Handbook of Statistics
Volume 53

Statistics in Industry and Government

Handbook of Statistics

Series Editors

C.R. Rao
C.R. Rao AIMSCS, University of Hyderabad Campus,
Hyderabad, India

Arni S.R. Srinivasa Rao
Medical College of Georgia, Augusta University, United States

Handbook of Statistics

Volume 53

Statistics in Industry and Government

Edited by

Ravindra Khattree
Distinguished University Professor of Applied Statistics
Co-Director, Institute of Data Science Participating Member,
Center for Biomedical Research Oakland University, MI, United States

Hon Yiu So
Department of Mathematics and Statistics,
Oakland University, Rochester, Michigan, United States

Arni S.R. Srinivasa Rao
Medical College of Georgia, Augusta, Georgia, United States

ELSEVIER

ACADEMIC PRESS

An imprint of Elsevier

Academic Press is an imprint of Elsevier
50 Hampshire Street, 5th Floor, Cambridge, MA 02139, United States
125 London Wall, London, EC2Y 5AS, United Kingdom

For accessibility purposes, images in electronic versions of this book are accompanied by alt text descriptions provided by Elsevier. For more information, see https://www.elsevier.com/about/accessibility.

Books and Journals published by Elsevier comply with applicable product safety requirements. For any product safety concerns or queries, please contact our authorised representative, Elsevier B.V., at productsafety@elsevier.com.

Publisher's note: Elsevier takes a neutral position with respect to territorial disputes or jurisdictional claims in its published content, including in maps and institutional affiliations.

Notices

Knowledge and best practice in this field are constantly changing. As new research and experience broaden our understanding, changes in research methods, professional practices, or medical treatment may become necessary.

Practitioners and researchers must always rely on their own experience and knowledge in evaluating and using any information, methods, compounds, or experiments described herein. In using such information or methods they should be mindful of their own safety and the safety of others, including parties for whom they have a professional responsibility.

To the fullest extent of the law, neither the Publisher nor the authors, contributors, or editors, assume any liability for any injury and/or damage to persons or property as a matter of products liability, negligence or otherwise, or from any use or operation of any methods, products, instructions, or ideas contained in the material herein.

ISBN: 978-0-443-31422-3
ISSN: 0169-7161

For information on all Academic Press publications
visit our website at https://www.elsevier.com/books-and-journals

Publisher: Zoe Kruze
Acquisition Editor: Mariana L. Kuhl
Editorial Project Manager: Sneha Apar
Production Project Manager: Abdulla Sait
Cover Designer: Arumugam Kothandan
Typeset by MPS Limited, India

Contents

Contributors

Demissie Alemayehu (207), Pfizer Inc., New York, NY, United States

Daniel Backenroth (207), Johnson & Johnson Innovative Medicine, Raritan, NJ, United States

John Buntin (353), Sociology & Criminology, University of Miami, Coral Gables, FL, United States

Sixia Chen (161), Department of Biostatistics and Epidemiology, The University of Oklahoma Health Sciences, Oklahoma City, OK, United States

Tianle Chen (207), Biogen, Inc., Cambridge, MA, United States

Erhard Cramer (81), Institute of Statistics, RWTH Aachen University, Aachen, Germany

Piyali Das (319), National Institute of Public Finance and Policy, New Delhi, India

Xinwei Deng (129), Department of Statistics, Virginia Tech, Blacksburg, VA, United States of America

Haoda Fu (207), Amgen Inc., Thousand Oaks, CA, United States

Chetan Ghate (319), Indian Statistical Institute, Delhi, India; Centre for Applied Macroeconomic Analysis (CAMA), Australian National University (ANU), Canberra, Australia

Subhadeep Halder (319), Indian Statistical Institute, Delhi, India; New York University, Abu Dhabi, United Arab Emirates

Takashi Isozaki (1), Tokyo Research, Sony Computer Science Laboratories, Inc.; Technology Strategy Department, Sony Group Corporation; Koozyt, Inc., Tokyo, Japan

Stephanie Kennedy (353), Sociology & Criminology, University of Miami, Coral Gables, FL, United States

Pandurang M. Kulkarni (207), Eli Lilly and Company, Indianapolis, IN, United States

C. Devon Lin (129), Department of Mathematics and Statistics, Queen's University, Kingston, ON, Canada

Man Ho Ling (51), Department of Mathematics and Information Technology, The Education University of Hong Kong, Tai Po, Hong Kong, SAR, P.R. China

James Livsey (243), US Census Bureau, Washington, DC, United States

Tucker McElroy (243), US Census Bureau, Washington, DC, United States

Brian Millen (207), Biogen, Inc., Cambridge, MA, United States

Kannan Natarajan (207), Pfizer Inc., New York, NY, United States

Hon Keung Tony Ng (81), Department of Mathematical Sciences, Bentley University, Waltham, MA, United States

Alex R. Piquero (353), Sociology & Criminology, University of Miami, Coral Gables, FL, United States

Karen Price (207), Eli Lilly and Company, Indianapolis, IN, United States

Yongming Qu (207), Eli Lilly and Company, Indianapolis, IN, United States

John K. Roman (353), Sociology & Criminology, University of Miami, Coral Gables, FL, United States

Satrajit Roychoudhury (207), Pfizer Inc., New York, NY, United States

Anita Shahrokhian (129), Department of Mathematics and Statistics, Queen's University, Kingston, ON, Canada

Wei Shen (207), Eli Lilly and Company, Indianapolis, IN, United States

Ram Tiwari (175), Bristol Myers Squibb, Berkeley Heights, NJ; Global StatSolutions, Reston, VA, United States

Wim Van Der Elst (207), Johnson & Johnson Innovative Medicine, Beerse, Belgium

Jixian Wang (175), Bristol Myers Squibb, Boudry, Switzerland

Changbao Wu (295), University of Waterloo, Waterloo, Canada

Chao Xu (161), Department of Biostatistics and Epidemiology, The University of Oklahoma Health Sciences, Oklahoma City, OK, United States

Yang Yu (81), Department of Statistics and Data Science, Southern Methodist University, Dallas, TX, United States

Hongtao Zhang (175), Merck & Co., Inc., North Wales, PA, United States

Preface

"Statistics is a gateway to knowledge and the progress a country makes is vitally dependent on the efficiency of its statistical system" C. R. Rao in Statistics, Statisticians and Public Policy Making. Sankhyā B 45 (2), 151–159, 1983.

References to the census of people and of agriculture as we know today can be found in the *Old Testament* (ca. 1500 BCE), in a Chinese book by *Kuan Tzu* (ca. 1000 BCE), and in the *Arthasastra* of Kautilya (ca. 321–297 BCE) written in Sanskrit. *Ain-i-Akbari (ca. 1590 CE)*, a book written in Persian provides detailed statistical data on the governance of early Mughal kings in India. Evidences of industrial applications involving statistical ideas are also apparent in ancient India. For example, without any formal notations and nomenclatures, *Varaha Mihira (ca. 505–587 CE)* in his book *Brihat Samhita*[1] had used experimental designs to specify the recipes for making optimum perfumes. (See Styan and Chu (2013)[2], Taneja (2010)[3] and Khattree (2020)[4]). In China, industrial experiments were conducted over a long period of time to improve the paper-making process by using different raw materials and techniques. Over time, this resulted in better quality paper at lower cost and in much bigger quantities. As a result, Chinese paper was exported to various countries along the Silk Road during the Song Dynasty (ca. 960–1279 CE).

Needless to say, the progress of a country heavily depends on how it is governed and how it strives to innovate by improving its existing industries and introducing the new ones. The two go hand in hand, and statistics in terms of correct information – either collected, extracted, derived, or created – plays an instrumental role in the process. And that is the rationale for bringing out this volume of Handbook of Statistics. This volume with an appropriate sub- title "*Statistics in Industry and Government*" sheds light on some of the current issues and cutting-edge techniques which play or can play important roles in

[1] *N. C. Iyer, The Brihat Samhita of Varaha Mihira, Tiruvadi Jotistantra Sabha, Madura, India (1884).*

[2] G. P. H. Styan and K. L. Chu, *An Illustrated Introduction to Some Old Magic Squares from India, In Combi- natorial Matrix Theory and Generalized Inverses of Matrices (Ed. R. B. Bapat, S. J. Kirkland, M. Prasad and S. Puntanen, Springer, New York, NY, (2013), 227–252.*

[3] I. J. Taneja, *Equivalent Versions of "Khajuraho" and "Lo-Shu" Magic Squares and the day 1st October 2010 (01.10.2010),* arXiv:1011.0451 (2010).

[4] R. Khattree, *On Construction of Constant Block-Sum Partially Balanced Incomplete Block Designs, Communications in Statistics-Theory and Methods, 49(11), (2020), 2585–2606.*

improving the workings of our industries and governmental organizations. Admittedly, any such volume, with constraints on its size, can never be exhaustive. Further, we are aware that with industrial and political rev- olutions that are currently taking place all across the world make it nearly impossible to predict what statistical techniques may have longer lasting impacts to the governance and industry. Thus, we have limited our selection to the topics and related statistical techniques which are likely to play an important role in our lives in the near future. Accordingly, we have chosen leading experts or teams of experts to write chapters of this book with the hope that readers will benefit from their expertise, experiences as well as opinions. The chapters do vary in content, style as well as in emphasis, be it mathematical or non-mathematical yet with technical issues pertaining to a chosen topic.

The Chapters 1 to 5 pertain to industry and address some of the more recent statistical techniques from which industries all across world can benefit. Topics cover various modeling issues as well as designing of complex experiments. They are all about supervised or unsuper- vised learning, either via observational data or by designed experiments.

Chapters 6 and 7 specifically address drug development related issues. This is an area where industry and government actively interact in the process of development and regulatory practices. The teams of experts chosen have considerable experience in working or dealing with industry as well as governmental regulatory agencies and include those with solid academic credentials in their previous years.

Chapters 8 to 11 are dedicated to the use of statistics in a variety of governmental agencies. Forecasting and prediction over time and of the state of the nation is an important component of any government, and Chapter 8 addresses related issues from a computational point of view. A related topic is that of data integration where various official data are to be combined in a suitable manner so that they not only make sense but can also be used for further analyses. This is the topic of Chapter 9. Chapter 10 deals with an important topic of public debt and provides important statistical insights into its analyses as well as interpretations. Finally, Chapter 11 is about a topic which is often overlooked but only to the detriment of our society. A better control on crime rates can be had and better or fairer justice can be delivered only when one can understand the present status via accurate and meaningful data and can do so in a timely manner. Thus, this chapter provides some deep insights into the American justice system in the contexts of occurrence of crimes. We believe that these two last chapters, although written in the contexts of two specific countries, mirror the situations for several other countries and thus will be able to provide general guidelines to improve governance.

Our sincere gratitude to Mariana Kuhl Leme, the acquisitions editor (Elsevier and North- Holland), for her overall administrative support during the preparation of this volume. We highly appreciate her valuable involvement in the project. Our thanks also go to Sneha Apar, the developmental editor

(Elsevier), for providing excellent support to the editors and for en- gaging with authors to address various technical queries throughout the preparation of the vol- ume, including during the proofing stage and production. We would like to extend our thanks to Sait Abdulla, the project manager of book production at RELX India Private Limited, Chennai, India, for leading the production, addressing several rounds of author queries, and providing assistance in printing activities. We are truly thankful to all the authors for writing their bril- liant chapters. Lastly, we would like to thank all our peer reviewers for their timely assessment of the chapters and their valuable recommendations for revision.

Overall, we as editors are very satisfied with the outcome of volume 53 of the Handbook of Statistics which focuses on industry and government applications.

Ravindra Khattree, Hon Yiu So, and Arni S.R. Srinivasa Rao

Chapter 1

The use of causal inference with structural models in industry

Takashi Isozaki[a,b,c,*]

[a]*Tokyo Research, Sony Computer Science Laboratories, Inc., Tokyo, Japan,* [b]*Technology Strategy Department, Sony Group Corporation, Tokyo, Japan,* [c]*Koozyt, Inc., Tokyo, Japan*

[*]*Corresponding author. e-mail address: isozaki@csl.sony.co.jp*

Chapter Outline

Handbook of Statistics, Vol. 53. https://doi.org/10.1016/bs.host.2025.04.007

Abstract

This article describes the use of probabilistic structural causal models in industry. With the accumulation of observational data and the development of inference methods in the models, it has become possible to infer causal relationships to a considerable extent if excellent inference algorithms are available. In industry, the key objective is not necessarily to infer causal relationships per se, but rather, to ensure that the inference of causal relationships leads to knowledge of the factors that have a direct or fundamental impact on the items to be improved, thereby leading to more efficient interventions. In this task, it is more important to discriminate between pseudo correlation and causation, unlike in prediction tasks, and it is also useful to distinguish between direct and indirect factors. This article first briefly reviews the theory, methods, and algorithms of causal inference that provide the foundations for its utilization, and then presents our approach to addressing a longstanding problem in this area. We then provide some discussion on the modeling and analytics of what we have been paying attention to when using the technologies in actual data analysis. After that, we introduce several publicly available examples of analyses using causal inference at both Sony and outside companies (e.g., companies involved in semiconductors, automobiles, plating and other manufacturing, product marketing, and analysis of machine learning in the music field). Our examples demonstrate that causal inference is effective in a broad range of business settings.

1 Introduction

There is no shortage of areas and systems that we humans wish to understand, manage, control, or intervene in. In the sciences, there are complex systems that are difficult to test with controlled experiments, such as evidenced in epidemiology, economics, brain and neuroscience, psychology, and other areas of the natural and social sciences; as such, most of the empirical sciences are to be found in such areas domains, (in addition to more fundamental disciplines such as physics, chemistry, and biology, where controlled experiments are the norm). On the other hand, businesses have similar demands in various data application fields. Examples include the pursuit of the causes of defects in manufacturing processes and the improvement of defects through controls, measures, and interventions based on this understanding; increasing the frequency of service use and billing based on an understanding of user behavior in marketing; intervention measures to keep users connected; analysis of human resource data to improve employee performance; and analysis of employees who have left the company to reduce job turnover. In complex systems, there is a need to understand in advance the cause-and-effect relationships that determine whether intervention and change can really take place in order to properly implement such controls and interventions, and if this is possible, it can be a business-critical process. Randomized controlled trials (RCTs) are known to be useful for understanding cause-effect relationships, but it is often difficult to conduct the desired controlled experiments in complex systems that have many variables because (1) there are too many variables to be controlled, or (2) there are many factors that cannot be controlled from practical, ethical, cost, or other perspectives. Therefore, we believe that there is a great need in industries that apply statistics and data science to infer some cause-effect relationships from observed data rather than from controlled experimental data. In fact, as described in this article, we have been actively promoting causal inference in data analysis for more than a decade, and have seen observed the need for this analysis in various data analysis scenarios.

A fundamental challenge in this context is that it is often difficult to substitute conventional correlation information for causal information. As an example to illustrate the difference between correlation and causation, Fig. 1 shows the results of an analysis using real data. The graph depicted in Fig. 1A, which shows whether there is a correlation or not in a pair of variables, is quite complex and difficult for humans to understand using domain knowledge. In contrast, the graph in (b) is much easier for humans to understand. Note that while there are 206 variables in both cases, there are 607 lines in (a) compared to just 251 lines in (b). In (a), one approach is to retain only highly correlated relationships from this complex set of relationships, but in this case, of course, the results of the selected relationships will change significantly with a slight difference in threshold values, but and more importantly, the highly correlated relationships are likely to be taken for granted due to the domain knowledge of the analysts, analysis team, or data acquirer/user, and are unlikely to lead to

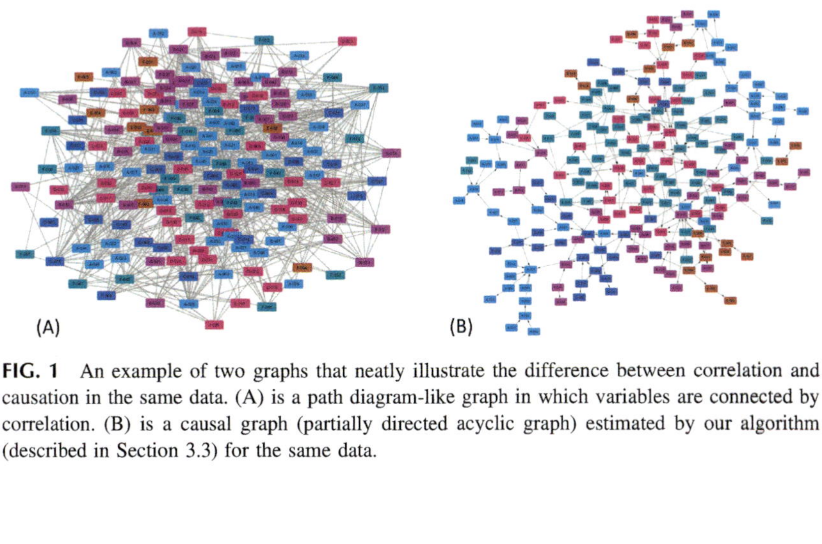

FIG. 1 An example of two graphs that neatly illustrate the difference between correlation and causation in the same data. (A) is a path diagram-like graph in which variables are connected by correlation. (B) is a causal graph (partially directed acyclic graph) estimated by our algorithm (described in Section 3.3) for the same data.

new discoveries, thus considerably reducing the value of the analysis. The purpose of analysis is partly to confirm domain knowledge, but it is also normal to expect new knowledge to be learned gleaned for the first time from the data. In addition, the determination of whether the relationship between variables is pseudo correlation or causation, and whether causation is a direct or indirect cause, relies on domain knowledge, which is not sufficient to provide the degree of objectivity expected of a data analysis. Therefore, causal inference, which enables analysis from a perspective other than simply the strength of the correlation and is able to explore causal relationships, has unique characteristics in the data analysis context.

The modeling of causality in statistics is said to have originated in the path diagrams first developed by Wright (1921). Currently, the structural causal model (SCM) (Pearl, 2000; Spirtes et al., 2000), which has been developed from both a graphical and probabilistic perspective, is considered its successor and has shown great promise. This model represents the human intuitive understanding of causality quite well. In this article, we mainly focus on the causal relationships based on SCMs and discuss their inference (i.e., causal discovery) methods and applications. Once the causal relationship is understood through the analysis, it is possible to devise and implement the control, intervention, and measures required to change the situation to a more favorable one based on the knowledge. For this purpose, Pearl (1995, 2000) created a framework for rigorous predictive calculation of the causal effects of virtual interventions on SCMs, which is also very useful as the two sides of a coin in causal inference, and we have implemented and used it as well (note, however, that we briefly touch on the causal effect inference in Section 3.4).

In Sections 2 and 3 of this article, we review the basics of causal modeling and inference, the challenges of inference, and related research. Next, we will present our approach to the problem, the theoretical and algorithmic approach, and our experimental results on their effectiveness. Section 4 discusses what is required of causal model representations and causal inference in actual data analysis, and discusses the relationship between the ideas and characteristics of data modeling and algorithmic modeling as classified by Breiman (2001) and causal inference. We will also discuss the attention that has been paid to disseminating causal analytics and the specific ways in which it has been disseminated. In Section 5, we will examine important concepts in causal inference from a more bird's-eye perspective. Section 6 presents examples of various data analyses we have conducted based on the science, technology, and methodology of causal inference described above. The author's company, Sony, is a conglomerate that covers electronics (cameras, TVs, audio devices, smartphones, entertainment robots, etc.), games, image sensor semiconductors, movies, music, finance (banks, life insurance, accident and disaster insurance), Internet service providers, sports support, entertainment content distribution, etc., and our causal inference and analysis software is used in most of these areas. We also provide software and contract analysis services to companies outside of Sony.[1] In this section, we will present a number of publicly available concrete examples from these various analysis cases, aiming to show how SCM-based causal inference can be effective in real data. We conclude the paper in Section 7 with a brief summary.

2 Foundation of structural causal models

2.1 Basic properties of markovianity for causal modeling

The underlying model throughout this article is the structural causal model (SCM), and of particular importance is its Markovian nature.[2] Markov models are well known in the treatment of time series data. All variables appearing below are assumed to be random variables unless otherwise noted. We give up on observing all events and inferring causality as a perfectly logical relationship, and will therefore treat causality among random variables as involving uncertainty. The Markovianity of a time-series variable is expressed as follows. Suppose a random variable X has state k in the time series, and denote this stochastic process as X_k. Given the current state X_n as a timeseries state $\{k = 1, 2, \ldots, n, n+1\}$, if the next state X_{n+1} is independent of $\{X_1, X_2, \ldots, X_{n-1}\}$, we say this stochastic process is Markovian. Then X_{n+1} is conditionally independent of $\{X_1, X_2, \ldots, X_{n-1}\}$ given X_n.

The Markovianity is also a powerful model in causal models. Let us consider the simplest causal chain model that demonstrates this. Suppose there are variables X, Y, and Z, and that there is a causal chain relationship from X as cause 1 to Y as result 1, and from Y as cause 2 to Z as result 2. In this case, the path diagram expression that draws an arrow line from the cause to the result is $X \rightarrow Y \rightarrow Z$. Then, as an expression of Markovianity, X and Z are said to be conditionally independent given Y and expressed as $X \perp\!\!\!\perp Z \mid Y$. The causal meaning of this model is that Z is caused by X and Y, but the direct cause to Z is only Y and thus Z is not directly affected by X. In causal relationships, as in this example, it is very often observed and recognized that there are other causes that can be said to be causes if they are traced back, but they are indirect causes, and there are other causes that have direct effects, and it is extremely important to express this in causal models. Of course, there are cases in which both X and Y have a direct effect on Z, and even in such cases, it is sufficient to use a model in which the two relationships $X \rightarrow Y \rightarrow Z$ and $X \rightarrow Z$ exist together.

---

[1] Until 2024, the project was implemented only in Japan, but starting this year, it is being implemented overseas as well.

[2] There are different approaches to modeling causality itself. Granger's causality (Granger, 1969) is a representative example, and the approach of Shimizu et al. (Shimizu et al., 2006) starting from a more statistical perspective may differ from the foundation we rely on, although it is a similar structural causality based on DAGs as a basis of representation. Note that the main topic of this article, causal discovery or causal search, is a different context and topic from Rubin's latent response model (Rubin, 1974, 1978) and Pearl's definition of causal effects with do-calculus (Pearl, 1995, 2000).

On the other hand, an important basic relationship is the so-called pseudo correlation, a relationship that is correlated but not causal (Simon, 1954). For example, suppose X is the common cause of Y and Z, but there is no causal relationship between Y and Z. In this case, correlation usually occurs between Y and Z, so the relationship between Y and Z is called pseudo correlation. This relationship is expressed as $Y \rightarrow X \rightarrow Z$ in the path diagram representation, and in this case, too, the conditional independence relationship $Y \perp\!\!\!\perp Z \mid X$ is considered to be established as a causal pattern with three variables. The characteristic of this path diagram is that no edge is drawn between Y and Z, and an edge is drawn between them only when there is a direct causal relationship. There is also a relationship $X \rightarrow Z \rightarrow Y$ if X is independent of Y in the three variables and both X and Z and Y and Z have no conditional independence (Wermuth, 1980; Reichenbach, 1956). The SCM is a structural model with Markovian conditions, described below, that develops our recognition and knowledge of such naive causal relationships.

2.2 Markovian structural causal models

Since the structural model of causality is usually represented as a directed acyclic graph (DAG) (Pearl, 2000; Spirtes et al., 2000), this article also discusses a model that assumes a DAG.[3] Since DAG is a graph-theoretic term that, as the name implies, refers to directed graphs and does not include Markovianity, it is better to call the structural causal models treated here causal Markovian DAGs, conditionally independent DAG models, or Markovian structural causal models.[4] However, in this article, we will simply refer to them as DAG models or SCMs, unless otherwise emphasized.

First, we will organize the graph-theoretic terminology used to graphically represent structural causal models with correspondence to random variables. Suppose a finite set of vertices $\boldsymbol{V}$, directed edges $\boldsymbol{E}$, and a directed graph $G = \langle \boldsymbol{V}, \boldsymbol{E} \rangle$, where each element of $\boldsymbol{V}$ corresponds to a random variable. Suppose that the graph represents the causal relationships and the directed edges represent the direction of causation, as in the path diagrams (Wright, 1921). When the directed edge from vertex X to vertex Y is $X \rightarrow Y$, X is said to be a parent of Y, and Y is said to be a child of X. In this case X and Y are also said to be adjacent to each other, and the vertex set adjacent to X is denoted by $\mathrm{Adj}_G(X)$. Also, the causal interpretation is that X is the direct cause of Y.

[3] Although research has been conducted on the usage of cyclic graphs (see e.g. (Spirtes, 1995; Richardson and Spirtes, 1999; Lacerda et al., 2008)), it has not yet reached the level of mainstream. We therefore assume DAGs as hidden SCMs in this article.

[4] The absence of edges in a graph-theoretic graph structure called a DAG does not immediately imply that it is Markovian because there could be a DAG model, such as ignoring edges with small weights in the edges above a certain threshold and not representing them. For example, a Chow-Liu tree (Chow and Liu, 1968), which represents only the relations with the greatest weights, is a DAG model if the direction of edges is given.

Denote the parent set by Pa_G (X) and the child set by Ch_G (X). In G with information only about its adjacency, i.e., the presence or absence of edges, it is said to be a skeleton. Also, a graph is said to be complete when all two vertices are adjacent. For a sequence of adjacent vertices $\langle V_1, V_2, \ldots, V_n\rangle$ in G, a path is the sequence, i.e., V_{k-1} and V_k are adjacent for $k = 2, 3, \ldots, n$, and a path such that V_{i-1} is the parent of V_i is called a directed path. If $X = Y$ or a directed path from X to Y exists for two vertices X and Y, we say X is an ancestor of Y and Y is a descendant of X. Ans_G (Y) and Des_G (X) denote the ancestor set of Y and the descendant set of X in the graph G, respectively.

In this article, it is assumed that there is a causal Markovian DAG model as an SCM and sample data generated from the relational model. As an axiom that generally links the Markovianity and causality of multivariate probability distributions, we present the causal Markov condition (or assumption) that defines the SCM, followed by the causal minimality condition and the causal faithfulness condition as related conditions and assumptions (Spirtes et al., 2000; Pearl, 2000).

Definition 1. (causal Markov condition) (Spirtes et al., 2000) *Let G be a causal graph with vertex set* **V** *and P* (**V**) *be a probability distribution over the vertices in* **V** *generated by the causal structure represented by G. G and P* (**V**) *satisfy the causal Markov condition if and only if for every X in* **V**, *X is independent of* **V** $\backslash(Des_G\ (X) \cup Pa_G\ (X))$ *given* Pa_G (*X*).

Definition 2. (causal minimality condition) (Spirtes et al., 2000) *Let G be a causal graph with vertex set* **V** *and P be a probability distribution on* **V** *generated by G.* $\langle G, P\rangle$ *satisfies the causal minimality condition if and only if for every proper subgraph H of G with vertex set* **V**, *the pair* $\langle H, P\rangle$ *does not satisfy the causal Markov condition.*

Definition 3. (causal faithfulness condition) (Spirtes et al., 2000) *Let G be a causal graph with vertex set* **V** *and P be a probability distribution on* **V** *generated by G.* $\langle G, P\rangle$ *satisfies the causal faithfulness condition if and only if every conditional independence relation true in P is entailed by the causal Markov condition applied to G.*

It should be noted that these axioms are not independent. A probability distribution P is faithful to the corresponding graph G if and only if $\langle G, P\rangle$ satisfies both the Markov and faithfulness conditions. The faithfulness and Markov conditions encompass the minimality, but the minimality and Markovianity do not encompass faithfulness (Spirtes et al., 2000).[5] The causal faithfulness condition is what Pearl calls stability[6] (Pearl, 2000), which requires that the cause-and-effect relationship be of a stable Markovian nature

[5] There is also a logical construction that takes the causal Markov condition as a theorem (Pearl, 2000), but we adopt this as our axiom.

[6] Strictly speaking, stability is a more stringent criterion (Andersen, 2013).

and not one that appears to be a Markovian relationship due to quantitative relationships between special parameters. We have mentioned above that the causal relationships we can infer from the data are limited to those that have such relationships (especially the causal Markov condition (CMC) and the causal faithfulness condition (CFC)), and there may be complex causal relationships that cannot be expressed by these axioms. However, this SCM is also called a model of the process of data and information generation, and it is recognized to a considerable extent as having validity as a model of this process (Spirtes et al., 2000; Pearl, 2000; Glymour and Cooper, 1999; Hausman and Woodward, 1999). On the other hand, we believe that there are points to be noted in the application of this model in real data analysis, which will be discussed in Section 4.

Next, as a more realistic setting, consider a causal model in which some of the constituent random variables $\boldsymbol{W}$ may not be observed. Let $\boldsymbol{V}$ be the set of variables that are observed (i.e., $\boldsymbol{V} \subseteq \boldsymbol{W}$). Let $\boldsymbol{L}$ be a set of unobserved variables and also a set of selection variables $\boldsymbol{S}$ whose values are fixed and not observed. These are characterized by the fact that they include all settings that are said to produce correlations. Let $\boldsymbol{V}, \boldsymbol{L}, \boldsymbol{S}$ be mutually disjoint sets, and let $\boldsymbol{W} = \boldsymbol{V} \cup \boldsymbol{L} \cup \boldsymbol{S}$ denote the relationship. It appears that infinite dimensions must be considered in a model that also includes unobserved variables, which may seem to be a very difficult situation. However, a model has been developed to handle this issue successfully (Verma and Pearl, 1990; Spirtes et al., 1995), called *ancestral graph model* (Richardson and Spirtes, 2002), and in this article, we introduce it and discuss it as the fundamental causal model. Zhang (2008b) described the characteristics of the ancestral graph models as follows:

A main attraction of ancestral graphs is that, without explicitly including latent variables, they can represent conditional independence relations and causal relations among observed variables when the underlying data generating process involves latent confounders and/or selection variables. This of course requires a richer syntax than that of DAGs.

The following sections describe mixed and ancestral graphs according to Richardson and Spirtes (2002) and Zhang (2008b). A *mixed graph* is a graph consisting of vertices and edges and may have three types of edges: directed ($\rightarrow$), bi-directed ($\leftrightarrow$) and undirected ($-$), and like DAG, there shall be at most one edge between two vertices. Given a mixed graph G and its two adjacent vertices X and Y, if $X \leftrightarrow Y$, we say X is a spouse of Y (and Y is also a spouse of X), and if $X - Y$, we say X is a neighbor of Y (and Y is also a neighbor of X). For the tip of an edge, if there is an arrow, it is called an arrowhead, and the part without an arrow is called a tail. At a non-end point V of a path π, V is said to be a collider if the two arrowheads both face V; otherwise, it is said to be a non-collider. When X and Z are adjacent in a path $\langle X, Y, Z\rangle$ consisting of three vertices X, Y, Z, we say that the path is shielded and the triple is a shielded triple. On the other hand, if X and Z are not adjacent on the path, the path is said to be unshielded, and the triple is called an unshielded triple. Also, if there is an arrowhead from X to Y and an arrowhead from Z to Y on that path, Y is called an unshielded collider. A vertex X is said to be anterior to a vertex Y if there is a path μ on which every edge is either of the form $T - U$ or $T \rightarrow U$ with U between T and Y, or $X = Y$. It means that there are no edges $T \leftrightarrow U$ and no edges $T \rightarrow U$ pointing toward X. Such a path is said to be an anterior path from X to Y, and $X \in \mathrm{Ant}_G(Y)$. In a mixed graph G, G is said to have a directed cycle if $X \rightarrow Y$ and $X \in \mathrm{Des}_G(Y)$, and if $X \leftrightarrow Y$ and $X \in \mathrm{Des}_G(Y)$, then G is said to have an almost directed cycle.

Definition 4. (ancestral graph) (Zhang, 2008b) *A mixed graph is ancestral if the following three conditions hold: (i) there is no directed cycle, (ii) there is no almost directed cycle, and (iii) for any undirected edge $X - Y$, X and Y have no parents or spouses.*

Directed acyclic graphs (DAGs) are a special case of ancestral graphs, and the first condition above is exactly the same as the condition for a DAG. The second and third above are unnecessary for a DAG, since in a DAG, an edge consists only of directed edges.

Ancestral graphs have the following properties (Richardson and Spirtes, 2002).

Proposition 1. *If G is an ancestral graph then, for three vertices X, Y, Z in G, the configurations $X - Y \leftrightarrow Z$ and $X - Y \rightarrow Z$ do not occur (regardless of whether X and Z are adjacent).*

Corollary 1. *In an ancestral graph, an anterior path from X to Y takes one of three forms: $X - \cdots - Y$, $X \rightarrow \cdots \rightarrow Y$, or $X - \cdots - \rightarrow \cdots \rightarrow Y$.*

Proposition 2. *If G is an ancestral graph, and X, Y are adjacent vertices in G, then each edge type has the following equivalence relationships: (i) $X - Y \Leftrightarrow X \in Ant_G(Y)$, $Y \in Ant_G(X)$, (ii) $X \rightarrow Y \Leftrightarrow X \in Ant_G(Y)$, $Y \notin \mathrm{Ant}_G(X)$, and (iii) $X \leftrightarrow Y \Leftrightarrow X \notin Ant_G(Y)$, $Y \notin Ant_G(X)$.*

When a DAG is a graph that expresses conditional independence as a Markov model, d-separation is defined as a property that is consistent with the causal Markov condition from a graphical perspective (Pearl, 1988). The following m-separation is an extension of this definition to the case of ancestral graphs (Richardson and Spirtes, 2002; Zhang, 2008b).

Definition 5. (m-separation) (Zhang, 2008b; Richardson and Spirtes, 2002) *In an ancestral graph, a path π between vertices X and Y is active (m-connecting) relative to a (possibly empty) set of vertices $\mathbf{Z}$ (X, $Y \notin \mathbf{Z}$) if.*

(i) *every non-collider on π is not a member of $\mathbf{Z}$, and.*
(ii) *every collider on π has a descendant in $\mathbf{Z}$.*

$\mathbf{X}$ and $\mathbf{Y}$ are said to be m-separated by $\mathbf{Z}$ if there is no active path between any vertex in $\mathbf{X}$ and any vertex in $\mathbf{Y}$ relative to $\mathbf{Z}$.

The properties of m-separation include the following (Richardson and Spirtes, 2002).

Proposition 3. *If π is a path m-connecting X and Y given* $\mathbf{Z}$ *in an ancestral graph G then every vertex on π is in* Ant_G $(\{X, Y\} \cup \mathbf{Z})$.

Proposition 4. *In an ancestral graph G, there is a path* $\boldsymbol{\mu}$ *between X and Y on which no non-collider is in a set* $\mathbf{Z}$ $(X, Y \notin \mathbf{Z})$ *and every collider is in* Ant_G $(\{X, Y\} \cup \mathbf{Z})$ *if and only if there is a path m-connecting X and Y given* $\mathbf{Z}$ *in G.*

Similar to d-separation in DAGs, ancestral graphs also have the following property called the global Markov property. That is, if ***X*** and ***Y*** are m-separated by ***Z***, then ***X*** and ***Y*** are conditionally independent given ***Z***. The set ***Z*** that m-separates *X* and *Y* is denoted by M-SEP(*X*, *Y*). On the other hand, the following property: when two vertices are not adjacent, there is a subset of other vertices that d-separates the two vertices, holds for DAGs, but not necessarily for ancestral graphs. This leads to the following definition.

Definition 6. (maximal ancestral graph) (Zhang, 2008b; Richardson and Spirtes, 2002) *An ancestral graph is said to be maximal if for any two nonadjacent vertices, there is a set of vertices that m-separates them.*

Of course, the DAGs are always maximal, and an ancestral graph is maximal in the sense that no new edges can be added without changing all conditionally independent relations. The existence of ***L*** and ***S*** gives rise to a complex relation between vertex (i.e., observed variable) sets and non-adjacent vertices in the m-separating maximal ancestral graph. For this purpose, we define the following notions.

Definition 7. (Zhang, 2008b) *In an ancestral graph, let X, Y be any two vertices, and* $\mathbf{L}$, $\mathbf{S}$ *be two disjoint sets of vertices not containing X, Y. A path π between X and Y is called an inducing path relative to* $\langle \mathbf{L}, \mathbf{S} \rangle$ *if every non-endpoint vertex on π is either in* $\mathbf{L}$ *or a collider, and every collider on π is an ancestor of either X, Y, or a member of* $\mathbf{S}$*. When* $\mathbf{L} = \mathbf{S} = \emptyset$, *$\pi$ is called a primitive inducing path between X and Y.*

This is a generalization by Richardson and Spirtes (2002) of the definition introduced by Verma and Pearl (1990). The primitive inducing path has the following properties (Zhang, 2008b).

Proposition 5. *An ancestral graph G is maximal if and only if there is no primitive inducing path between any two non-adjacent vertices in G.*

The following relationship exists between an ancestral graph and a maximal ancestral graph (MAG) (Richardson and Spirtes, 2002).

Proposition 6. *If* G_0 *is an ancestral graph, there exists a unique maximal ancestral graph G formed by adding* $\leftrightarrow$ *edges to* G_0 *such that the conditional independence relation set remains invariant.*

The causal meaning of an edge in MAG is as follows (Zhang, 2008b). The relationship $X - Y$ means that X is the cause of Y or is the cause of some selection variable, and Y is the cause of X or is the cause of some selection variable. In a situation where there are no selection variables in the data, which can be assumed in many cases in data analysis, in the relation $X \rightarrow Y$, it is clear that X is the cause of Y, and in the relation $X \leftrightarrow Y$, X is not the cause of Y and Y is not the cause of X, meaning that X and Y are pseudo correlations due to latent confounding variables.

Two MAGs may be indistinguishable from conditional independence relations, and both graphs are said to be Markov equivalent. The definition of Markov equivalence in MAGs is as follows.

Definition 8. (Markov equivalence) (Zhang, 2008b) *Two MAGs G_1, G_2 (with the same set of vertices) are Markov equivalent if for any three disjoint sets of vertices* $\mathbf{X}$, $\mathbf{Y}$, $\mathbf{Z}$, $\mathbf{X}$ *and* $\mathbf{Y}$ *are m-separated by* $\mathbf{Z}$ *in* G_1 *if and only if* $\mathbf{X}$ *and* $\mathbf{Y}$ *are m-separated by* $\mathbf{Z}$ *in* G_2.

The Markov equivalence in MAGs is slightly different from that in DAGs, and in that it requires additional conditions. The following definition is for that purpose (Zhang, 2008b; Spirtes and Richardson, 1996).

Definition 9. (Zhang, 2008b) *In a MAG, a path π between X and Y: $\pi = \langle X, \ldots, W, V, Y\rangle$ is a discriminating path for V if (i) π includes at least three edges, (ii) V is a non-endpoint vertex on π, and is adjacent to Y on π, and X is not adjacent to Y, and every vertex between X and V is a collider on π and is a parent of Y.*

The following proposition is proved by Spirtes and Richardson (1996) regarding Markov equivalence in MAGs.

Proposition 7. (Zhang, 2008b) *Two MAGs over the same set of vertices are Markov equivalent if and only if (i) they have the same adjacencies, (ii) they have the same unshielded colliders, and (iii) if a path π is a discriminating path for vertex V in both graphs, then V is a collider on the path in one graph if and only if it is a collider on the path in the other.*

MAGs in the same class are not distinguishable from each other from a probability distribution, thus from the pattern of correlations in inference from the data, and such a class is said to be a Markov equivalence class. Therefore, it is common for us to be able to identify up to this class without being able to identify up to the MAG when we infer causality from observational data. In this case, there are arrowheads and tails that cannot be deduced and are undetermined, so they are represented by circles ($\circ$). Just as the graph representation of the Markov equivalence class of a DAG is called a PDAG, that of a MAG is called a partial ancestral graph, or a PAG for short. The definition of PAGs is as follows.

Definition 10. (partial ancestral graph) (Zhang, 2008b) *Let $[G]$ be the Markov equivalence class of a MAG G. A partial ancestral graph (PAG) for*

[G] is a graph P with possibly three kinds of marks and hence six kinds of edges: $-$, $\rightarrow$, $\leftrightarrow$, $\circ\!\rightarrow$, $\circ\!-\!\circ$, *and* $\circ\!-$, *such that (1) P has the same adjacencies as G and any member of [G] does, and (2) every non-circle mark in P is an invariant mark in [G]. If it is furthermore true that (3) every circle in P corresponds to a variant mark in [G], P is called the maximally informative PAG for [G].*

As an example of PAGs, an expert-designed DAG model (Beinlich et al., 1989) called the Alarm network is transformed into a PAG using direction rules (Zhang, 2008b) that will be mentioned in Section 3, as shown in Fig. 2 (in which $\rightarrow$ is represented by $\rightarrowtail$) (Isozaki and Kuroki, 2017).

2.3 Discussion: Significance of the structural causal models

We discuss here the significance of being structural in the Markovian structural causality model introduced as a framework in this article. The fact that the causal models are Markovian structural models allows us to distinguish the order of causation, and the axiom of the causal Markov condition allows us to clearly reason out which variables are the direct causes and which are the indirect causes for a given variable of interest for the model being inferred. This is important when considering control and intervention, and whether to intervene in a direct factor that seems to have a large effect and significantly change the variable of interest, or whether to intervene in a root factor that affects multiple direct factors and eliminates the problem all together. In other words, when there are multiple causal variables for a given variable of interest,

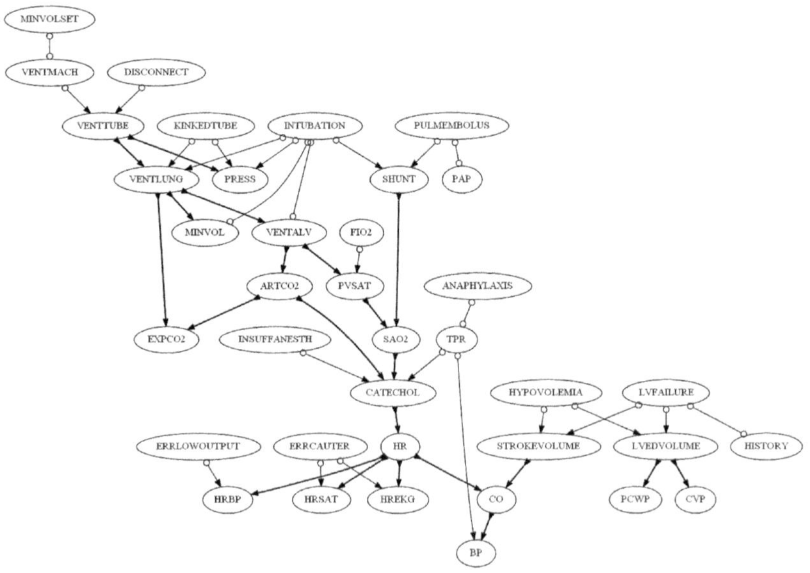

FIG. 2 An example of PAG representation: alarm network.

they can be further decomposed to make more detailed sense of them, and these suggestions can be utilized to implement the next action to be taken more effectively and efficiently. This is the major significance of considering structural models. In fact, we will see this in the specific case of semiconductors discussed in Section 6.

2.4 A brief review of historical developments of SCMs

In this subsection, we briefly review the historical development of structural causal models up to the DAG models. Wright's path diagram (Wright, 1921) was a pioneer in causal modeling, and the path coefficient utilizing the partial correlation coefficient was also defined by Wright (Wright, 1934), laying the foundation for subsequent path analysis and structural equation models. While research has started from the perspective of how to linguistically describe the causal relationships themselves (Burks, 1951; Simon, 1952), attempts to infer causal relationships from data based on a more analytical perspective have been initiated by Simon (1953, 1954) and Blalock (1962), Simon (1954) explored the possibility of identifying causal relationships using partial correlation coefficients under the assumption of linear causality in three variables, and discussed the causal interpretation of pseudo correlation and its identification, while Blalock (1962) extended and developed the discussion to four variables. Reichenbach (1956) modeled causality and discussed its relationship to conditional independence, clarifying the pseudo correlation associated with the common cause principle in a way that remains relevant today. In this early stage, several different fields— genetics, sociology, logic, and physical philosophy—each developed their own models and arguments for causal reasoning. The Markovian nature of matter has long been discussed in terms of statistical physics by Gibbs, the physicist who pioneered physical chemistry, and this led to the development of Markov models in statistics, including the model featuring undirected graphs with multiple variables (Miyakawa, 1999). This development appears to have extended to temporal relationships which, in conjunction with early investigations, led to the development of Markov models of causality based on acyclic directed graphs. Kiivery et al. first showed that Markov conditions can be derived from the decomposition representation of probability distributions in directed graphs (Kiivery et al., 1984; Hausman and Woodward, 1999; Pearl, 2000), and Pearl and Verma later specified the Markov condition as the more fundamental axiomatic entity that defines causality (Pearl and Verma, 1991; Pearl, 2000). This relationship, called the causal Markov condition, is now recognized as the basis for Markovian structural causal models. Pearl et al. investigated various properties of Markovian acyclic directed graphs, and in particular introduced the concept of d-separation (Pearl, 1988), which combines graph structure and conditional independence, concepts that are important in SCMs, and revealed various important properties (Geiger and Pearl, 1988; Verma and Pearl, 1988).

3 Inference of the causal models from observed data

In this section, we review how to infer the MAGs and PAGs defined in the previous section from the data. Rather than saying that structural causal models are simply Markovian, we believe it is their asymmetric Markovian nature that should be emphasized. It is already evident in the simplest case, where the conditional independence relation is different in the three-variable case with two causes, and in the same three-variable case with two outcomes. That is, it is asymmetric because the association between outcome variables can be decomposed by the causal variables and the association between causal variables cannot be decomposed by the outcome variables. This asymmetric Markovian nature of cause and effect is a key feature of the Markov causality models. Therefore, one of the basic policies of causal discovery methodology is to identify this asymmetry in the data.

3.1 Basic methodology: Constraint-based approach

In the inference of Markovian structural causal models, the early idea of inferring their properties directly from the data, i.e., finding a set of conditionally independent relational variables, remains the basic structure of the algorithm (Pearl and Verma, 1991; Pearl, 2000; Spirtes et al., 2000). According to this policy, one can examine the adjacency between vertices in a graph by examining conditional independence, and at the same time, constraints due to that set arise, from which further inferences can be layered, thereby making it possible to infer edge orientation (i.e., causal direction) to a significant degree (Pearl and Verma, 1991; Pearl, 2000; Spirtes et al., 2000). Constraint-based causal inference rules for DAG/PDAG (partially DAG) were completed by Verma and Pearl (1992) and Meek (1995a), and the rules for MAG/PAG were completed by Zhang (2008b). In total, there are about ten rules that are applied to PAGs. This methodology/approach is called constraint-based (CB) algorithm/approach.

The Fast Causal Inference (FCI) algorithm is one of the earliest and most prominent algorithms in PAG inference (Spirtes et al., 2000, 1995, 1999; Zhang, 2008b). It is an extension to PAG inference of the PC algorithm (Spirtes and Glymour, 1991; Spirtes et al., 2000), which is still considered the baseline algorithm for PDAG inference today. The wildcard (*) appearing below indicates that it is an arbitrary type of arrowhead/tail at the edge endpoints, and $|\boldsymbol{Z}|$ denotes the number of variables in set $\boldsymbol{Z}$. The basic algorithmic structure of the CB approach, which is also the basis of FCI, is as follows.

A1) Construct a complete graph G with the observed variables as vertices. Each edge consists of $\circ - \circ$.

A2) For any two vertices X and Y on G, search for a condition set $\boldsymbol{S}$ such that X and Y are independent ($X, Y \in\!\!/ \boldsymbol{S}$) in ascending order of $|\boldsymbol{S}|$ and using adjacency information found. If $\boldsymbol{S}$ is found, it is stored in the set *Sepset* (X, Y) and each edge forms a skeleton consisting of $\circ - \circ$.

A3) For a discovered unshielded triple $\langle X, Z, Y\rangle$, if and only if Z does not belong to *Sepset* (X, Y), form the following unshieled collider: $X \ast\!\!\rightarrow Z \leftarrow\!\!\ast\, Y$.

A4) Execute Zhang's ten direction rules to determine arrowheads and tails as much as possible.

In PAG inference that considers the presence of latent variables, compared to the PDAG case, a new set called Possible-D-Sep is defined (Spirtes et al., 1999) and required in Step A2, and a new path called the definite discriminating path is also defined, which is required to determine some directions of edges (Zhang, 2008b).

Estimation of the conditional independence relationship from the data can be done by means of statistical hypothesis tests. Following the standard method of this approach, in conducting a statistical hypothesis test, conditional independence is tested as the null hypothesis and conditional dependence as the alternative hypothesis, and edges are removed only when the null hypothesis is not rejected (Spirtes et al., 2000). As a specific test method, if the random variable is assumed to follow a Gaussian distribution and a linear structural equation model, the following Fisher's Z-transform is typically used:

$$z(\rho_{XY.\mathbf{Z}}, n) = \frac{1}{2}\sqrt{n - |\mathbf{Z}| - 3}\,\ln\left(\left|\frac{1 + \rho_{XY.\mathbf{Z}}}{1 - \rho_{XY.\mathbf{Z}}}\right|\right), \tag{1}$$

where $\rho_{XY.\mathbf{Z}}$ is the population partial correlation between X and Y under a given $\mathbf{Z}$ (Spirtes et al., 2000). On the other hand, for discrete probability distributions, the χ^2 test quantity

$$\chi^2 = \sum \frac{(Observed - Expected)^2}{Expected} \tag{2}$$

can be used for the performance of the independence tests, where *Observed* denotes observed frequency and *Expected* denotes expected frequency (Spirtes et al., 2000). More sophisticated kernel independence test methods have also been developed that are applicable to continuous variables and can capture nonlinear relationships and then be applied to causal inference (Sun et al., 2007; Fukumizu et al., 2008; Gretton et al., 2008; Tillman et al., 2009; Zhang et al., 2011).

3.2 Critical issues, other approaches, and proposals

The CB approach in causal discovery for the latent variable models, starting with the IC* algorithm (Pearl, 2000), however, does not work so well in practice, and much research continues (e.g., (Zhang and Spirtes, 2008; Triantafillou and Tsamardinos, 2016; Zhalama et al., 2017; Marx et al., 2021; Sadeghi and Soo, 2022; Shiragur et al., 2024)), including several proposed algorithms to address the related problems (e.g., (Spirtes et al., 1995, 1999; Ramsey et al., 2006; Ogarrio et al., 2016)). On the other hand, in the FCI algorithm (Spirtes et al., 1999), it was proved that the underlying PAG structure can be correctly inferred if all the conditional independence tests are correct under the CMC and CFC (Spirtes et al., 1999), as is the case for the fully observable model (Spirtes et al., 2000), so the approach is promising from the theoretical perspective. As to why it does not work, it is becoming increasingly recognized that the CFC breaks down, and a fundamental solution has remained elusive. Although research by Spirtes et al. (2000) and Meek (1995b) indicates that the CFC is likely not probabilistically broken, it is recognized that situations frequently occur where the CFC must be considered statistically broken due to lack of statistical power of detection and discrimination in a finite sample size (Ramsey et al., 2006; Zhang and Spirtes, 2008; Isozaki, 2014), and research on other approaches has been advanced to overcome this problem. One is the score-based approach, led by Greedy Equivalent Search (GES) (Chickering, 2002), which uses some model selection criterion (e.g., the Bayesian Information Criterion (BIC) (Schwarz, 1978)) to select a model with the highest scores instead of conducting conditional independence (CI) tests. Although research from the viewpoint of causal models has been slowing down due to the problems of scalability and generality, such as the difficulty of inferring MAG/PAG type-SCMs, GES, which can take the Markov equivalence into account, is still considered a promising approach. In fact, a hybrid algorithm using GES and FCI (or PC) has been proposed (Zhalama et al., 2017; Ogarrio et al., 2016). A new score-based algorithm has also been developed, but it does not match the conservative FCI described below (Triantafillou and Tsamardinos, 2016) in terms of the standard structural Hamming distance criterion (Isozaki and Kuroki, 2017). The conservative FCI does not solve the FCI problem of inaccurate inferences about the presence of edges due to the problem of the CFC violations. Another approach, LiNGAM (Shimizu et al., 2006), has the basic constraint of linear causality with numerical variables, infers linear structural equation models based on independent component analysis (ICA), and exploits the degree of non-Gaussianity under the assumption that causality is an additive relationship. This approach is effective in the inference of causal orders thanks to taking advantage of the assumption that causalities are additive relationships, and it has been studied vigorously. To add, an inference algorithm that takes into account latent confounding variables has also been proposed (Maeda and Shimizu, 2020).

Although the CB approach has a major problem in that the accuracy of inference is not easily improved due to the CFC breakdown, it also has several major practical advantages that seem not to have been mentioned much so far. The strength of structural causal models is that they are highly interpretable and understandable in many cases, and if it were not necessary to understand the model in the first place, there would be little reason for the existence of these SCMs. Interpretability is an essential aspect of the model, as the analysts or users of the results must understand the inferred SCMs and then consider appropriate interventions to help alleviate or resolve the problem in focus. In understanding this structural model, the interpretation of the presence or absence of edges is first and foremost paramount. On this point, the CB approach can give a clear-cut answer: that is, the reason there is no edge for two given variables is that it was inferred from the data that conditional independence is established with a certain set of conditional variables. It is also explicit as to which set of conditional variables did it. Conversely, the reason there is an edge in two variables is also clear, and the answer is understandable: the conditional independence relationship was not found by examining various sets of conditions. This is a major advantage of this approach, which directly and rigorously infers the CMC relationships; other methods, such as those using scores, do not seem able to provide it. When using the SCMs for data analysis, it may be necessary to explain the inference results to various parties, and this property of being able to explain the results to others has become a vital element in industry. More fundamentally, the CB approach guarantees that the inferred results are Markovian causal models based on the CMC at least from the statistical standpoint. There is a suspicion that inference by other approaches, such as score-based and ICA-based ones, does not theoretically guarantee that it is CMC in finite sample sizes.[7] To add to this, since the models inferred by the presence or absence of conditional independence are inferred to possess Markovianity, Markovianity clearly decomposes the edges representing some kind of correlation into direct causality, indirect causality, and pseudo correlation due to hidden confounding variables or selection bias. This inherits an important aspect in the path diagram and path analysis (Kuroki, 2017). Therefore, the meaning of the edges is in principle clearer, and this also makes it easier for analysts and users to understand the inferred results, which makes the transition to intervention and other actions in response to the results of the analysis smoother.

The CB approach has drawbacks in terms of the inference accuracy although it has some practical and theoretical advantages as discussed above, and several studies of the approach thus have attempted to tackle this issue. Some of which are presented here. Ramsey et al. (2006) decomposed CFC into adjacency faithfulness and orientation faithfulness, and proposed a conservative PC (CPC) algorithm applied to DAG/PDAG as a measure to mitigate the orientation faithfulness problem in particular. This algorithm reduces orientation errors by carefully orienting and leaving some edges undirected.

[7] In other words, while the models inferred by these approaches are commonly termed DAG models, it seems to be a subtle question whether the Markovian DAG model, which assumes CMC, is common among these approaches. This suggests that it may be time to clarify the terminology that distinguishes them. We thus often emphasize that we are dealing with the Markovian causal models in this article.

The policy has been extended and applied to FCI for MAG/PAG inference (called conservative FCI (CFCI); note that the output is not maximal informative PDAG/PAG), and several studies related to this stream have followed (e.g., (Lemeire et al., 2012; Zhalama et al., 2017). Extensive research on the CFC is ongoing, including discussion and proposals intended to relax the concept (Zhang, 2013; Zhalama et al., 2017; Zhang and Spirtes, 2016; Marx et al., 2021). Under the hypothesis that the more variables in the set of conditional variables for a CI test, the greater the likelihood of insufficient data and thus the higher the rate of testing errors, algorithms and frameworks have been proposed to avoid tests involving such multiple variables by splitting large causal models into smaller parts (Yehezkel and Lerner, 2009; Xie and Geng, 2008; Kocaoglu, 2023; Shiragur et al., 2024). However, with the exception of Kocaoglu (2023), there has been no extension to MAG/PAG and it is limited to DAG/PDAG (with the further limitation that the study of Shiragur et al. (2024) also uses intervention data). On the other hand, algorithms that aim to be applied to causal model inference for large sets of variables by greatly reducing computation time, with the same level of accuracy as FCI, include the MBCS* algorithm (Pellet and Elisseeff, 2008) and the RFCI algorithm, an approximation algorithm that skips the tests related to Possible D-Sep (Colombo et al., 2012).

3.3 Combining stage algorithm against faithfulness violations

In response to the larger problem that the CFC in the CB approach must be considered weakly broken statistically, we propose an algorithm for inferring DAG/PDAG called the Combining Stage (CS) algorithm (Isozaki, 2014). Later, we extend it for MAG/PAG inference and call it the CS* algorithm (Isozaki and Kuroki, 2017). Here, we provide an overview of the concept and algorithm.

The basic strategy of the CS/CS* algorithm is to avoid unnecessary CI tests. To this end, we proceed with inference without conducting the tests by CB inference as much as possible, assuming that there are some statistical faithfulness violations, i.e., correlation patterns that make it difficult to identify whether or not the null hypothesis can be statistically rejected. This seems a similar but the strategy is quite different to the PC algorithm (Spirtes and Glymour, 1991) that dramatically improved the early algorithms that were computationally intractable such as IC (Verma and Pearl, 1990) and SGS (Spirtes et al., 1990), and we pursue it from the perspective of accuracy rather than computational complexity. It is also an attempt to solve this problem directly, not from the perspective of careful management of the current algorithm or by using other approaches such as (Ramsey et al., 2006; Lemeire et al., 2012; Zhalama et al., 2017; Ogarrio et al., 2016); essentially, we want to take advantage of the CB approach as it is. We define the minimal blocker condition (MBC) in order to proceed. Before we do so, let us give another equivalent definition of m-separation.

Definition 11. *In an ancestral graph G, a path* **p** *between two vertices X and Y is called blocked by a vertex set* **Z** *(X, Y* $\notin$ **Z***) that has the possibility of being the empty set if the following conditions hold: (i) a non-collider is in* **Z** *on* **p***, or (ii) a collider and its descendants are not in* **Z** *on* **p***. If all the paths between X and Y are blocked by* **Z***, X and Y are m-separated by* **Z***.*

Definition 12. (minimal blocker condition) (Isozaki, 2014; Isozaki and Kuroki, 2017) *In* Definition 11, if **Z** *(possibly empty) blocks a path between two vertices X and Y, each element of* **Z** *is called a blocker of the path. Then, among the sets of blockers of a path, the minimal size set of blockers is called the minimal blockers. In a MAG, a blocker set* **S** *satisfies the Minimal Blocker Condition for a path between a pair of vertices X and Y if the path contains no colliders.*

The following two propositions are valid in MAGs, the same as in DAGs (Isozaki, 2014).

Proposition 8. (Isozaki and Kuroki, 2017) *A collider on path* $\boldsymbol{\mu}$ *between two vertices X and Y in a MAG is not a blocker on the path for the pair.*

Proposition 9. (Isozaki and Kuroki, 2017) *If path* $\boldsymbol{\mu}$ *between two vertices X and Y in a MAG has colliders, any vertex Z on* $\boldsymbol{\mu}$ *is not in the minimal blocker on* $\boldsymbol{\mu}$ *for the pair.*

Using the MBC, we have designed the CS/CS* algorithm to significantly reduce the number of CI tests while maintaining theoretical rigor. Its overview is as follows.

C1) Construct a complete graph G with the observed variables as vertices. Each edge consists of $\circ - \circ$.

C2) For any two vertices X and Y on G, search for a condition set $\boldsymbol{S}$ such that X and Y are independent ($X, Y \notin \boldsymbol{S}$) in ascending order of $|\boldsymbol{S}|$ and using adjacency information found. Skip the test for the candidate set that does not satisfy the MBC. If $\boldsymbol{S}$ is found, it is stored in the set *Sepset* (X, Y).

C3) For a discovered unshielded triple $\langle X, Z, Y\rangle$, if and only if Z does not belong to *Sepset* (X, Y), form the following directed collision: $X * \!\!\rightarrow Z \leftarrow\!\! * Y$.

C4) Repeat C2) and C3) until no changes are found.

C5) Execute Zhang's ten directedness rules to determine arrowheads and tails as much as possible, and the output graph is a maximal informative PAG at the end.

The soundness of the above algorithm is guaranteed by the following theorem proved by the authors (Isozaki and Kuroki, 2017).

Theorem 1. (Isozaki and Kuroki, 2017) *Suppose that an algorithm searches for a PAG consisting of a vertex set* **V***. The algorithm starts from a complete graph, performs CI tests in ascending order of conditioning sets* **S** *starting from* $|\mathbf{S}| = 0$ *for any currently connecting pair X, Y* $\in$ **V***, and removes the edge between X and Y if it detects a set* **S** *such that* $X \perp\!\!\!\perp Y \mid \mathbf{S}$*. In such an algorithm, if a conditioning set* **S** *has a condition* $|\mathbf{S}| > 0$, and **S** does not satisfy MBC, **S** is not an m-separator candidate at the order of CI tests in the algorithm.

TABLE 1 Average relative error evaluations of missing edges in each learned PAG and each sample size (1000, 2000, 5000, and 10,000) used in experiments. Smaller values than 1.00 indicate better results than those of CS* algorithm.

Dataset	Algorithm	1000	2000	5000	10,000
	CS*	1.00	1.00	1.00	1.00
Alarm	FCI	1.65	2.11	2.45	2.71
	RFCI	1.65	1.55	1.79	1.58
Carpo	FCI	1.73	2.03	2.28	2.42
	RFCI	1.81	1.95	2.00	1.94
Hailfinder	FCI	1.79	1.81	2.12	2.13
	RFCI	1.55	1.53	1.84	1.86
Insurance	FCI	1.42	1.48	1.73	1.90
	RFCI	1.44	1.39	1.24	1.07
Water	FCI	1.04	1.03	1.04	1.03
	RFCI	1.04	1.02	1.02	1.01

The CS/CS* algorithm[8] is designed to check whether or not the minimal blocker condition is satisfied by utilizing the information of directed edges in the process of inferring adjacency, and to determine whether or not to perform the statistical CI tests. This reduces not only the number of independence tests given the condition set of a large number of variables but also the number of tests even at the lowest order of $|\boldsymbol{S}|$ for which partial correlations are calculated, thereby reducing inference errors due to statistical breaks in faithfulness. In fact, Isozaki and Kuroki (2017) showed that the CS* algorithm was able to suppress errors in inferring adjacent vertices in all cases except the Water dataset (see Table 1), as the authors aimed, i.e., it is more robust against adjacency faithfulness violations[9] in comparison with the prominent FCI and RFCI algorithms. It is particularly interesting to note that even with relatively large sample sizes, which should increase the accuracy of the test, the effect of the CS* algorithm is significantly large in some cases. Also, as shown in Table 2, CS* achieved about 1/3, 1/2, and 2/3 the number of errors compared to FCI and about 1/2, 1/2, and 2/3 the number of errors compared to RFCI in the standard structural Hamming distance index, which represents the total error in the Alarm, Carpo, and Hailfinder datasets in a sample size of 10,000, which is not small.

[8] See the papers (Isozaki, 2014; Isozaki and Kuroki, 2017) for other originalities further included.
[9] The accuracy difference from FCI/RFCI is naturally smaller in Water since there are very few cases where the number of CI tests is reduced by MBC.

TABLE 2 Comparison of relative inference accuracy of CS*, FCI, and RFCI algorithms on several benchmark datasets. Comparison of the average structural Hamming distance using ten different datasets with sample sizes of 5000 (top row) and 10,000 (bottom row). The ratios are calculated using CS* as the standard (1.00). Smaller values indicate better results.

		Alarm	Carpo	Hailfinder	Insurance	Water
	CS*	1.00	1.00	1.00	1.00	1.00
5000	FCI	1.62	1.86	1.59	1.71	0.95
↑	RFCI	1.61	1.79	1.48	1.23	0.96
10,000	FCI	2.94	2.15	1.51	1.79	0.95
↑	RFCI	2.09	1.91	1.44	1.24	0.96

3.4 Other topics

The author previously proposed a variation on the statistic of the CI tests using a concept in physics and demonstrated its utility (Isozaki and Ueno, 2009). This is an expansion of the idea used in the parameter estimation method (Isozaki et al., 2008, 2009), which the author proposed earlier, and is aimed at a physics-like generalization of the maximum likelihood principle to apply even to small data. It is formulated in the motif of a certain similarity between the free energy minimization principle in physics (especially thermodynamics) and estimation/testing/model-selection in statistics. Helmholtz's principle of minimum free energy is the principle that for a quantity $F = U - H/\beta$ expressed in terms of free energy F, internal energy U, entropy H, and inverse temperature β, nature tries to realize a state with minimum F.[10] By defining the data-(inverse) temperature β here $(0 < \beta < 1)$, the author can determine the corresponding inverse temperature for a given number of data, and under these conditions, the amount of Kullback-Leibler divergence between the true likelihood and the likelihood estimated from finite data is regarded as information distortion energy, U, and Shannon entropy H. The formulation for applying this to the G^2 statistic is proposed as follows. The test quantity, with the random variables X, Y, $\mathbf{Z}$, the finite number of data N, and the corresponding finite inverse temperature β, is derived from Helmholtz's minimum free energy principle, which is newly defined as G_β^2 in the following formula.

$$G_\beta^2 = G^2 - 2NI_\beta(X; Y \mid \mathbf{Z})(1 - \beta)/\beta. \tag{3}$$

In Eq. (3), $I_\beta(X; Y|\mathbf{Z})$ is the conditional mutual information quantity (Cover and Thomas, 2006) estimated by the Boltzmann probability distributions for X and Y subject to $\mathbf{Z}$ in the formulation with β of Isozaki et al. (2009). This test quantity is called G_β^2 test quantity by the authors. Naturally, $N \to \infty$ corresponds to $G_\beta^2 \to G^2$, i.e., at the limit of the infinite number of data corresponding to absolute zero, G_β^2 is consistent with G^2, a kind of likelihood ratio test that was proved to be the strongest test by Neyman-Pearson. The G_β^2 test can be said to be a tempered likelihood test by the free energy minimum principle involving inverse temperature β explicitly related to the number of data. Isozaki and Ueno (2009) have shown that the G_β^2 test is more accurate than the G^2 test with respect to edge orientation for the artificial datasets used in the experiment, up to about 500 to 5000 samples. The inverse temperature β formulation was subsequently refined and derived from a more fundamental definition (Isozaki, 2012, 2015).

We will present an overview of the inference of causal effects, which we have not mentioned so far in this article, and then also mention our research. The theory of the inference of causal effects in SCMs was formed by Pearl in a new mathematical framework called the do-calculus (Pearl, 1995, 2000). Pearl

[10] Helmholtz free energy was also noted by neural network giant Le Cun (LeCun and Huang, 2005).

proposed the back door/front door criterion and discussed the identifiability of causal effects when a Markovian SCM is formed. In the case where the SCM is known in the form of a DAG consisting of a vertex set $V = \{X, Y, Z_1, \ldots Z_k\}$, the causal effect on the variable $Y = y$ of an intervention to fix the value $X = x$ for some variable X is defined with the probability density function f as follows:

$$f(y \mid do(X = x)) = \int \cdots \int \frac{f_V(x, y, z_1, \ldots, z_k)}{f_{X \cdot Pa}(x \mid pa(x))} dz_1 \cdots dz_k, \tag{4}$$

where lowercase letters denote specific values of the case variables and $f_{X \cdot Pa}(x \mid pa(x))$ denotes that it is a probability density function conditional on the parent set of X.

What is important in this calculation is the identification of confounding variables, which are variables that affect the variable for which we want to see change (response variable) and that are related to, but not affected by, the intervention variable (also called the treatment variable). The confounding variables are also called adjustment variables. The adjustment variables that satisfy the back door criterion are not uniquely fixed and can be chosen. Noda and Isozaki (2024, 2025) showed that inference accuracy may decrease depending on which set of adjustment variables is selected for a finite sample size, and proposed the Correlation-based Adjustment Variable Selection (CAVS) algorithm to mitigate this decrease for DAGs and completed PDAGs (which can be called maximal informative PDAGs). The CAVS algorithm uses information of both the graph structure and the data to assign a preference between sets of variables that satisfy the backdoor criterion, and is practically effective for not small datasets. Fig. 3 shows its effectiveness for quasi-real-data (Abramson et al., 1996) in their experiments, demonstrating that the CAVS algorithm can select the best adjustment variable set (MountainFcst) to keep estimation accuracy for small sample sizes (Noda and Isozaki, 2024, 2025). In the underlying theory of the inference of intervention effects in MAG/PAGs, recent further developments following Zhang (2008a)'s early work have been remarkable (e.g., (Jaber et al., 2018, 2019, 2022)).

4 Toward fully demonstrating the potential of causal inference in real data analysis

4.1 On the realistic causal structure modeling

In industrial data application scenarios, when people want to determine a cause-and-effect relationship, their motivation is typically to solve, mitigate, or address some problem for improvement. In addition to prediction and identification tasks, for which AI and machine learning are now actively utilized, there is a need to find clues to solve various issues and problems, and to have data analysis support decision making for such improvement activities. Thus, if causal relationships can be determined from the data, they should be able to aid in these

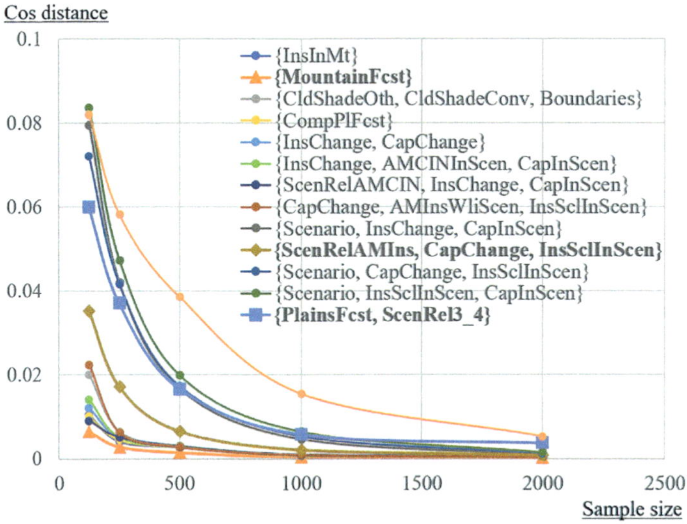

FIG. 3 Experimental results on the accuracy of causal effect estimation with adjusted variable selection on pseudo-real data. The horizontal axis represents the sample size and the vertical axis represents the cosine distance between the correct and estimated values. The CAVS algorithm has been able to select the best choice, MountainFcst, among the many candidates of the adjustment variables, and has been able to avoid a significant drop in estimation accuracy even with very small amounts of data.

considerations (as the case studies discussed below demonstrate). Moreover, now that there is a large amount of uncontrolled observational data available, more and more companies and departments are becoming equipped to do so.

In this situation, what is needed to infer causality through data analysis? The user acquiring the data often has some knowledge in the area, so it is quite possible that they already know the causal relationship to some extent, albeit to a different degree. Therefore, it is difficult to conclude that the analysis was worthwhile if the causal relationships obtained as a result of the analysis were mostly already known to the people who acquired the data or to the users who conducted the analysis. Of course, one of the major values of causal data analysis is that the causal relationships that were somehow already assumed also really emerge, and then the results provide a knowledge boost and convince people to take action with increased objectivity. Also of value is the fact that causal relationships one was previously unaware of can be noticed for the first time through analysis. In particular, a relationship that becomes explicit for the first time through analysis, and that is understandable and satisfactory, is of great value.[11]

[11] Conversely, if the relationship is known for the first time but the sense of conviction is low, there may still be a high hurdle to take action, unless the intervention is expected to have a large effect.

For the above reasons, in the task of inferring causal relationships from data, it is vital to be able to discover new ones, and accordingly, to not miss important relationships that users may be unaware of. In this context, what are the relationships that users know about, and what are the relationships they do not know about? It is of course a flipside relationship, but while it is relatively easy to notice monotone causal relationships, it is often quite difficult to perceive the non-monotonic and nonlinear nature of the relationships. This is because non-monotonic and nonlinear relationships will often seem random to human intuitive understanding and difficult to correlate in a small sample. Indeed, even a light analysis with a small number of samples is more likely to look like a random scatterplot than a monotonic relationship. Moreover, many users are likely unaware of relationships that are likely to occur only due to a combination of factors. This is because it is also not simple that one can understand the relationships immediately from its limited experiences, but at the same time, this is the information one expects to be discovered in the analysis. As an example, think back to a time when you were in a dangerous situation like a car accident. In such a situation, it is easy to imagine that many automobile accidents would have occurred when multiple conditions happened to co-occur. In other words, in causal inference, the ability to capture non-monotonic and nonlinear relationships as well as co-occurring relationships would make the analysis more valuable, and these would be important features. Also, of course, not all variables that are minimally necessary to describe all causal relationships will be observed. Therefore, it is also necessary to clarify how unobserved variables are positioned in the model, and furthermore, to have causal inference techniques that take them into account (this is why we introduced MAG/PAGs, which are causal models to identify pseudo correlations produced by latent variables). We are thus working on the assumption of the following requirements for causal models to be inferred: (1) the consideration of non-observables, (2) the discovery of non-linear causality, and (3) the discovery of non-additive (i.e., co-occurring) causality. In fact, in many analyses, new and important findings are obtained from data by identifying complex causal relationships in (2) and (3).

We recently recognized that these ideas aligned somewhat with the methodology of causal effect inference proposed by Ogburn and Shpitser (2021) corresponding to the criticisms of statistics in the well-known paper by Breiman (2001), and we had made a causal discovery analytics independently. Breiman (2001) believes that the emphasis of traditional statistics is what can be called a data modeling culture, based on parametric modeling. On the other hand, there is what could be called an algorithmic modeling culture, which emphasizes algorithms for prediction and identification in machine learning. His insights on these two cultures can be summarized as follows.

- The data modeling culture based on statistics has placed too much trust in overly parametric models, which may not achieve good modeling.

- The idea is that it is important to achieve the objective and more emphasis should be placed on algorithms, even if they are black-box-like, although the focus seems to be primarily on forecasting and identification tasks and it is often not necessary to be able to understand the data in the first place.

The greater the emphasis on data modeling, the more assumptions and suppositions tend to be made, such as probability distribution functions, the specification of input-output functional forms (e.g., logistics type, linearity, additivity, etc.), and so on. This increases the possibility of incorrect representation of real-world data. In this respect, as we have already mentioned, we agree with Breiman's criticism in many parts. On the other hand, the excessive bias toward algorithms, as Breiman argues, is not compatible with the idea that causal modeling is originally a type of modeling for human understanding, as already discussed in Section 2, and therefore, the black box nature should be eliminated as much as possible. Similar ideas have already been expressed by Ogburn and Shpitser (2021) in the context of causal effect inference. Regarding causal modeling, they call the (supervised) machine learning approach to inference of causal effects (a position almost identical to the algorithmic modeling approach described by Breiman) the optimists, and warn that excessive optimism can cause problems, a position the author also holds.

Interestingly, however, algorithms are intrinsically important to causal data modeling, as we have already noted. As mentioned in the previous section, our research and other studies have shown that the accuracy of inferred models relies on algorithms, not just on the reduction of computational efficiency. Therefore, the emphasis on algorithms with respect to causal discovery inference is an essential aspect in the research domain. On the other hand, if the algorithm is too much of a black box, the reasons for the composition of the output model may not be understood, so the idea of a data modeling culture is still important. In this sense, as Ogburn and Shpitser (2021) state, a balanced integration of data modeling and algorithmic modeling is important even in the causal discovery inference domain.

4.2 On the validity of the SCMs and data to be entered

The validity of the structural causal models based on the causal Markov condition as a model of causality has long been debated (e.g. (Pearl, 2000; Spirtes et al., 2000; Hausman and Woodward, 1999; Cartwright, 2001; Hausman and Woodward, 2004; Dawid, 2010)). Since the causal faithfulness condition has already been discussed in Sections 3.2 and 3.3, we here examine the causal Markov condition (CMC) from the standpoint and experience of those engaged in actual data analysis.

The CMC that the SCM assumes must be composed of appropriate variables, as the SCM is said to be a model that describes the data generation process among variables. We pay particular attention to the criticism that the CMC assumption may not hold for arbitrary random variables (e.g., (Cartwright, 2001)). To begin with, a review of the history of causal modeling in the early days of Wright (1921),

Simon (1954), Blalock (1962), and others shows that SCM is a model of naive cause and effect chain processes. Therefore, we believe there are some variable sets that do not conform to this model (in particular, variable sets that are inclusive, such as the coexistence of a variable calculated from a set of variables and its original variable, as Cartwright says, are dangerous), and we recommend that analysis be conducted with care in variable selection and variable creation. In other words, the set of variables for which CMC is valid is not the set of all random variables but rather the set of *natural* variables that can form causal chain processes, and we recognize that this is a problem that requires further consideration and discussion before it can be generalized. Therefore, instructors who train analysts need to be well versed in understanding the origins of SCMs and the CMC.

At the same time, we believe that users on the business side need not be so concerned about the criticism that SCMs based on the CMC may not be able to express all causal relationships, as Cartwright (2001) has stated. What users should do is infer causal relationships from the data to the extent that current science is able to do so, and recognize that this does not need be perfect; if it is valid as a suggestion, it will be useful enough for business decision making. In fact, as discussed in Section 6, many valid cases have been found.

4.3 Training causal modelers and promoting the dissemination of causal inference

When using causal inference in actual data analysis, it is extremely risky to use analytical tools without understanding the overview of causal models, as erroneous conclusions may be rampant; the value of causal inference will eventually be greatly undermined if this happens.[12] Our recommendations for avoiding such erroneous conclusions are as follows.

- Examine whether the input variables are appropriate for causal modeling based on the CMC, based as per on the discussion in the previous subsection, and develop know-how through trial and error and experience.
- Note that inferred causality is not proven and is not 100 % accurate.
- Therefore, examine the possible interpretations of the model using as much domain knowledge as possible.
- The mentor should convey the meaning of causality and pseudo correlation based on the foundational properties of the Markovian SCMs.
- Interpret the results based on the basic algorithmic features for Markovian inference.
- The above should be taught to users by researchers or mentors familiar with the theory and algorithms, and analysts should be trained accordingly.
- Know-how should be accumulated and gradually clarified to make future training more efficient and effective.

[12] Thanks to Prof. Kuroki for his discussions on the tool and its social implementation.

While consolidating these ideas, the author has expanded the practice of causal inference as a project, which began with the author and a promotion member, and continues to manage the project and train analysts. We have been making efforts to spread the use adoption of causal inference by introducing the existence and use of our own tools to new employees of Sony Group companies in Japan, and have also been cultivating analysts in less mature stages by introducing the basic concept of causal inference, our tools, and case studies while increasing the number of project members. It is now widely used by employees of Sony Group companies around the world as an important leading option for data analysis.

5 Discussion from the viewpoint of another discipline

The following is an attempt to clarify the significance of causal inference from the perspective of metascience in contrast to another discipline. The importance of partial correlation in causal models has already been described in the works of Wright (1921), Simon (1954), and Blalock (1962) et al. In fact, examining conditional independence is nothing more than examining partial correlation in a broad sense, but when nonlinearities and non-parametrics are also included, the term "partial correlation" seems to be used mostly in the framework of the linear causal models only. However, the importance of the concept of partial correlation itself still stands out and should be emphasized in causal inference. This subsection discusses this point in terms of contrast with physics.

The theoretical roots important for modeling the world in the major areas of physics are typically expressed as partial differential equations, such as the Euler-Lagrange equation in analytical mechanics,[13] Maxwell equations in electromagnetism,[14] Schr¨odinger equation in quantum mechanics,[15] Navier-Stokes.

equation in fluid mechanics, and Einstein 's equation in general relativity. In other words, the universal importance of partial derivatives stands out more than that of the total derivatives of multivariable for modeling a world consisting of many variables, as a total differential ($df(x_1, \ldots, x_n)$) is decomposed into the sum of partial differentiations as follows.

$$df(x_1, \ldots, x_n) = \sum_{k=1}^{n} \frac{\partial f}{\partial x_k} dx_k.$$

Analogous to this fact, the concept of partial correlation, which has a closer correspondence to the partial derivative, may be just as important in statistics as in physics, rather than the concept of correlation, which is a mixture of various relationships that correspond to the total derivative. Unlike partial derivatives, however, partial correlations are difficult to utilize except when the

---

[13] $\partial L(q(t), \dot{q}(t), t)/\partial q_i - d/dt(\partial L(q(t), \dot{q}(t), t)/\partial \dot{q}_i) = 0.$

[14] e.g., $\nabla \times \boldsymbol{E}(\boldsymbol{r}, t) = -\partial \boldsymbol{B}(\boldsymbol{r}, t)/\partial t.$

[15] $i\hbar\, \partial\psi(\boldsymbol{r}, t)/\partial t = \hat{H}\, \psi(\boldsymbol{r}, t).$

relationship is linear, as it is challenging to identify the variables involved. Causal inference analysts perform the task of searching for input variables by calculating partial correlations. In this sense, we believe that the combination of causal modeling and inference is a significant analysis methodology from the viewpoint of statistical modeling because it enables partial correlation analysis to be performed, which is desirable from the viewpoint of world modeling but has been difficult to achieve so far.

We would like to add a little more scientific perspective on the subject. Specifically, the theory and algorithms of Markovian SCMs decompose correlation into direct causation, indirect causation, and pseudo correlation due to latent confounding variables and selection bias. This decomposition is certainly effective in terms of extracting, understanding, and utilizing information from observation data (see Section 3.2). Science has developed in the direction of decomposing what can be decomposed, and from the viewpoint of scientific development, this trend of decomposing correlations seems a natural and correct direction.

6 Case studies

We have implemented the causal inference (both discovery and intervention inference) methods and algorithms that we have researched and developed as a software named Causight,[16] and have provided it to various users. In addition to the original algorithms and methods introduced in this article and the previous papers (Isozaki, 2014; Isozaki and Kuroki, 2017), Causight also implements several proprietary technologies that are not yet in the public domain to effectively perform causal analysis, such as more improved inference algorithm from CS*, methods that integrate various data formats and variable types, effective preprocessing methods, and efficient and effective ways to display and deepen the analysis results.

Sony is a conglomerate company that operates a wide variety of businesses spanning electronics, semiconductors, games, movies, music, life insurance, accident and disaster insurance, banking, Internet providers, sports support systems, and more. Causight has been utilized in most of its domains worldwide. At Sony, causal analysis is becoming widely recognized as an important analysis method for factor analysis. Causight is also provided as a software tool to external companies, and it is currently being used by many major companies in Japan. In this section, we will clarify that causal inference is effective in a wide variety of fields, including manufacturing (automotive, plating, panel manufacturing, semiconductors), marketing (finance, electronics, music), and education.

[16] This software had previously only been available in Japan under the name CALC, but it has been expanded to North America and Europe under the new name Causight from 2025.

6.1 Data analysis activities at Mazda's automotive manufacturing plant

The General Incorporated Association Value Chain Initiative (IVI) (https://ivi.org/en/en-top/) organization in Japan separates the parts of the manufacturing industry that should compete from the parts that should cooperate with each other, and opens up issues that are difficult for companies to solve alone, encouraging the participating companies to work together to solve them (https://ivi.org/en/organizations/). We here present an example of an issue that the automaker Mazda Motor Corporation (https://www.mazda.com) tackled within the IVI, and show how Causight contributed to solving the problem. The following results and photos are based on a Japanese-language material released by IVI.[17]

In the casting process of cylinder head parts, which are also used for engine parts (Fig. 4) that are implemented in Mazda's proprietary SKYACTIV engine technology,[18] there was a defect in the hole that had been a problem for many years (Fig. 5). This defect was believed to lead to product defects such as engine strength degradation and oil leakage, and despite years of investigation to solve the problem, the true cause was not found.

Causal analysis was carried out using historical data of the casting process for parts containing molten metal components made of various substances and materials. The direct influence of a component, which had been unknown until then, was found by Causight, and it was also suggested that the amount of components could be related to the problem. Subsequent offline experimental verification confirmed that the components had a real influence, and also showed that the degree of influence differed depending on the amount of components. In this way, an important factor affecting the quality failure of automobile engine parts was discovered by causal analysis, and causal inference could significantly contribute to solving the problem. The results of this activity were commended by IVI, receiving IVI's Excellence Award in 2020.

6.2 Manufacturing data analysis activities of Tsukada Riken industry

Tsukada Riken Industry Co., Ltd. (https://www.tukada-riken.co.jp/en/) is a manufacturing company in Japan that manufactures and sells a variety of plastic plating and other plating products. Fig. 6 shows the blue plating products of this company. In the plating manufacturing process of this company, it is known that there are several modes in which defective products occur. In order to eliminate one of the defect modes, historical data of the manufacturing process was analyzed by Causight. Small defects of several tens of microns sometimes occur in products using satin plating, which is a plating method for diffusing light on the plating surface. It is often used for the design of automobile exteriors and faucet

[17] https://iv-i.org/wp/wp-content/uploads/2020/03/symposium2020-spring14.pdf.

[18] https://www.mazda.co.jp/beadriver/dynamics/skyactiv/.

FIG. 4 A Mazda car with a SKYACTIV engine and its cylinder head.

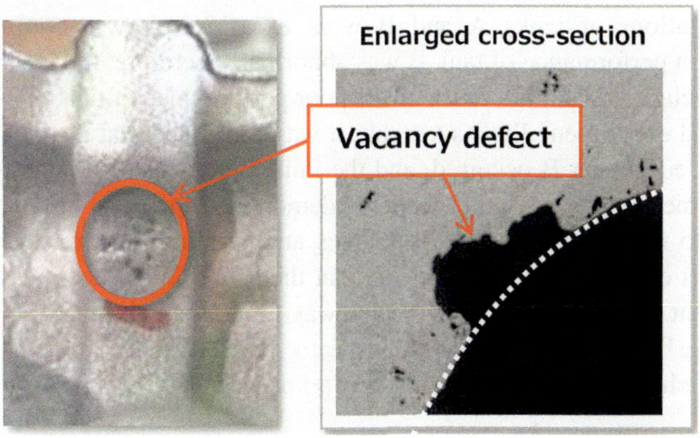

FIG. 5 View of a cylinder head in engine parts and its defects.

FIG. 6 Blue plating products manufactured by Tsukada Riken.

fittings, and in this case, especially strict control is required. Satin plating is formed by stacking layers of copper, nickel, and chromium on a plastic material in order. In this process, the plastic materials are immersed in multiple tanks filled with plating solution. Data were accumulated for each production lot from production to inspection and then analyzed by Causight. As a result, the temperature of one of the tanks, called tank B, was inferred as a variable affecting the failure rate in that mode. An increase in the temperature of tank B indicated that tank B was used, and therefore, it was suggested that there was a problem in tank B. When investigating the cause of this variable further back, the date-time information variable appeared, which is information indicating the date and time at which tank B was used. A causal chain was thus suggested (i.e., date-time → temperature of tank B → failure rate), and the users' sense of conviction increased because they were able to understand the causal relationship. When the plating solutions of tanks A and B were examined, it was confirmed that the filtration performance of tank B was abnormal. Therefore, the location of the cause was clarified, and the cause of the poor performance of tank B was further investigated experimentally. As a result, it was confirmed that a slight backflow of water used in tank B occurred, and the failure rate was significantly reduced by taking measures to prevent this. This demonstrates that the causal inference was able to narrow down which variables among various correlated variables had a direct effect on the problem, and that the action to dig deep was effective and efficient. In addition, another factor was found, namely, that the judgment results were biased depending on the inspector. This led to the implementation of employee education, and as a result, this point was also improved.

6.3 Manufacturing data analysis activities at JEMS

Japan E.M. Solutions Co., Ltd. (JEMS) is a Japanese company that manufactures various types of monitors and wearable devices, and is promoting the use of factory big data (https://www.j-em-s.com/en/index.html). At the time of the analysis, the company was already promoting standardization and automation of manufacturing processes and had achieved nearly 100 % automation of factory lines, but the sheer volume of factory data was taking time to analyze for improvement. Therefore, the company decided to use Causight for the purpose of efficient improvement of the manufacturing process (Fig. 7). As a result, the company was able to achieve overwhelming improvements in inspection accuracy, inspection time reduction, and the false alarm rate of inspection. Also, with regard to predictive maintenance of consumables, it was found that the number of NGs increased according to a certain range of values of a particular automatic machine transfer pallet, which clarified when the pallet should be replaced and made it possible to use the tool for predictive maintenance. Other helpful results were obtained as well, such as the discovery of many false alarms for certain operators, which clarified the points for work guidance (Isozaki et al., 2024).

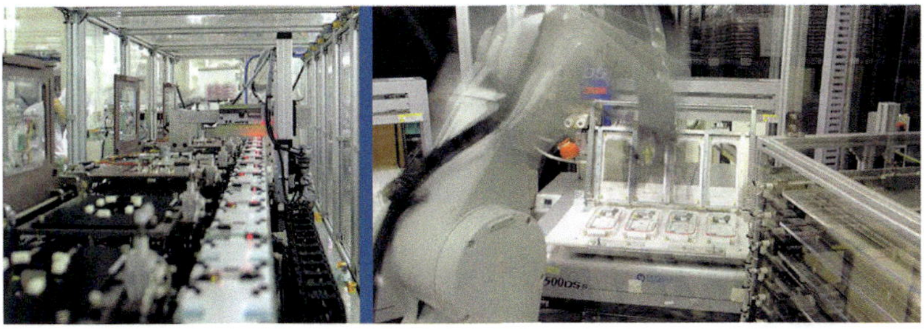

FIG. 7 JEMS manufacturing process.

6.4 Data analysis activities at Sony's semiconductor manufacturing plant

Sony Semiconductor Manufacturing Corporation, a semiconductor affiliate of the Sony group companies, is engaged in the mass production development and manufacturing of various semiconductor products, including CMOS image sensors. Image sensors are semiconductors that convert light into electrical signals and serve as a basic device for image processing in digital cameras, cameras mounted in automobiles and drones, security cameras, etc., as shown in Fig. 8. These sensors are also implemented in smart-phones, making them one of the most important electronic devices in modern times. Sony supplies these image sensors not only to its own company but also to many others: according to a report by Yole (https://www.yolegroup.com/), Sony's market share of image sensor semiconductors was 42 % in 2022 and 45 % in 2023, remaining at the top of the world.[19] This semiconductor was also utilized in the camera that captured and provided an image of the decisive moment when the Japanese football player Kaoru Mitoma didn't break the line, called Mitoma's 1-mm (Fig. 9), which was a hot topic at the 2022 FIFA World Cup in Qatar.

The manufacturing process of image sensor semiconductors consists of hundreds of processes, each of which is accompanied by many parameters such as processing dimensions and equipment parameters. As a result, it is a huge and complex process with thousands to tens of thousands of variables extracted as data. In addition, since the preceding process often affects the following processes, pseudo correlation is likely to occur, which makes it difficult to determine which variables to control to improve the manufacturing yield. Conventionally, domain knowledge based on past cases and the experience of engineers have been relied on to a large extent since the relationship between many variables is complicated, nonlinear causality is assumed, and various forms of data must be handled uniformly. Therefore, it is difficult to distinguish cause and effect by conventional analysis alone.

---

[19] https://www.yolegroup.com/press-release/sony-is-set-to-capture-50-of-the-cis-market/.

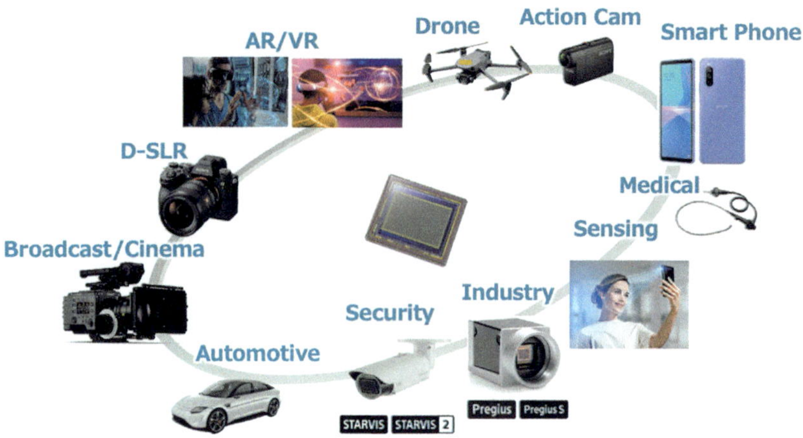

FIG. 8 Applications of Sony's image sensor.

FIG. 9 Application of Sony's image sensor in the camera that took the famous photo of Japanese football player Kaoru Mitoma as *Mitoma's 1-mm* at the 2022 FIFA World Cup in Qatar.

At this production site, the defect rate is predicted by machine learning, and the actual defect rate results are also obtained afterwards. The defect rate has been decreasing for a certain time, and although LightGBM (Ke et al., 2017), a Gradient Boosting Machine (Friedman, 2001) method that is basically an advanced form of the decision tree utilized in the machine learning method, was able to predict the decrease (see the left side of Fig. 10), the cause of the decrease in the defect rate was unknown. The analysts at the plant therefore analyzed the variables that have a large contribution to the predicted values in the prediction model using SHapley Additive exPlanations (SHAP) (Lundberg and Lee, 2017), which is a well-known Explainable AI (XAI, see e.g. (Arrieta et al., 2020))

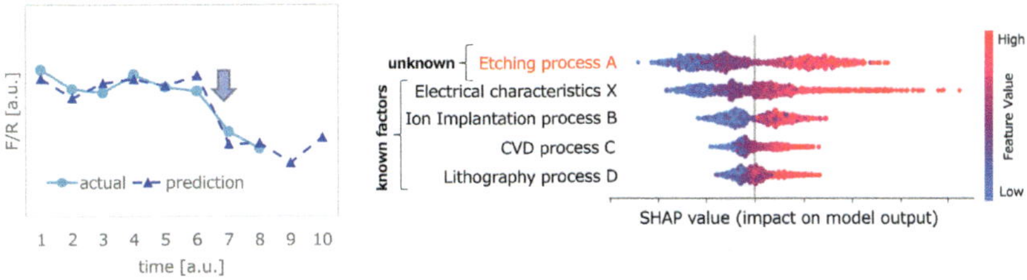

FIG. 10 Comparison of predicted and actual measurement in the left side. Analysis result of SHAP, which represents the basis of the prediction in the right side.

method. They found that a measured value of the performance in etching process A (called EP-A as a variable from now on) had a large influence on the predicted defect rate. The result of SHAP including the association of the variable EP-A is shown on the right side of Fig. 10. However, the mechanism of why this variable influenced the defect rate remained unclear, which made it impossible to determine whether it influenced the defect rate indirectly[20] (i.e., there was a causal relationship) or whether it only fluctuated in conjunction with other factors (i.e., there was no causal relationship but rather a pseudo correlation). Since the decrease in the defect rate was not a controlled result, it was necessary to confirm whether it was possible to reduce and stabilize the defect rate by controlling EP-A. In other words, clarifying the mechanism of the influence of EP-A on the defect rate (also called false rate) was a key task in the manufacturing process of this device.

Therefore, they inferred the causality behind the data of the manufacturing processes using Causight. Setting values, information about environments and equipment units, and various measured values were recorded as data and then utilized as variables. The inferred causal graph is partially shown in Fig. 11 (here, we often used a simplified representation graph of the latent variable model that we set up independently). As indicated in this graph, it was inferred that the variable EP-A, which we had been focusing on, affected three groups of variables belonging to the measurement values of electrical characteristics Y, lithography process F, and CVD process E, and these directly affected the false rate (F/R), which is the variable of interest in this analysis, and those relationships formed causal paths, as shown in Fig. 11. A simplified model of this result is shown in Fig. 12, where EP-A (shown on the left), whose role had been unknown, is now known to affect three groups, and they all affected F/R. In short, EP-A was interpreted as a more fundamental causal variable of F/R. Physical interpretation was also performed, and the interpretation was convincing among the members. This relationship could be understood in the time-series diagram of the process on the right, and it was judged that controlling

[20] The officials were already aware that EP-A was not the direct cause of the defect rate.

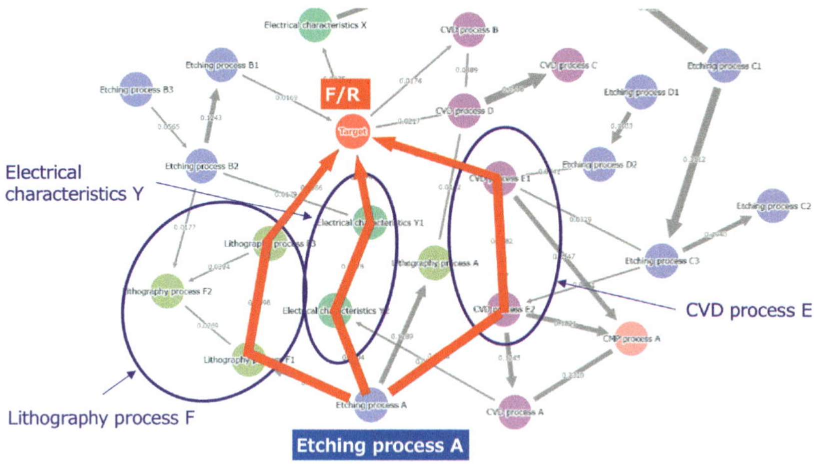

FIG. 11 Inferred partial causal graph. The red arrows represent the causal chains.

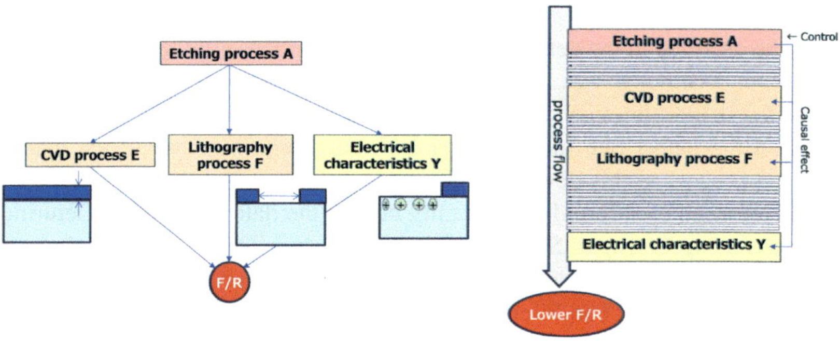

FIG. 12 The mechanism discovered by the causal analysis and interpreted by the officials. Intervention and control were performed accordingly.

EP-A could also control F/R. Therefore, when they actually intervened and controlled this fundamental factor, they were able to keep F/R stably low, and analysis by causal inference made a significant contribution. The significance of using the causal inference tool was then recognized. As a result, Causight is now linked to the unified analysis system of multiple plants, and well over several dozen trained production site workers utilize it regularly. What is noteworthy here is that the combination of causal models with machine learning and XAI was effective. The prediction task itself is important, and machine learning has been used, and in recent years, XAI tools also are beginning to be implemented in industry. However, it is often difficult to decide whether to immediately intervene on important variables identified by

XAI tools. This case clearly shows that causal analysis has enabled us to clarify the causal role of the variables and to make efficient and effective decisions about intervention.

6.5 Contribution to Sony assurance's marketing activities

Sony Assurance Inc. is an accident and disaster insurance company operating in Japan, providing products such as automobile insurance and fire insurance. One of its key sales methods is direct marketing. Data analysis is an active part of its marketing activities, and Causight has been used for many years. Here, we will present an example of an analysis of a direct mail promotion to obtain contracts (Isozaki et al., 2024).

Sony Assurance has been sending direct mail (DM) to existing customers who are known to own more than one car and encouraging them to sign up for a second car, with the aim of acquiring new policies. One of the issues in this activity has been how to improve the application rate for automobile insurance after the DMs are sent. Therefore, Causight analyzed past DM response data to search for variables that directly contribute to the variable of whether or not a new policy is obtained, and the variable of "age" emerged from other many variables. This indicates that there is a difference in the subscription rate for the second vehicle depending on the age of the customer.

In many cases, it is difficult to narrow down the variables by the magnitude of correlation alone, since there are many correlated variables, and in many cases, the line is drawn based on a slight difference in the magnitude of the correlation. However, by understanding the difference between direct and indirect effects, or whether the correlation is large but inferred to be a pseudo correlation, Causight analysis makes it easier to narrow down the variables and facilitate the transition to the next business action, which is how to implement measures and interventions for those variables. Its business associates then considered how to absorb this age difference and decided to take the action of changing the content of their traditional DMs for the older age groups that had not responded well. The previous DM design mainly emphasized price and discounts regardless of age, but for the older age group, the design was changed to visually emphasize the benefits of consolidating insurance for the whole family (Fig. 13).

When AB testing was conducted on these two designs, the application rate increased 1.5 times for Group B with the new design compared to Group A with the conventional design (Fig. 14). By narrowing the focus to the variables that Causight analysis inferred had a more direct impact, the business activity efficiency was improved by moving quickly to the activity of considering action, and the actual contract rate was increased by 1.5 times.

6.6 Contribution to marketing activities at Sony sales company

A data analysis using causal inference was conducted on the sales activities of a Sony group company located outside of Japan. The challenge was how to develop

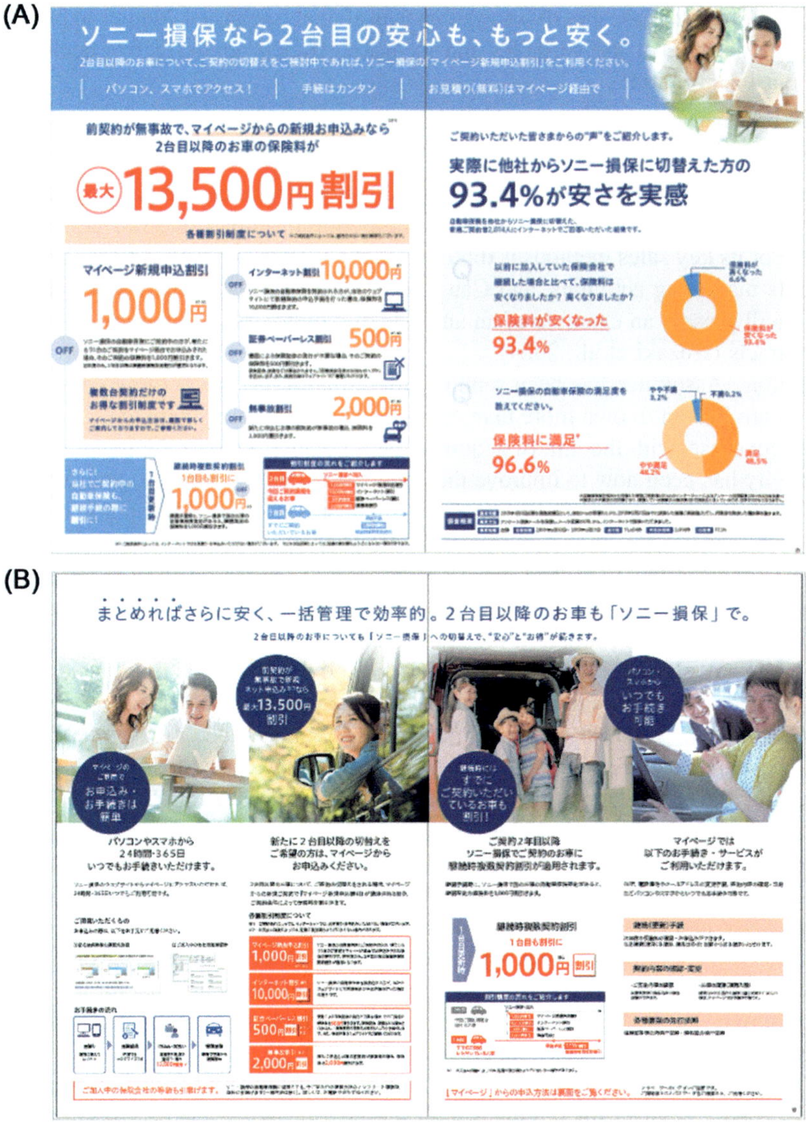

FIG. 13 Direct mail advertisement featuring (A) the conventional design and the design based on the results of causal inference analysis. While ad (A) offers a discount of up to 13,500 yen, (B) visually emphasizes the benefits of combining insurance for a family.

effective marketing activities. The marketing team found it difficult to come up with customer segments that would lead to sales. Therefore, data on customer attributes and purchase history were analyzed utilizing causal inference to determine which attributes and behavioral characteristics loyal customers possess.

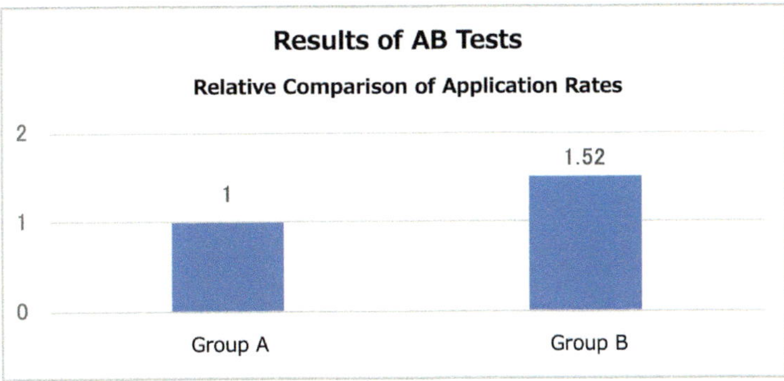

FIG. 14 Increasing ratio of new subscribers acquired. Compared to the conventional design A, the new design B produced a significant sales effect, with a 1.5-fold increase in the contract rate.

Although the team knew that various factors were correlated, causal analysis allowed them to narrow down the factors that were directly related.

The team focused on a variable that had not previously been on their radar among the inferred direct causes: the type of email address used by the customer. The analysis results suggested that the type of email address clearly differentiates the degree of loyalty. Fig. 15 shows the conditional probabilities of this relationship, where the horizontal axis is the type of email address used by the customers, which denotes the condition, and the vertical axis is the probability distribution of each. Blue represents loyal customers, red represents normal customers, and orange represents customers who have drifted away. Each of the classes was defined by the team. Among the email types A, B, C, and D, we can see that there are many normal customers, especially in A and B, but still, the degree of loyalty varies considerably.

The team then further validated the results of this analysis through AB testing. When the team distributed coupons for Sony's audio products via direct mail during Black Friday to one group of email addresses (Group A) and another group (Group B′), Group A was twice as large as Group B′ in terms of both click-through rate and sales. The result was that the company was able to identify loyal customers and effectively utilize management resources, which was reflected in the sales activities.

6.7 Analysis of learning machines for music hit predictions at Sony music publishing

Sony Music Publishing Inc. (SMP) is a music publishing company that manages, develops, and promotes copyrights for music around the world. Early prediction of the degree to which SMP label's song will be a hit is an important component of management operations at SMP. At the same time, digital information such as streaming data is easy to collect, and indicators such as

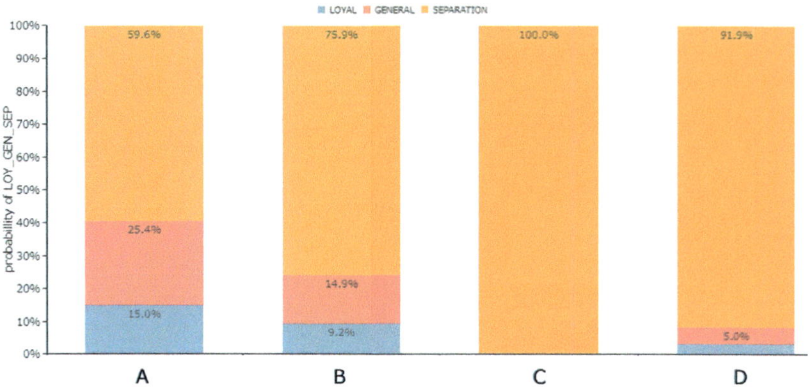

FIG. 15 Conditional probability chart showing differences in loyalty probability by email address types A, B, C, and D.

artists' popularity and influence can be easily obtained as information related to SNS applications, making the music field a good match for data analysis. After a song is released, machine learning is utilized to immediately perform predictive analysis, but predictions can naturally fail. Therefore, the team used causal inference to analyze the causes of the mistaken predictions. This is the act of analyzing the learning machines, so to speak. Such analysis is not limited to this team at Sony. Since we, as researchers and developers of causal inference methods, algorithms, and tools, encourage this type of analysis, several teams have actually performed it on data from a variety of domains. As a result of this analysis, the analysts were able to identify the causes of difficult predictions among various correlated factors and update the machine learning predictors accordingly, leading to a significant improvement in the accuracy of the predictions. This result became well known to many.

6.8 Contribution to the education system through the development of Sony Network Communications Inc

Sony Network Communications Inc. (SNC) formed a joint research project with Saitama Prefecture, a local government in Japan, and the Ministry of Education, Culture, Sports, Science and Technology (MEXT) to conduct AI-based student learning support system research. In this activity, SNC used the causal inference tool Causight in the background to compile and analyze the results of academic achievement tests in arithmetic, mathematics, and Japanese language for approximately 300,000 elementary and junior high school students in Saitama Prefecture, as well as survey data on non-cognitive skills related to daily life, study habits, and attitude toward learning, to develop a learning support system for each individual student. This system visualizes each student's stumbling points and is designed to allow students to obtain information on their weak points, review materials for those

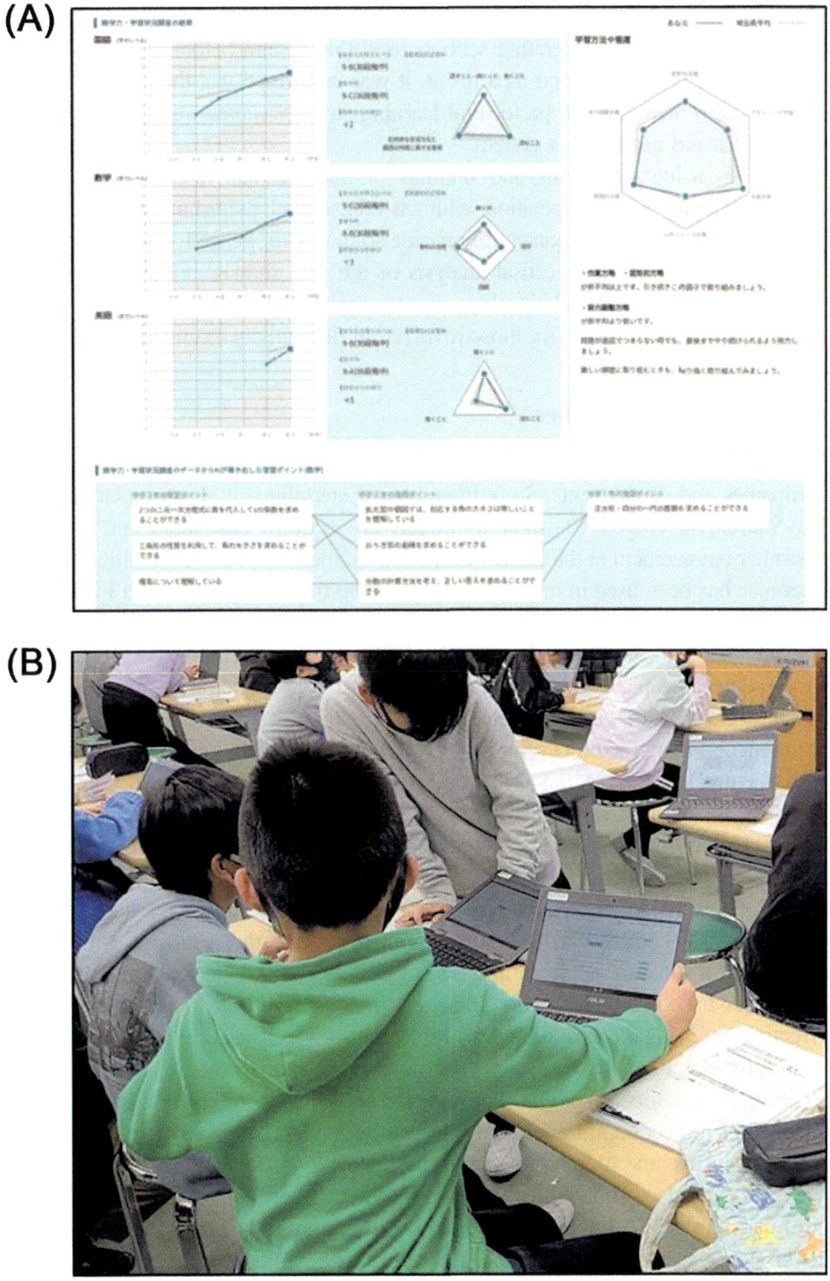

FIG. 16 SNC's practical research activities on AI utilization. (A) Part of the learning support system developed by SNC. (B) Elementary school students accessing the system through tablets.

weak points, and access advice via the network from tablets distributed at each school (Fig. 16). The cooperating schools used this system before morning classes and for home study, etc., and as a result, it was confirmed that the growth in academic achievement and the prefectural learning strategy scores were higher than in schools that did not use this system.

SNC is using the results and findings of this practical research project and commercializing it in cooperation with an educational publishing company that provides learning material content to provide solutions for general schools and cram schools. The results of the causal analysis of the data showed nonlinear and non-additive (co-occurrence) relationships among various variables, which were evaluated as useful information for those involved in this activity to utilize in the future.

6.9 Other case studies

To close this section, we briefly present a few other applications in several user companies and departments. Sony Interactive Entertainment, the company behind PlayStation, leveraged Causight to effectively acquire customers and increase customer engagement at the launch of the PlayStation Network. In addition, causal inference has been used in research surveys at the firm Tokyu Research Institute to analyze the preferences and behavior of visitors to large downtown areas in the metropolitan area in Japan, and by the PERSOL Research Institute to analyze survey data on Japanese work styles. Seiko Epson Corporation (a machine manufacturer), AGC (a glass manufacturer), Murata Manufacturing (an electronic components manufacturer), Sumitomo Heavy Industries (a heavy machinery manufacturer), and ANA (an airline company) are among the companies that have disclosed their use of Causight for causal inference as user companies. Dentsu, an advertising company in Japan, also conducts data analysis and consulting services that focus on non-financial values. We should also mention that the results of causal analysis using Causight provided important suggestions for deciding the content of TV commercials for some companies, and the results of a user log analysis of a software application gave suggestions on how to make the software more user-friendly, which were reflected in the UI design and proved to be effective. Inferring the causal effects of interventions is also being used, although it is unfortunate that there are no specific examples that can be made public. However, we can confirm that there are indeed some specific use cases, such as analysis of the economic effects of interventions in certain businesses, which we believe will become more common in the business world.

Causight also contributes to social science, epidemiology, and public administration. We have conducted questionnaire data analysis on positive psychology with social psychologist professors Takashi Maeno, Toshiyuki Yasui, and Makiko Takao (Kitashima et al., 2021; Yasui et al., 2021; Takao et al., 2023; Yasui et al., 2024, 2025). Additionally, we are working with Japanese medical universities to conduct a causal analysis of a large-scale disease database in Japan.

6.10 Summary of case studies

In this article, we have presented examples of the use of causal inference in a wide range of businesses of Sony group companies, including manufacturing, marketing (consumer electronic and insurance products), music, and education, as well as examples of non-Sony manufacturing companies. In these cases, the data analysis using causal inference was able to efficiently determine both the direct and indirect causes of the problem, which would have been difficult or time-consuming using conventional methods, leading to prompt interventional actions in business. These examples demonstrate that causal inference has had significant effects in the real world. Thus, if workers have a dataset with a certain sample size and set of variables, and have mentors who are well versed in the causal theory and methods, they can now take important business actions (e.g., interventions and measures) that are transparent, efficient, and effective, with some objective prospects and by persuading those around them more easily.

7 Conclusion

This article has described the industrial applications of data analysis using the science and technology of causal inference based on Markovian structural causal models. As a background, we reviewed the representations, theories, and algorithms of the causal models, and discussed our research on the difficulties of inference with finite data and how to alleviate them. Next, we described the forms of the model representation and some debate on the validity of the model and data in order to make the analysis useful for real-world data. Those are related to causal modeling methodology in response to criticisms in statistical modeling. We also mentioned methods for dissemination of the causal analytics and for training analysts. We then presented case studies of the inference analyses using Causight, our causal inference tool for both discovery and intervention, which is being utilized in various companies and divisions at Sony, a conglomerate with a wide range of businesses including electronics products, entertainment, financial services, and more. We presented examples of applications of the tool, which we also sell to external customers, in the automotive, panel manufacturing, and plating industries. These examples demonstrate how the inference of cause-and-effect relationships from data leads to rapid transition and effective implementation of business-critical actions such as measures and interventions.

Acknowledgments

The author would like to thank a number of people for their help and support with this work.

Hiroaki Kitano, President of Sony Computer Science Laboratories, Inc. (Sony CSL), and Shinichi Furuya, Head of Tokyo Research of Sony CSL, for various foundational supports.

Tetsu Natsume of Sony CSL, who managed the entire project with the author as well as managed the promotion department for disseminating the results of the research both within and outside of Sony.

Atsushi Noda, Masahiro Yamamoto, and Ryonosuke Koda, who helped with research, product development, data analysis practice, analysis support, sales support, education, and analyst training activities.

Prof. Manabu Kuroki for his research collaboration.

Dr. Shin Takagi for his many valuable comments.

The Sony CSL staff, including Yoichi Ueda, Ikuko Matsusaka, and Asuka Tejima, for performing various tasks such as sales, contracts, and public relations with Sony group companies.

The team of Yoshifumi Ueno, Ryo Mukaiyama, and Kou Kobayashi, who collaborated on the causal analytics project at Sony Group Corp. and promoted analysis practices, support, and SaaS development, along with the team of Noriyuki Sakai, Yoshiyuki Nakamura, and Hironori Morita, who mainly promoted analysis practices.

Members Yuki Nakata, Tetsuaki Fukushima, Hirokazu Matsuyama, Katsuyuki Ogiwara, and Seiji Yanagida of Dentsu Soken Inc., and members Takahiko Sueyoshi, Yasushi Miyajima, Yasuhiko Terashita, Hideyuki Ono, and Katsuhiro Takematsu of Koozyt, Inc., for collaborating on causal analysis services.

The various staff at Sony Group Corp. and Sony group companies, especially Koichiro Masuda, Taiwa Komatsu, and Naoki Hideshima of Sony Semiconductor Manufacturing Corporation, and Seigo Taniguchi of Sony Network Communications Inc., as well as many people at other companies, for their help in introducing the case studies.

Declaration of generative AI and AI-assisted technologies in the writing process

Statement: During the preparation of this work the author used DeepL in order to improve readability

References

Abramson, B., Brown, J., Winkler, R.L., 1996. Hailfinder: a Bayesian system for forecasting severe weather. Int. J. Forecast 12, 57–71.

Andersen, H., 2013. When to expect violations of causal faithfulness and why it matters. Philosophy Sci. 80, 672–683.

Arrieta, A., D´ıaz-Rodr´ıguez, N., Ser, J., Bennetot, A., Tabik, S., Barbado, A., et al., 2020. Explainable artificial intelligence (XAI): concepts, taxonomies, opportunities and challenges toward responsible AI. Inf. Fusion 58, 82–115.

Beinlich, I., Suermondt, H., Chavez, R. Cooper, G., 1989, The ALARM monitoring system: a case study with two probabilistic inference techniques for belief networks. In: Proceedings of European Conference on Artificial Intelligence in Medicine (AIME-89), pp. 247–256.

Blalock Jr, H.M., 1962. Four-variable causal models and partial correlations. Am. J. Sociol. 68 (2), 182–194.

Breiman, L., 2001. Statistical modeling: the two cultures. Stat. Sci. 16 (3), 199–231.

Burks, A.W., 1951. The logic of causal propositions. Mind 60 (239), 363–382.

Cartwright, N., 2001. What is wrong with Bayes nets? Monist 84 (2), 242–264.

Chickering, D.M., 2002. Optimal structure identification with greedy search. J. Mach. Learn. Res. 3, 507–554.

Chow, C.K., Liu, C.N., 1968. Approximating discrete probability distributions with dependence trees. IEEE Trans. Inform. Theory 14, 462–467.

Colombo, D., Maathuis, M.H., Kalisch, M., Richardson, T.S., 2012. Learning high-dimensional directed acyclic graphs with latent and selection variables. Ann. Stat. 40 (1), 294–321.

Cover, T.M., Thomas, J.A., 2006. Elements of Information Theory, Second ed. John Wiley & Sons, Hoboken, NJ.

Dawid, A.P., 2010. Beware of the DAG!. Proceedings of Workshop on Causality: Objectives and Assessment at NIPS 2008. PMLR, Whistler, pp. 59–86.

Friedman, J., 2001. Greedy function approximation: a gradient boosting machine. Ann. Stat. 29, 1189–1232.

Fukumizu, K., Gretton, A., Sun, X., Schölkopf, B., 2008. Kernel measures of conditional dependence. Proceeding of 20th Advances in Neural Information Processing Systems (NIPS-08). MIT Press, pp. 489–496.

Geiger, D., Pearl, J., 1988. On the logic of causal models. Proceeding of Workshop on Uncertainty in Artificial Intelligence (UAI-88). St Paul, Minneapolis.

Glymour, C., Cooper, G.F. (Eds.), 1999. Computation, Causation, and Discovery. AAAI/MIT Press, Cambridge, MA.

Granger, C.W.J., 1969. Investigating causal relations by econometric models and cross-spectral methods. Econometrica 37 (3), 424–438.

Gretton, A., Fukumizu, K., Teo, C.H., Song, L., Schölkopf, B., 2008. A kernel statistical test of independence. Proceeding of 20th Advances in Neural Information Processing Systems (NIPS-08). MIT Press, pp. 585–592.

Hausman, D.M., Woodward, J., 1999. Independence, invariance and the causal Markov condition. Br. J. Philosophy Sci. 50 (4), 521–583.

Hausman, D.M., Woodward, J., 2004. Modularity and the causal Markov condition: a restatement. Br. J. Philosophy Sci. 55 (1), 147–161.

Isozaki, T., 2012. A thermodynamical approach for probability estimation. arXiv Prepr arXiv:1201.1384v2.

Isozaki, T., 2014. A robust causal discovery algorithm against faithfulness violation. Trans. Japanese Soc. Artif. Intell. 29 (1), 137–147.

Isozaki, T., 2015. Fluctuation-model-based discrete probability estimation for small samples. Open. J. Stat. 5, 465–474.

Isozaki, T., 2024. Understanding the reasons for AI decisions using the causal inference technology CALC (in Jpn), in Technical Information Institute, Co, Ltd. (ed.), How Advance AI Mach. Learn. Less Data, Improve Accuracy, Dev. Explainable AI (Jpn.), Technical Information Institute, Co., Ltd., Tokyo, Japan, pp. 288–294.

Isozaki, T., Kato, N., Ueno, M., 2008. Minimum free energies with "data temperature" for parameter learning of Bayesian networks. In: Proceeding of IEEE International Conference on Tools with Artificial Intelligence (ICTAI 2008), pp. 371–378.

Isozaki, T., Kato, N., Ueno, M., 2009. "Data temperature" in minimum free energies for parameter learning of Bayesian networks. Int. J. Artif. Intell. Tools 18 (5), 653–671.

Isozaki, T., Kuroki, M., 2017. Learning causal graphs with latent confounders in weak faithfulness violations. N. Gener. Comput. 35, 29–45.

Isozaki, T., Ueno, M., 2009. Minimum free energy principle for constraint-based learning Bayesian networks. In: Proceeding of European Conference on Machine Learning and Principles and Practice of Knowledge Discovery in Databases (ECML-PKDD-09), Part1, LNAI5781, pp. 612–627.

Jaber, A., Ribeiro, A.H., Zhang, J., Bareinboim, E., 2022. Causal identification under Markov equivalence: calculus, algorithm, and completeness. In: Proceeding of 36th Annual Conference on Neural Information Processing Systems (NeurIPS-22), pp. 3679–3690.

Jaber, A., Zhang, J., Bareinboim, E., 2018. Causal identification under Markov equivalence. In: Proceeding of Conference on Uncertainty in Artificial Intelligence (UAI-18), pp. 978–987.

Jaber, A., Zhang, J., Bareinboim, E., 2019. Causal identification under Markov equivalence: completeness results. In: Proceeding of International Conference on Machine Learning (ICML-19), pp. 2981–2989.

Ke, G., Meng, Q., Finley, T., Wang, T., Chen, W., Ma, W., et al., 2017. LightGBM: a highly efficient gradient boosting decision tree. In: Proceedings of the 31st Advances in Neural Information Processing Systems (NeurIPS-17), pp. 3146–3154.

Kiivery, H., Speed, T.P., Carlin, J.B., 1984. Recursive causal models. J. Australian Math. Soc. (Ser. A) 36, 30–52.

Kitashima, C., Asukai, M., Sueyoshi, T., Isozaki, T., Maeno, T., 2021. Structuralizing and analysis of subjectivewell-being -psychological factors using factor analysis and partialcorrelation- (in Jpn). Japanese Journal of Affective Engineering 20(2), 129–139.

Kocaoglu, M., 2023. Characterization and learning of causal graphs with small conditioning sets. In: Proceeding of 37th Annual Conference on Neural Information Processing Systems (NeurIPS-23), pp. 74140– 74179.

Kuroki, M., 2017. An Introduction to Structural Causal Models (in Jpn). Kyoritsu Shuppan Co., Ltd, Tokyo, Japan.

Lacerda, G., Spirtes, P., Ramsey, J., Hoyer, P.O., 2008. Discovering cyclic causal models by independent components analysis. Proceeding of Conference on Uncertainty in Artificial Intelligence (UAI-08), pp. 366–374.

LeCun, Y., Huang, F.J., 2005. Loss functions for discriminative training of energy- based models. In: Proceeding of International Workshop on Artificial Intelligence and Statis- tics (AISTATS-05), pp. 206–213.

Lemeire, J., Meganck, S., Cartella, F., Liu, T., 2012. Conservative independence- based causal structure learning in absence of adjacency faithfulness. Int. J. Approx. Reason. 53, 1305–1325.

Lundberg, S., Lee, S., 2017. A unified approachto interpreting model predictions. In: Proceedings of the 31st Advances in NeuralInformation Processing Systems (NeurIPS-17), pp. 4765–4774.

Maeda, T.N., Shimizu, S., 2020. RCD: repetitive causal discovery of linear non-gaussian acyclic models with latent confounders. In: Proceedings of International Conference on Artificial Intelligence and Statistics (AISTATS-20), pp. 735–745.

Marx, A., Gretton, A., Mooij, J.M., 2021. A weaker faithfulness assumption based on triple interactions. In: Proceedings of Conference on Uncertainty in Artificial Intelligence (UAI-21), pp. 451–460.

Meek, C., 1995a. Causal inference and causal explanation with background knowledge. In: Proceedings of Conference on Uncertainty in Artificial Intelligence (UAI-95), pp. 403–410.

Meek, C., 1995b. Strong completeness and faithfulness in Bayesian networks. In: Proceedings of Conference on Uncertainty in Artificial Intelligence (UAI-95), pp. 411–418.

Miyakawa, M., 1999. Graphical Modeling (in Jpn). Asakura Publishing Co., Ltd, Tokyo, Japan.

Noda, A., Isozaki, T., 2024. Adjustment variable selection for intervention estimation robust to data size. In: Proceedings of AAAI 2024 Workshop on Graphs and More Complex Structures for Learning and Reasoning.

Noda, A., Isozaki, T., 2025. Practically effective adjustment variable selection in causal inference. J. Phys. Complex 6, 015001.

Ogarrio, J.M., Spirtes, P., Ramsey, J., 2016. A hybrid causal search algorithm for latent variable models. In: Proceedings of International Conference on Probabilistic Graphical Models, pp. 368–379.

Ogburn, E.L., Shpitser, I., 2021. Causal modelling: the two cultures. Observational Stud. 7, 179–183.

Pearl, J., 1988. Probabilistic Reasoning in Intelligent Systems. Morgan Kaufmann, San Mateo, CA.

Pearl, J., 1995. Causal diagrams for empirical research. Biometrika 82, 669–688.

Pearl, J., 2000. Causality, Models, Reasoning, and Inference. Cambridge University Press, New York, NY.

Pearl, J., Verma, T., 1991. A theory of inferred causation. In: Proceedings of Conference on Principles of Knowledge Representation and Reasoning, pp. 441–452.

Pellet, J., Elisseeff, A., 2008. Finding latent causes in causal networks: an efficient approach based on Markov blankets. In: Proceedings of Advances in Neural Information Processing Systems 21 (NIPS 08), pp. 1249 1256.

Ramsey, J., Spirtes, P., Zhang, J., 2006. Adjacency-faithfulness and conservative causal inference. In: Proceedings of Conference on Uncertainty in Artificial Intelligence (UAI- 06), pp. 401–408.

Reichenbach, H., 1956. The Direction of Time. Dover Publications, Mineola, NY Republication of the work published by Universityof California Press, Berkely.

Richardson, T., Spirtes, P., 1999. Automated discovery of linear feedback models. In: Glymour, C., Cooper, G.F. (Eds.), Computation, Causation, and Discovery. AAAI/MIT Press, Cambridge, MA, pp. 253–302.

Richardson, T., Spirtes, P., 2002. Ancestral graph Markov models. Ann. Stat. 30 (4), 962–1030.

Rubin, D.B., 1974. Estimating causal effects of treatments in randomized and non- randomized studies. J. Educ. Psychol. 66, 688–701.

Rubin, D.B., 1978. Bayesian inference of causal effects: the role of randomization. Ann. Stat. 6, 34–58.

Sadeghi, K., Soo, T., 2022. Conditions and assumptions for constraint-based causal structure learning. J. Mach. Learn. Res. 23, 1–34.

Schwarz, G., 1978. Estimating the dimension of a model. Ann. Stat. 6, 461–464.

Shimizu, S., Hoyer, P.O., Hyv¨arinen, A., Kerminen, A., 2006. A linear non- gaussian acyclic model for causal discovery. J. Mach. Learn. Res. 7, 2003–2030.

Shiragur, K., Zhang, J., Uhler, C., 2024. Causal discovery with fewer conditional independence tests. In: Proceedigs of International Conference on Machine Learning (ICML- 24), pp. 45060–45078.

Simon, H.A., 1952. On the definition of the causal relation. J. Philosophy 49 (16), 517–528.

Simon, H.A., 1953. Causal ordering and identifiability. In: Hood, W.C., Koopmans, T.C. (Eds.), Studies in Econometric Method. John Wiley & Sons, NY, pp. 49–74.

Simon, H.A., 1954. Spurious correlation: a causal interpretation. J. Am. Stat. Assoc. 49, 467–479.

Spirtes, P., 1995. Directed cyclic graphical representation of feedback models. In: Proceedings of Conference on Uncertainty in Artificial Intelligence (UAI-95), pp. 491–498.

Spirtes, P., Glymour, C., 1991. An algorithm for fast recovery of sparse causal graphs. Soc. Sci. Computer Rev. 9 (1), 62–72.

Spirtes, P., Glymour, C., Scheines, R., 1990. Causality from probability. In: Tiles, J., McKee, G., Dean, G. (Eds.), Evolving Knowledge in the Natural and Behavioral Sciences. Pitman Publishing, London.

Spirtes, P., Glymour, C., Scheines, R., 2000. Causation, Prediction and Search, Second ed. MIT Press, Cambridge, MA.

Spirtes, P., Meek, C., Richardson, T., 1995. Causal inference in the presence of latent variables and selection bias. In: Proceedings of Conference on Uncertainty in Artificial Intelligence (UAI-95), pp. 499–506.

Spirtes, P., Meek, C., Richardson, T., 1999. An algorithm for causal inference in the presenceof latent variables and selection bias. In: Glymour, C., Cooper, G.F. (Eds.), Computation, Causation, and Discovery. AAAI/MIT Press, Cambridge, MA, pp. 211–252.

Spirtes, P., Richardson, T., 1996. A polynomial time algorithm for determining DAG equivalence in the presence of latent variables and selection bias. In: Proceedings of International Workshop on Artificial Intelligence and Statistics (AISTATS-96).

Sun, X., Janzing, D., Schölkopf, B., Fukumizu, K., 2007. A kernel-based causal learning algorithm. In: Proceedings of International Conference on Machine Learning (ICML- 07), pp. 855–862.

Takao, M., Sueyoshi, T., Egami, H., Isozaki, T., Yasui, T., 2023. Pathways through local currency enhances subjective well-being by altruism and connections with residents: a relationship-oriented local currency platform as validation field (in Jpn). J. Jpn. Assoc. Regional Dev. Vitalization 19, 11–19.

Tillman, R., Gretton, A., Spirtes, P., Nonlinear directed acyclic structure learning with weakly additive noise models. In: Proceedings of 22nd Advances in Neural Information Processing Systems (NIPS-09), pp. 1847–1855.

Triantafillou, S., Tsamardinos, I., 2016. Score based vs constraint based causal learning in the presence of confounders. In: Proceedings of UAI 2016 Workshop on Causation: Foundation to Application.

Verma, T., Pearl, J., 1988. Causal networks: semantics and expressiveness. In: Proceedings of Workshop on Uncertainty in Artificial Intelligence (UAI-88), pp. 352–359.

Verma, T., Pearl, J., 1990. Equivalence and synthesis of causal models. In: Proceedings of Conference on Uncertainty in Artificial Intelligence (UAI-90), pp. 220–227.

Verma, T., Pearl, J., 1992. An algorithm for deciding if a set of observed independencies has a causal explanation. In: Proceedings of Conference on Uncertainty in Artificial Intelligence (UAI-92), pp. 323–330.

Wermuth, N., 1980. Linear recursive equations, covariance selection and path analysis. J. Am. Stat. Assoc. 75 (372), 963–972.

Wright, S., 1921. Correlation and causation. J. Agric. Res. 20 (7), 557–585.

Wright, S., 1934. The method of path coefficients. Ann. Math. Stat. 5, 161–215.

Xie, X., Geng, Z., 2008. A recursive method for structural learning of directed acyclic graphs. J. Mach. Learn. Res. 9, 459–483.

Yasui, T., Sueyoshi, T., Isozaki, T., Asukai, M., Yamakawa, A., Egami, H., et al., 2021. Social system design of community currency to enhance subjective well-being, by identifying direct correlation factors between money spending and subjective well-being (in Jpn). J. Syst. Des. Soc. Jpn. 1 (1), 43–57.

Yasui, T., Sueyoshi, T., Isozaki, T., Takao, M., 2024. Identifying pathways where pro-environmental behavior affects residents 'subjective well-being through nature connectedness and place attachment by exploratory factor analysis: empirical study in two Japanese cities. In: Proceedings of European Conference on Positive Psychology (ECPP- 24).

Yasui, T., Sueyoshi, T., Miyajima, Y., Takematsu, K., Isozaki, T., Takao, M., 2025. Identifying key factors which pro-environment behaviors correlate subjective well- being of residents by exploratory factor and Markovian modeling analysis: empirical and comparative research in regenerative city. In: Proceedings World Congress on Positive Psychology (IPPA-25).

Yehezkel, R., Lerner, B., 2009. Bayesian network structure learning by recursive autonomy identification. J. Mach. Learn. Res. 10, 1527–1570.

Zhalama, Zhang, J., Mayer, W., 2017. Weakening faithfulness: some heuristic causal discovery algorithms. Int. J. Data Sci. Analytics 3, 93–104.

Zhang, J., 2008a. Causal reasoning with ancestral graphs. J. Mach. Learn. Res. 9, 1437–1474.

Zhang, J., 2008b. On the completeness of orientation rules for causal discovery in the presence of latent confounders and selection bias. Artif. Intell. 172, 1873–1896.

Zhang, J., 2013. A comparison of three Occam's razors for Markovian causal models. Br. J. Philosophy Sci. 64 (2), 423–448.

Zhang, J., Spirtes, P., 2008. Detection of unfaithfulness and robust causal inference. Minds Mach. 18, 239–271.

Zhang, J., Spirtes, P., 2016. The three faces of faithfulness. Synthese 193, 1011–1027.

Zhang, K., Peters, J., Janzing, D., Scholköpf, B., 2011. Kernel-based conditional independence test and application in causal discovery. In: Proceedings of Conference on Un- certainty in Artificial Intelligence (UAI-11), pp. 804–813.

Chapter 2

An overview of models for one-shot device testing data analysis

Man Ho Ling*
Department of Mathematics and Information Technology, The Education University of Hong Kong, Tai Po, Hong Kong, SAR, P.R. China
**Corresponding author. e-mail address: amhling@eduhk.hk*

Chapter Outline

Abstract

One-shot devices are a unique class of products designed to function only once and then become unusable or discarded. These specialized devices are found in various industries, such as consumer electronics, aerospace, and military applications. The reliability of one-shot devices is of paramount importance, as they must perform their critical functions correctly the first and only time they are used in emergency or safety-critical situations and their failure can lead to catastrophic consequences. This chapter provides an overview of the latest developments and research contributions in modeling one-shot device testing data. It covers a range of topics, including popular lifetime distribution models, reliability analysis of devices with multiple failure modes, semi-parametric approaches, step-stress accelerated life testing, and manufacturing defects on one-shot devices. By synthesizing the key findings and insights from the recent literature, this chapter aims to open up more potential avenues for future investigations.

1 Introduction

One-shot devices are a unique class of products that are designed to operate or function only once, after which they become unusable or must be discarded.

Handbook of Statistics, Vol. 53. https://doi.org/10.1016/bs.host.2025.04.004

These specialized devices are commonly found in various industries, including consumer electronics, aerospace systems, and military applications (Bain and Engelhardt, 1991; Vintr and Valis, 2007; Newby, 2008; Fan et al., 2009).

For example, in automobiles, airbag deployment systems are designed to be used only once. When a collision is detected, a small explosive charge is ignited, which inflates the airbag to protect the vehicle's occupants. After the airbag has been deployed, the system must be replaced. In certain military and aerospace applications, primer circuits are utilized to trigger the firing of rockets or missiles. These circuits are engineered to be activated a single time, ensuring the reliable and safe launch of the projectile. Fire extinguishers represent another common example of one-shot devices. When the fire extinguisher is activated, the pressurized agent is released through a nozzle, allowing the user to direct the agent onto the fire. Once the extinguisher has been fully discharged, it cannot be recharged and must be replaced with a new unit. Electro-explosive devices, also known as pyrotechnic initiators, are one-shot devices employed to initiate the detonation or activation of other explosive or pyrotechnic systems. These devices typically consist of a small explosive charge that is triggered by the application of an electrical current or static discharge. After the explosion, the electro-explosive devices cannot be reused. In addition, one-shot device testing data can arise from destructive testing (Morris, 1987; Zheng et al., 2018), where binary responses (success/failure) can be observed. When devices fail to operate during testing, their lifetimes are known to be before the test time, resulting in left-censored observations. Conversely, when devices successfully pass the test, their lifetimes are known to be beyond the test time, resulting in right-censored observations.

The reliability of one-shot devices is of paramount importance, as they must perform their critical functions correctly the first and only time they are used. For instance, both fire extinguishers and electro-explosive devices are examples of one-shot devices that are designed to perform a critical function in emergency or safety-critical situations. Failure of these devices can lead to catastrophic consequences. However, evaluating the reliability of one-shot devices poses unique challenges due to the nature of their failure data. Unlike conventional products, the actual lifetimes of one-shot devices are completely censored, either left-censored (when the device fails before the end of the test) or right-censored (when the device successfully completes the test). This type of censoring data presents significant obstacles in reliably assessing the failure behavior and lifetime characteristics of these devices.

In recent years, researchers have devoted considerable attention to addressing the reliability modeling and analysis of one-shot devices. Various studies have focused on developing advanced reliability models, optimizing inspection schedules and preventive maintenance strategies, and enhancing supply chain management for these specialized products. Researchers have extensively studied the reliability modeling and optimization of one-shot units, tackling a variety of challenges in this domain. A series of works have explored different optimization problems to maximize the operational use of one-shot units (Cheng and Elsayed, 2016, 2017a, 2017b, 2018). Another study addressed the issue of inspection schedule problems when maintenance resources are limited (Zhao and Yun, 2018). Other researchers have focused on determining the optimum timing for regular inspections and replacements (Yun et al., 2014), as well as identifying the optimal preventive maintenance periods (Azimian et al., 2021a) and warranty lengths (Wu et al., 2023) for one-shot devices. The research in this area has also delved into supply chain management and maintenance strategy considerations. Several studies have explored these aspects, providing valuable insights for practitioners (Azimian et al., 2021b, 2022a, 2022b, 2022c). Additionally, a comprehensive overview of the reliability analysis and data collection methodologies for one-shot devices has served as a valuable resource for researchers and engineers working in this field (Balakrishnan et al., 2021a). Given the breadth and depth of the existing literature, this chapter aims to provide an extended overview of the latest developments and research contributions in the field of one-shot device reliability. It covers a range of topics, including popular lifetime distribution models, reliability analysis of devices with multiple failure modes, semi-parametric and non-parametric approaches, step-stress accelerated life testing, and the impact of manufacturing defects on one-shot device reliability.

The remainder of the chapter is organized as follows. Section 2 presents various popular lifetime distribution models for reliability studies. Section 3 discusses one-shot devices with multiple failure modes. Section 4 discusses the proportional hazard model. Section 5 focuses on step-stress accelerated life testing experiments. One-shot devices with manufacturing defects are discussed in Section 6. Finally, some concluding remarks are given in Section 7.

2 Lifetime distributions for reliability studies

Due to increasing customer demands and the need for highly reliable products with extended lifetimes, the use of accelerated life tests (ALTs) has become a popular approach for reliability assessment (Nelson, 2009; Meeker and Escobar, 2014). In these tests, product units are subjected to higher-than-normal stress levels, such as temperature, voltage, or pressure, to induce failures more rapidly than under normal operating conditions. By observing the failures at these elevated stress levels, researchers can then extrapolate the lifetime and reliability of the product under normal use conditions. Reliability analysis of one-shot device testing data obtained from ALTs has thus received significant attention in the past decade. Researchers have extensively discussed the estimation and inferential issues under different parametric models in this context.

In a typical constant-stress ALT (CSALT) setup for one-shot devices, the test units are divided into I groups, each subjected to a specific stress level x_i. For the i-th test group, N_i devices are examined at time τ_i, and the number of failures n_i is recorded. The observed dataset can be summarized as $\mathbf{z} = \{\tau_i, N_i, n_i, x_i, i = 1,2,...,I\}$, as shown in Table 1.

TABLE 1 One-shot device testing data under a CSALT with I test groups.

Test group	Inspection time	Stress level	# Of tested devices	# Of failures
1	τ_1	x_1	N_1	n_1
2	τ_2	x_2	N_2	n_2
$\vdots$	$\vdots$	$\vdots$	$\vdots$	$\vdots$
I	τ_I	x_I	N_I	n_I

Here, let $\boldsymbol{\theta}$ be the model parameters, and the lifetime distribution has a cumulative distribution function (CDF) $F(t; \boldsymbol{\theta})$. The likelihood function for the dataset $\mathbf{z}$ is then given by

$$L(\boldsymbol{\theta}; \mathbf{z}) = C \prod_{i=1}^{I} [F(\tau_i; \boldsymbol{\theta})]^{n_i} [1 - F(\tau_i; \boldsymbol{\theta})]^{N_i - n_i},$$

where C is the normalizing constant.

In this chapter, we will explore four of the most widely used lifetime distribution models for reliability analysis of one-shot devices, including the exponential, Weibull, gamma, and log-normal distributions. Readers interested in exploring a wider range of lifetime distributions applicable to reliability analysis may refer to two excellent resources (Johnson et al., 1994, 1995). These references provide in-depth coverage of numerous lifetime distributions and their properties, offering a valuable reference for researchers and practitioners in the field.

2.1 Exponential lifetime distribution

The exponential distribution is a simple and commonly applied model for its memoryless property, which can be suitable for certain types of one-shot devices. Abundant research on exponential distribution for one-shot devices has appeared in the reliability literature. Fan et al (Fan et al., 2009). assumed that the device lifetimes $T_{i,k}$ are independent and identically distributed exponential random variables with failure rate $\lambda_i > 0$, for $i = 1, ..., I$, $k = 1, 2, ..., N_i$. The probability density function (PDF) and CDF are given by

$$f(t) = \lambda_i \exp(-\lambda_i t), \quad t > 0 \tag{1}$$

and

$$F(t) = 1 - \exp(-\lambda_i t), \quad t > 0, \tag{2}$$

respectively. Here, we can relate the failure rate at the i-th test group to the stress levels through a log-linear link function, that is

$$\lambda_i = \exp(a_0 + a_1 x_i). \tag{3}$$

Thus, the model parameters $\boldsymbol{\theta} = (a_0, a_1)$, and the corresponding reliability at a mission time t and the mean-time-to-failure (MTTF) under normal operating conditions x_0 are given by

$$R_0(t) = \exp(-t\exp(a_0 + a_1 x_0)) \tag{4}$$

and

$$E[T_0] = \exp(-a_0 - a_1 x_0), \tag{5}$$

respectively.

Given the complete censoring of actual lifetimes in one-shot device testing, researchers have developed various statistical techniques to estimate the reliability and MTTF. An expectation-maximization (EM) algorithm was developed to obtain the maximum likelihood estimates (MLEs) of these key reliability metrics (Balakrishnan and Ling, 2012a, 2012b). This approach proved useful in handling the censored nature of the data. Considering the Bayesian perspective, a zero-failure reliability demonstration test for electro-explosive devices was introduced under the exponential distribution assumption (Fan and Chang, 2009). This Bayesian framework allowed for reliability assessment even in the absence of observed failures. Building upon the Bayesian approach, a two-stage hierarchical Bayes technique, also under the exponentiality assumption, was recently employed to study the reliability of electro-explosive devices (Sharma and Upadhyay, 2018). This hierarchical modeling approach can capture the potential heterogeneity in the underlying device populations. In addition to the maximum likelihood and Bayesian techniques, alternative robust inference methods have been explored for one-shot devices with exponential lifetimes (Balakrishnan et al., 2019a, 2020). Minimum density power divergence estimators (MDPDEs) and weighted MDPDEs (WMDPDEs) were developed to provide more resilient reliability estimates in the presence of data contamination.

2.2 Weibull lifetime distribution

The more flexible Weibull distribution is also widely employed, as it can capture a variety of failure rate behaviors, from increasing to decreasing and constant, allowing it to effectively model early failures, aging failures, and exponential failure behavior, respectively. For $i = 1, 2, ..., I$, $k = 1, 2, ..., N_i$, Balakrishnan and Ling (Balakrishnan and Ling, 2013) assumed that the device lifetimes $T_{i,k}$ have Weibull distribution with corresponding PDF and CDF as

$$f_T(t) = \frac{\eta_i t^{\eta_i - 1}}{\alpha_i^{\eta_i}} \exp\left(-\left(\frac{t}{\alpha_i}\right)^{\eta_i}\right), \quad t > 0,$$

and

$$F_T(t) = 1 - \exp\left(-\left(\frac{t}{\alpha_i}\right)^{\eta_i}\right), \quad t > 0,$$

where $\alpha_i > 0$ and $\eta_i > 0$ are the scale and shape parameters, respectively. We assume that both these parameters relate to the stress levels in log-linear forms as

$$\alpha_i = \exp(a_0 + a_1 x_i) \quad \mathit{and} \quad \eta_i = \exp(b_0 + b_1 x_i).$$

Hence, the model parameters is $\theta = (a_0, a_1, b_0, b_1)$. Then, the corresponding reliability at a mission time t and MTTF under normal operating conditions x_0 are given by

$$R_0(t) = \exp\left(-\left(\frac{t}{\alpha_0}\right)^{\eta_0}\right) \tag{6}$$

and

$$E[T_0] = \alpha_0 \Gamma\left(1 + \frac{1}{\eta_0}\right), \tag{7}$$

respectively, where $\alpha_0 = \exp(a_0 + a_1 x_0)$, $\eta_0 = \exp(b_0 + b_1 x_0)$, and $\Gamma(\cdot)$ denotes the complete gamma function.

For one-shot devices with Weibull distributed lifetimes, researchers have developed various inference and optimization techniques to address the challenges in reliability analysis as well as ALT planing. Regarding the CSALT design, a procedure was proposed to determine the optimal test plan that minimizes the asymptotic variance of the maximum likelihood estimator for the reliability at a specific mission time (Balakrishnan and Ling, 2014a). This optimization approach considered budget and time constraints, providing practitioners with guidance on the item allocation, inspection frequency, and termination time at each stress level. By optimizing the CSALT design, this method helps to enhance the precision of reliability estimates under the Weibull lifetime model. Moving beyond the classical maximum likelihood approach, Bayesian and robust inference methods have also been investigated for the Weibull distribution in the context of one-shot device reliability prediction. The Bayesian framework allows for the incorporation of prior knowledge and can be particularly useful when the available data is limited (Mun et al., 2019). Moreover, minimum density power divergence approaches have been explored as an alternative to the maximum likelihood method, offering more resilient reliability estimates when data are contaminated (Balakrishnan et al., 2019a; Castilla, 2024).

2.3 Gamma lifetime distribution

The gamma distribution provides additional flexibility by introducing a shape parameter, allowing for more diverse failure time characteristics. Its hazard function can also be increasing, decreasing, or constant. Balakrishnan and Ling (Balakrishnan and Ling, 2014b) assumed that the device lifetimes $T_{i,k}$ follow gamma distribution with shape parameter $\alpha_i > 0$ and scale parameter $\beta_i > 0$ for $i = 1, 2, ..., I$, $k = 1,2, ..., N_i$. The corresponding PDF and CDF are given by

$$f(t) = \frac{t^{\alpha_i - 1}}{\Gamma(\alpha_i)\beta_i^{\alpha_i}} \exp\left(-\frac{t}{\beta_i}\right), \quad t > 0,$$

and

$$F(t) = \int_0^t \frac{y^{\alpha_i - 1}}{\Gamma(\alpha_i)\beta_i^{\alpha_i}} \exp\left(-\frac{y}{\beta_i}\right) dy, \quad t > 0,$$

respectively. The CDF $F(t)$ can be readily expressed in terms of the lower incomplete gamma ratio as

$$F(t) = \int_0^{\frac{t}{\beta_i}} \frac{y^{\alpha_i - 1}}{\Gamma(\alpha_i)} \exp(-y) dy = \gamma\left(\alpha_i, \frac{t}{\beta_i}\right).$$

Within each test group, we assume that both shape and scale parameters are related to the stress factors in log-linear forms as

$$\alpha_i = \exp(a_0 + a_1 x_i) \quad and \quad \beta_i = \exp(b_0 + b_1 x_i).$$

Hence, the model parameters is $\theta = (a_0, a_1, b_0, b_1)$, and the corresponding reliability at a mission time t and the MTTF under normal operating conditions x_0 are given by

$$R_0(t) = \Gamma\left(\alpha_0, \frac{t}{\beta_0}\right) \tag{8}$$

and

$$E[T_0] = \alpha_0 \beta_0, \tag{9}$$

respectively, where $\alpha_0 = \exp(a_0 + a_1 x_0)$, $\beta = \exp(b_0 + b_1 x_0)$, and $\Gamma(u, v) = 1 - \gamma(v, u)$ is the upper incomplete gamma ratio.

The reliability analysis of one-shot devices has seen further advancements with the development of robust estimators under alternative lifetime distributions, such as the gamma distribution. Researchers have explored the application of robust statistical methods, specifically MDPDEs, in the context of gamma distributed lifetimes (Balakrishnan et al., 2019b). This work was motivated by a tumor toxicological dataset involving mice, where it was observed that the incidence of tumors was related to the dosage of benzidine dihydrochloride, and that male mice exhibited higher expected lifetimes compared to females. It is important to note that numerous lifetime distributions are available for modeling one-shot device testing data, and the selection of an appropriate model is crucial for accurate reliability evaluation. Studies have been conducted to analyze the impact of model misspecification, specifically between the Weibull and gamma lifetime distribution models (Ling and Balakrishnan, 2017). These investigations have revealed that the consequences of model misspecification can be non-negligible, underscoring the importance of employing appropriate statistical tools, such as goodness-of-fit tests, to ensure the selection of the most suitable lifetime distribution for the one-shot device data at hand. In this regard, researchers have investigated the performance of various goodness-of-fit tests to assist practitioners in identifying the most suitable lifetime distribution for a given set of one-shot device data (Balakrishnan and Chimitova, 2017). By carefully evaluating the fit of the chosen model, reliability engineers can enhance the accuracy and reliability of their assessments for one-shot devices. A holistic approach, which encompasses both robust inference methods and model selection diagnostics, provides a more comprehensive framework for the reliable evaluation of one-shot device lifetimes and their associated failure risks.

2.4 Log-normal lifetime distribution

The log-normal distribution is often used when the failure times exhibit a skewed, positive-valued distribution. It has increasing-decreasing hazard behavior for units that experience early failure (resulting in increasing hazard) and then stabilize over time in terms of performance (resulting in decreasing hazard). Balakrishnan and Castilla (Balakrishnan and Castilla, 2023) assumed that the device lifetimes $T_{i,k}$ have log-normal distribution with corresponding PDF and CDF as

$$f_T(t) = \frac{1}{\sqrt{2\pi}\sigma_i t}\exp\left(-\frac{(\ln(t) - \mu_i)^2}{2\sigma_i^2}\right), \quad t > 0,$$

and

$$F_T(t) = \Phi\left(\frac{\ln(t) - \mu_i}{\sigma_i}\right), \quad t > 0,$$

where $\Phi(\cdot)$ is the CDF of the standard normal distribution, and $\mu_i > 0$ and $\sigma_i > 0$ are the scale and shape parameters, respectively. We assume that both these parameters relate to the stress levels in a linear form and a log-linear form, respectively, as

$$\mu_i = a_0 + a_1 x_i \quad and \quad \sigma_i = \exp(b_0 + b_1 x_i).$$

Hence, the model parameters is $\boldsymbol{\theta} = (a_0, a_1, b_0, b_1)$. Then, the corresponding reliability at a mission time t and MTTF under normal operating conditions x_0 are given by

$$R_0(t)=1 - \Phi\left(\frac{\ln(t) - \mu_0}{\sigma_0}\right) \tag{10}$$

and

$$E[T_0] = \exp\left(\mu_0 + \frac{\sigma_0^2}{2}\right), \tag{11}$$

respectively, where $\mu_0 = a_0 + a_1 x_0$, $\sigma_0 = \exp(b_0 + b_1 x_0)$.

By leveraging robust estimators, such as MDPDEs, under the log-normal distribution model has been developed to enhance the resilience of the analysis and helps to prevent biased or misleading conclusions, due to data contamination (Balakrishnan and Castilla, 2023). Researchers have explored the development of optimal CSALT designs when the underlying lifetime distribution follows a log-normal model (Balakrishnan and Castilla, 2022). These optimization techniques aim to determine the most effective test plan, in terms of item allocation, inspection frequency, and termination time at each stress level, while considering budget and test duration constraints. By optimizing the CSALT design under the log-normal assumption, reliability engineers can enhance the precision of the estimated reliability metrics for one-shot devices. Building upon these advancements, studies have also investigated the impact of model misspecification on the optimal CSALT design (Balakrishnan et al., 2022). Specifically, researchers have examined the implications of assuming different lifetime distributions, such as gamma, Weibull, and log-normal, when the true underlying distribution may differ. This analysis provides valuable insights into the potential consequences of selecting an inappropriate lifetime model.

2.5 Other lifetime distributions

The reliability analysis of one-shot devices has continued to evolve, with researchers exploring the application of increasingly flexible lifetime distribution models to better capture the unique characteristics of these systems. Beyond the well-established distributions, such as the Weibull, gamma, and log-normal, the reliability literature has witnessed the introduction of other highly flexible models for analyzing one-shot device testing data. These include Birnbaum-Saunders distribution (Zhu et al., 2021) generalized exponential distribution (Wu et al., 2020), generalized gamma distribution (Zhu and Liu, 2022), and logistic-exponential distribution (Baghel and Mondal, 2024). The generalized gamma distribution is a highly flexible model encompasses several well-known lifetime models, such as the exponential, Weibull, gamma and log-normal distributions, as special cases, providing reliability engineers with a more comprehensive modeling approach. The generalized exponential and logistic-exponential distributions offer greater flexibility in terms of the shape of the failure rate function, which can be increasing, decreasing, or even bathtub-shaped. This versatility is particularly beneficial when modeling one-shot device failures that may exhibit complex hazard rate patterns.

3 Devices with multiple failure modes

The reliability analysis of one-shot devices has primarily focused on scenarios with a single failure mode. However, as products become increasingly complex, incorporating multiple components and subsystems, the reliability community has recognized the need to expand the analytical approaches to capture these more intricate failure mechanisms.

Modern one-shot devices often comprise complex structures with multiple interconnected components, each contributing to the overall system's functionality and reliability. For instance, a fire extinguisher contains components such as a cylinder, a valve, and various chemicals, while an automobile airbag system includes a crash sensor, an inflator, and the airbag itself. In these complex one-shot devices, the lifetime and failure behavior of individual components can provide valuable insights that may not be readily apparent from examining the system-level performance alone. By focusing on the component-level reliability, researchers and engineers can gain a more nuanced understanding of the failure modes and their relative contributions to the overall system reliability. This granular approach enables the identification of critical components, the quantification of their respective reliability characteristics, and the development of more targeted mitigation strategies. Furthermore, the component-level reliability analysis can lead to more precise inferences and reliability estimates for the one-shot device as a whole. By incorporating the component-level lifetime data and failure mechanisms, reliability models can be refined to better reflect the intricate interdependencies and interactions within the system. This enhanced modeling approach can, in turn, result in more accurate predictions of the system's performance and service life, crucial for making informed decisions in safety-critical applications.

The shift towards component-level reliability analysis for one-shot devices represents a significant advancement in the field. It acknowledges the growing complexity of modern products and the need for a more comprehensive understanding of failure modes to ensure reliable and safe performance. By embracing this multi-component perspective, researchers and practitioners can further enhance the reliability evaluation and optimization of one-shot devices, ultimately leading to improved design, maintenance, and decision-making processes. This section will cover three popular approaches for analyzing one-shot device testing data with multiple failure modes, including competing risks modeling, copula modeling, and gamma frailty modeling.

3.1 Competing risks modeling

In the context of reliability analysis for one-shot devices, the data collected from life tests often includes valuable information about the device's status and the specific failure modes responsible for malfunctions. Through careful follow-up investigations, researchers can identify the underlying causes of failure, providing a nuanced understanding of the system's reliability characteristics.

Consider, for example, the case of testing electro-explosive devices. A successfully detonated device is typically considered a surviving unit, while a device that fails to explode is labeled as a malfunctioning device. The specific failure mode, such as burnout of resistance wires or leakage of organic fuel, can be identified through detailed post-failure analyses. In a CSALT setup, researchers may investigate one-shot devices with multiple failure modes. Suppose a CSALT study is conducted with I test groups for one-shot devices with two failure modes, including burnout of resistance wires (failure mode 1) and leakage of organic fuel (failure mode 2). The observed dataset can be described as follows:

For each test group $i = 1, ..., I$,

1. N_i devices are placed on test at a temperature level of x_i and inspected at time τ_i.
2. the number of malfunctioning devices with the m-th failure mode is observed and denoted as $n_{i,m}$, for $m = 1, 2$;
3. the number of surviving devices is observed and denoted by $n_{i,\varnothing} = N_i - n_{i,1} - n_{i,2}$.

The observed dataset is represented as $\mathbf{z} = \{x_i, \tau_i, n_{i,\varnothing}, n_{i,1}, n_{i,2}, i = 1, ..., I\}$ and shown in Table 2.

Let us denote the time to the m-th failure mode as $T_{i,m}$ for $m = 1, 2$ and $i = 1, ..., I$. Balakrishnan et al (Balakrishnan et al., 2015a). assumed that the lifetime distribution of $T_{i,m}$ is the exponential with rate parameter $\lambda_{i,m}$. The lifetimes to each failure mode are independent. The reliability, the probability with failure mode 1, and the probability with failure mode 2 at time τ_i are given by

$$p_{i,\varnothing} = \prod_{m=1}^{2} (1 - F_{T_{i,m}}(\tau_i)) = \exp\left(-\sum_{m=1}^{2} \lambda_{i,m}\tau_i\right), \tag{12}$$

$$p_{i,1} = \left(\frac{\lambda_{i,1}}{\lambda_{i,1} + \lambda_{i,2}}\right)\left(1 - \exp\left(-\sum_{m=1}^{2} \lambda_{i,m}\tau_i\right)\right) \tag{13}$$

$$p_{i,2} = \left(\frac{\lambda_{i,2}}{\lambda_{i,1} + \lambda_{i,2}}\right)\left(1 - \exp\left(-\sum_{m=1}^{2} \lambda_{i,m}\tau_i\right)\right), \tag{14}$$

where the rate parameters are related to temperature levels through log-linear functions, i.e., $\lambda_{i,m} = \exp(a_{0m} + a_{1m}x_i)$.

Then, the log-likelihood function for the given data is

$$\ell(\boldsymbol{\theta}) \propto \sum_{i=1}^{I} n_{i,\varnothing} \ln(p_{i,\varnothing}) + n_{i,1} \ln(p_{i,1}) + n_{i,2} \ln(p_{i,2}), \tag{15}$$

where $\boldsymbol{\theta} = (a_{01}, a_{02}, a_{11}, a_{12})$ is the model parameters.

TABLE 2 One-shot device testing data with two failure modes for competing risks models.

			# Of devices		
Test group	Stress level	Inspection time	∅	1	2
1	x_1	τ_1	$n_{1,\emptyset}$	$n_{1,1}$	$n_{1,2}$
2	x_2	τ_2	$n_{2,\emptyset}$	$n_{2,1}$	$n_{2,2}$
⋮	⋮	⋮	⋮	⋮	⋮
I	x_I	τ_I	$n_{I,\emptyset}$	$n_{I,1}$	$n_{I,2}$

The reliability analysis of one-shot devices with multiple failure modes has been studied in the literature, with researchers developing a range of statistical inference approaches to address the competing risk scenarios. These include likelihood-based, Bayesian, and robust inference methods (Balakrishnan et al., 2015a, 2015b, 2015c, 2023a; Balakrishnan and Castilla, 2024), each with its own advantages and applications.

3.2 Copula modeling

The preceding reliability analysis of one-shot devices with multiple failure modes relies on the assumption of independence between the failure modes, which is made for mathematical convenience. However, in some real-world scenarios, the lifetimes to each failure mode may be correlated. In such cases, more advanced modeling approaches, such as copula models, can provide a more accurate representation of the underlying reliability characteristics. Ling et al (Ling et al., 2021a). considered the case of CSALTs with multiple higher-than-normal temperatures and multiple inspection times. In their setup, $N_{i,j}$ devices are placed at temperature x_i and inspected at time τ_j, for $i = 1, ..., I$ and $j = 1, ..., J$. The observed dataset $\mathbf{z}$ consists of the following information and can be summarized as Table 3.

1. The number of devices without any failures, $n_{i,j,\emptyset}$;
2. The number of devices with only failure mode 1, $n_{i,j,1}$;
3. The number of devices with only failure mode 2, $n_{i,j,2}$;
4. The number of devices with both failure modes, $n_{i,j,12}$.

It is important to note that in this setup, the functions of both components of the device are triggered at the inspection time, and both failure modes may be identified simultaneously, even though the times to failure modes are not the same. This differs from the typical competing risks model, where only the root cause of failure is identified.

TABLE 3 One-shot device testing data with two failure modes for copula models.

			# Of devices			
Test group	Stress level	Inspection time	∅	1	2	12
1	x_1	τ_1	$n_{1,1\varnothing}$	$n_{1,1,1}$	$n_{1,1,2}$	$n_{1,1,12}$
⋮	⋮	⋮	⋮	⋮	⋮	⋮
J	x_1	τ_J	$n_{1,J,\varnothing}$	$n_{1,J,1}$	$n_{1,J,2}$	$n_{1,J,12}$
$J+1$	x_2	τ_1	$n_{2,1,\varnothing}$	$n_{2,1,1}$	$n_{2,1,2}$	$n_{2,1,12}$
⋮	⋮	⋮	⋮	⋮	⋮	⋮
IJ	x_I	τ_J	$n_{I,J,\varnothing}$	$n_{I,J,1}$	$n_{I,J,2}$	$n_{I,J,12}$

To address the potential dependence between the failure modes, copula models can be a useful approach. Copula models, first introduced by Sklar (Sklar, 1959), provide a flexible way to model multivariate distributions by capturing the dependence structure between random variables through a copula function involving a single dependence parameter α. Using the copula function $C_\alpha(U, V)$, which characterizes the dependence between the standard uniformly distributed random variables U and V, we can derive the following joint and marginal probabilities:

$$
\begin{aligned}
P(U \le u, V \le v) &= C_\alpha(u, v);\\
P(U \le u, V > v) &= P(U \le u) - P(U \le u, V \le v) = P(U \le u) - C_\alpha(u, v);\\
P(U > u, V \le v) &= P(V \le v) - P(U \le u, V \le v) = P(V \le v) - C_\alpha(u, v);\\
P(U > u, V > v) &= 1 - P(U \le u) - P(V \le v) + P(U \le u, V \le v)\\
&= 1 - P(U \le u) - P(V \le v) + C_\alpha(u, v).
\end{aligned}
$$

These expressions demonstrate the flexibility of the copula-based approach in modeling the dependence between the failure modes, which can lead to more accurate reliability estimation and enhanced failure mode identification for one-shot devices.

The Gumbel-Hougaard copula is a popular copula model that can effectively characterize the positive dependence among random variables. Suppose we have failure times $T_{i,m}$ with marginal probability distributions $F_{i,m}(t)$, where i represents the temperature and m represents the failure mode. It is assumed that the correlation between the times to failure for the two modes changes with temperature x_i, and this relationship can be expressed as

$$y_i = \ln(\alpha_i - 1) = d_0 + d_1 x_i, \quad i = 1, 2, ..., I,$$

where $\boldsymbol{\theta} = (d_0, d_1)$ is the vector of model parameters. This indicates that the correlation between the failure times increases with increasing temperature when $d_0 > 0$. The joint cdf of $T_{i,1}$ and $T_{i,2}$ under the Gumbel-Hougaard copula can be written as

$$\begin{aligned} C_{\alpha_i(\theta)}(t_1, t_2) &= P(T_{i,1} \leq t_1, T_{i,2} \leq t_2) \\ &= \exp\left(-\left[(-\ln F_{i,1}(t_1))^{\alpha_i} + (-\ln F_{i,2}(t_2))^{\alpha_i}\right]^{1/\alpha_i}\right). \end{aligned}$$

When $\alpha_i = 1$, we have $C_{\alpha_i(\theta)}(t_1, t_2) = F_{i,1}(t_1)F_{i,2}(t_2)$, implying that $T_{i,1}$ and $T_{i,2}$ are independent at temperature x_i. Moreover, the dependence parameter remains unchanged over temperatures when $a_1 = 0$.

Since the occurrence of any one of the failure modes causes the device to fail, the lifetime distribution of the device at temperature x_i is denoted by $T_i = \min T_{i,1}, T_{i,2}$. The reliability of the device at a mission time t is then given by

$$R_i(t) = P(T_{i,1} > t, T_{i,2} > t) = 1 - F_{T_{i,1}}(t) - F_{T_{i,2}}(t) + C_{\alpha_i(\theta)}(t, t).$$

When $\alpha_i = 1$, the reliability simplifies to

$$R_i(t) = 1 - F_{T_{i,1}}(t) - F_{T_{i,2}}(t) + F_{T_{i,1}}(t)F_{T_{i,2}}(t) = R_{T_{i,1}}(t)R_{T_{i,2}}(t),$$

indicating independent failure modes. However, when α_i tends to ∞, the failure modes are highly and positively correlated, and the device is more likely to fail due to the failure mode with a higher probability of failure. Devices with highly and positively correlated failure modes can also have longer lifetimes than those with independent failure modes, as a longer time to one failure mode can provide information about the likelihood of the other failure mode not occurring soon.

The Frank copula is another popular copula model, where the dependence parameter α_i is assumed to relate to temperature x_i in a linear form

$$y_i = \alpha_i = d_0 + d_1 x_i, \quad i = 1, 2, ..., I,$$

where $\theta = (d_0, d_1)$ is the vector of model parameters.

The joint cdf of $T_{i,1}$ and $T_{i,2}$ under the Frank copula can be written as

$$\begin{aligned} C_{\alpha_i(\theta)}(t_1, t_2) &= P(T_{i,1} \leq t_1, T_{i,2} \leq t_2) \\ &= -\frac{1}{\alpha_i} \ln\left(1 + \frac{(\exp(-\alpha_i F_{T_{i,1}}(t_1)) - 1)(\exp(-\alpha_i F_{T_{i,2}}(t_2)) - 1)}{(\exp(-\alpha_i) - 1)}\right). \end{aligned}$$

When α_i approaches ∞, the failure modes are highly and positively correlated, while when α_i approaches $-\infty$, the failure modes are highly and negatively correlated. Devices with independent failure modes have longer lifetimes than those with highly and negatively correlated failure modes, as a failure mode with a longer time can provide information that the time to the other failure mode will likely be short, leading to a shorter device lifetime.

The log-likelihood function based on the observed data $\mathbf{z}$ is given by

$$\ell(\theta) = \sum_{i=1}^{I} \sum_{j=1}^{J} n_{i,j,12} \ln(p_{i,j,12}) + n_{i,j,1} \ln(p_{i,j,1}) + n_{i,j,2} \ln(p_{i,j,2}) + n_{i,j,\varnothing} \ln(p_{i,j,\varnothing}).$$

where $p_{i,j,12}$, $p_{i,j,1}$, $p_{i,j,2}$, and $p_{i,j,\varnothing}$ represent the probabilities of both failure modes, failure mode 1 only, failure mode 2 only, and reliability, respectively, that is,

$$\begin{aligned}
p_{i,j,12} &= C_{\alpha_i(\theta)}(\tau_j, \tau_j), \\
p_{i,j,1} &= F_{T_{i,1}}(\tau_j) - C_{\alpha_i(\theta)}(\tau_j, \tau_j), \\
p_{i,j,2} &= F_{T_{i,2}}(\tau_j) - C_{\alpha_i(\theta)}(\tau_j, \tau_j), \\
p_{i,j,\varnothing} &= 1 - F_{T_{i,1}}(\tau_j) - F_{T_{i,2}}(\tau_j) + C_{\alpha_i(\theta)}(\tau_j, \tau_j).
\end{aligned}$$

A two-step procedure has been developed for finding the MLEs of the model parameters, where the marginal distributions are first estimated by the proportions of failures (Ling et al., 2021a). Recently, the Bayesian and the maximum likelihood approaches for one-shot device testing data with correlated failure modes using copula models have been proposed (Ashkamini, Sharma and Upadhyay, 2023; Salem and Salah, 2024; Salah and Salem, 2025), and the issue of copula model misspecification for one-shot devices has also been discussed in the literature (Prajapati et al., 2023). By relaxing the independence assumption and adopting copula models, researchers can gain a better understanding of the underlying dependence structure and its impact on the reliability of one-shot devices with multiple failure modes, leading to improved design, maintenance, and decision-making processes.

3.3 Gamma frailty modeling

The gamma frailty model is an alternative approach for modeling the dependence between failure modes. Frailty, in this context, refers to the influence of common, but unobservable, covariates on the hazard function as a random effect in a model. Frailty models provide an easily understandable interpretation of this latent heterogeneity, in contrast to the copula models. The key idea behind the gamma frailty model is to interpret the frailty as a random proportionality factor that modifies the hazard functions of the component failure times. This describes the dependence between the two components within each device, which arises from common environmental or operational factors.

Suppose that the data consist of N systems, each with two components, which are divided into I test groups. In the i-th test group, N_i systems are exposed to a higher-than-normal stress level x_i and examined at a pre-specified time τ_i. There are four possible combinations of failure modes observed from the life test. Let $n_{i,X}$ denote the number of systems with observed failure modes $X \in \{\varnothing, 1, 2, 12\}$ at time τ_i. We use $\mathbf{z}_i = \{x_i, \tau_i, n_{i,\varnothing}, n_{i,1}, n_{i,2}, n_{i,12}\}$, for $i = 1, ..., I$, to denote the observed data for the i-th test group, and $\mathbf{z} = \{\mathbf{z}_1, ..., \mathbf{z}_I\}$ to denote all observed data. Table 4 presents the form of one-shot device testing data with two failure modes under a CSALT with I test groups.

Ling et al (Ling et al., 2021b). assumed that, conditioned on an unobserved frailty variable $\gamma > 0$, the failure time $T_{i,m}$ of the m-th component in the i-th

TABLE 4 One-shot device testing data with two failure modes for gamma frailty models.

			# Of devices			
Test group	Stress level	Inspection time	$\emptyset$	1	2	12
1	x_1	τ_1	$n_{1,\emptyset}$	$n_{1,1}$	$n_{1,2}$	$n_{1,12}$
2	x_2	τ_2	$n_{2,\emptyset}$	$n_{2,1}$	$n_{2,2}$	$n_{2,12}$
⋮	⋮	⋮	⋮	⋮	⋮	⋮
I	x_I	τ_I	$n_{I,\emptyset}$	$n_{I,1}$	$n_{I,2}$	$n_{I,12}$

device follows an exponential distribution with a rate parameter of $\gamma\lambda_{i,m} > 0$. The corresponding conditional CDF and reliability function are

$$F_{T_{i,m}}(t|\gamma) = P(T_{i,m} \leq t|\gamma) = 1 - \exp(-\gamma\lambda_{i,m}t), \quad t > 0,$$
$$R_{T_{i,m}}(t|\gamma) = 1 - F_{T_{i,m}}(t|\gamma) = \exp(-\gamma\lambda_{i,m}t), \quad t > 0.$$

The failure rate $\lambda_{i,m}$ of the m-th component is assumed to be related to the temperature x_i through a link function: $\lambda_{i,m} = \exp(a_{m0} + a_{m1}x_i)$, where $\theta = (a_{10}, a_{11}, a_{20}, a_{21}, \beta)$ represents the model parameters. The frailties γ are assumed to be common among components within the same device, but different across devices. They are modeled as following a gamma distribution with a scale parameter $\beta > 0$ and a shape parameter $1/\beta > 0$. The mean of the frailty is 1, and its variance is β, which ensures model identifiability (Liu, 2012). The PDF of γ is given by

$$f_\gamma(y) = \frac{1}{\Gamma\left(\frac{1}{\beta}\right)\beta^{\frac{1}{\beta}}} y^{\frac{1}{\beta}-1} \exp\left(-\frac{y}{\beta}\right), \quad y > 0, \ \beta > 0.$$

For a device in the i-th test group, conditioned on the frailty γ, the failure times of the two components are independent. The joint reliability function of $(T_{i,1}, T_{i,2})$ is given by

$$R_{T_{i,1},T_{i,2}}(t_1, t_2) = P(T_{i,1} > t_1, T_{i,2} > t_2) = (1 + \beta(\lambda_{i,1}t_1 + \lambda_{i,2}t_2))^{-\frac{1}{\beta}}. \tag{16}$$

This joint reliability function captures the dependence between the failure times of the two components within the same device, which is introduced by the shared frailty variable γ. However, constraints on β are imposed for a finite mean and variance of the component lifetime.

Furthermore, the failure time distribution of the i-th component under the m-th stress level, $T_{i,m}$, is modeled using a Lomax (or Pareto Type II) distribution, which has the CDF

$$F_{T_{i,m}}(t) = P(T_{i,m} \leq t) = 1 - (1 + \beta\lambda_{i,m}t)^{-\frac{1}{\beta}}, \quad t > 0, \tag{17}$$

This distribution has a scale parameter of $(\beta\lambda_{i,m})^{-1}$ and a shape parameter of $1/\beta$. The mean and variance of $T_{i,m}$ are given by

$$\mu_{T_{i,m}} = \frac{1}{\lambda_{i,m}(1-\beta)}, \quad (\beta < 1),$$
$$\sigma^2_{T_{i,m}} = \frac{1}{\lambda^2_{i,m}(1-\beta)^2(1-2\beta)}, \quad \left(\beta < \frac{1}{2}\right).$$

The failure times of the components are dependent when $\beta \neq 0$, and a larger value of β indicates stronger dependence between the components. When β tends to zero, the components become independent, and the frailties of all devices are equal to one, resulting in independent failure times.

The joint reliability function can also be rewritten in the form of the Clayton survival copula

$$P(T_{i,1} > t_1, T_{i,2} > t_2) = \left(\left(R_{T_{i,1}}(t_1) + R_{T_{i,2}}(t_2)\right)^{-\beta} - 1\right)^{-\frac{1}{\beta}}, \tag{18}$$

where $R_{T_{i,m}}(t) = 1 - F_{T_{i,m}}(t)$ is the reliability function of $T_{i,m}$. The mean lifetime of the component, $\mu_{T_{i,m}}$, is increasing with β, indicating that stronger dependence between the components results in a larger mean component lifetime.

The log-likelihood function of the model parameters $\theta = (a_{10}, a_{11}, a_{20}, a_{21}, \beta)$, based on the observed data $\mathbf{z}$, is as follows:

$$\ell(\theta) = \sum_{i=1}^{I} n_{i,12}\ln(p_{i,12}) + n_{i,1}\ln(p_{i,1}) + n_{i,2}\ln(p_{i,2}) + n_{i,\varnothing}\ln(p_{i,\varnothing}),$$

where

$$p_{i,\varnothing} = P(T_{i,1} > \tau_i, T_{i,2} > \tau_i) = g_i(12, \tau_i),$$
$$p_{i,1} = P(T_{i,1} \leq \tau_i, T_{i,2} > \tau_i) = g_i(2, \tau_i) - g_i(12, \tau_i),$$
$$p_{i,2} = P(T_{i,1} > \tau_i, T_{i,2} \leq \tau_i) = g_i(1, \tau_i) - g_i(12, \tau_i),$$
$$p_{i,12} = P(T_{i,1} \leq \tau_i, T_{i,2} \leq \tau_i) = 1 - g_i(1, \tau_i) - g_i(2, \tau_i) + g_i(12, \tau_i),$$
$$g_i(m, t) = P(T_{i,m} > t) = (1 + \beta\lambda_{i,m}t)^{-\frac{1}{\beta}}, \quad m = 1, 2,$$
$$g_i(12, t) = P(T_{i,1} > t, T_{i,2} > t) = (1 + \beta(\lambda_{i,1} + \lambda_{i,2})t)^{-\frac{1}{\beta}}.$$

An EM algorithm was developed to find the MLEs of the model parameters for the gamma frailty model (Ling et al., 2021b). Building on this, the optimal CSALT design was studied for one-shot devices with correlated failure modes (Ling, 2022). Ling et al (Ling et al., 2024a). further extended the gamma frailty model, wherein the component lifetimes are assumed to follow gamma distributions.

4 Proportional hazards model

The models discussed in the preceding sections are parametric models, where the device lifetimes are assumed to be fully described by a probability distribution consisting of a number of model parameters. While this parametric setup allows for inference development, it can be quite sensitive to departures from the assumed parametric distribution. As an alternative, semi-parametric models that consist of both parametric and non-parametric components can be a good choice for model fitting. One such semi-parametric model that has been studied for analyzing one-shot device testing data is the proportional hazards model. In the proportional hazards model, a parametric link function is adopted to extrapolate the data from elevated stress conditions to normal operating conditions. In additional, the proportional hazards assumption is made, which means that the hazard rate changes in a non-parametric way and does not belong to any particular parametric distribution.

Consider a CSALT experiment with I test groups and J inspection times. For the i-th test group, N_i devices are placed under stress level x_i, of which $N_{i,j}$ are tested at the j-th inspection time τ_j, with $N_i = \sum_{j=1}^{J} N_{i,j}$. The number of devices that have failed by each inspection time are recorded, denoted as $n_{i,j}$. One-shot device testing data obtained from such a life-test can then be represented as $\mathbf{z} = (x_i, \tau_j, N_{i,j}, n_{i,j})$, for $i = 1, ..., I, j = 1, ..., J$.

Under the proportional hazards assumption ((Cox, 1972), (Cox, 2018)), the cumulative hazard function for the i-th test group is given by

$$H_i(t; \boldsymbol{\eta}, a) = H_0(t; \boldsymbol{\eta})\lambda_i(a), \tag{19}$$

where $H_0(t; \boldsymbol{\eta})$ is the baseline cumulative hazard function, the parameter $\boldsymbol{\eta}$ measures the changes in the baseline, and the parameter a captures the influence of the stress factor. The model in (19) is thus composed of two independent components. Ling et al (Ling et al., 2015). assumed a log-linear link function to relate the stress levels to the failure times of the units to obtain the cumulative hazard function from (19) to be

$$H_i(t; \boldsymbol{\eta}, a) = H_0(t; \boldsymbol{\eta})\exp(ax_i); \tag{20}$$

the corresponding reliability function is then

$$R_i(t; \boldsymbol{\eta}, a) = \exp(-H_i(t; \boldsymbol{\eta}, a)) = R_0(t; \boldsymbol{\eta})^{\exp(ax_i)}, \tag{21}$$

where $R_0(t; \boldsymbol{\eta}) = \exp(-H_0(t; \boldsymbol{\eta}))$ is the baseline reliability function. It is interesting to note that the baseline reliability function is a decreasing function over time, that is, $0 < R_0(\tau_J; \eta) < R_0(\tau_{J-1}; \eta) < \cdots < R_0(\tau_2; \eta) < R_0(\tau_1; \eta) < 1$. This requires the reliability functions to be parameterized in terms of the model parameters, such as

$$\gamma(\eta_J) = 1 - R_0(\tau_J; \boldsymbol{\eta}) = 1 - \exp(-\exp(\eta_J)), \tag{22}$$

and, for $j = 1, ..., J - 1$,

$$\gamma(\eta_j) = \frac{1 - R_0(\tau_j; \boldsymbol{\eta})}{1 - R_0(\tau_{j+1}; \boldsymbol{\eta})} = 1 - \exp(-\exp(\eta_j)). \tag{23}$$

Hence, $\boldsymbol{\theta} = (a, \eta_1, ..., \eta_J)$ is the vector of model parameters.

Here the η_j's capture the probabilities of failure between successive inspection times. It can be seen that $R_0(\tau_j; \boldsymbol{\eta}) \approx R_0(\tau_{j+1}; \boldsymbol{\eta})$ when $\eta_j \to +\infty$, and $R_0(\tau_j; \boldsymbol{\eta})$ is much larger than $R_0(\tau_{j+1}; \boldsymbol{\eta})$ when $\eta_j \to -\infty$. The value of η_j gives us an idea about the probability of failure between successive times τ_j and τ_{j+1}. It is observed that few devices will fail between τ_j and τ_{j+1} when η_j tends to $+\infty$, and many devices will fail between τ_j and τ_{j+1} when η_j tends to $-\infty$.

Recent works have also explored Bayesian and robust inferences under the proportional hazards model for one-shot device testing data (Balakrishnan et al., 2021b; Sharma et al., 2021). Additionally, maximum likelihood approaches using EM-algorithms and Nelson-Aalen based estimation have been developed to make non-parametric inferences (Zhu and Balakrishnan, 2022).

5 Step-stress accelerated life tests

Step-stress accelerated life tests (SSALTs) are a type of ALT where the stress level applied to the devices under test is increased at pre-specified time points, unlike CSALTs where each device is subjected to a single, predefined stress level. Compared to CSALTs, SSALTs typically require fewer samples and are more efficient, as well as less costly to collect lifetime data. The step-stress approach has proven to be a valuable tool in accelerated life testing, providing researchers and engineers with a more efficient and cost-effective method for obtaining lifetime data (Sha and Pan, 2014; Haghighi and Bae, 2015; Han and Bai, 2020; Bai et al., 2022; Kateri and Nikolov, 2024). This makes SSALTs an attractive alternative that has gained significant attention in the research literature.

In a SSALT experiment for one-shot devices, the test begins with N devices placed under a higher-than-normal stress level x_1. At the first inspection time τ_1, N_1 devices are randomly selected from the N devices and tested, with n_1 failures recorded. The remaining $N - N_1$ devices are then subjected to an even higher stress level x_2. This process continues, with N_i devices randomly selected from the remaining devices at the i-th inspection time τ_i, tested, and the number of failures n_i recorded. The remaining devices are then placed under the next higher stress level x_{i+1}. Finally, at the last inspection time τ_I, the remaining $N_I = N - \sum_{k=1}^{I-1} N_k$ devices, which have experienced the full sequence of I stress levels, are examined, and the number of failures n_I is recorded. The one-shot device test data is recorded as $\mathbf{z} = \{\tau_i, N_i, n_i, x_i, i = 1, ..., I\}$, as summarized in Table 5. It is noted that Table 1 for CSALTs and Table 5 for SSALTs are the same. The key difference between SSALTs and CSALTs is that in SSALTs, the devices have gone through various stress levels by the time they are examined at each inspection point, while in CSALTs, the devices have only experienced a

TABLE 5 One-shot devices test data under a SSALT with I stages.

Stage	Inspection time	Stress level	# Of tested devices	# Of failures
1	τ_1	x_1	N_1	n_1
2	τ_2	x_2	N_2	n_2
$\vdots$	$\vdots$	$\vdots$	$\vdots$	$\vdots$
I	τ_I	x_I	N_I	n_I

single stress level. This brings different modeling approaches and inferential methods for CSALTs and SSALTs.

The research literature describes three fundamental models for understanding the effect of increased stress levels on the lifetime distribution of a device, including the tampered random variable model (DeGroot and Goel, 1979; Koley et al., 2023; Sultana and Dewanji, 2023; Ahmad et al., 2024), the tampered failure-rate model (Bhattacharyya and Soejoeti, 1989; Madi, 1993; Wang and Fei, 2003; Bobotas and Kateri, 2015), and the cumulative exposure model (Nelson, 1980; Miller and Nelson, 1983; Alhadeed and Yang, 2005; Balakrishnan et al., 2007). This chapter focuses on the cumulative exposure model, which is the most widely adopted approach for analyzing lifetime data obtained from SSALT experiments. The key premise of the CE model is that the remaining lifespan of test units depends solely on the total cumulative exposure they have experienced, regardless of how this exposure was accumulated over time. This means that at any fixed stress level in the SSALT experiment, the test units fail according to the corresponding CDF for that specific stress level. The continuity of the lifetime CDF across stress level changes is a critical feature that allows for effective data analysis and lifetime prediction under the CE model. This modeling approach has thus become the dominant method for analyzing data from SSALT experiments, as it provides a comprehensive framework for understanding the relationship between increased stress levels and device lifetime.

Assuming that the lifetime of one-shot devices follows a Weibull distribution with scale parameter $\alpha_i(>0)$ and shape parameter $a_2(>0)$, the cumulative hazard function and the reliability function under the CE model is given by

$$H(t) = \begin{cases} \left(\frac{t}{\alpha_1}\right)^{a_2}, & 0 < t \le \tau_1 \\ \left(\sum_{j=1}^{i-1} \frac{\Delta_j}{\alpha_j} + \frac{t-\tau_{i-1}}{\alpha_i}\right)^{a_2}, & \tau_{i-1} < t \le \tau_i, \end{cases} \tag{24}$$

for $i = 2, \ldots, I$. The associated reliability function can also be obtained as

$$R(t) = \exp(-H(t))$$
$$= \begin{cases} \exp\left(-\left(\frac{t}{\alpha_1}\right)^{a_2}\right), & 0 < t \leq \tau_1 \\ \exp\left(-\left(\sum_{j=1}^{i-1} \frac{\Delta_j}{\alpha_j} + \frac{t-\tau_{i-1}}{\alpha_i}\right)^{a_2}\right), & \tau_{i-1} < t \leq \tau_i. \end{cases} \tag{25}$$

Furthermore, the Weibull scale parameter for each stage is modeled using a log-linear model with the corresponding stress level. Specifically, $\alpha_i = \exp(a_0 + a_1 x_i)$. Finally, the log-likelihood function for the SSALT data is given by

$$\ell(\boldsymbol{\theta}) \propto \sum_{i=1}^{I} n_i \ln(1 - R(\tau_i; \boldsymbol{\theta})) + (N_i - n_i)\ln(R(\tau_i; \boldsymbol{\theta})), \tag{26}$$

where $\boldsymbol{\theta} = (a_0, a_1, a_2)$ is the model parameters.

Significant advancements have been made in the research on the optimal design of simple SSALTs with two stress levels (I=2 stages). For the commonly used exponential and Weibull lifetime distributions, two recent studies have investigated the optimal design of such simple SSALT experiments (Ling, 2019; Ling and Hu, 2020). These works have provided valuable insights and guidelines for researchers and practitioners on how to efficiently plan and conduct two-stage SSALT experiments for one-shot devices. For instance, under the Weibull lifetime distribution, the SSALT data may not provide sufficient information to estimate all the model parameters simultaneously, as the number of model parameters exceeds the number of observed data points. In such cases, the variance-covariance matrix of the MLEs for the model parameters does not even exist. It is advised to fix one of the model parameters to overcome the estimation problem, such as the parameter related to the stress level a_1. This approach is highly advantageous for simple two-stage SSALTs, as the corresponding optimal designs generally require fewer test samples compared to designs with a fixed shape parameter a_2. Recently, Ling (Ling, 2025) demonstrated that a minimum of three stress levels are required for SSALTs to construct a non-singular variance-covariance matrix of the MLEs for the model parameters of the Weibull lifetime distribution. In addition to the optimal design problem, the literature has also explored robust statistical inference methods for non-destructive one-shot devices tested using SSALTs. In this context, researchers have developed advanced techniques for analyzing SSALT data under the exponential and log-normal lifetime distribution assumptions (Balakrishnan et al., 2023b, 2024).

6 One-shot devices with manufacturing defects

In manufacturing processes, it is possible that defective devices are inadvertently produced. These defective devices are unable to function properly at any time, and their lifetimes are effectively zero. However, as aforementioned, one-shot devices are designed for a single use and cannot be continuously monitored during their lifetime. As a result, any defects in these devices may not be detected until the scheduled inspection time. When a one-shot device fails during the inspection, it is difficult to determine whether the failure was due to a defect in the device or if it simply malfunctioned before the inspection time. This is because the destructive nature of one-shot devices means they cannot be further examined after a failure. The limited resources, capabilities, or time available to investigate each failed one-shot device can exacerbate this issue. Practitioners may not have the means to thoroughly analyze each malfunctioning device and determine the true nature of the failure. Without the ability to investigate each failed device, the root cause of the malfunction often remains unknown. This is known as the "masked" failure problem, where the practitioner cannot confidently distinguish between a defective device and a non-defective device that happened to fail prematurely. In summary, the reliability analysis of one-shot devices is complicated by the lack of continuous monitoring, the destructive nature of failures, and the potential for “masked” failures where the root cause of the malfunction is unclear. Shang et al (Shang et al., 2023). proposed a mixture model to address these challenges.

Suppose the manufacturing process has a defect rate of p, where p is a value between 0 and 1. To study the reliability of one-shot devices with manufacturing defects, a life test is conducted with a pre-specified masking rate, denoted as r_{masked}. In this test, K devices are randomly selected and divided into I test groups. At each inspection time τ_i, K_i devices are examined. If a device is malfunctioned, a careful investigation to identify the root cause of the malfunction for the device with probability of $1 - r_{masked}$. In other words, the probability for masked failure is r_{masked}. The observed data includes the number of surviving devices ($n_{S,i}$), the number of malfunctioning devices with unidentified causes ($n_{M,i}$), the number of identified defective devices ($n_{D,i}$), and the number of identified non-defective devices that failed before the inspection ($n_{F,i}$). It is noting that the number of malfunctioning devices $n_i = n_{F,i} + n_{M,i} + n_{D,i}$. The observed data for the i-th inspection is denoted as $z_i = \{n_{S,i}, n_{M,i}, n_{D,i}, n_{F,i}\}$, and the observed data for all I inspections can be summarized as $z = (z_1, ..., z_I)$, as shown in Table 6.

To analyze the data of one-shot devices with manufacturing defects, the lifetime of the j-th device in the i-th test group is denoted as T_{ij}. If a device fails before the inspection time τ_i, its lifetime is left-censored ($t_{ij} < \tau_i$). If a device performs properly, its lifetime is right-censored ($t_{ij} \geq \tau_i$). It is assumed that the lifetimes of non-defective devices follow an absolutely continuous lifetime distribution with CDF of $F_X(x; \boldsymbol{\theta}_X)$, where $\boldsymbol{\theta}_X$ is the vector of parameters. Defective devices, on the other hand, have a lifetime of zero. The possible

TABLE 6 One-shot device testing data with devices survived, masked, defective, and normally failed.

Test group	Inspection time	# Of surviving devices	# Of devices masked	# Of defective devices	# Of devices normally failed
1	τ_1	$n_{S,1}$	$n_{M,1}$	$n_{D,1}$	$n_{F,1}$
2	τ_2	$n_{S,2}$	$n_{M,2}$	$n_{D,2}$	$n_{F,2}$
$\vdots$	$\vdots$	$\vdots$	$\vdots$	$\vdots$	$\vdots$
I	τ_I	$n_{S,I}$	$n_{M,I}$	$n_{D,I}$	$n_{F,I}$

lifetime distributions include the exponential, Weibull, gamma, log-normal, or generalized gamma distributions. For a simple random sample, the probability that a device is defective is p, which represents the defect rate. The probability that a non-defective device is selected and its lifetime is less than t is $(1-p)F_X(t;\,\theta_X)$. Combining these probabilities, the corresponding CDF for the lifetime T is given by

$$F_T(t) = \Pr(T < t) = p + (1-p)F_X(t;\,\theta_X). \tag{27}$$

This model accounts for the presence of defective devices in the population, which is a critical consideration in the reliability analysis of one-shot devices. When we further assume mutual independence between masking, censoring, and defect rate, the likelihood function of the parameter vector $\theta = (\theta_X, p)$ for the i-th test group can then be expressed as

$$\begin{aligned} L_i(\theta;\,z_i) &\propto P_{S,i}^{n_{S,i}} P_{M,i}^{n_{M,i}} P_D^{n_{D,i}} P_{F,i}^{n_{F,i}} \\ &\propto [p]^{n_{D,i}}[1-p]^{n_{S,i}+n_{F,i}}[p+(1-p)F_X(\tau_i;\,\theta_X)]^{n_{M,i}} \\ &\quad [F_X(\tau_i;\,\theta_X)]^{n_{F,i}}[1-F_X(\tau_i;\,\theta_X)]^{n_{S,i}}. \end{aligned} \tag{28}$$

The analysis of one-shot devices with manufacturing defects, considering the masking effect, is an important topic in reliability engineering, as it helps to accurately assess the performance and reliability of these devices. Recent studies have shown that ignoring the existence of defective devices in reliability analysis can lead to serious biases in the estimation of the mean lifetime (Shang et al., 2023). Subsequently, a Bayesian inference approach for the defect rate has been developed, which can provide more useful insights when the defect rate is small (Ling et al., 2024b).

7 Concluding remarks

This chapter has provided an extensive overview of the research and developments in the field of one-shot device reliability. The key challenges in evaluating the reliability of one-shot devices stem from the nature of destructive experiments, which leads to completely censored, either left-censored (when the device fails before the end of the test) or right-censored (when the device successfully completes the test). This type of data presents significant obstacles in reliably assessing the failure behavior and lifetime characteristics of these devices. To address these challenges, researchers have devoted considerable attention to developing advanced reliability models for one-shot devices. The chapter has covered a range of topics, including popular lifetime distribution models, reliability analysis of devices with multiple failure modes, semi-parametric and non-parametric approaches, step-stress accelerated life testing, and the impact of manufacturing defects on one-shot device reliability. After the comprehensive reviews on modeling for one-shot devices from the recent literature, this chapter has aimed to open up more potential avenues for future investigations in the field of one-shot device reliability. The research in this area continues to evolve, and further advancements are expected to enhance the understanding, design, and management of these critical and specialized devices.

One of the critical challenges in the reliability analysis of one-shot devices is the potential for model misspecification. Researchers have proposed a variety of parametric and non-parametric models to characterize the failure behavior of one-shot devices, including Weibull, lognormal, and exponential distributions, as well as semi-parametric and non-parametric approaches. However, the underlying failure mechanisms and data-generating processes for one-shot devices may not always conform to the assumptions of these models. Model misspecification can lead to biased parameter estimates, inaccurate reliability predictions. Future research should explore the robustness of existing reliability models to departures from their underlying assumptions. This may involve investigating the sensitivity of model parameters and reliability estimates to violations of assumptions, such as the presence of competing failure modes, time-varying covariates, or non-constant failure rates. Additionally, developing diagnostic tools and goodness-of-fit tests specifically tailored to one-shot device data would enable researchers and practitioners to assess the appropriateness of the selected models. This could include model comparison techniques and Bayesian model selection methods to identify the most suitable reliability models for a given one-shot device application.

Additionally, research into optimal experimental designs for one-shot device reliability testing could yield significant benefits. Due to the high cost and limited availability of one-shot devices, researchers should investigate innovative experimental designs that can maximize the information gained from a limited number of test units or accelerated test conditions. Adaptive and sequential designs have not been explored for one-shot devices. Developing experimental protocols that can dynamically adjust the test conditions, sample sizes, or inspection schedules based on the observed data. This can help identify the most informative test conditions and efficiently allocate resources. Bayesian experimental designs is also one of the potential directions for experimental design research. Incorporating prior information about one-shot device reliability, manufacturing processes, and failure modes into the design of experiments. Bayesian approaches can help balance the exploration of new reliability models and the exploitation of existing knowledge.

References

Ahmad, H.H., Ramadan, D.A., Almetwally, E.M., 2024. Tampered random variable analysis in step-stress testing: modeling, inference, and applications. Mathematics 12 (8), 1248.

Alhadeed, A.A., Yang, S.S., 2005. Optimal simple step-stress plan for cumulative exposure model using log-normal distribution. IEEE Trans. Reliab. 54 (1), 64–68.

Ashkamini, Sharma, R., Upadhyay, S.K., 2023. Bayes analysis of one-shot device testing data with correlated failure modes using copula models. Commun. Stat. - Simul. Comput. 1–20.

Azimian, M., Karbasian, M., Atashgar, K., 2021b. Designing a multi-objective mathematical model for determining the reliable suppliers of one-shot systems based on risk assessment. Res. Prod. Oper. Manag. 12 (2), 47–70.

Azimian, M., Karbasian, M., Atashgar, K., 2022a. A multi-objective mathematical model for selecting reliable suppliers for one-shot systems. Expert. Syst. Appl. 207, 117858.

Azimian, M., Karbasian, M., Atashgar, K., 2022c. Developing a novel mathematical approach toward minimizing sustainable circular economy costs of one-shot systems. Prod. Eng. 16 (5), 627–634.

Azimian, M., Karbasian, M., Atashgar, K., Kabir, G., 2021a. A new approach to select the reliable suppliers for one-shot devices. Prod. Eng. 15 (3), 371–382.

Azimian, M., Karbasian, M., Atashgar, K., Kabir, G., 2022b. Appropriate strategy selection for reliability-centered maintenance of one-shot systems using fuzzy model. J. Qual. Maint. Eng. 28 (3), 634–652.

Baghel, S., Mondal, S., 2024. Analysis of one-shot device testing data under logistic-exponential lifetime distribution with an application to SEER gallbladder cancer data. Appl. Math. Model 126, 159–184.

Bai, X., Shi, Y., Ng, H.K.T., 2022. Statistical inference of Type-I progressively censored step-stress accelerated life test with dependent competing risks. Commun. Stat. - Theory Methods 51 (10), 3077–3103.

Bain, L.J., Engelhardt, M., 1991. Reliability test plans for one-shot devices based on repeated samples. J. Qual. Technol. 23 (4), 304–311.

Balakrishnan, N., Castilla, E., 2022. EM-based likelihood inference for one-shot device test data under log-normal lifetimes and the optimal design of a CSALT plan. Qual. Reliab. Eng. Int. 38 (2), 780–799.

Balakrishnan, N., Castilla, E., 2023. Robust estimation based on one-shot device test data under log-normal lifetimes. Statistics 57 (5), 1061–1086.

Balakrishnan, N., Castilla, E., 2024. Robust inference for destructive one-shot device test data under Weibull lifetimes and competing risks. J. Comput. Appl. Math. 437, 115452.

Balakrishnan, N., Castilla, E., Jaenada, M., Pardo, L., 2023b. Robust inference for nondestructive one-shot device testing under step-stress model with exponential lifetimes. Qual. Reliab. Eng. Int. 39 (4), 1192–1222.

Balakrishnan, N., Castilla, E., Ling, M.H., 2022. Optimal designs of constant-stress accelerated life-tests for one-shot devices with model misspecification analysis. Qual. Reliab. Eng. Int. 38 (2), 989–1012.

Balakrishnan, N., Castilla, E., Martín, N., Pardo, L., 2019a. Robust estimators and test statistics for one-shot device testing under the exponential distribution. IEEE Trans. Inf. Theory 65 (5), 3080–3096.

Balakrishnan, N., Castilla, E., Martín, N., Pardo, L., 2019a. Robust inference for one-shot device testing data under Weibull lifetime model. IEEE Trans. Reliab. 69 (3), 937–953.

Balakrishnan, N., Castilla, E., Martín, N., Pardo, L., 2019b. Robust estimators for one-shot device testing data under gamma lifetime model with an application to a tumor toxicological data. Metrika 82 (8), 991–1019.

Balakrishnan, N., Castilla, E., Martín, N., Pardo, L., 2020. Robust inference for one-shot device testing data under exponential lifetime model with multiple stresses. Qual. Reliab. Eng. Int. 36 (6), 1916–1930.

Balakrishnan, N., Castilla, E., Martín, N., Pardo, L., 2021b. Divergence-based robust inference under proportional hazards model for one-shot device life-test. IEEE Trans. Reliab 70 (4), 1355–1367.

Balakrishnan, N., Castilla, E., Martin, N., Pardo, L., 2023a. Power divergence approach for one-shot device testing under competing risks. J. Comput. Appl. Math. 419, 114676.

Balakrishnan, N., Chimitova, E., 2017. Goodness-of-fit tests for one-shot device accelerated life testing data. Commun. Stat. - Simul. Comput. 46 (5), 3723–3734.

Balakrishnan, N., Jaenada, M., Pardo, L., 2024. Non-destructive one-shot device test under step-stress experiment with lognormal lifetime distribution. J. Comput. Appl. Math. 437, 115483.

Balakrishnan, N., Kundu, D., Ng, K.T., Kannan, N., 2007. Point and interval estimation for a simple step-stress model with Type-II censoring. J. Qual. Technol. 39 (1), 35–47.

Balakrishnan, N., Ling, M.H., 2012a. EM algorithm for one-shot device testing under the exponential distribution. Comput. Stat. & Data Anal. 56 (3), 502–509.

Balakrishnan, N., Ling, M.H., 2012b. Multiple-stress model for one-shot device testing data under exponential distribution. IEEE Trans. Reliab. 61 (3), 809–821.

Balakrishnan, N., Ling, M.H., 2013. Expectation maximization algorithm for one shot device accelerated life testing with Weibull lifetimes, and variable parameters over stress. IEEE Trans. Reliab. 62 (2), 537–551.

Balakrishnan, N., Ling, M.H., 2014a. Best constant-stress accelerated life-test plans with multiple stress factors for one-shot device testing under a Weibull distribution. IEEE Trans. Reliab. 63 (4), 944–952.

Balakrishnan, N., Ling, M.H., 2014b. Gamma lifetimes and one-shot device testing analysis. Reliab. Eng. & Syst. Saf. 126, 54–64.

Balakrishnan, N., Ling, M.H., So, H.Y., 2021a. Accelerated Life Testing of One-Shot Devices: Data Collection and Analysis. John Wiley & Sons, Hoboken, New Jersey.

Balakrishnan, N., So, H.Y., Ling, M.H., 2015a. EM algorithm for one-shot device testing with competing risks under exponential distribution. Reliab. Eng. & Syst. Saf. 137, 129–140.

Balakrishnan, N., So, H.Y., Ling, M.H., 2015b. A Bayesian approach for one-shot device testing with exponential lifetimes under competing risks. IEEE Trans. Reliab. 65 (1), 469–485.

Balakrishnan, N., So, H.Y., Ling, M.H., 2015c. EM algorithm for one-shot device testing with competing risks under Weibull distribution. IEEE Trans. Reliab. 65 (2), 973–991.

Bhattacharyya, G.K., Soejoeti, Z., 1989. A tampered failure rate model for step-stress accelerated life test. Commun. Stat.-Theory Methods 18 (5), 1627–1643.

Bobotas, P., Kateri, M., 2015. The step-stress tampered failure rate model under interval monitoring. Stat. Methodol 27, 100–122.

Castilla, E., 2024. A new estimation approach based on phi-divergence measures for one-shot device accelerated life testing. Qual. Reliab. Eng. Int. 40 (4), 2048–2066.

Cheng, Y., Elsayed, E.A., 2016. Reliability modeling and prediction of systems with mixture of units. IEEE Trans. Reliab. 65 (2), 914–928.

Cheng, Y., Elsayed, E.A., 2017a. Optimal sequential ALT plans for systems with mixture of one-shot units. IEEE Trans. Reliab. 66 (4), 997–1011.

Cheng, Y., Elsayed, E.A., 2017b. Reliability modeling of mixtures of one-shot units under thermal cyclic stresses. Reliab. Eng. & Syst. Saf. 167, 58–66.

Cheng, Y., Elsayed, E.A., 2018. Reliability modeling and optimization of operational use of one-shot units. Reliab. Eng. & Syst. Saf. 176, 27–36.

Cox, D.R., 1972. Regression models and life-tables. J. R. Stat. Society: Ser. B (Methodol 34 (2), 187–202.

Cox, D.R., 2018. Analysis of Survival Data. Chapman and Hall/CRC.

DeGroot, M.H., Goel, P.K., 1979. Bayesian estimation and optimal designs in partially accelerated life testing. Nav. Res. Logist. Q. 26 (2), 223–235.

Fan, T.H., Balakrishnan, N., Chang, C.C., 2009. The Bayesian approach for highly reliable electro-explosive devices using one-shot device testing. J. Stat. Comput. Simul. 79 (9), 1143–1154.

Fan, T.H., Chang, C.C., 2009. A Bayesian zero-failure reliability demonstration test of high quality electro-explosive devices. Qual. Reliab. Eng. Int. 25 (8), 913–920.

Haghighi, F., Bae, S.J., 2015. Reliability estimation from linear degradation and failure time data with competing risks under a step-stress accelerated degradation test. IEEE Trans. Reliab. 64 (3), 960–971.

Han, D., Bai, T., 2020. Design optimization of a simple step-stress accelerated life test-contrast between continuous and interval inspections with non-uniform step durations. Reliab. Eng. & Syst. Saf. 199, 106875.

Johnson, N.L., Kotz, S., Balakrishnan, N., 1994. Continuous Univariate Distributions, Second ed. vol. 1 John Wiley & Sons, New York.

Johnson, N.L., Kotz, S., Balakrishnan, N., 1995. Continuous Univariate Distributions, Second ed. vol. 2 John Wiley & Sons, New York.

Kateri, M., Nikolov, N.I., 2024. Product of spacings estimation in step-stress accelerated life testing: an alternative to maximum likelihood. IEEE Trans. Reliab (in press).

Koley, T., Sultana, F., Dewanji, A., 2023. Parametric analysis of tampered random variable model for multiple step-stress life test. J. Stat. Theory Pract. 17 (2), 28.

Ling, M.H., 2019. Optimal design of simple step-stress accelerated life tests for one-shot devices under exponential distributions. Probability Eng. Informational Sci. 33 (1), 121–135.

Ling, M.H., 2022. Optimal constant-stress accelerated life test plans for one-shot devices with components having exponential lifetimes under gamma frailty models. Mathematics 10 (5), 840.

Ling, M.H., 2025. Optimal designs of multiple step-stress accelerated life tests for one-shot devices with Weibull lifetime distributions. IEEE Trans. Reliab (in press).

Ling, M.H., Balakrishnan, N., 2017. Model mis-specification analyses of Weibull and gamma models based on one-shot device test data. IEEE Trans. Reliab. 66 (3), 641–650.

Ling, M.H., Balakrishnan, N., Bae, S.J., 2024a. On the application of inverted Dirichlet distribution for reliability inference of completely censored components with dependence structure. Computers & Ind. Eng. 196, 110452.

Ling, M.H., Balakrishnan, N., Yu, C., So, H.Y., 2021b. Inference for one-shot devices with dependent k-out-of-M structured components under gamma frailty. Mathematics 9 (23), 3032.

Ling, M.H., Chan, P.S., Ng, H.K.T., Balakrishnan, N., 2021a. Copula models for one-shot device testing data with correlated failure modes. Commun. Stat. - Theory Methods 50 (16), 3875–3888.

Ling, M.H., Hu, X.W., 2020. Optimal design of simple step-stress accelerated life tests for one-shot devices under Weibull distributions. Reliab. Eng. & Syst. Saf. 193, 106630.

Ling, M.H., Ng, H.K.T., Shang, X., Bae, S.J., 2024b. Efficient Bayesian inference for a defect rate based on completely censored data. Appl. Math. Model 128, 123–136.

Ling, M.H., So, H.Y., Balakrishnan, N., 2015. Likelihood inference under proportional hazards model for one-shot device testing. IEEE Trans. Reliab. 65 (1), 446–458.

Liu, X., 2012. Planning of accelerated life tests with dependent failure modes based on a gamma frailty model. Technometrics 54 (4), 398–409.

Madi, M.T., 1993. Multiple step-stress accelerated life test: the tampered failure rate model. Commun. Stat.–Theory Methods 22 (9), 295–306.

Meeker, W.Q., Escobar, L.A., 2014. Statistical Methods for Reliability Data. John Wiley & Sons, New York.

Miller, R., Nelson, W., 1983. Optimum simple step-stress plans for accelerated life testing. IEEE Trans. Reliab. 32 (1), 59–65.

Morris, M.D., 1987. A sequential experimental design for estimating a scale parameter from quantal life testing data. Technometrics 29 (2), 173–181.

Mun, B.M., Lee, C., Jang, S.G., Ryu, B.T., Bae, S.J., 2019. A Bayesian approach for predicting functional reliability of one-shot devices. Int. J. Ind. Eng. 26 (1).

Nelson, W., 1980. Accelerated life testing-step-stress models and data analyses. IEEE Trans. Reliab. 29 (2), 103–108.

Nelson, W.B., 2009. Accelerated Testing: Statistical Models, Test Plans, and Data Analysis. John Wiley & Sons, Hoboken, New Jersey.

Newby, M., 2008. Monitoring and maintenance of spares and one shot devices. Reliab. Eng. & Syst. Saf. 93 (4), 588–594.

Prajapati, D., Ling, M.H., Chan, P.S., Kundu, D., 2023. Misspecification of copula for one-shot devices under constant stress accelerated life-tests. Proc. Inst. Mech. Engineers, Part. O: J. Risk Reliab 237 (4), 725–740.

Salah, R.N., Salem, M., 2025. Point and interval estimation for one-shot devices under Weibull distribution with dependent failure modes using copulas. Commun. Stat.-Simulation Comput. 1–16.

Salem, M., Salah, R.N., 2024. Bayesian inference for one-shot devices with Weibull dependent failure modes using copulas. J. Indian. Soc. Probability Stat. 1–23.

Sha, N., Pan, R., 2014. Bayesian analysis for step-stress accelerated life testing using Weibull proportional hazard model. Stat. Pap. 55, 715–726.

Shang, X., Ng, H.K.T., Ling, M.H., 2023. On reliability analysis of one-shot devices with manufacturing defects. Qual. Eng. 35 (1), 79–94.

Sharma, R., Srivastava, R., Upadhyay, S.K., 2021. A hierarchical Bayes analysis and comparison of PH Weibull and PH exponential models for one-shot device testing experiment. Int. J. Reliability, Qual. Saf. Eng. 28 (05), 2150036.

Sharma, R., Upadhyay, S.K., 2018. A hierarchical Bayes analysis for one-shot device testing experiment under the assumption of exponentiality. Commun. Stat. - Simul. Comput. 47 (5), 1297–1314.

Sklar, A., 1959. Fonctions de repartition an dimensions et leurs marges. Publ. de.... L'Institut de Statistique de L'Universit de Paris 8, 229–231.

Sultana, F., Dewanji, A., 2023. Tampered random variable modeling for multiple step-stress life test. Commun. Stat.-Theory Methods 52 (15), 5387–5406.

Vintr, Z., Valis, D., 2007, January. Modeling and analysis of the reliability of systems with one-shot items. In: 2007 Annual Reliability and Maintainability Symposium, IEEE, pp. 380–385.

Wang, R.H., Fei, H.L., 2003. Uniqueness of the maximum likelihood estimate of the Weibull distribution tampered failure rate model. Commun. Stat.-Theory Methods 32 (12), 2321–2338.

Wu, S.J., Hsu, C.C., Huang, S.R., 2020. Optimal designs and reliability sampling plans for one-shot devices with cost considerations. Reliab. Eng. & Syst. Saf. 197, 106795.

Wu, S.J., Huang, S.R., Wang, J.H., 2023. Determination of warranty length for one-shot devices with Rayleigh lifetime distribution. Commun. Stat. - Theory Methods 52 (5), 1400–1416.

Yun, W.Y., Han, Y.J., Kim, H.W., 2014. Simulation-based inspection policies for a one-shot system in storage over a finite time span. Commun. Stat. - Simul. Comput. 43 (8), 1979–2003.

Zhao, Q.Q., Yun, W.Y., 2018. Determining the inspection intervals for one-shot systems with support equipment. Reliab. Eng. & Syst. Saf. 169, 63–75.

Zheng, J., Li, Y., Wang, J., Shiju, E., Li, X., 2018. Accelerated thermal aging of grease-based magnetorheological fluids and their lifetime prediction. Mater. Res. Express 5 (8), 085702.

Zhu, X., Balakrishnan, N., 2022. One-shot device test data analysis using non-parametric and semi-parametric inferential methods and applications. Reliab. Eng. & Syst. Saf. 221, 108319.

Zhu, X., Liu, K., 2022. Reliability of one-shot device with generalized gamma lifetime under cyclic accelerated life-test. Proc. Inst. Mech. Engineers, Part. O: J. Risk Reliab 236 (6), 1007–1023.

Zhu, X., Liu, K., He, M., Balakrishnan, N., 2021. Reliability estimation for one-shot devices under cyclic accelerated life-testing. Reliab. Eng. & Syst. Saf. 212, 107595.

Chapter 3

Optimal experimental planning for experiments based on coherent systems with industrial applications

Yang Yu[a], **Hon Keung Tony Ng**[b,*], **and Erhard Cramer**[c]
[a]*Department of Statistics and Data Science, Southern Methodist University, Dallas, TX, United States,* [b]*Department of Mathematical Sciences, Bentley University, Waltham, MA, United States,* [c]*Institute of Statistics, RWTH Aachen University, Aachen, Germany*
[*]*Corresponding author. e-mail address: tng@bentley.edu*

Chapter Outline

Abstract

In industrial engineering and manufacturing, assessing the reliability of a product or system is crucial. Life-testing and reliability experiments are commonly employed to

Handbook of Statistics, Vol. 53. https://doi.org/10.1016/bs.host.2025.04.006

gain robust insights into product or system lifetime distributions. During these tests, successive times to failure are recorded, and lifetime data are collected. Life-testing is useful across various industrial environments, including the automobile, materials, telecommunications, and electronics sectors. Statisticians focus on two key problems: statistical inference based on life-test data and the effective planning of life-testing experiments under specific constraints. While the experimental design for life tests has been extensively studied, the planning of experiments involving the integration of test units into systems has not been thoroughly explored. This chapter addresses the optimal experimental planning problem when test units are incorporated into coherent systems for experimentation. Our focus is on planning progressively Type-II censored life-testing experiments, with discussions on other experimental schemes. Utilizing the notion of system signature - a structural reliability tool representing the configuration of a coherent system - we derive maximum likelihood estimators for the model parameters and the expected Fisher information matrix under various experimental settings, models, and assumptions. Optimal experimental plans are derived based on different optimality criteria, such as D-optimality, A-optimality, and V-optimality, under varying conditions. Numerical and Monte Carlo simulation studies are conducted to demonstrate the advantages and disadvantages of incorporating systems in life-testing experiments.

1 Introduction

Assessing the reliability of products and systems is a critical aspect of industrial engineering and manufacturing. Life-testing and reliability experiments are commonly employed to gather insights into the lifetime distributions of products or systems. Typically, a sample of items is subjected to stresses and environmental conditions that simulate normal operating environments. Throughout the life test, failure times are recorded, and lifetime data are collected. This approach is widely useful across various industries, including automotive, materials, telecommunications, and electronics.

Statistical inference for lifetime data and the efficient planning of life-testing experiments under constraints are key challenges of interest to both statisticians and engineers. While the design of life-testing experiments has been studied extensively, there has not been a significant focus on experimental planning that incorporates placing test units into systems for life-testing. In many cases, life-testing experiments can be conducted using individual items or by testing systems composed of those items. For instance, in a reliability study of joints used in case furniture, as presented in Klos et al. (2018), the experiment was conducted using 10 joints connected in series, with the number of working cycles recorded until the failure of the 10-component series system. This approach helps determine the durability and safety of the joints. Testing systems, rather than individual components, can save significant experimental time, as the lifetime of a system is typically shorter than the longest lifetime of its individual components. Therefore, it is crucial to develop experimental plans that consider both system-based and individual item-based testing approaches. In this chapter, we address the problem of optimal experimental design in life-testing experiments.

In an ideal scenario, researchers would have a complete sample in a life-testing experiment without any censoring, meaning they observe all the failure times of the items in the experiment (cf. Ng, 2010). Nevertheless, this is generally difficult to achieve in practice due to the limitations of time and experimental resources. Therefore, it is important to consider statistical analysis and inference based on censored life-testing experiments. This chapter focuses on progressively Type-II right-censored life testing experiments, a flexible and efficient censoring scheme that reduces the cost and duration of life testing while maintaining the ability to estimate different reliability metrics precisely. The progressive Type-II right-censoring scheme can be described as follows (for details, see Balakrishnan and Aggarwala, 2000; Balakrishnan and Cramer, 2014; Balakrishnan and Cramer, 2023): Consider a life-testing experiment involving m units tested simultaneously. When the first failure occurs, R_1 experimental units are randomly removed from the remaining $(m-1)$ surviving units. At the time of the second failure, R_2 experimental units are randomly removed from the remaining $(m-R_1-1)$ surviving units. This process continues similarly until the r-th failure, at which point all remaining $(m-R_1-R_2-\ldots-R_{r-1}-1)$ surviving units are removed. The censoring plan $(R_1, R_2, \ldots, R_r)$ is predetermined before the study begins.

The literature has explored the problem of optimal progressive censoring after recognizing that progressive Type-II censoring is a flexible scheme that allows for the adjustment of $(R_1, R_2, \ldots, R_r)$. In one notable study, Balakrishnan and Aggarwala (2000) defined objective functions to identify optimal progressive censoring plans based on the variances and covariances of the best linear unbiased estimators (BLUEs) of parameters in location-scale distributions. They implemented analytical and computational methods to calculate these functions for the location-scale family of distributions, where moments of the usual order statistics are accessible. Similarly, Ng et al. (2004) investigated optimal progressive censoring plans for experimental units following the Weibull distribution. Meanwhile, Burkschat et al. (2006) determined optimal progressive Type-II censoring schemes for exponential, uniform, and Pareto distributions, utilizing the variance of the BLUE for one-parameter models. For two-parameter models, they applied two optimality criteria: minimizing the trace and the determinant of the variance-covariance matrix of the BLUEs. Further, Burkschat et al. (2007) developed optimal censoring schemes using p-criteria from experimental design applied to the variance-covariance matrix of the BLUEs and discussed the monotonicity properties of the trace and determinant of the variance-covariance matrix of the maximum likelihood estimators (MLEs) relative to sample size and the initial number of experimental units. In Cramer and Davies (2018), restrictions on the censoring plans have been additionally imposed, leading to constrained optimization problems. Dahmen et al. (2012) established a representation of the Fisher information matrix of the MLEs in terms of the hazard rate of the baseline distribution and identified A- and D-optimal progressively Type-II censoring plans across various distributions. Additionally, Salemi et al. (2019)

examined the relationship between the missing information matrix and the Fisher information matrix, presenting a straightforward method for determining optimal progressive censoring plans from a specific class of one-step censoring plans under A-optimality and D-optimality criteria for models with multiple parameters. In Bhattacharya and Balakrishnan (2024), the authors proposed a probabilistic approach for determining optimal progressive censoring plans by defining a probability structure on feasible solutions, facilitating the computation of updated solutions. Research has also explored optimal censoring schemes based on non-parametric statistical inference. For instance, Balakrishnan and Han (2007) investigated optimal censoring schemes for non-parametric confidence intervals of population quantiles in the context of progressive Type-II right censoring.

In life testing experiments, the test units can be expensive since they are usually products under development. Moreover, the time allowed for the experiment is always a constraint. Observing the lifetimes of systems formed by different numbers of components makes it possible for researchers to save time, save surviving items for future use, and improve the accuracy of parameter estimation. For example, for progressively censored life testing experiments, Wu and Kus (2009), Hermanns and Cramer (2017) and Hermanns and Cramer (2018) used the series systems, the parallel systems, and the k-out-of-n systems, respectively, for the life testing experiments. These authors also developed the statistical inference under those experiments with different kinds of systems. Note that another connection of coherent system data to progressive censoring schemes has been presented in Cramer and Navarro (2015) and Cramer and Navarro (2016).

Although statistical inference for system-level progressively censored data and optimal progressive censoring schemes have been studied in the literature, a comprehensive study of the optimal experimental planning for progressively Type-II censored experiments based on coherent systems has not been done to our knowledge. Therefore, this chapter aims to develop the algorithm to obtain the optimal progressive Type-II censoring schemes when the components can be put into coherent systems for the life testing experiment.

The remainder of the paper is organized as follows. In Section 2, we introduce the notation of system structure, coherent systems, and system signatures. The literature on statistical inference based on system lifetime data is also reviewed in Section 2. In Section 3, we describe the model for progressively Type-II censored experiments with coherent systems and outline different optimality criteria. Specifically, we review the notation for the system signature and the location-scale family of distributions. In Section 4, we derive the MLEs of the model parameters and the corresponding Fisher information matrix for the location-scale family of distributions. Section 5 discusses various optimality criteria, including commonly used criteria such as D-optimality, A-optimality, and V-optimality, as well as a comprehensive criterion and different settings for optimal experimental planning. In Section 6, we consider multiple cases involving different types of systems used in experiments with progressively censored samples. Section 7 presents computational formulas for lognormal and Weibull distributed components to illustrate the methodology developed in this chapter, with numerical examples for n values up to 4 or 5 to demonstrate the advantages and disadvantages of the approaches. In Section 8, a practical example of insulating fluids testing is used to illustrate the methodologies for planning a future progressively Type-II censored experiment, followed by concluding remarks and discussions about future research directions in Section 9.

2 System structure, coherent system and system signature

2.1 System structure and coherent system

In engineering, a system can be described as an assemblage of elements organized to fulfill specific purposes. Consider a configuration comprising n distinct elements, where each element can exist in one of two conditions: operational or non-operational. We can denote the condition of the i-th element within this configuration using a binary indicator β_i, referred to as the *state variable*. This indicator is defined as follows:

$$\beta_i = \begin{cases} 1, & \text{if the i} - \text{th component is functioning,} \\ 0, & \text{if the i} - \text{th component is failed.} \end{cases}$$

Then, $\beta = (\beta_1, \beta_2, \ldots, \beta_n)$, denotes the *state vector* of the system and the state of the system can be described by the *structure function* defined as

$$\phi(\beta) = \phi(\beta_1, \beta_2, \ldots, \beta_n).$$

In Fig. 1, we illustrate the series and parallel systems with structure functions

$$\phi(\beta) = \beta_1\beta_2 \cdots \beta_n = \prod_{i=1}^{n} \beta_i,$$

and

$$\phi(\beta) = 1 - \prod_{i=1}^{n} (1 - \beta_i),$$

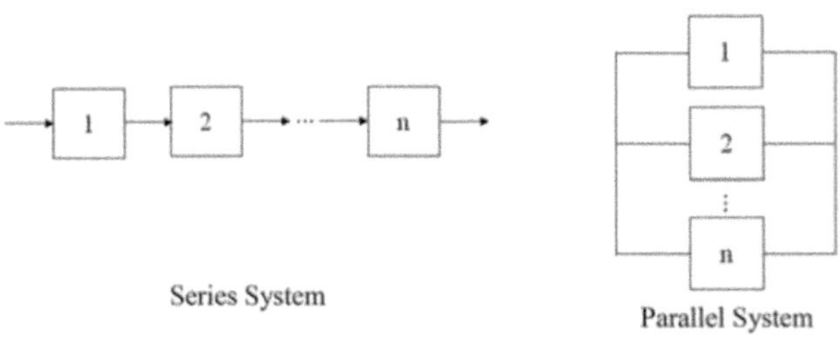

FIG. 1 Series and parallel n-component systems.

respectively. The state of the series system is $\phi(\beta) = 1$ (i.e., functioning) if and only if all the n components in the system are functioning (i.e., $\beta_i = 1$ for all $i = 1, 2, ..., n$). The state of the parallel system is $\phi(\beta) = 1$ (i.e., functioning) if at least one of the n components in the system is functioning (i.e., $\beta_i = 1$ for at least one $i = 1, 2, ..., n$).

A component within a system is considered relevant if the system's status relies on all its components. A system structure function is classified as monotone if repairing a failed component cannot worsen the system's condition. A coherent system is defined as one that is both relevant and possesses a monotone structure function. For instance, a system equipped with a spare component does not qualify as coherent, as the system's status does not depend on the spare. Beyond the structure function $\phi(\beta)$, various representations exist for illustrating a system's structure, such as minimal path sets and minimal cut sets (see, for example, Proschan and Barlow, 1965; Rausand Hoyland, 2004). For more details on system reliability theory, one may refer to Rausand et al. (2020) and Navarro (2022). In this chapter, we focus on a coherent system with n independent and identically distributed (i.i.d.) components, represented through system signature and minimal signature.

2.2 System signature

Consider a coherent system with n components, each represented by i.i.d. lifetimes denoted as $X_1, ..., X_n$. The ordered lifetimes are expressed as $X_{1:n} < X_{2:n} < ... < X_{n:n}$, with the overall lifetime of the n-component system defined as $T = T(X_{1:n}, ..., X_{n:n})$. We assume that the component lifetimes adhere to a shared absolutely continuous distribution characterized by the cumulative distribution function (CDF) $F_X(\cdot)$, the probability density function (PDF) $f_X(\cdot)$, and the survival function (SF) given by $\bar{F}_X(\cdot) = 1 - F_X(\cdot)$. Additionally, we denote the CDF, PDF, and SF of the i-th order statistics as $F_{i:n}(\cdot)$, $f_{i:n}(\cdot)$, and $\bar{F}_{i:n}(\cdot)$, respectively. The CDF, PDF, and SF for the n-component coherent system are represented as $F_T(\cdot)$, $f_T(\cdot)$, and $\bar{F}_T(\cdot)$, respectively.

Let I be a random variable that indicates the corresponding ordered component lifetimes of the system failure, i.e., $I = i$ if the n-component system fails at the i-th ordered component failure. In this case, we have $T = X_{i:n}$. For some systems, the value of I is fixed (i.e., $\Pr(I = i) = 1$). For example, for a k-out-of-n system $\Pr(I = k) = 1$, where $k = 1$ corresponds to the series system, and $k = n$ corresponds to the parallel system (see, Fig. 1). Suppose the random variable I is not observable, then we consider the probability mass function of I as $\Pr(I = i) = \Pr(T = X_{i:n}) = s_i$, $0 \le s_i \le 1$, $i = 1, 2, ..., n$, with $\sum_{i=1}^{n} s_i = 1$. The n-dimensional probability vector $s = (s_1, s_2, ..., s_n)$ is called the system signature, proposed by Samaniego (1985) (see also, Samaniego, 2007), which is a nonparametric representation of the system structure that does not depend on the underlying lifetime distribution of the components. For the series and parallel systems illustrated in Fig. 1), the system signatures are one point distributions $s = (1, 0 ..., 0)$ and $s = (0, ..., 0, 1)$, respectively. Note that the number of failed components leading to system failure will usually be random. In the following, we denote this number by R. Then, the expected number of failed components in a single coherent system with signature vector s is given by the expected value of the signature vector, that is,

$$E(R) = \sum_{i=1}^{n} i s_i. \tag{1}$$

From Samaniego (1985), the PDF and SF of system lifetime T can be expressed in terms of the CDF and PDF of the component lifetime as

$$f_T(t) = \sum_{i=1}^{n} s_i \binom{n}{i} i f_X(t) [F_X(t)]^{i-1} [\bar{F}_X(t)]^{i-1} \tag{2}$$

and

$$\bar{F}_T(t) = \sum_{i=1}^{n} s_i \sum_{j=0}^{i-1} \binom{n}{j} [F_X(t)]^j [\bar{F}_X(t)]^{n-j}. \tag{3}$$

Navarro et al. (2007) considered another representation called *minimal signature* of the system by expressing the SF of the system lifetime T as a generalized mixture of the SFs of i-component series system lifetimes $(i = 1, 2, ..., n)$, i.e. $\bar{F}_{1:i}(\cdot)$:

$$\bar{F}_T(t) = \sum_{i=1}^{n} a_i \bar{F}_{1:i}(t) = \sum_{i=1}^{n} a_i [\bar{F}(t)]^i, \tag{4}$$

for some integers (which can be positive or non-positive) $a_1, a_2, ..., a_n$ that do not depend on the underlying lifetime distribution of the components with $\sum_{i=1}^{n} a_i = 1$. The n-dimensional vector $a = (a_1, a_2, ..., a_n)$ is called the minimal signature of the system. For example, the minimal signature of a n-component series system is $a = (0, ..., 0, 1)$. The minimal signature of a system can be obtained from its system signature and vice versa (see Navarro et al., 2007).

2.3 Statistical inference based on system lifetime data

When n-component systems are used in a life testing experiment, there have been several studies on the estimation of parameters of the lifetime distributions of components based on system lifetime data. With known signatures and Type-II censored data, Balakrishnan et al. (2011b) discussed linear inference for the lifetime distributions of components and derived the best linear unbiased estimators (BLUE) for the parameters. Fan and Wang (2011) established a cumulative exposure model for series system with masking and Type-I censoring, and gave inferences on functions of the unknown parameters of the assumed Weibull distribution. Aslett et al. (2015) studied Bayesian inference for systems and networks based on signature. Zhang et al. (2015) studied the statistical inference and estimation of parameters using the maximum likelihood estimation method and regression-based method. Yang et al. (2016) proposed the stochastic expectation-maximization (SEM) algorithm to estimate the model parameters with complete and censored system lifetimes. Similarly, with unknown system structure, Yang et al. (2019) adopted the expectation-maximization (EM) algorithm to do inference and gave illustration based on a two-parameter Weibull distribution. Hermanns et al. (2020) developed EM algorithms for ordered and censored system lifetime data under a proportional hazard rate model. Fallah et al. (2021) developed the tools for statistical inference of component lifetime distribution based on system lifetime data under a proportional reversed hazard model. Tavangar and Asadi (2020) considered different estimation methods for the reliability of components of the system when progressively censored system failure times are observed.

In addition to parametric inference, nonparametric statistical inference has been developed in the literature. For example, Bhattacharya and Samaniego (2010) considered the nonparametric maximum likelihood estimate of the CDF of component lifetimes based on the observed system failure times and characterized its asymptotic behavior. Balakrishnan et al. (2011a) developed exact nonparametric inference for component lifetime distribution in the form of confidence intervals for population quantiles and tolerance limits, and Coolen and Al-nefaiee (2012) discussed nonparametric predictive inference for system lifetimes with exchangeable components. Hall et al. (2015) and Jin et al. (2017) studied the nonparametric estimation of a common component reliability function using independent samples from coherent systems when the system design is known and unknown, respectively. Qu et al. (2024) and Qu and Ng (2024) developed and evaluated nonparametric, distribution-free hypothesis testing procedures for assessing the homogeneity of component lifetime distributions based on system lifetime data, using empirical likelihood methods, for complete and censored samples, respectively.

3 Progressively type-II censored experiments with coherent systems

Suppose we have $N = m \times n$ components available for a progressively Type-II censored experiment, and these components can be placed as m coherent n-component systems with system signature $s = (s_1, s_2, ..., s_n)$ for the life testing experiment, we plan to fail $r \leq m$ systems (i.e., r is the effective sample size and m is the total sample size in terms of systems) with progressive censoring plan $\boldsymbol{R} = (R_1, R_2, ..., R_r)$. We use $X_1, X_2, ..., X_n$ to denote the i.i.d. component lifetimes, and $T_1, T_2, ..., T_m$ to denote the m system lifetimes. For the special case when $n = 1$, the life testing experiment is done at the component level in which $m = N$ components are placed on the life test, and the test is terminated as soon as the r-th component failure with progressive censoring plan $\boldsymbol{R} = (R_1, R_2, ..., R_r)$ is observed.

In this chapter, we consider a general model that the lifetimes of the components are i.i.d. with a distribution that belongs to the log-location-scale family of distributions, then the log-transformed component lifetimes, denoted as $U = \ln X$, follow a distribution belong to the location-scale family of distributions with location parameter $\mu \in (-\infty, \infty)$ and scale parameter $\sigma \in (0, \infty)$. Suppose $f_U(\cdot; \theta)$, $F_U(\cdot; \theta)$ and $\bar{F}_U(\cdot; \theta)$ are the PDF, CDF, and SF of the log-lifetimes of components in the systems (i.e., $U = \ln X$) with parameter vector $\theta = (\mu, \sigma) \in \Theta = (-\infty, \infty) \times (0, \infty)$, then we have

$$\begin{aligned} f_U(u; \mu, \sigma) &= \frac{1}{\sigma} f^*\left(\frac{u-\mu}{\sigma}\right), \\ F_U(u; \mu, \sigma) &= F^*\left(\frac{u-\mu}{\sigma}\right), \\ \bar{F}_U(u; \mu, \sigma) &= 1 - F^*\left(\frac{u-\mu}{\sigma}\right), \end{aligned}$$

where $f^*(\cdot)$ and $F^*(\cdot)$ are, respectively, the PDF and CDF of the standardized distribution (i.e., $\mu = 0$ and $\sigma = 1$) of the corresponding distribution in the location-scale family. The PDF and SF of the log-transformed system lifetimes (i.e., $V = \ln T$) can be expressed in terms of the PDF, CDF, and SF of the log-transformed component lifetimes as

$$\begin{aligned} f_V(v; \theta) &= \sum_{i=1}^{n} s_i \binom{n}{i} i f_U(v; \theta) [F_U(v; \theta)]^{i-1} [\bar{F}_U(v; \theta)]^{i-1} \\ &= \frac{1}{\sigma} f^*\left(\frac{v-\mu}{\sigma}\right) \sum_{i=1}^{n} s_i \binom{n}{i} i \left[F^*\left(\frac{v-\mu}{\sigma}\right)\right]^{i-1} \left[\bar{F}^*\left(\frac{v-\mu}{\sigma}\right)\right]^{n-i} \\ &= \frac{1}{\sigma} f^*(z) \sum_{i=1}^{n} s_i \binom{n}{i} i [F^*(z)]^{i-1} [\bar{F}^*(z)]^{n-i}, \end{aligned} \tag{5}$$

and

$$\begin{aligned} \bar{F}_V(v; \theta) &= \sum_{i=1}^{n} s_i \sum_{j=0}^{i-1} \binom{n}{i} [F_U(v; \theta)]^j [\bar{F}_U(v; \theta)]^{n-j} \\ &= \sum_{i=1}^{n} s_i \sum_{j=0}^{i-1} \binom{n}{j} \left[F^*\left(\frac{v-\mu}{\sigma}\right)\right]^j \left[\bar{F}^*\left(\frac{v-\mu}{\sigma}\right)\right]^{n-j} \\ &= \sum_{i=1}^{n} s_i \sum_{j=0}^{i-1} \binom{n}{j} [F^*(z)]^j [\bar{F}^*(z)]^{n-j}, \end{aligned} \tag{6}$$

where

$$z = \frac{v-\mu}{\sigma}.$$

The location-scale family contains many commonly used distributions, including the normal, smallest extreme value, largest extreme value, and logistic distributions. These distributions are often used for modeling log-transformed lifetime data. For example, if the component lifetime X follows a lognormal distribution, then $U = \ln X$ follows the normal distribution, which belongs to the location-scale family. If the component lifetime X follows a Weibull distribution, then $U = \ln X$ follows the smallest extreme value (SEV) distribution, which belongs to the location-scale family.

4 Maximum likelihood estimation

4.1 Maximum likelihood estimation of model parameters

To obtain a progressively Type-II censored sample with the progressive censoring plan $R = (R_1, R_2, ..., R_r)$, R_j operating systems are randomly removed from the life test when the ℓ-th failure is observed, $\ell = 1, 2, ..., r$. Thus, the pre-fixed progressive Type-II censoring plan $(R_1, ..., R_r)$ and the pre-fixed number of failures r satisfy the constraint $\Sigma_{\ell=1}^{r} R_\ell + r = m$, and the ordered log-transformed observed progressively Type-II censored sample with coherent systems is denoted as $V_{1:r:m} < V_{2:r:m} < ... < V_{r:r:m}$. The likelihood function and the log-likelihood function based on the observed system-level progressively Type-II censored sample can be expressed as

$$L(\theta) = \left(\prod_{\ell=1}^{r} \gamma_\ell\right) \prod_{\ell=1}^{r} f_V(v_{\ell:r:m};\, \theta)[\bar{F}_V(v_{\ell:r:m};\, \theta)]^{R_\ell}, \qquad v_{1:r:m} < v_{2:r:m} < ... < v_{r:r:m},$$

and

$$l(\theta) = C + \sum_{\ell=1}^{r} \ln f_V(v_{\ell:r:m};\, \theta) + \sum_{\ell=1}^{r} R_\ell \ln[\bar{F}_V(v_{\ell:r:m};\, \theta)],\ v_{1:r:m} < v_{2:r:m} < ... < v_{r:r:m},$$

respectively, where $\gamma_\ell = \Sigma_{j=\ell}^{r}(R_j + 1) = m - R_1 - R_2 - ... - R_{\ell-1} - \ell + 1$ denotes the number of items remaining in the experiment before the ℓ-th failure for $\ell = 1, 2, ..., r$, and $C = \Sigma_{\ell=1}^{r} \ln \gamma_\ell$ is a constant that does not depend on the parameter vector θ.

We can obtain the likelihood equations by taking the first partial derivatives of the log-likelihood function and setting them to zero, i.e.,

$$\frac{\partial l(\theta)}{\partial \mu} = \sum_{\ell=1}^{r} \frac{1}{f_V(v_{\ell:r:m};\, \theta)} \frac{\partial f_V(v_{\ell:r:m};\, \theta)}{\partial \mu} + \sum_{\ell=1}^{r} \frac{R_\ell}{\bar{F}_V(v_{\ell:r:m};\, \theta)} \frac{\partial \bar{F}_V(v_{\ell:r:m};\, \theta)}{\partial \mu} = 0, \tag{7}$$

$$\frac{\partial l(\theta)}{\partial \sigma} = \sum_{\ell=1}^{r} \frac{1}{f_V(v_{\ell:r:m};\, \theta)} \frac{\partial f_V(v_{\ell:r:m};\, \theta)}{\partial \sigma} + \sum_{\ell=1}^{r} \frac{R_\ell}{\bar{F}_V(v_{\ell:r:m};\, \theta)} \frac{\partial \bar{F}_V(v_{\ell:r:m};\, \theta)}{\partial \sigma} = 0. \tag{8}$$

The MLEs of the parameters μ and σ, denoted as $\hat{\mu}$ and $\hat{\sigma}$, respectively, can be obtained by solving Eqs. (7) and (8) simultaneously.

The first derivatives of the SF and the PDF of the log-transformed system lifetime with respect to the parameters μ and σ can be expressed in terms of the log-transformed component lifetime distributions as

$$\frac{\partial \bar{F}_V(v_{\ell:r:m};\boldsymbol{\theta})}{\partial \mu} = \left[-\frac{1}{\sigma}f^*(z_{\ell:r:m})\right]\sum_{i=1}^{n} s_i \sum_{j}^{i-1}\binom{n}{j}[\,j\,[F^*(z_{\ell:r:m})]^{j-1}\,[\bar{F}^*(z_{\ell:r:m})]^{n-j}$$
$$-\,[F^*(z_{\ell:r:m})]^{j}\,(n-j)\,[\bar{F}^*(z_{\ell:r:m})]^{n-j-1}], \tag{9}$$

$$\frac{\partial \bar{F}_V(v_{\ell:r:m};\boldsymbol{\theta})}{\partial \sigma} = \left[-\frac{z_{\ell:r:m}}{\sigma}f^*(z_{\ell:r:m})\right]\sum_{i=1}^{n} s_i \sum_{j}^{i-1}\binom{n}{j}[j\,[F^*(z_{\ell:r:m})]^{j-1}[\bar{F}^*(z_{\ell:r:m})]^{n-j}$$
$$-\,[F^*(z_{\ell:r:m})]^{j}\,(n-j)\,[\bar{F}^*(z_{\ell:r:m})]^{n-j-1}], \tag{10}$$

and

$$\frac{\partial f_V(v_{\ell:r:m};\boldsymbol{\theta})}{\partial \mu} = \frac{1}{\sigma}\frac{\partial f^*(z_{\ell:r:m})}{\partial \mu}\sum_{i=1}^{n} s_i\binom{n}{i} i\,[F^*(z_{\ell:r:m})]^{i-1}[\bar{F}^*(z_{\ell:r:m})]^{n-i}$$
$$-\,\frac{1}{\sigma^2}(f^*(z_{\ell:r:m}))^2\sum_{i=1}^{n} s_i\binom{n}{i} i\,[(i-1)[F^*(z_{\ell:r:m})]^{i-2}[\bar{F}^*(z_{\ell:r:m})]^{n-i}$$
$$-\,[F^*(z_{\ell:r:m})]^{i-1}(n-i)\,[\bar{F}^*(z_{\ell:r:m})]^{n-i-1}], \tag{11}$$

$$\begin{aligned}\frac{\partial f_V(v_{\ell:r:m};\boldsymbol{\theta})}{\partial \sigma} &= \left[-\frac{1}{\sigma^2}f^*(z_{\ell:r:m}) + \frac{1}{\sigma}\frac{\partial f^*(z_{\ell:r:m})}{\partial \sigma}\right]\\ &\quad\times \sum_{i=1}^{k} s_i\binom{n}{i} i\,[F^*(z_{\ell:r:m})]^{i-1}[\bar{F}^*(z_{\ell:r:m})]^{n-i}\\ &\quad-\,\frac{z_{\ell:r:m}}{\sigma^2}(f^*(z_{\ell:r:m}))^2\sum_{i=1}^{n} s_i\binom{n}{i} i\,[(i-1)\\ &\quad[F^*(z_{\ell:r:m})]^{i-2}[\bar{F}^*(z_{\ell:r:m})]^{n-i}\\ &\quad-\,[F^*(z_{\ell:r:m})]^{i-1}(n-i)[\bar{F}^*(z_{\ell:r:m})]^{n-i-1}].\end{aligned} \tag{12}$$

4.2 Fisher information and asymptotic variance-covariance matrices

The expected Fisher information matrix of the parameter vector $\boldsymbol{\theta} = (\mu, \sigma)$ can be expressed as

$$\boldsymbol{I}(\mu,\sigma) = -\begin{bmatrix} E\left(\frac{\partial^2 l(\boldsymbol{\theta})}{\partial \mu^2}\right) & E\left(\frac{\partial^2 l(\boldsymbol{\theta})}{\partial \mu \partial \sigma}\right) \\ E\left(\frac{\partial^2 l(\boldsymbol{\theta})}{\partial \sigma \partial \mu}\right) & E\left(\frac{\partial^2 l(\boldsymbol{\theta})}{\partial \sigma^2}\right)\end{bmatrix}. \tag{13}$$

The second partial derivatives of the log-likelihood function involved in the Fisher information matrix in Eq. (13) can be expressed in terms of the PDF and CDF of the log-transformed component lifetimes as described in Section 4.1.

Then, the asymptotic variance-covariance matrix of the MLEs of μ and σ can be obtained as the inverse of the expected Fisher information matrix as

$$\boldsymbol{V}(\mu, \sigma) = \boldsymbol{I}^{-1}(\mu, \sigma) = \begin{bmatrix} Var(\hat{\mu}) & Cov(\hat{\mu}, \hat{\sigma}) \\ Cov(\hat{\mu}, \hat{\sigma}) & Var(\hat{\sigma}) \end{bmatrix}. \tag{14}$$

For a particular standardized location-scale distribution, $f^*(\cdot)$, $F^*(\cdot)$, and $\bar{F}^*(\cdot)$ are specified and the normal equations in Eqs. (7) and (8), the expected Fisher information matrix in Eq. (13) and the asymptotic variance-covariance matrix in Eq. (14) can be obtained.

5 Experimental planning and optimality criteria

5.1 Settings for optimal experimental planning

This subsection explores various settings and scenarios for planning the progressively Type-II censored experiments described in Section 1. Suppose N components are available for a progressively Type-II censored experiment, and these components can either be tested on a life test individually or as part of systems formed by them. We consider the case where the system structures used in the experiments are identical. Specifically, the N components are arranged into M (= $\lfloor N/n \rfloor$, *where* $\lfloor a \rfloor$ *is the largest integer less than* a) n-component systems with system signature $s = (s_1, \ldots, s_n)$. These M systems are placed on a life test with a progressive censoring plan $R = (R_1, R_2, \ldots, R_r)$ such that r failures are observed. In a progressively Type-II censored experiment, we assume that the systems used are identically structured, meaning they share the same system signature. However, systems used in different experiments may vary in structure. To facilitate comparison across different progressively Type-II censored experiments, we consider the following two scenarios:

Scenario I. Specifying the expected number of failed components in the failed systems:

In this scenario, we fix the expected number of failed components in the failed systems, denoted as (cf. Eq. (1))

$$E(R) = \left[\sum_{i=1}^{n} i s_i \right] r.$$

For example, if we have $N = 20$ components available for the experiment and use ten 2-component parallel systems (i.e., $s = (0, 1)$) for the progressively Type-II censored experiment with an effective sample size $r = 4$, then $E(R) = 2 \times 4 = 8$. Note that $E(R)$ is not the total number of failed components in the progressively censored experiment, as the censored systems can contain different numbers of failed components, except for series systems. In the aforementioned example, with ten 2-component systems and $r = 4$, the six censored systems can contain 0 or 1 failed component.

Scenario II. Specifying the maximum number of failed components in both failed and censored systems:

To take into account the component failures in the censored systems, we consider the maximum possible number of failed components in both failed and censored systems, denoted as $R_{\max}$. In the aforementioned example, with ten 2-component systems and $r = 4$, the four failed systems contain two failed components in each system, and the six censored systems can contain at most one failed component in each system. Therefore, the maximum possible number of failed components in both failed and censored systems is

$$R_{\max} = (2 \times 4) + (1 \times 6) = 14.$$

Note that this scenario is considered the worst-case scenario.

When the values of $N = n \times m$ (i.e., the number of components available for the experiment) and $E(R)$ or $R_{\max}$ are fixed in advance according to the availability of resources, we aim to determine the optimal experimental plan under progressive Type-II censoring with respect to an objective function (optimality criterion). In other words, we aim to determine the number of components in each system n, the system signature $s = (s_1, s_2, ..., s_n)$, and the progressive censoring plan $R = (R_1, R_2, ..., R_r)$.

To obtain the optimal progressive censoring schemes, we consider the following settings when N and r ($E(R)$ in Scenario I or $R_{\max}$ in Scenario II) are specified.

[M1] Fix $R_\ell = m - r$, $R_{\ell'} = 0$, for $\ell, \ell' = 1, 2, ..., r$ and $\ell' \neq \ell$, then we determine the optimal number of components in the system (denoted as $n^{(\ell)}$) and the system signature (denoted as $s^{(\ell)}$) that optimize the specific objective function. The progressive censoring plan that censored all $m - r$ systems at a certain observed failure is known as the progressive Type-II extremal censoring scheme (see, for example, Ng et al., 2017). In this setting, for each progressive Type-II extremal censoring scheme $(0, ..., 0, R_\ell = m - r, 0, ..., 0)$ for $\ell = 1, 2, ..., r$, we search for the optimal values of n and s. These extremal plans are denoted "one-step censoring plans" in the literature see, for example, Balakrishnan et al., 2008). It is pointed out in Balakrishnan and Cramer (2014) that these plans are the vertices of the simplex of admissible censoring plans (see also Cramer and Davies, 2018).

[M2] Fix $\sum_{\ell=1}^{r} R_\ell = m - r$, and then determine the values of R_ℓ (number of censored systems at each observed failure), the values of n (number of components in the system) and s (system structure) that optimize the specific objective function. In this setting, we either fix the expected number of failed components in those failed systems (i.e., Scenario I) or the maximum possible number of failed components in the experiment (i.e., Scenario II). We do not put any restriction on the censoring plan $(R_1, R_2, ..., R_r)$.

5.2 Optimality criteria

Suppose we are interested in estimating the model parameters or function of the model parameters based on the settings presented in Section 5.1, or controlling the experimental time and cost, we plan the optimal progressively Type-II censored experiment by considering the following optimality criteria:

[C1] D-optimality: Maximizing the determinant of the expected Fisher information matrix, **I**, i.e., maximizing the differential Shannon information contained in the MLEs and minimizing the volume of the Wald-type joint confidence region for the model parameter (μ, σ). Specifically, we aim to maximize

$$\det[\boldsymbol{I}(\mu, \sigma)] = \left| E\left(-\frac{\partial^2 l(\theta)}{\partial \mu^2}\right) E\left(-\frac{\partial^2 l(\theta)}{\partial \sigma^2}\right) - \left[E\left(-\frac{\partial^2 l(\theta)}{\partial \mu \partial \sigma}\right)\right]^2 \right|.$$

[C2] A-optimality: Minimizing the trace of the asymptotic variance-covariance matrix of the MLEs, $V(\mu, \sigma)$, i.e., minimizing the sum of the asymptotic variances of the MLEs of μ and σ. Specifically, we aim to minimize

$$Var(\hat{\mu}) + Var(\hat{\sigma}).$$

For notational convenience, we denote $V_\mu = Var(\hat{\mu})$ and $V_\sigma = Var(\hat{\sigma})$.

[C3] V-optimality: Minimizing the variance of the estimator of μ, $Var(\hat{\mu})$. This criterion minimizes the variance of the estimate of the slope parameter.

[C4] Minimization of the expected total time of the experiment, denoted as TTE, which is defined as

$$E(TTE) = E\{[\exp(V_{r:r:m})]\}.$$

Using the relationship between $Z_{\ell:r:m}$ and $V_{\ell:r:m}$:

$$Z_{\ell:r:m} = \frac{V_{\ell:r:m} - \mu}{\sigma},$$

the TTE can be expressed in terms of $Z_{\ell:r:m}$, which is the ℓ-th ordered standardized log-transformed failure time, as

$$E(TTE) = \exp(\mu) E[\exp(\sigma Z_{r:r:m})]. \tag{15}$$

Based on the progressive Type-II censored experiments with coherent systems described in Section 3, the experiment terminates at the last observed system failure. This criterion aims to minimize the termination time of the experiment.

[C5] Minimization of the expected maximum total time for all the components spending on the experiment, denoted as TTT, which is defined as

$$\begin{aligned} E(TTT) &= nE\left\{\sum_{\ell=1}^{r} (R_\ell + 1)\exp(V_{\ell:r:m})\right\} \\ &= \exp(\mu) n \sum_{\ell=1}^{r} (R_\ell + 1) E[\exp(\sigma Z_{\ell:r:m})]. \end{aligned}$$

Since the components that did not fail in the life testing experiment can be used in the future for other purposes, the shorter the time those components spent in the experiment, the longer they can be used for other purposes. Therefore, this criterion aims to minimize the maximum expected total time the components spend in the experiment.

[C6] Minimization of a cost function: For experimental planning purposes, we consider the optimal progressive censoring scheme by including the cost in the life testing experiment due to failures and time and the cost in the inaccuracy of parameter estimation. Specifically, for specific values of N, m, n, and system signature s, we minimize the objective function

$$\psi(\mathbf{R}) = c_1 E(R) + c_2 E(TTE) + c_3(Var(\hat{\mu}) + Var(\hat{\sigma})), \tag{16}$$

where c_1, c_2 and c_3 are the cost per unit of failed components, the cost per unit of time for the duration of the experiment, and the cost of variance of the estimates of the unknown model parameters, respectively.

6 Specific coherent systems being considered

Many coherent systems can be considered for the progressively Type-II censored experiment involving systems, especially when N and n are large. In practice, it may be more feasible to use coherent systems with a small number of components since using systems with a large number of components reduces the flexibility of the progressively censored experiments. For example, with $N = 120$, if we consider using 3-component coherent systems (i.e., $n = 3$), we have 40 systems for the progressively Type-II censored experiment. However, if we consider using 10-component coherent systems (i.e., $n = 10$), we only have 12 systems for the progressively Type-II censored experiment. Therefore, to simplify the optimal experimental planning problem and for illustrative purposes, we restrict our discussion to $n \leq 3$. Note that the methodologies proposed in this chapter can be extended to $n > 3$.

For $n = 1$, the experiment reduces to a component-wise progressively Type-II censored experiment, in which the optimal experimental schemes have been discussed under different optimality criteria in the literature. For $n = 2$, two possible coherent systems can be considered: the series system with system signature $s = (1, 0)$ and the parallel system with system signature $s = (0, 1)$. For $n = 3$, five possible coherent systems can be considered, including the series system with system signature $s = (1, 0, 0)$, parallel system, 2-out-of-3 system with system signature $s = (0, 1, 0)$, parallel-series system with system signature $s = (0, 2/3, 1/3)$, and series-parallel system with system signature $s = (1/3, 2/3, 0)$ (see Shaked and Suarez-Llorens, 2003). Among those different types of systems, series systems, and parallel systems are just special cases of k-out-of-n systems for $k = 1$ and $k = n$, respectively; hence, we consider k-out-on-n, parallel-series, and series-parallel systems for $n \leq 3$. All coherent systems of order 4 and 5 can be found in Shaked and Suarez-Llorens (2003) and Navarro and Rubio (2009), respectively.

The following subsections provide the formulas required for obtaining the normal equations in Eqs. (7) and (8) for specific n-component coherent systems with $n \leq 3$.

6.1 k-out-of-n systems

For a k-out-of-n system, the system signature is $s = (0, ..., 0, s_k = 1, 0, ..., 0)$. Thus, the SF and PDF of the log-transformed system lifetimes can be expressed as

$$\begin{aligned}\bar{F}_V(v;\,\theta) &= \sum_{i=1}^{n} s_i \sum_{j=0}^{i-1} \binom{n}{j} [F^*(z)]^j [\bar{F}^*(z)]^{n-j} \\ &= \sum_{j=0}^{k-1} \binom{n}{j} [F^*(z)]^j [\bar{F}^*(z)]^{n-j}, \\ \textit{and}\ f_V(v;\,\theta) &= \frac{1}{\sigma} f^*(z) \sum_{i=1}^{n} s_i \binom{n}{i} i\, [F^*(z)]^{i-1} [\bar{F}^*(z)]^{n-i} \\ &= \frac{1}{\sigma} f^*(z) \binom{n}{k} l\, [F^*(z)]^{k-1} [\bar{F}^*(z)]^{n-k}.\end{aligned}$$

The log-likelihood functions in Eq. (7) can be simplified as

$$l(\theta) = C + \sum_{\ell=1}^{r} \ln f_V(v_{\ell:r:m};\,\theta) + \sum_{\ell=1}^{r} R_\ell \ln[\bar{F}_V(v_{\ell:r:m};\,\theta)],$$

$$v_{1:r:m} < v_{2:r:m} < \ldots < v_{r:r:m},$$

where

$$\begin{aligned}\ln f_V(v_{\ell:r:m};\,\theta) &= \ln\left\{\frac{1}{\sigma} f^*(z_{\ell:r:m}) \binom{n}{k} l\, [F^*(z_{\ell:r:m})]^{k-1} [\bar{F}^*(z_{\ell:r:m})]^{n-k}\right\} \\ \textit{and}\ \ln \bar{F}_V(v_{\ell:r:m};\,\theta) &= \ln\left\{\sum_{j=0}^{k-1} \binom{n}{j} [F^*(z_{\ell:r:m})]^j [\bar{F}^*(z_{\ell:r:m})]^{n-j}\right\}.\end{aligned}$$

Then, we can obtain the first partial derivatives of the log-likelihood function with respect to μ and σ. respectively, as

$$\begin{aligned}\frac{\partial l(\theta)}{\partial \mu} &= \sum_{\ell=1}^{r} \left[\frac{\partial \ln f_V(v_{\ell:r:m};\,\theta)}{\partial \mu} + R_\ell \frac{\partial \ln \bar{F}_V(v_{\ell:r:m};\,\theta)}{\partial \mu}\right] \\ \textit{and}\ \frac{\partial l(\theta)}{\partial \sigma} &= \sum_{\ell=1}^{r} \left[\frac{\partial \ln f_V(v_{\ell:r:m};\,\theta)}{\partial \sigma} + R_\ell \frac{\partial \ln \bar{F}_V(v_{\ell:r:m};\,\theta)}{\partial \sigma}\right].\end{aligned}$$

6.1.1 Series system: $k = 1$

For $k = 1$, we have the series system. With progressive Type-II censoring, the model is called "first-failure censoring" (cf. Balakrishnan and Cramer, 2014, p. 529), which was introduced in Wu and Kus (2009). This model can be viewed as a standard progressive censoring model with a modified censoring plan $(R_1^*, R_2^*, ..., R_r^*)$, where $R_\ell^* = n(R_\ell + 1) - 1$, $\ell = 1, 2, ..., r$. Thus, the PDF and SF of the log-transformed system lifetimes can be further simplified as

$$f_V(v; \boldsymbol{\theta}) = \frac{n}{\sigma} f^*(z)[\bar{F}^*(z)]^{n-1} \text{ and } \bar{F}_V(v; \boldsymbol{\theta}) = [\bar{F}^*(z)]^n, \quad n = 1, 2, 3.$$

Then, we have

$$\begin{aligned}
\ln f_V(v_{\ell:r:m}; \boldsymbol{\theta}) &= \ln\left\{\frac{n}{\sigma} f^*(z_{\ell:r:m})[\bar{F}^*(z_{\ell:r:m})]^{n-1}\right\} \\
&= \ln n - \ln\sigma + \ln f^*(z_{\ell:r:m}) + (n-1)\ln \bar{F}^*(z_{\ell:r:m}) \\
\text{and } \ln \bar{F}_V(v_{\ell:r:m}; \mathbf{q}) &= \ln\{[\bar{F}^*(z_{\ell:r:m})]^n\} = n \ln \bar{F}^*(z_{\ell:r:m}).
\end{aligned}$$

Thus, we can obtain the first partial derivatives of the PDF and SF as

$$\begin{aligned}
\frac{\partial \ln f_V(v_{\ell:r:m}; \boldsymbol{\theta})}{\partial\mu} &= \frac{\partial \ln f^*(z_{\ell:r:m})}{\partial\mu} + (n-1)\frac{\partial \ln \bar{F}^*(z_{\ell:r:m})}{\partial\mu} \\
\frac{\partial \ln f_V(v_{\ell:r:m}; \boldsymbol{\theta})}{\partial\sigma} &= -\frac{1}{\sigma} + \frac{\partial \ln f^*(z_{\ell:r:m})}{\partial\sigma} + (n-1)\frac{\partial \ln \bar{F}^*(z_{\ell:r:m})}{\partial\sigma},
\end{aligned}$$

and

$$\begin{aligned}
\frac{\partial \ln \bar{F}_V(v_{\ell:r:m}; \boldsymbol{\theta})}{\partial\mu} &= n\left(\frac{\partial \ln \bar{F}^*(z_{\ell:r:m})}{\partial\mu}\right) \\
\frac{\partial \ln \bar{F}_V(v_{\ell:r:m}; \boldsymbol{\theta})}{\partial\sigma} &= n\left(\partial\frac{\ln \bar{F}^*(z_{\ell:r:m})}{\partial\sigma}\sigma\right).
\end{aligned}$$

Hence, the normal equations can be expressed as

$$\begin{aligned}
\frac{\partial l(\boldsymbol{\theta})\mu}{\partial\mu} &= \sum_{\ell=1}^{r}\left[\frac{\partial \ln f^*(z_{\ell:r:m})}{\partial\mu} + (R_\ell n + n - 1)\frac{\partial \ln \bar{F}^*(z_{\ell:r:m})}{\partial\mu}\right] \\
&= \sum_{\ell=1}^{r}\left[\frac{1}{f^*(z_{\ell:r:m})}\frac{\partial f^*(z_{\ell:r:m})}{\partial\mu} + (R_\ell n + n - 1)\frac{1}{\bar{F}^*(z_{\ell:r:m})}\frac{\partial \bar{F}^*(z_{\ell:r:m})}{\partial\mu}\right] = 0
\end{aligned} \tag{17}$$

$$\begin{aligned}
\frac{\partial l(\boldsymbol{\theta})}{\partial\sigma} &= \sum_{\ell=1}^{r}\left[-\frac{1}{\sigma} + \frac{\partial \ln f^*(z_{\ell:r:m})}{\partial\sigma} + (R_\ell n + n - 1)\frac{\partial \ln \bar{F}^*(z_{\ell:r:m})}{\partial\sigma}\right] \\
&= \sum_{\ell=1}^{r}\left[-\frac{1}{\sigma} + \frac{1}{f^*(z_{\ell:r:m})}\frac{\partial f^*(z_{\ell:r:m})}{\partial\sigma} + (R_\ell n + n - 1)\frac{1}{\bar{F}^*(z_{\ell:r:m})}\frac{\partial \bar{F}^*(z_{\ell:r:m})}{\partial\sigma}\right] = 0.
\end{aligned} \tag{18}$$

6.1.2 Parallel system: $k = n$

For $k = n$, we have the parallel system with system signature $s = (0, 0, ..., 1)$. The PDF and SF of the log-transformed system lifetimes can be expressed as

$$f_V(v;\boldsymbol{\theta}) = \frac{n}{\sigma} f^*(z)[F^*(z)]^{n-1},$$

$$\text{and } \bar{F}_V(v;\boldsymbol{\theta}) = \sum_{j=0}^{n-1} \binom{n}{j} [F^*(z)]^j [\bar{F}^*(z)]^{n-j}$$

$$= 1 - [F^*(z)]^n,$$

respectively. Then, we have

$$\ln f_V(v_{\ell:r:m};\boldsymbol{\theta}) = \ln\left\{\frac{n}{\sigma} f^*(z)[F^*(z)]^{n-1}\right\}$$

$$= \ln n - \ln\sigma + \ln f^*(z_{\ell:r:m}) + (n-1)\ln F^*(z_{\ell:r:m})$$

$$\text{and } \ln \bar{F}_V(v_{\ell:r:m};\boldsymbol{\theta}) = \ln\{1 - [F^*(z_{\ell:r:m})]^n\}.$$

Thus, we can obtain the first partial derivatives of the PDF and SF as

$$\begin{aligned}\frac{\partial \ln f_V(v_{\ell:r:m};\boldsymbol{\theta})}{\partial\mu} &= \frac{\partial \ln f^*(z_{\ell:r:m})}{\partial\mu} + (n-1)\frac{\partial \ln F^*(z_{\ell:r:m})}{\partial\mu} \\ &= \frac{1}{f^*(z_{\ell:r:m})}\frac{\partial f^*(z_{\ell:r:m})}{\partial\mu} + \frac{n-1}{F^*(z_{\ell:r:m})}\frac{\partial F^*(z_{\ell:r:m})}{\partial\mu}\end{aligned} \tag{19}$$

$$\begin{aligned}\frac{\partial \ln f_V(v_{\ell:r:m};\boldsymbol{\theta})}{\partial\sigma} &= -\frac{1}{\sigma} + \frac{\partial \ln f^*(z_{\ell:r:m})}{\partial\sigma} + (n-1)\frac{\partial \ln F^*(z_{\ell:r:m})}{\partial\sigma} \\ &= -\frac{1}{\sigma} + \frac{1}{f^*(z_{\ell:r:m})}\frac{\partial f^*(z_{\ell:r:m})}{\partial\sigma} + \frac{n-1}{F^*(z_{\ell:r:m})}\frac{\partial F^*(z_{\ell:r:m})}{\partial\sigma},\end{aligned} \tag{20}$$

and.

$$\frac{\partial \ln \bar{F}_V(v_{\ell:r:m};\boldsymbol{\theta})}{\partial\mu} = -\frac{n[F^*(z_{\ell:r:m})]^{n-1}}{1-[F^*(z_{\ell:r:m})]^n}\frac{\partial F^*(z_{\ell:r:m})}{\partial\mu}$$

$$\frac{\partial \ln \bar{F}_V(v_{\ell:r:m};\boldsymbol{\theta})}{\partial\sigma} = -\frac{n[F^*(z_{\ell:r:m})]^{n-1}}{1-[F^*(z_{\ell:r:m})]^n}\frac{\partial F^*(z_{\ell:r:m})}{\partial\sigma}.$$

The normal equations can be obtained as

$$\frac{\partial l(\boldsymbol{\theta})}{\partial \mu} = \sum_{\ell=1}^{r} \left[\frac{1}{f^*(z_{\ell:r:m})} \frac{\partial f^*(z_{\ell:r:m})}{\partial \mu} + \frac{(n-1) - (R_\ell n + n - 1)[F^*(z_{\ell:r:m})]^n}{F^*(z_{\ell:r:m})(1 - [F^*(z_{\ell:r:m})]^n)} \frac{\partial F^*(z_{\ell:r:m})}{\partial \mu} \right] = 0$$

$$\frac{\partial l(\boldsymbol{\theta})}{\partial \sigma} = \sum_{\ell=1}^{r} \left[-\frac{1}{\sigma} + \frac{1}{f^*(z_{\ell:r:m})} \frac{\partial f^*(z_{\ell:r:m})}{\partial \sigma} + \frac{(n-1) - (R_\ell n + n - 1)[F^*(z_{\ell:r:m})]^n}{F^*(z_{\ell:r:m})(1 - [F^*(z_{\ell:r:m})]^n)} \frac{\partial F^*(z_{\ell:r:m})}{\partial \sigma} \right] = 0.$$

6.1.3 2-out-of-n system

For 2-out-of-n ($n > 2$) system, the system signature is $s = (0, 1, 0, ..., 0)$, and the PDF and SF of the log-transformed system lifetimes are

$$f_V(v; \boldsymbol{\theta}) = \frac{1}{\sigma} f^*(z) \{ n(n-1) [F^*(z)] [\bar{F}^*(z)]^{n-2} \}$$

$$\text{and } \bar{F}_V(v; \boldsymbol{\theta}) = [\bar{F}^*(z)]^n + n [F^*(z)] [\bar{F}^*(z)]^{n-1}.$$

Then, we have

$$\ln f_V(v_{\ell:r:m}; \boldsymbol{\theta}) = \ln n(n-1) - \ln \sigma + \ln f^*(z_{\ell:r:m}) + \ln \{ [F^*(z_{\ell:r:m})] [\bar{F}^*(z_{\ell:r:m})]^{n-2} \}$$

$$\text{and } \ln \bar{F}_V(v_{\ell:r:m}; \boldsymbol{\theta}) = \ln \{ [\bar{F}^*(z_{\ell:r:m})]^n + n [F^*(z_{\ell:r:m})] [\bar{F}^*(z_{\ell:r:m})]^{n-1} \}.$$

Thus, we can obtain the first partial derivatives of the PDF and SF with respect to μ and σ as

$$\frac{\partial \ln f_V(v_{\ell:r:m}; \boldsymbol{\theta})}{\partial \mu} = \frac{\partial \ln f^*(z_{\ell:r:m})}{\partial \mu} + \frac{\ln \{ [F^*(z_{\ell:r:m})] [\bar{F}^*(z_{\ell:r:m})]^{n-2} \}}{\partial \mu}$$

$$\frac{\partial \ln f_V(v_{\ell:r:m}; \boldsymbol{\theta})}{\partial \sigma} = -\frac{1}{\sigma} + \frac{\partial \ln f^*(z_{\ell:r:m})}{\partial \sigma} + \frac{\ln \{ [F^*(z_{\ell:r:m})] [\bar{F}^*(z_{\ell:r:m})]^{n-2} \}}{\partial \sigma}$$

and

$$\frac{\partial \ln \bar{F}_V(v_{\ell:r:m}; \boldsymbol{\theta})}{\partial \mu} = \frac{\partial \ln \{ [\bar{F}^*(z_{\ell:r:m})]^n + n [F^*(z_{\ell:r:m})] [\bar{F}^*(z_{\ell:r:m})]^{n-1} \}}{\partial \mu}$$

$$\frac{\partial \ln \bar{F}_V(v_{\ell:r:m}; \boldsymbol{\theta})}{\partial \sigma} = \frac{\partial \ln \{ [\bar{F}^*(z_{\ell:r:m})]^n + n [F^*(z_{\ell:r:m})] [\bar{F}^*(z_{\ell:r:m})]^{n-1} \}}{\partial \sigma}.$$

The normal equations can be obtained as

$$\frac{\partial l(\boldsymbol{\theta})}{\partial \mu} = \sum_{\ell=1}^{r} \left[\frac{1}{f^*(z_{\ell:r:m})} \frac{\partial f^*(z_{\ell:r:m})}{\partial \mu} + \frac{\partial \ln\{[F^*(z_{\ell:r:m})][\bar{F}^*(z_{\ell:r:m})]^{n-2}\}}{\partial \mu} \right.$$
$$\left. + R_\ell \frac{\partial \ln\{[\bar{F}^*(z_{\ell:r:m})]^n + n[F^*(z_{\ell:r:m})][\bar{F}^*(z_{\ell:r:m})]^{n-1}\}}{\partial \mu} \right] = 0, \tag{21}$$

$$\frac{\partial l(\boldsymbol{\theta})}{\partial \sigma} = \sum_{\ell=1}^{r} \left[-\frac{1}{\sigma} + \frac{1}{f^*(z_{\ell:r:m})} \frac{f^*(z_{\ell:r:m})}{\partial \sigma} + \frac{\partial \ln\{[F^*(z_{\ell:r:m})][\bar{F}^*(z_{\ell:r:m})]^{n-2}\}}{\partial \sigma} \right.$$
$$\left. + R_\ell \frac{\partial \ln\{[\bar{F}^*(z_{\ell:r:m})]^n + n[F^*(z_{\ell:r:m})][\bar{F}^*(z_{\ell:r:m})]^{n-1}\}}{\partial \sigma} \right] = 0. \tag{22}$$

6.1.4 3-out-of-n

For 3-out-of-n ($n > 3$) system, the system signature is $s = (0, 0, 1, 0, \ldots, 0)$, and the PDF and SF of the log-transformed system lifetimes are

$$f_V(v; \boldsymbol{\theta}) = \tfrac{1}{\sigma} f^*(z) \left\{ \tfrac{n(n-1)(n-2)}{2} [F^*(z)]^2 [\bar{F}^*(z)]^{n-3} \right\}$$
$$\text{and } \bar{F}_V(v; \boldsymbol{\theta}) = [\bar{F}^*(z)]^n + n[F^*(z)][\bar{F}^*(z)]^{n-1} + \tfrac{n(n-1)}{2}[F^*(z)]^2[\bar{F}^*(z)]^{n-2}.$$

Then, we have

$$\ln f_V(v_{\ell:r:m}; \boldsymbol{\theta}) = \ln\left[\tfrac{n(n-1)(n-2)}{2}\right] - \ln\sigma + \ln f^*(z_{\ell:r:m}) + \ln\{[F^*(z_{\ell:r:m})]^2[\bar{F}^*(z_{\ell:r:m})]^{n-3}\}$$
$$\text{and} \quad \ln \bar{F}_V(v_{\ell:r:m}; \boldsymbol{\theta}) = \ln\left\{ [\bar{F}^*(z_{\ell:r:m})]^n + n[F^*(z_{\ell:r:m})][\bar{F}^*(z_{\ell:r:m})]^{n-1} + \tfrac{n(n-1)}{2}[F^*(z)]^2[\bar{F}^*(z)]^{n-2} \right\}.$$

Thus, we can obtain the first partial derivatives of the PDF and SF with respect to μ and σ as

$$\frac{\partial \ln f_V(v_{\ell:r:m}; \boldsymbol{\theta})}{\partial \mu} = \frac{\partial \ln f^*(z_{\ell:r:m})}{\partial \mu} + \frac{\partial \ln\{[F^*(z_{\ell:r:m})]^2[\bar{F}^*(z_{\ell:r:m})]^{n-3}\}}{\partial \mu}$$
$$\frac{\partial \ln f_V(v_{\ell:r:m}; \boldsymbol{\theta})}{\partial \sigma} = -\frac{1}{\sigma} + \frac{\partial \ln f^*(z_{\ell:r:m})}{\partial \sigma} + \frac{\partial \ln\{[F^*(z_{\ell:r:m})]^2[\bar{F}^*(z_{\ell:r:m})]^{n-3}\}}{\partial \sigma}$$

and

$$\frac{\partial \ln \bar{F}_V(v_{\ell:r:m};\boldsymbol{\theta})}{\partial \mu} = \frac{\ln\left\{[\bar{F}^*(z_{\ell:r:m})]^n + n[F^*(z_{\ell:r:m})][\bar{F}^*(z_{\ell:r:m})]^{n-1} + \frac{n(n-1)}{2}[F^*(z)]^2[\bar{F}^*(z)]^{n-2}\right\}}{\partial \mu}$$

$$\frac{\partial \ln \bar{F}_V(v_{\ell:r:m};\boldsymbol{\theta})}{\partial \sigma} = \frac{\partial \ln\left\{[\bar{F}^*(z_{\ell:r:m})]^n + n[F^*(z_{\ell:r:m})][\bar{F}^*(z_{\ell:r:m})]^{n-1} + \frac{n(n-1)}{2}[F^*(z)]^2[\bar{F}^*(z)]^{n-2}\right\}}{\partial \sigma}.$$

The normal equations can be obtained as

$$\frac{\partial l(\boldsymbol{\theta})}{\partial \mu} = \sum_{\ell=1}^{r}\left[\frac{1}{f^*(z_{\ell:r:m})}\frac{\partial f^*(z_{\ell:r:m})}{\partial \mu} + \frac{\partial \ln\{[F^*(z_{\ell:r:m})]^2[\bar{F}^*(z_{\ell:r:m})]^{n-3}\}}{\partial \mu} + \frac{\partial \ln\left\{[\bar{F}^*(z_{\ell:r:m})]^n + n[F^*(z_{\ell:r:m})][\bar{F}^*(z_{\ell:r:m})]^{n-1} + \frac{n(n-1)}{2}[F^*(z)]^2[\bar{F}^*(z)]^{n-2}\right\}}{\partial \mu}\right] = 0, \tag{23}$$

$$\frac{\partial l(\boldsymbol{\theta})}{\partial \sigma} = \sum_{\ell=1}^{r}\left[-\frac{1}{\sigma} + \frac{1}{f^*(z_{\ell:r:m})}\frac{\partial f^*(z_{\ell:r:m})}{\partial \sigma} + \frac{\partial \ln\{[F^*(z_{\ell:r:m})]^2[\bar{F}^*(z_{\ell:r:m})]^{n-3}\}}{\partial \sigma} + \frac{\partial \ln\left\{[\bar{F}^*(z_{\ell:r:m})]^n + n[F^*(z_{\ell:r:m})][\bar{F}^*(z_{\ell:r:m})]^{n-1} + \frac{n(n-1)}{2}[F^*(z)]^2[\bar{F}^*(z)]^{n-2}\right\}}{\partial \sigma}\right] = 0. \tag{24}$$

6.2 Series-parallel and parallel-series systems for $n = 3$

Illustrations of the following systems can be found in Table 1 of Shaked and Suarez-Llorens (2003).

6.2.1 Series-parallel systems for $n = 3$

For $n = 3$, the system signature for series-parallel systems is $s = (1/3, 2/3, 0)$, and the PDF and SF of the log-transformed system lifetimes are

$$\begin{aligned} f_V(v;\boldsymbol{\theta}) &= \frac{1}{\sigma}f^*(z)\{[\bar{F}^*(z)]^2 + 2[F^*(z)][\bar{F}^*(z)]\} \\ &= \frac{1}{\sigma}f^*(z)\{1 - [F^*(z)]^2\} \\ \textit{and}\ \ \bar{F}_V(v;\boldsymbol{\theta}) &= [\bar{F}^*(z)]^3 + 2[F^*(z)][\bar{F}^*(z)]^2 \\ &= [F^*(z)]^3 - [F^*(z)]^2 - F^*(z) + 1, \end{aligned}$$

respectively. Then, we have

$$\ln f_V(v_{\ell:r:m}; \Theta) = -\ln\sigma + \ln f^*(z_{\ell:r:m}) + \ln\{1 - [F^*(z_{\ell:r:m})]^2\}$$

$$\textit{and} \quad \ln \bar{F}_V(v_{\ell:r:m}; \Theta) = \ln\{[F^*(z)]^3 - [F^*(z)]^2 - F^*(z) + 1\}.$$

Thus, we can obtain the first partial derivatives of the PDF and SF with respect to μ and σ as

$$\frac{\partial \ln f_V(v_{\ell:r:m}; \boldsymbol{\theta})}{\partial\mu} = \frac{\partial \ln f^*(z_{\ell:r:m})}{\partial\mu} + \frac{\partial \ln\{1 - [F^*(z_{\ell:r:m})]^2\}}{\partial\mu}$$

$$\frac{\partial \ln f_V(v_{\ell:r:m}; \boldsymbol{\theta})}{\partial\sigma} = -\frac{1}{\sigma} + \frac{\partial \ln f^*(z_{\ell:r:m})}{\partial\sigma} + \frac{\partial \ln\{1 - [F^*(z_{\ell:r:m})}{\partial\sigma}$$

and

$$\frac{\partial \ln \bar{F}_V(v_{\ell:r:m}; \boldsymbol{\theta})}{\partial\mu} = \frac{\partial \ln\{[F^*(z)]^3 - [F^*(z)]^2 - F^*(z) + 1\}}{\partial\mu}$$

$$\frac{\partial \ln \bar{F}_V(v_{\ell:r:m}; \boldsymbol{\theta})}{\partial\sigma} = \frac{\partial \ln\{[F^*(z)]^3 - [F^*(z)]^2 - F^*(z) + 1\}}{\partial\sigma}$$

The normal equations can be obtained as

$$\begin{aligned}\frac{\partial l(\boldsymbol{\theta})}{\partial\mu} = \sum_{\ell=1}^{r}\Bigg[&\frac{1}{f^*(z_{\ell:r:m})}\frac{\partial f^*(z_{\ell:r:m})}{\partial\mu} + \frac{\partial \ln\{1 - [F^*(z_{\ell:r:m})]^2\}}{\partial\mu} \\ &+ \frac{\partial \ln\{[F^*(z)]^3 - [F^*(z)]^2 - F^*(z) + 1\}}{\partial\mu}\Bigg] = 0.\end{aligned} \tag{25}$$

$$\begin{aligned}\frac{\partial l(\boldsymbol{\theta})}{\partial\sigma} = \sum_{\ell=1}^{r}\Bigg[&-\frac{1}{\sigma} + \frac{1}{f^*(z_{\ell:r:m})}\frac{\partial f^*(z_{\ell:r:m})}{\partial\sigma} + \frac{\partial \ln\{1 - [F^*(z_{\ell:r:m})]^2\}}{\partial\sigma} \\ &+ \frac{\partial \ln\{[F^*(z)]^3 - [F^*(z)]^2 - F^*(z) + 1\}}{\partial\sigma}\Bigg] = 0.\end{aligned} \tag{26}$$

6.2.2 Parallel-series systems for $n = 3$

For $n = 3$, the system signature for parallel-series systems is $s = (0, 2/3, 1/3)$. Thus, the PDF and SF of the log-transformed system lifetimes are,

$$\begin{aligned} f_V(v; \boldsymbol{\theta}) &= \frac{1}{\sigma}f^*(z)\left\{2[F^*(z)][\bar{F}^*(z)] + \frac{1}{3}[F^*(z)]^2\right\} \\ &= \frac{1}{\sigma}f^*(z)\left\{2[F^*(z)] - \frac{5}{3}[F^*(z)]^2\right\} \\ \textit{and} \quad \bar{F}_V(v; \boldsymbol{\theta}) &= [\bar{F}^*(z)]^3 + 3[F^*(z)][\bar{F}^*(z)]^2 + [F^*(z)]^2[\bar{F}^*(z)], \\ &= [F^*(z)]^3 - 2[F^*(z)]^2 + 1, \end{aligned}$$

respectively. Thus, we have

$$\ln f_V(v_{\ell:r:m};\boldsymbol{\theta}) = -\ln\sigma + \ln f^*(z_{\ell:r:m}) + \ln\left\{2[F^*(z)] - \frac{5}{3}[F^*(z)]^2\right\}$$

and $\ln \bar{F}_V(v_{\ell:r:m};\boldsymbol{\theta}) = \ln\{[F^*(z)]^3 - 2[F^*(z)]^2 + 1\}.$

Then, we can obtain the first partial derivatives of the PDF and SF with respect to μ and σ as

$$\frac{\partial \ln f_V(v_{\ell:r:m};\boldsymbol{\theta})}{\partial\mu} = \frac{\partial \ln f^*(z_{\ell:r:m})}{\partial\mu} + \frac{\partial \ln\left\{2[F^*(z_{\ell:r:m})] - \frac{5}{3}[F^*(z_{\ell:r:m})]^2\right\}}{\partial\mu}$$

$$\frac{\partial \ln f_V(v_{\ell:r:m};\boldsymbol{\theta})}{\partial\sigma} = -\frac{1}{\sigma} + \frac{\partial \ln f^*(z_{\ell:r:m})}{\partial\sigma} + \frac{\partial \ln\left\{2[F^*(z_{\ell:r:m})] - \frac{5}{3}[F^*(z_{\ell:r:m})]^2\right\}}{\partial\sigma}$$

and

$$\frac{\partial \ln \bar{F}_V(v_{\ell:r:m};\boldsymbol{\theta})}{\partial\mu} = \frac{\partial \ln\{[F^*(z_{\ell:r:m})]^3 - 2[F^*(z_{\ell:r:m})]^2 + 1}{\partial\mu}$$

$$\frac{\partial \ln \bar{F}_V(v_{\ell:r:m};\boldsymbol{\theta})}{\partial\sigma} = \frac{\partial \ln\{[F^*(z_{\ell:r:m})]^3 - 2[F^*(z_{\ell:r:m})]^2 + 1}{\partial\sigma}$$

The normal equations can be obtained as

$$\frac{\partial l(\boldsymbol{\theta})}{\partial\mu} = \sum_{\ell=1}^{r}\left[\frac{1}{f^*(z_{\ell:r:m})}\frac{\partial f^*(z_{\ell:r:m})}{\partial\mu} + \frac{\partial \ln\left\{2[F^*(z_{\ell:r:m})] - \frac{5}{3}[F^*(z_{\ell:r:m})]^2\right\}}{\partial\mu} + \frac{\partial \ln\{[F^*(z_{\ell:r:m})]^3 - 2[F^*(z_{\ell:r:m})]^2 + 1\}}{\partial\mu}\right] = 0, \tag{27}$$

$$\frac{\partial l(\boldsymbol{\theta})}{\partial\sigma} = \sum_{\ell=1}^{r}\left[-\frac{1}{\sigma} + \frac{1}{f^*(z_{\ell:r:m})}\frac{\partial f^*(z_{\ell:r:m})}{\partial\sigma} + \frac{\partial \ln\left\{2[F^*(z_{\ell:r:m})] - \frac{5}{3}[F^*(z_{\ell:r:m})]^2\right\}}{\partial\sigma} + \frac{\partial \ln\{[F^*(z_{\ell:r:m})]^3 - 2[F^*(z_{\ell:r:m})]^2 + 1\}}{\partial\sigma}\right] = 0. \tag{28}$$

7 Numerical Illustrations

7.1 Lognormal distributed components

Consider the lifetimes of the components follow a two-parameter lognormal distribution with CDF and PDF

$$F_{LN}(x;\mu,\sigma) = \Phi_{nor}\left[\frac{\ln(x)-\mu}{\sigma}\right],\ x>0,$$

$$f_{LN}(x;\mu,\sigma) = \frac{1}{\sigma x}\phi_{nor}\left(\frac{\ln(x)-\mu}{\sigma}\right) = \frac{1}{\sqrt{2\pi}\sigma x}\exp\left[-\frac{1}{2}\left(\frac{\ln(x)-\mu}{\sigma}\right)^2\right]x>0,$$

respectively, where $\exp(\mu)$ is the scale parameter, $\sigma > 0$ is the shape parameter, and Φ_{nor} and ϕ_{nor} are the CDF and PDF for the standard normal distribution:

$$\Phi_{nor}(z) = \int_{-\infty}^{z} \phi_{nor}(w)dw, \quad \bar{\Phi}_{nor}(z)=1 - \Phi_{nor}(z),$$
$$\phi_{nor}(z) = \frac{1}{\sqrt{2\pi}} \exp\left(-\frac{z^2}{2}\right), \ -\infty < z < \infty.$$

Then, the log-transformed component lifetimes follow a normal distribution with location parameter μ and scale parameter σ with CDF, SF and PDF

$$F_U(u; \mu, \sigma) = \Phi_{nor}\left(\frac{u-\mu}{\sigma}\right),$$
$$\bar{F}_U(u; \mu, \sigma) = 1 - \Phi_{nor}\left(\frac{u-\mu}{\sigma}\right) = \bar{\Phi}_{nor}\left(\frac{u-\mu}{\sigma}\right),$$
$$\text{and } f_U(u; \mu, \sigma) = \frac{1}{\sigma}\phi_{nor}\left(\frac{u-\mu}{\sigma}\right), \quad -\infty < u < \infty,$$

respectively.

We consider the experimental planning for a fixed value of N and r when various types of coherent systems are used in progressively Type-II censored experiments. In Tables 1 and 2, we present the optimal progressive censoring schemes for n-component coherent systems with lognormal distributed components under settings **[M1]** and **[M2]**. In Table 3, we present the optimal progressive censoring plans for n-component coherent systems with lognormal distributed components ($n = 1, 2,$ and 3) by fixing $R_{\max} = 8$ under setting **[M1]** and **[M2]**. The optimal values are highlighted in bold font in these tables.

7.2 Weibull distributed components

Suppose the lifetimes of the components follow a two-parameter Weibull distribution with CDF and PDF

$$F_{WE}(x; \eta, \beta) = 1 - \exp\left[-\left(\frac{x}{\eta}\right)^{\beta}\right], \quad x > 0,$$
$$f_{WE}(x; \eta, \beta) = \frac{\beta}{\eta}\left(\frac{x}{\eta}\right)^{\beta-1} \exp\left[-\left(\frac{x}{\eta}\right)^{\beta}\right], \quad x > 0,$$

respectively, where $\eta > 0$ is the scale parameter and $\beta > 0$ is the shape parameter. Then, the log-transformed component lifetimes follow a SEV distribution (also known as Gumbel Type I distribution) with location parameter μ and scale parameter σ, i.e.,

$$F_U(u; \mu, \sigma) = \Phi_{sev}\left(\frac{u-\mu}{\sigma}\right), \ -\infty < u < \infty,$$
$$f_U(u; \mu, \sigma) = \frac{1}{\sigma}\phi_{sev}\left(\frac{u-\mu}{\sigma}\right), \ -\infty < u < \infty,$$

TABLE 1 Optimal progressive censoring schemes for n-component systems with lognormal distributed components when $N = 30$, by fixing $n = 1$, 2, and 3 under setting [M1] and [M2]. The corresponding optimal value for a criterion is highlighted in bold.

								[C1]	[C2]	[C3]	[C4]	[C5]	[C6] with (c_1, c_2, c_3)						
System type	n	Setting	m	r	$E(R)$	R_{max}	R	$\|I\|$	$V\mu + V\sigma$	V_μ	$\overline{TTE}$	$\overline{TTT}$	(1,1,10)	(1,0,10)	(1,1,0)	(0,1,10)	(0,1,5)	(0,1,20)	(0,1,30)
Series	1	[M1]	30	6	6	6	(0,24,0,0,0,0)	194.48	0.1984	0.1420	27.91	92.08	35.90	7.98	33.91	29.90	28.91	31.88	33.87
							(24,0,0,0,0,0)	172.23	0.1943	0.1419	30.56	89.61	33.5	2.94	31.56	32.50	31.53	34.45	36.39
							(0,0,0,0,0,24)	143.63	0.2905	0.1818	3.13	87.90	12.04	8.91	9.13	6.04	4.58	8.94	11.85
		[M2]	30	6	6	6	(0,24,0,0,0,0)	194.48	0.1984	0.1420	27.91	92.08	35.90	7.98	33.91	29.90	28.91	31.88	33.87
							(23,0,0,0,0,1)	171.21	0.1933	0.1369	16.60	84.33	24.54	7.93	22.60	18.54	17.57	20.47	22.40
							(20,0,0,0,0,4)	166.61	0.2000	0.1330	8.35	81.47	16.35	8.00	14.35	10.35	9.35	12.35	14.35
							(0,0,0,0,0,24)	143.63	0.2905	0.1818	3.13	87.90	12.04	8.90	9.13	6.04	4.58	8.94	11.85
							(5,0,0,0,0,19)	147.64	0.2678	0.1671	3.52	86.20	12.19	8.68	9.52	6.19	4.85	8.87	11.55
							(6,0,0,0,0,18)	148.27	0.2640	0.1650	3.61	85.72	12.25	8.64	9.61	6.25	4.93	8.89	11.53
	2	[M1]	15	6	6	6	(0,9,0,0,0,0)	183.59	0.207	0.1404	12.13	86.15	20.20	8.07	18.13	14.20	13.17	16.27	18.34
							(9,0,0,0,0,0)	171.43	0.1985	0.1348	12.84	83.33	20.83	7.99	18.84	14.83	13.83	16.81	18.80
							(0,0,0,0,0,9)	146.57	0.2796	0.1752	3.36	87.25	12.16	8.80	9.36	6.16	4.76	8.95	11.75
		[M2]	15	6	6	6	(0,9,0,0,0,0)	183.59	0.207	0.1404	12.13	86.15	20.20	8.07	18.13	14.20	13.17	16.27	18.34
							(9,0,0,0,0,0)	171.43	0.1985	0.1348	12.84	83.33	20.83	7.99	18.84	14.83	13.83	16.81	18.80
							(0,0,0,0,0,9)	146.57	0.2796	0.1752	3.36	87.25	12.16	8.80	9.36	6.16	4.76	8.95	11.75
							(8,0,0,0,0,1)	167.13	0.2058	0.1357	8.17	82.30	16.23	8.06	14.17	10.23	9.20	12.29	14.34
							(3,0,0,0,0,6)	152.09	0.2521	0.1581	4.03	85.17	12.55	8.52	10.03	6.55	5.29	9.07	11.59
	3	[M1]	10	6	6	6	(0,4,0,0,0,0)	174.29	0.2189	0.1443	8.28	85.23	16.47	8.19	14.28	10.47	9.37	12.66	14.85
							(4,0,0,0,0,0)	168.06	0.2113	0.1385	8.57	83.25	16.68	8.11	14.57	10.68	9.63	12.8	14.91
							(0,0,0,0,0,4)	149.97	0.2682	0.1684	3.70	86.61	12.38	8.68	9.7	6.38	5.04	9.06	11.75
		[M2]	10	6	6	6	(0,4,0,0,0,0)	174.29	0.2189	0.1443	8.28	85.23	16.47	8.19	14.28	10.47	9.37	12.66	14.85

Continued

Table 1 Optimal progressive censoring schemes for n-component systems with lognormal distributed components when $N = 30$, by fixing $n = 1, 2,$ and 3 under setting [M1] and [M2]. The corresponding optimal value for a criterion is highlighted in bold. —Cont'd

								[C1]	[C2]	[C3]	[C4]	[C5]	[C6] with (c_1, c_2, c_3)						
System type	n	Setting	m	r	$E(R)$	R_{max}	R	$\|I\|$	$V\mu + V\sigma$	V_μ	$\overline{TTE}$	$\overline{TTT}$	(1,1,10)	(1,0,10)	(1,1,0)	(0,1,10)	(0,1,5)	(0,1,20)	(0,1,30)
							(4,0,0,0,0,0)	168.06	**0.2113**	**0.1385**	**8.57**	**83.25**	**16.68**	**8.11**	**14.57**	**10.68**	**9.63**	**12.80**	**14.91**
							(0,0,0,0,0,4)	149.97	0.2682	0.1684	**3.70**	**86.61**	**12.38**	**8.68**	**9.70**	**6.38**	**5.04**	**9.06**	**11.75**
							(1,0,0,0,0,3)	153.23	0.254	0.1598	4.11	85.58	12.65	8.54	10.11	6.65	5.38	9.19	**11.73**
Parallel	2	[M1]	**15**	**3**	**6**	**18**	**(0,12,0)**	**70.43**	**0.2595**	**0.1078**	**20.73**	**177.68**	**29.33**	**8.60**	**26.73**	**23.33**	**22.03**	**25.92**	**28.52**
							(12,0,0)	69.79	**0.2465**	**0.1261**	**28.01**	**169.73**	**36.47**	**8.47**	**34.01**	**30.47**	**29.24**	**32.94**	**35.40**
							(0,0,12)	61.01	0.3010	**0.0992**	**6.23**	**178.90**	**15.24**	**9.01**	**12.23**	**9.24**	**7.73**	**12.25**	**15.26**
		[M2]	**15**	**3**	**6**	**18**	**(0,12,0)**	**70.43**	**0.2595**	**0.1078**	**20.73**	**177.68**	**29.33**	**8.60**	**26.73**	**23.33**	**22.03**	**25.92**	**28.52**
							(12,0,0)	69.79	**0.2465**	**0.1261**	**28.01**	**169.73**	**36.47**	**8.47**	**34.01**	**30.47**	**29.24**	**32.94**	**35.40**
							(1,0,11)	61.39	0.2971	**0.0991**	**6.40**	**177.85**	**15.37**	**8.97**	**12.40**	**9.37**	**7.89**	**12.34**	**15.31**
							(0,0,12)	61.01	0.3010	0.0992	**6.23**	**178.90**	**15.24**	**9.01**	**12.23**	**9.24**	**7.73**	**12.25**	**15.26**
							(11,0,1)	69.22	0.2481	0.1164	15.96	**165.62**	**24.44**	**8.48**	**21.96**	**18.44**	**17.20**	**20.92**	**23.40**
	3	[M1]	**10**	**2**	**6**	**22**	**(8,0)**	**39.33**	**0.3295**	**0.1258**	**25.04**	**240.98**	**34.33**	**9.30**	**31.04**	**28.33**	**26.69**	**31.63**	**34.92**
							(0,8)	36.03	0.3845	**0.0980**	**8.65**	**252.04**	**18.50**	**9.84**	**14.65**	**12.50**	**10.57**	**16.34**	**20.19**
		[M2]	**10**	**2**	**6**	**22**	**(8,0)**	**39.33**	**0.3295**	**0.1258**	**25.04**	**240.98**	**34.33**	**9.30**	**31.04**	**28.33**	**26.69**	**31.63**	**34.92**
							(0,8)	36.03	0.3845	**0.0980**	**8.65**	**252.04**	**18.50**	**9.84**	**14.65**	**12.50**	**10.57**	**16.34**	**20.19**
							(7,1)	38.86	0.3398	0.1176	15.30	**239.28**	**24.70**	**9.40**	**21.30**	**18.70**	**17.00**	**22.10**	**25.50**
2-out-of-n	3	[M1]	**10**	**3**	**6**	**13**	**(7,0,0)**	**80.81**	**0.2513**	**0.1232**	**13.28**	**127.92**	**21.79**	**8.51**	**19.28**	**15.79**	**14.54**	**18.31**	**20.82**
							(0,7,0)	80.04	0.2783	**0.1227**	**10.71**	**135.80**	**19.49**	**8.78**	**16.71**	**13.49**	**12.1**	**16.28**	**19.06**
							(0,0,7)	69.76	0.3273	0.1271	**4.96**	**139.41**	**14.23**	**9.27**	**10.96**	**8.23**	**6.60**	**11.51**	**14.78**
		[M2]	**10**	**3**	**6**	**13**	**(7,0,0)**	**80.81**	**0.2513**	**0.1232**	**13.28**	**85.28**	**21.79**	**8.51**	**19.28**	**15.79**	**14.54**	**18.31**	**20.82**

							(5,0,2)	76.09	0.2753	0.1181	7.24	131.18	15.99	8.75	13.24	9.99	8.62	12.75	15.50
							(0,0,7)	69.76	0.3273	0.1271	4.96	139.41	14.23	9.27	10.96	8.23	6.60	11.51	14.78
							(2,0,5)	71.53	0.3092	0.1227	5.46	136.50	14.55	9.09	11.46	8.55	7.01	11.64	14.74
Series-parallel	3	[M1]	10	4	6 2/3	14	(0,6,0,0)	79.86	0.2573	0.1458	11.93	115.16	21.17	9.24	18.60	14.50	13.22	17.08	19.65
							(6,0,0,0)	77.68	0.2471	0.1467	13.32	110.36	22.46	9.14	19.99	15.79	14.56	18.26	20.73
							(0,0,0,6)	65.04	0.3143	0.1515	4.48	118.08	14.29	9.81	11.15	7.62	6.05	10.77	13.91
		[M2]	10	4	6 2/3	14	(0,6,0,0)	79.86	0.2573	0.1458	11.93	115.16	21.17	9.24	18.60	14.50	13.22	17.08	19.65
							(6,0,0,0)	77.68	0.2471	0.1467	13.32	110.36	22.46	9.14	19.99	15.79	14.56	18.26	20.73
							(4,0,0,2)	72.46	0.2644	0.1381	6.83	112.25	16.14	9.31	13.50	9.47	8.15	12.12	14.76
							(0,0,0,6)	65.04	0.3143	0.1515	4.48	118.08	14.29	9.81	11.15	7.62	6.05	10.77	13.91
							(1,0,0,5)	66.59	0.3014	0.1465	4.81	116.84	14.49	14.68	16.48	12.82	6.32	10.84	13.85
Parallel-series	3	[M1]	10	3	7	23	(0,7,0)	63.56	0.2836	0.1226	16.37	172.26	26.21	9.84	23.37	19.21	17.79	22.04	24.88
							(7,0,0)	62.22	0.2706	0.1364	21.42	169.30	31.13	9.71	28.42	24.13	22.77	26.83	29.54
							(0,0,7)	58.26	0.3128	0.1146	6.04	169.05	16.17	10.13	13.04	9.17	7.60	12.30	15.42
		[M2]	10	3	7	23	(0,7,0)	63.56	0.2836	0.1226	16.37	172.26	26.21	9.84	23.37	19.21	17.79	22.04	24.88
							(7,0,0)	62.22	0.2706	0.1364	21.42	169.30	31.13	9.71	28.42	24.13	22.77	26.83	29.54
							(1,0,6)	58.73	0.3066	0.1144	6.35	167.77	16.42	10.07	13.35	9.42	7.88	12.48	15.55
							(0,0,7)	58.26	0.3128	0.1146	6.04	169.05	16.17	10.13	13.04	9.17	7.60	12.30	15.42
							(5,0,2)	61.72	0.2779	0.1197	9.46	162.73	19.24	9.78	16.46	12.24	10.85	15.02	17.80

[a] For criterion **[C6]**, the objective function is $\psi(\mathbf{R}) = c_1E(R) + c_2E(TTE) + c_3(Var(\hat{\mu}) + Var(\hat{\sigma}))$.

TABLE 2 Optimal progressive censoring schemes for n-component systems with lognormal distributed components when $N = 24$, by fixing $n = 1, 2, 3$, and 4 under setting [M1] and [M2].

								[C1]	[C2]	[C3]	[C4]	[C5]	[C6] with (c_1, c_2, c_3)						
System type	n	Setting	m	r	$E(R)$	R_{max}	R	$\|I\|$	$V\mu + V\sigma$	V_μ	$\overline{TTE}$	$\overline{TTT}$	(1,1,10)	(1,0,10)	(1,1,0)	(0,1,10)	(0,1,5)	(0,1,20)	(0,1,30)
Series	1	[M1]	24	4	4	4	(0,20,0,0)	94.62	0.3162	0.2241	19.61	64.56	26.77	7.16	23.61	22.77	21.19	25.93	29.10
							(20,0,0,0)	88.84	0.2904	0.2128	23.75	62.94	30.65	6.90	27.75	26.65	25.20	29.56	32.46
							(0,0,0,20)	70.15	0.4529	0.2935	2.74	62.68	11.27	8.53	6.74	7.27	5.00	11.80	16.33
		[M2]	24	4	4	4	(0,20,0,0)	94.62	0.3162	0.2241	19.61	64.56	26.77	7.16	23.61	22.77	21.19	25.93	29.10
							(20,0,0,0)	88.84	0.2904	0.2128	23.75	62.94	30.65	6.90	27.75	26.65	25.20	29.56	32.46
							(18,0,0,2)	85.78	0.2985	0.2067	8.94	58.34	15.93	6.99	12.94	11.93	10.43	14.91	17.90
							(0,0,0,20)	70.15	0.4529	0.2935	2.74	62.68	11.27	8.53	6.74	7.27	5.00	11.80	16.33
							(17,0,0,3)	84.24	0.3063	0.2084	7.23	58.13	14.29	7.06	11.23	10.29	8.76	13.36	16.42
							(2,0,0,18)	71.00	0.4385	0.2839	2.86	62.15	11.25	8.39	6.86	7.25	5.05	11.63	16.02
							(10,0,0,10)	76.06	0.3718	0.2415	3.79	59.81	11.51	7.72	7.79	7.51	5.65	11.23	14.94
							(12,0,0,8)	77.77	0.3535	0.2308	4.25	59.19	11.79	7.54	8.25	7.79	6.02	11.32	14.86
	2	[M1]	12	4	4	4	(0,8,0,0)	88.64	0.3322	0.2276	9.22	61.35	16.54	7.32	13.22	12.54	10.88	15.86	19.19
							(8,0,0,0)	86.23	0.3016	0.2089	10.48	58.94	17.50	7.02	14.48	13.50	11.99	16.51	19.53
							(0,0,0,8)	71.24	0.4401	0.2855	2.87	62.27	11.27	8.40	6.87	7.27	5.07	11.67	16.07
		[M2]	12	4	4	4	(0,8,0,0)	88.64	0.3322	0.2276	9.22	61.35	16.54	7.32	13.22	12.54	10.88	15.86	19.19
							(8,0,0,0)	86.23	0.3016	0.2089	10.48	58.94	17.50	7.02	14.48	13.50	11.99	16.51	19.53
							(0,0,0,8)	71.24	0.4401	0.2855	2.87	62.27	11.27	8.40	6.87	7.27	5.07	11.67	16.07
							(7,0,0,1)	83.02	0.3180	0.2135	6.50	58.49	13.68	7.18	10.50	9.68	8.09	12.86	16.04
							(4,0,0,4)	76.13	0.3745	0.2435	3.82	60.01	11.57	7.75	7.82	7.57	5.69	11.31	15.06
							(5,0,0,3)	78.03	0.3559	0.2326	4.31	59.42	11.87	7.56	8.31	7.87	6.09	11.43	14.99
	3	[M1]	8	4	4	4	(0,4,0,0)	84.55	0.3469	0.2331	6.59	60.83	14.06	7.47	10.59	10.06	8.32	13.53	17.00
							(4,0,0,0)	83.54	0.3200	0.2154	7.20	58.86	14.40	7.20	11.20	10.40	8.80	13.60	16.80

							(0,0,0,4)	72.09	0.4277	0.2772	3.05	61.90	11.32	8.28	7.05	7.32	5.19	11.60	15.88
		[M2]	8	4	4	4	(0,4,0,0)	84.55	0.3469	0.2331	6.59	60.83	14.06	7.47	10.59	10.06	8.32	13.53	17.00
							(4,0,0,0)	83.54	0.3200	0.2154	7.20	58.86	14.40	7.20	11.20	10.40	8.80	13.60	16.80
							(0,0,0,4)	72.09	0.4277	0.2772	3.05	61.90	11.32	8.28	7.05	7.32	5.19	11.60	15.88
							(2,0,0,2)	76.23	0.3768	0.2451	3.88	60.21	11.65	7.77	7.88	7.65	5.76	11.42	15.18
	4	[M1]	6	4	4	4	(0,2,0,0)	81.60	0.3597	0.2389	5.40	60.69	12.99	7.60	9.40	8.99	7.20	12.59	16.19
							(2,0,0,0)	81.25	0.3400	0.2255	5.70	59.33	13.10	7.40	9.70	9.10	7.40	12.50	15.90
							(0,0,0,2)	73.48	0.4131	0.2682	3.28	61.48	11.42	8.13	7.28	7.42	5.35	11.54	15.67
		[M2]	6	4	4	4	(0,2,0,0)	81.60	0.3597	0.2389	5.40	60.69	12.99	7.60	9.40	8.99	7.20	12.59	16.19
							(2,0,0,0)	81.25	0.3400	0.2255	5.70	59.33	13.10	7.40	9.70	9.10	7.40	12.50	15.90
							(0,0,0,2)	73.48	0.4131	0.2682	3.28	61.48	11.42	8.13	7.28	7.42	5.35	11.54	15.67
							(1,0,0,1)	76.56	0.3782	0.2465	3.97	60.36	11.75	7.78	7.97	7.75	5.86	11.53	15.32
Parallel	2	[M1]	12	2	4	14	(10,0)	34.81	0.3636	0.1716	19.86	124.24	27.49	7.64	23.86	23.49	21.67	27.13	30.76
							(0,10)	30.16	0.4430	0.1526	5.55	129.76	13.98	8.43	9.55	9.98	7.76	14.41	18.84
		[M2]	12	2	4	14	(10,0)	34.81	0.3636	0.1716	19.86	124.24	27.49	7.64	23.86	23.49	21.67	27.13	30.76
							(4,6)	31.09	0.4194	0.1513	6.32	126.98	14.52	8.19	10.32	10.52	8.42	14.71	18.91
							(0,10)	30.16	0.4430	0.1526	5.55	129.76	13.98	8.43	9.55	9.98	7.76	14.41	18.84
							(9,1)	33.98	0.3723	0.1604	11.41	122.52	19.14	7.72	15.41	15.14	13.27	18.86	22.58
	3	[M1]	8	2	6	18	(6,0)	34.54	0.3530	0.1418	25.55	217.32	35.08	9.53	31.55	29.08	27.32	32.61	36.14
							(0,6)	32.18	0.3996	0.1151	9.68	223.38	19.68	10.00	15.68	13.68	11.68	17.67	21.67
		[M2]	8	2	6	18	(6,0)	34.54	0.3530	0.1418	25.55	217.32	35.08	9.53	31.55	29.08	27.32	32.61	36.14
							(0,6)	32.18	0.3996	0.1151	9.68	223.38	19.68	10.00	15.68	13.68	11.68	17.67	21.67
							(5,1)	33.99	0.3651	0.1333	15.73	214.94	25.38	9.65	21.73	19.38	17.55	23.03	26.68
	4	[M1]	6	2	8	20	(4,0)	33.70	0.4021	0.1786	30.37	311.70	42.39	12.02	38.37	34.39	32.38	38.41	42.43
							(0,4)	32.66	0.4477	0.1684	13.93	316.65	26.40	12.48	21.93	18.40	16.17	22.88	27.36
		[M2]	6	2	8	20	(4,0)	33.70	0.4021	0.1786	30.37	311.70	42.39	12.02	38.37	34.39	32.38	38.41	42.43
							(0,4)	32.66	0.4477	0.1684	13.93	316.65	26.40	12.48	21.93	18.40	16.17	22.88	27.36

Continued

Table 2 Optimal progressive censoring schemes for n-component systems with lognormal distributed components when $N = 24$, by fixing $n = 1, 2, 3$, and 4 under setting [M1] and [M2]. —Cont'd

								[C1]	[C2]	[C3]	[C4]	[C5]	[C6] with (c_1, c_2, c_3)						
System type	n	Setting	m	r	$E(R)$	R_{max}	R	$\|I\|$	$V\mu + V\sigma$	V_μ	TTE	TTT	(1,1,10)	(1,0,10)	(1,1,0)	(0,1,10)	(0,1,5)	(0,1,20)	(0,1,30)
							(3,1)	33.36	0.4207	0.1764	19.58	308.97	31.79	12.21	27.58	23.79	21.69	28.00	32.20
2-out-of-n	3	[M1]	8	2	6	13	(6,0)	39.83	0.3859	0.1855	10.13	95.15	17.99	7.86	14.13	13.99	12.06	17.85	21.71
							(0,6)	34.38	0.4909	0.2015	4.41	101.76	13.31	8.91	8.41	9.31	6.86	14.22	19.13
		[M2]	8	2	6	13	(6,0)	39.83	0.3859	0.1855	10.13	95.15	17.99	7.86	14.13	13.99	12.06	17.85	21.71
							(5,1)	37.90	0.4118	0.1840	6.80	96.33	14.92	8.12	10.80	10.92	8.86	15.04	19.16
							(0,6)	34.38	0.4909	0.2015	4.41	101.76	13.31	8.91	8.41	9.31	6.86	14.22	19.13
							(2,4)	35.23	0.4664	0.1940	4.82	100.04	13.48	8.66	8.82	9.48	7.15	14.15	18.81
							(3,3)	35.85	0.4510	0.1900	5.17	99.02	13.68	8.51	9.17	9.68	7.42	14.19	18.70
Series-parallel	3	[M1]	8	4	6 2/3	14	(6,0)	25.54	0.5022	0.3085	8.34	36.74	16.69	8.36	11.67	13.36	10.85	18.38	23.41
							(0,6)	20.69	0.6750	0.3748	3.09	39.33	13.18	10.08	6.43	9.84	6.47	16.59	23.34
		[M2]	8	4	6 2/3	14	(6,0)	25.54	0.5022	0.3085	8.34	36.74	16.69	8.36	11.67	13.36	10.85	18.38	23.41
							(0,6)	20.69	0.6750	0.3748	3.09	39.33	13.18	10.08	6.43	9.84	6.47	16.59	23.34
							(2,4)	21.51	0.6304	0.3509	3.46	38.65	13.10	9.64	6.80	9.77	6.62	16.07	22.37
							(4,2)	22.91	0.5721	0.3242	4.28	37.72	13.33	9.05	7.61	10.00	7.14	15.72	21.44
							(5,1)	24.07	0.5360	0.3115	5.27	37.16	13.97	8.69	8.61	10.63	7.95	15.99	21.35
Parallel-series	3	[M1]	10	3	7	23	(6,0)	31.18	0.4026	0.1923	15.35	121.77	24.04	8.69	20.02	19.38	17.36	23.40	27.43
					.		(0,6)	28.89	0.4646	0.1771	5.31	122.37	14.63	9.31	9.98	9.96	7.64	14.61	19.25
		[M2]	10	3	7	23	(6,0)	31.18	0.4026	0.1923	15.35	121.77	24.04	8.69	20.02	19.38	17.36	23.40	27.43
							(3,3)	29.83	0.4356	0.1753	6.38	119.85	15.41	9.02	11.05	10.74	8.56	15.10	19.45
							(0,6)	28.89	0.4646	0.1771	5.31	122.37	14.63	9.31	9.98	9.96	7.64	14.61	19.25
							(5,1)	30.89	0.4108	0.1802	8.95	118.56	17.72	8.77	13.61	13.05	11.00	17.16	21.27
							(1,5)	29.10	0.4566	0.1762	5.55	121.56	14.79	9.23	10.22	10.12	7.84	14.68	19.25

[a] For criterion **[C6]**, the objective function is $\psi(\mathbf{R}) = c_1E(R) + c_2E(TTE) + c_3(Var(\hat{\mu}) + Var(\hat{\sigma}))$.

TABLE 3 Optimal progressive censoring schemes for n-component systems with lognormal distributed components when $N = 18$, by fixing $n = 1, 2,$ and 3, $R_{max} = 8$ under setting [M1] and [M2]. The corresponding optimal value for a criterion is highlighted in bold.

								[C1]	[C2]	[C3]	[C4]	[C5]	[C6] with (c_1, c_2, c_3)						
System type	n	Setting	m	r	$E(R)$	R_{max}	R	$\|I\|$	$V\mu + V\sigma$	V_μ	$\overline{TTE}$	$\overline{TTT}$	(1,1,10)	(1,0,10)	(1,1,0)	(0,1,10)	(0,1,5)	(0,1,20)	(0,1,30)
	1	[M1]	18	8	8	8	**(0,0,10,0,0,0,0,0)**	**217.5**	**0.1552**	**0.1022**	**32.96**	**104.27**	**42.51**	**9.55**	**40.96**	**34.51**	**33.74**	**36.06**	**37.62**
							(0,10,0,0,0,0,0,0)	214.69	0.1551	0.105	34.81	105.17	44.36	9.55	42.81	36.36	35.59	37.91	39.46
							(0,0,0,0,0,0,0,10)	182.19	0.1725	0.0965	6.27	92.55	16.00	9.73	14.27	8.00	7.13	9.72	11.45
		[M2]	18	8	8	8	**(0,0,10,0,0,0,0,0)**	**217.5**	**0.1552**	**0.1022**	**32.96**	**104.27**	**42.51**	**9.55**	**40.96**	**34.51**	**33.74**	**36.06**	**37.62**
							(0,8,2,0,0,0,0,0)	215.11	0.155	0.1038	34.06	104.67	43.61	9.55	42.06	35.61	34.84	37.16	38.71
							(3,0,0,0,0,0,0,7)	186.33	0.1649	0.0956	7.67	92.22	17.32	9.65	15.67	9.32	8.49	10.97	12.62
							(0,0,0,0,0,0,0,10)	182.19	0.1725	0.0965	6.27	92.55	16.00	9.73	14.27	8.00	7.13	9.72	11.45
							(0,9,0,1,0,0,0,0)	214.45	0.155	0.1037	33.78	104.34	43.33	9.55	41.78	35.33	34.56	36.88	38.43
	2	[M1] and [M2]	9	8	8	8	**(0,1,0,0,0,0,0,0)**	**200.68**	**0.1590**	**0.0976**	**14.76**	**95.39**	**24.35**	**9.59**	**22.76**	**16.35**	**15.56**	**17.94**	**19.53**
							(1,0,0,0,0,0,0,0)	198.19	0.1589	0.0976	14.85	95.11	24.44	9.59	22.85	16.44	15.65	18.03	19.62
							(0,0,0,0,0,0,0,1)	192.95	0.1633	0.0963	9.71	93.42	19.34	9.63	17.71	11.34	10.52	12.97	14.61
2-out-of-3	3	[M1]	6	2	4	8	**(4,0)**	**33.82**	**0.3933**	**0.1814**	**10.40**	**83.06**	**22.33**	**11.93**	**18.40**	**14.33**	**12.36**	**18.26**	**22.20**
							(0,4)	30.33	0.4657	0.1812	5.12	87.15	17.78	12.66	13.12	9.78	7.45	14.43	19.09
		[M2]	6	2	4	8	**(4,0)**	**33.82**	**0.3933**	**0.1814**	**10.40**	**83.06**	**22.33**	**11.93**	**18.40**	**14.33**	**12.36**	**18.26**	**22.20**
							(3,1)	32.30	0.4171	0.1767	7.09	84.05	19.26	12.17	15.09	11.26	9.17	15.43	19.60
							(1,3)	30.82	0.4520	0.1788	5.46	86.28	17.98	12.52	13.46	9.98	7.72	14.50	19.02
							(0,4)	30.33	0.4657	0.1812	5.12	87.15	17.78	12.66	13.12	9.78	7.45	14.43	19.09
Series-parallel	3	[M1]	6	2	3 1/3	8	**(4,0)**	**21.76**	**0.5036**	**0.2939**	**8.61**	**60.01**	**21.64**	**13.04**	**16.61**	**13.64**	**11.12**	**18.68**	**23.72**
							(0,4)	18.47	0.6191	0.3185	3.73	62.86	17.93	14.19	11.73	9.93	6.83	16.12	22.31
		[M2]	6	2	3 1/3	8	**(4,0)**	**21.76**	**0.5036**	**0.2939**	**8.61**	**60.01**	**21.64**	**13.04**	**16.61**	**13.64**	**11.12**	**18.68**	**23.72**

Continued

Table 3 Optimal progressive censoring schemes for n-component systems with lognormal distributed components when $N = 18$, by fixing $n = 1, 2$, and 3, $R_{max} = 8$ under setting [M1] and [M2]. The corresponding optimal value for a criterion is highlighted in bold. —Cont'd

								[C1]	[C2]	[C3]	[C4]	[C5]	[C6] with (c_1, c_2, c_3)						
System type	n	Setting	m	r	$E(R)$	R_{max}	R	$\|I\|$	$V\mu + V\sigma$	V_μ	$\overline{TTE}$	$\overline{TTT}$	(1,1,10)	(1,0,10)	(1,1,0)	(0,1,10)	(0,1,5)	(0,1,20)	(0,1,30)
							(3,1)	20.37	0.5376	**0.2927**	**5.53**	**60.52**	**18.90**	**13.38**	**13.53**	**10.90**	**8.22**	**16.28**	**21.66**
							(2,2)	19.47	0.5697	0.3004	4.54	61.38	18.24	13.70	12.54	10.24	7.39	**15.93**	**21.63**
							(0,4)	18.47	0.6191	0.3185	**3.73**	**62.86**	**17.93**	**14.19**	**11.73**	**9.93**	**6.83**	**16.12**	**22.31**

[a] For criterion **[C6]**, the objective function is $\psi(R) = c_1 R_{max} + c_2 E(TTE) + c_3(Var(\hat{\mu}) + Var(\hat{\sigma}))$.

where $\sigma = 1/\beta$, $\mu = \ln(\eta)$, and $\Phi_{sev}(z) = 1 - \exp[-\exp(z)]$ and $\phi_{sev}(z) = \exp[z - \exp(z)]$ are the CDF and PDF for the standard SEV distribution.

We consider the experimental planning for a fixed value of N and r when various types of coherent systems are used in the progressively Type-II censored experiments when the component lifetimes follow a Weibull distribution. In Tables 4 and 5, we present the optimal progressive censoring schemes for n-component coherent systems with Weibull distributed components under settings **[M1]** and **[M2]**. In Table 6, we present the optimal progressive censoring schemes for n-component coherent systems with Weibull distributed components (n = 1, 2, and 3) by fixing $R_{\max} = 8$ under setting **[M1]** and **[M2]**.

7.3 Discussions

Based on the numerical results presented in Table 1, for optimality criteria **[C1]** (D-optimality), **[C2]** (A-optimality) and **[C4]** ($\overline{TTE}$), we would choose to perform the life testing experiment using individual components (i.e., $n = 1$) over other types of systems. Specifically, for optimality criterion **[C1]** (D-optimality) that maximizes the determinant of the expected Fisher information matrix, when $N = 30$ and $r = 6$, we would censor all 24 components at the second observed failure. In addition, for optimality criterion **[C2]** (A-optimality) that minimizes the sum of the asymptotic variance of the MLEs, we would choose to perform a progressively Type-II censored experiment by censoring $R_1 = 23$ and $R_6 = 1$ components at the first and last observed failures, respectively. For optimality criterion **[C4]** ($\overline{TTE}$) that minimizes the expected total time on test, the optimal scheme would censor all of the 24 components at the last observed failure, which is the conventional Type-II censoring scheme. This result agrees with our intuition as the conventional Type-II censoring scheme gives the shortest experimental time among all the progressive censoring schemes.

When we consider $E(R) = 6$ and experimental schemes that give $E(R)$ close to 6, the optimal values of the objective functions for **[C1]**, **[C2]**, and **[C4]** when using series, parallel, 2-out-of-n, series-parallel, and parallel-series systems are ($|I| = 194.48$, $V_\mu + V_\sigma = 0.1933$, $\overline{TTE} = 3.13$), ($|I| = 70.43$, $V_\mu + V_\sigma = 0.2465$, $\overline{TTE} = 6.23$), ($|I| = 80.81$, $V_\mu + V_\sigma = 0.2513$, $\overline{TTE} = 4.96$), ($|I| = 79.86$, $V_\mu + V_\sigma = 0.2471$, $\overline{TTE} = 4.48$) and ($|I| = 63.56$, $V_\mu + V_\sigma = 0.2706$, $\overline{TTE} = 6.04$), respectively. If one considers the optimality criterion **[C6]** that minimizes a combined cost function involving the costs due to failures and time and inaccuracy of parameter estimation, based on the choices of c_1, c_2, and c_3, which are the cost per unit if failed components, the cost per unit of time for the duration of the experiment, and the cost of variance of the estimates of the unknown model parameters, respectively, putting individual components in the experiments still produce best performance compared to other settings. Note that for series-parallel and parallel-series

TABLE 4 Optimal progressive censoring schemes for n-component systems with Weibull distributed components when $N = 30$, by fixing $n = 1, 2$, and 3 under setting [M1] and [M2]. The corresponding optimal value for a criterion is highlighted in bold.

								[C1]	[C2]	[C3]	[C4]	[C5]	[C6] with (c_1, c_2, c_3)						
System type	n	Setting	m	r	$E(R)$	R_{max}	R	$\|I\|$	$V\mu + V\sigma$	V_μ	$\overline{TTE}$	$\overline{TTT}$	(1,1,10)	(1,0,10)	(1,1,0)	(0,1,10)	(0,1,5)	(0,1,20)	(0,1,30)
Series	1	[M1]	**30**	**6**	**6**	**6**	**(0,24,0,0,0,0)**	**118.21**	**0.2257**	**0.1749**	**15.89**	**44.33**	**24.15**	**8.26**	**21.89**	**18.15**	**17.02**	**20.40**	**22.66**
							(24,0,0,0,0,0)	108.74	**0.2220**	**0.1668**	**17.12**	**44.33**	**25.34**	**8.22**	**23.12**	**19.34**	**18.23**	**21.56**	**23.78**
							(0,0,0,0,0,24)	43.19	0.6115	0.4726	**1.62**	**44.33**	**13.73**	**12.12**	**7.62**	**7.73**	**4.68**	**13.85**	**19.97**
		[M2]	**30**	**6**	**6**	**6**	**(0,24,0,0,0,0)**	**118.21**	**0.2257**	**0.1749**	**15.89**	**44.33**	**24.15**	**8.26**	**21.89**	**18.15**	**17.02**	**20.40**	**22.66**
							(24,0,0,0,0,0)	108.74	**0.2220**	**0.1668**	**17.12**	**44.33**	**25.34**	**8.22**	**23.12**	**19.34**	**18.23**	**21.56**	**23.78**
							(0,0,0,0,0,24)	43.19	0.6115	0.4726	**1.62**	**44.33**	**13.73**	**12.12**	**7.62**	**7.73**	**4.68**	**13.85**	**19.97**
							(12,0,0,0,0,12)	54.69	0.4157	0.3060	2.73	44.33	**12.89**	**10.16**	**8.73**	**6.89**	**4.81**	**11.04**	**15.20**
							(6,0,0,0,0,18)	47.67	0.5168	0.3909	2.01	44.33	13.18	11.17	8.01	7.18	**4.59**	**12.35**	**17.51**
							(16,0,0,0,0,8)	62.20	0.3464	0.2500	3.66	44.33	13.13	9.46	9.66	7.13	5.39	**10.59**	**14.05**
							(18,0,0,0,0,6)	67.60	0.3121	0.2233	4.46	44.33	13.58	9.12	10.46	7.58	6.02	10.70	**13.82**
	2	[M1]	**15**	**6**	**6**	**6**	**(0,9,0,0,0,0)**	**87.96**	**0.2759**	**0.2077**	**8.21**	**44.33**	**16.97**	**8.76**	**14.21**	**10.97**	**9.59**	**13.73**	**16.49**
							(9,0,0,0,0,0)	84.47	**0.2600**	**0.1890**	**8.68**	**44.33**	**17.28**	**8.60**	**14.68**	**11.28**	**9.98**	**13.88**	**16.48**
							(0,0,0,0,0,9)	45.28	0.5713	0.4388	**1.81**	**44.33**	**13.52**	**11.71**	**7.81**	**7.52**	**4.66**	**13.23**	**18.95**
		[M2]	**15**	**6**	**6**	**6**	**(0,9,0,0,0,0)**	**87.96**	**0.2759**	**0.2077**	**8.21**	**44.33**	**16.97**	**8.76**	**14.21**	**10.97**	**9.59**	**13.73**	**16.49**
							(9,0,0,0,0,0)	84.47	**0.2600**	**0.1890**	**8.68**	**44.33**	**17.28**	**8.60**	**14.68**	**11.28**	**9.98**	**13.88**	**16.48**
							(0,0,0,0,0,9)	45.28	0.5713	0.4388	**1.81**	**44.33**	**13.52**	**11.71**	**7.81**	**7.52**	**4.66**	**13.23**	**18.95**
							(4,0,0,0,0,5)	53.45	0.4389	0.3267	2.63	44.33	**13.02**	**10.39**	**8.63**	**7.02**	**4.83**	**11.41**	**15.8**
							(1,0,0,0,0,8)	46.88	0.5394	0.4114	1.95	44.33	13.35	11.39	7.95	7.35	**4.65**	**12.74**	**18.13**
							(6,0,0,0,0,3)	60.46	0.3687	0.2695	3.51	44.33	13.20	9.69	9.51	7.20	5.36	**10.89**	**14.58**
							(7,0,0,0,0,2)	65.54	0.3330	0.2414	4.28	44.33	13.61	9.33	10.28	7.61	5.95	10.94	**14.27**
	3	[M1]	**10**	**6**	**6**	**6**	**(0,4,0,0,0,0)**	**73.65**	**0.3251**	**0.2436**	**5.65**	**44.33**	**14.90**	**9.25**	**11.65**	**8.90**	**7.28**	**12.15**	**15.40**

							(4,0,0,0,0,0)	72.31	0.3091	0.2261	5.87	44.33	14.96	9.09	11.87	8.96	7.42	12.05	15.14
							(0,0,0,0,0,4)	48.04	0.5268	0.4019	2.08	44.33	13.35	11.27	8.08	7.35	4.71	12.62	17.88
		[M2]	10	6	6	6	(0,4,0,0,0,0)	73.65	0.3251	0.2436	5.65	44.33	14.90	9.25	11.65	8.90	7.28	12.15	15.40
							(4,0,0,0,0,0)	72.31	0.3091	0.2261	5.87	44.33	14.96	9.09	11.87	8.96	7.42	12.05	15.14
							(0,0,0,0,0,4)	48.04	0.5268	0.4019	2.08	44.33	13.35	11.27	8.08	7.35	4.71	12.62	17.88
							(2,0,0,0,0,2)	55.62	0.4235	0.3156	2.94	44.33	13.17	10.23	8.94	7.17	5.06	11.41	15.64
							(3,0,0,0,0,1)	61.80	0.3683	0.2712	3.82	44.33	13.50	9.68	9.82	7.50	5.66	11.19	14.87
Parallel	2	[M1]	15	3	6	18	(12,0,0)	60.87	0.2615	0.1557	15.95	97.65	24.57	8.62	21.95	18.57	17.26	21.18	23.80
							(0,12,0)	54.21	0.2891	0.1552	12.4	110.03	21.29	8.89	18.4	15.29	13.85	18.18	21.07
							(0,0,12)	33.42	0.4327	0.1940	4.15	118.10	14.48	10.33	10.15	8.48	6.31	12.80	17.13
		[M2]	15	3	6	18	(12,0,0)	60.87	0.2615	0.1557	15.95	97.65	24.57	8.62	21.95	18.57	17.26	21.18	23.80
							(10,0,2)	47.95	0.2975	0.1499	8.45	103.94	17.43	8.98	14.45	11.43	9.94	14.40	17.38
							(0,0,12)	33.42	0.4327	0.1940	4.15	118.10	14.48	10.33	10.15	8.48	6.31	12.80	17.13
							(4,0,8)	36.36	0.3880	0.1744	4.86	113.99	14.74	9.88	10.86	8.74	6.80	12.62	16.50
							(7,0,5)	40.11	0.3475	0.1600	5.88	109.87	15.36	9.48	11.88	9.36	7.62	12.83	16.31
	3	[M1]	10	2	6	22	(8,0)	39.77	0.3194	0.1420	14.60	153.70	23.79	9.19	20.60	17.79	16.20	20.99	24.18
							(0,8)	26.33	0.4489	0.1245	6.02	174.83	16.51	10.49	12.02	10.51	8.26	15.00	19.49
		[M2]	10	2	6	22	(8,0)	39.77	0.3194	0.1420	14.60	153.70	23.79	9.19	20.60	17.79	16.20	20.99	24.18
							(2,6)	27.34	0.4313	0.1238	6.45	172.03	16.76	10.31	12.45	10.76	8.61	15.08	19.39
							(0,8)	26.33	0.4489	0.1245	6.02	174.83	16.51	10.49	12.02	10.51	8.26	15.00	19.49
							(3,5)	28.05	0.4203	0.1240	6.74	170.21	16.94	10.20	12.74	10.94	8.84	15.15	19.35
2-out-of-*n*	3	[M1]	10	3	6	13	(7,0,0)	53.54	0.3004	0.1786	8.95	73.94	17.95	9.00	14.95	11.95	10.45	14.96	17.96
							(0,0,7)	32.70	0.5300	0.2971	3.12	85.94	14.42	11.30	9.12	8.42	5.77	13.72	19.02
		[M2]	10	3	6	13	(7,0,0)	53.54	0.3004	0.1786	8.95	73.94	17.95	9.00	14.95	11.95	10.45	14.96	17.96
							(0,0,7)	32.70	0.5300	0.2971	3.12	85.94	14.42	11.30	9.12	8.42	5.77	13.72	19.02
							(2,0,5)	34.86	0.4804	0.2664	3.53	83.75	14.33	10.80	9.53	8.33	5.93	13.14	17.94

Continued

Table 4 Optimal progressive censoring schemes for n-component systems with Weibull distributed components when $N = 30$, by fixing $n = 1, 2$, and 3 under setting [M1] and [M2]. The corresponding optimal value for a criterion is highlighted in bold. —Cont'd

								[C1]	[C2]	[C3]	[C4]	[C5]	[C6] with (c_1, c_2, c_3)						
System type	n	Setting	m	r	$E(R)$	R_{max}	R	$\|I\|$	$V\mu + V\sigma$	V_μ	$\overline{TTE}$	$\overline{TTT}$	(1,1,10)	(1,0,10)	(1,1,0)	(0,1,10)	(0,1,5)	(0,1,20)	(0,1,30)
							(5,0,2)	41.45	0.3835	0.2124	4.93	79.22	14.77	9.84	10.93	8.77	6.85	**12.60**	**16.44**
Series-parallel	3	[M1]	**10**	**4**	**6 2/3**	**14**	**(6,0,0,0)**	**43.58**	**0.3166**	**0.1964**	**8.98**	**62.30**	**18.81**	**9.83**	**15.65**	**12.15**	**10.56**	**15.31**	**18.48**
							(0,6,0,0)	43.01	0.3381	0.2152	8.06	64.96	18.11	10.05	14.73	11.44	9.75	14.82	**18.2**
							(0,0,0,6)	24.70	0.5605	0.3377	**2.73**	**69.08**	**15**	**12.27**	**9.4**	**8.34**	**5.53**	**13.94**	**19.55**
		[M2]	**10**	**4**	**6 2/3**	**14**	**(6,0,0,0)**	**43.58**	**0.3166**	**0.1964**	**8.98**	**62.30**	**18.81**	**9.83**	**15.65**	**12.15**	**10.56**	**15.31**	**18.48**
							(0,0,0,6)	24.70	0.5605	0.3377	**2.73**	**69.08**	**15.00**	**12.27**	**9.40**	**8.34**	**5.53**	**13.94**	**19.55**
							(2,0,0,4)	27.67	0.4807	0.282	3.35	67.73	**14.82**	**11.47**	**10.02**	**8.16**	**5.75**	**12.96**	**17.77**
							(4,0,0,2)	32.68	0.3967	0.2301	4.6	65.68	15.23	10.63	11.27	8.57	6.58	**12.53**	**16.50**
Parallel-series	3	[M1]	**10**	**3**	**7**	**23**	**(7,0,0)**	**47.19**	**0.3037**	**0.1824**	**12.89**	**109.98**	**22.93**	**10.04**	**19.89**	**15.93**	**14.41**	**18.96**	**22.00**
							(0,0,7)	30.62	0.4589	0.2274	**4.00**	**107.06**	**15.59**	**11.59**	**11.00**	**8.59**	**6.29**	**13.18**	**17.77**
		[M2]	**10**	**3**	**7**	**23**	**(7,0,0)**	**47.19**	**0.3037**	**0.1824**	**12.89**	**109.98**	**22.93**	**10.04**	**19.89**	**15.93**	**14.41**	**18.96**	**22.00**
							(6,0,1)	41.50	0.3287	**0.1798**	**8.28**	**100.83**	**18.57**	**10.29**	**15.28**	**11.57**	**9.92**	**14.85**	**18.14**
							(0,0,7)	30.62	0.4589	0.2274	**4.00**	**107.06**	**15.59**	**11.59**	**11.00**	**8.59**	**6.29**	**13.18**	**17.77**
							(2,0,5)	32.57	0.4217	0.2096	4.56	107.58	15.78	11.22	11.56	8.78	6.67	**12.99**	**17.21**
							(4,0,3)	35.63	0.3786	0.1920	5.59	104.73	16.38	10.79	12.59	9.38	7.48	13.16	**16.95**

[a] For criterion **[C6]**, the objective function is $\psi(\mathbf{R}) = c_1 E(R) + c_2 E(TTE) + c_3(Var(\hat{\mu}) + Var(\hat{\sigma}))$.

TABLE 5 Optimal progressive censoring schemes for n-component coherent systems with Weibull distributed components when $N = 24$, by fixing $n = 1, 2, 3$, and 4 under settings [M1] and [M2]. The corresponding optimal value for a criterion is highlighted in bold.

								[C1]	[C2]	[C3]	[C4]	[C5]	[C6] with (c_1, c_2, c_3)						
System type	n	Setting	m	r	$E(R)$	R_{max}	R	$\|I\|$	$V\mu + V\sigma$	V_μ	$\overline{TTE}$	$\overline{TTT}$	(1,1,10)	(1,0,10)	(1,1,0)	(0,1,10)	(0,1,5)	(0,1,20)	(0,1,30)
Series	1	[M1]	**24**	**4**	**4**	**4**	**(20,0,0,0)**	**53.81**	**0.3297**	**0.2553**	**13.85**	**29.56**	**21.15**	**7.30**	**17.85**	**17.15**	**15.5**	**20.45**	**23.74**
							(0,20,0,0)	52.8	0.373	0.2972	11.71	29.56	19.44	7.73	15.71	15.44	13.58	19.17	**22.9**
							(0,0,20,0)	40.07	0.5025	0.4026	8.35	29.56	17.38	9.02	12.35	13.38	10.87	**18.4**	**23.43**
							(0,0,0,20)	20.09	0.9841	0.7849	**1.32**	**29.56**	**15.16**	**13.84**	**5.32**	**11.16**	**6.24**	**21.00**	**30.84**
		[M2]	**24**	**4**	**4**	**4**	**(20,0,0,0)**	**53.81**	**0.3297**	**0.2553**	**13.85**	**29.56**	**21.15**	**7.30**	**17.85**	**17.15**	**15.50**	**20.45**	**23.74**
							(0,0,0,20)	20.09	0.9841	0.7849	**1.32**	**29.56**	**15.16**	**13.84**	**5.32**	**11.16**	**6.24**	**21.00**	**30.84**
							(14,0,0,6)	29.77	0.5492	0.4149	3.11	29.56	**12.60**	**9.49**	**7.11**	**8.60**	**5.85**	**14.09**	**19.59**
							(10,0,0,10)	25.20	0.6899	0.5312	2.16	29.56	13.06	10.90	6.16	9.06	**5.61**	**15.96**	**22.86**
							(16,0,0,4)	33.58	0.4750	0.3559	4.07	29.56	12.82	8.75	8.07	8.82	6.45	**13.57**	**18.32**
							(17,0,0,3)	36.25	0.4375	0.3271	4.86	29.56	13.24	8.37	8.86	9.24	7.05	13.61	**17.99**
	2	[M1]	**12**	**4**	**4**	**4**	**(8,0,0,0)**	**41.02**	**0.3984**	**0.3009**	**7.08**	**29.56**	**15.07**	**7.98**	**11.08**	**11.07**	**9.07**	**15.05**	**19.03**
							(0,0,8,0)	32.15	0.5982	0.4738	4.71	29.56	**14.69**	**9.98**	**8.71**	**10.69**	**7.70**	**16.67**	**22.65**
							(0,0,0,8)	20.74	0.9377	0.7448	**1.42**	**29.56**	**14.8**	**13.38**	**5.42**	**10.8**	**6.11**	**20.18**	**29.55**
		[M2]	**12**	**4**	**4**	**4**	**(8,0,0,0)**	**41.02**	**0.3984**	**0.3009**	**7.08**	**29.56**	**15.07**	**7.98**	**11.08**	**11.07**	**9.07**	**15.05**	**19.03**
							(0,0,0,8)	20.74	0.9377	0.7448	**1.42**	**29.56**	**14.80**	**13.38**	**5.42**	**10.80**	**6.11**	**20.18**	**29.55**
							(6,0,0,2)	29.82	0.5549	0.4208	3.20	29.56	**12.75**	**9.55**	**7.20**	**8.75**	**5.98**	**14.30**	**19.85**
							(4,0,0,4)	25.18	0.6966	0.5378	2.19	29.56	13.16	10.97	6.19	9.16	**5.67**	**16.12**	**23.09**
							(7,0,0,1)	33.83	0.4787	0.3605	4.31	29.56	13.10	8.79	8.31	9.10	6.70	**13.88**	**18.67**
	3	[M1]	**8**	**4**	**4**	**4**	**(4,0,0,0)**	**34.79**	**0.4738**	**0.3588**	**4.82**	**29.56**	**13.56**	**8.74**	**8.82**	**9.56**	**7.19**	**14.30**	**19.04**
							(0,0,0,4)	21.56	0.8869	0.7013	**1.56**	**29.56**	**14.43**	**12.87**	**5.56**	**10.43**	**6.00**	**19.30**	**28.17**

Continued

Table 5 Optimal progressive censoring schemes for n-component coherent systems with Weibull distributed components when $N = 24$, by fixing $n = 1, 2, 3$, and 4 under settings [M1] and [M2]. The corresponding optimal value for a criterion is highlighted in bold. —Cont'd

								[C1]	[C2]	[C3]	[C4]	[C5]	[C6] with (c_1, c_2, c_3)						
System type	n	Setting	m	r	$E(R)$	R_{max}	R	$\|I\|$	$V\mu + V\sigma$	V_μ	$\overline{TTE}$	$\overline{TTT}$	(1,1,10)	(1,0,10)	(1,1,0)	(0,1,10)	(0,1,5)	(0,1,20)	(0,1,30)
		[M2]	**8**	**4**	**4**	**4**	**(4,0,0,0)**	**34.79**	**0.4738**	**0.3588**	**4.82**	**29.56**	**13.56**	**8.74**	**8.82**	**9.56**	**7.19**	**14.30**	**19.04**
							(0,0,0,4)	21.56	0.8869	0.7013	**1.56**	**29.56**	**14.43**	**12.87**	**5.56**	**10.43**	**6.00**	**19.30**	**28.17**
							(3,0,0,1)	28.60	0.5931	0.4532	2.98	29.56	**12.91**	**9.93**	**6.98**	**8.91**	**5.94**	**14.84**	**20.77**
							(2,0,0,2)	25.27	0.7001	0.5418	2.24	29.56	13.24	11.00	6.24	9.24	**5.74**	**16.24**	**23.24**
	4	[M1]	**6**	**4**	**4**	**4**	**(2,0,0,0)**	**30.94**	**0.5481**	**0.4188**	**3.69**	**29.56**	**13.18**	**9.48**	**7.69**	**9.18**	**6.43**	**14.66**	**20.14**
							(0,0,0,2)	22.61	0.8303	0.6534	**1.75**	**29.56**	**14.06**	**12.30**	**5.75**	**10.06**	**5.91**	**18.36**	**26.66**
		[M2]	**6**	**4**	**4**	**4**	**(2,0,0,0)**	**30.94**	**0.5481**	**0.4188**	**3.69**	**29.56**	**13.18**	**9.48**	**7.69**	**9.18**	**6.43**	**14.66**	**20.14**
							(0,0,0,2)	22.61	0.8303	0.6534	**1.75**	**29.56**	**14.06**	**12.30**	**5.75**	**10.06**	**5.91**	**18.36**	**26.66**
							(1,0,0,1)	25.47	0.7000	0.5429	2.31	29.56	13.31	11.00	6.31	9.31	**5.81**	**16.31**	**23.31**
Parallel	2	[M1]	**12**	**2**	**4**	**14**	**(10,0)**	**28.18**	**0.3930**	**0.2292**	**11.89**	**72.51**	**19.82**	**7.93**	**15.89**	**15.82**	**13.85**	**19.75**	**23.68**
							(0,10)	16.14	0.6471	0.3113	**3.60**	**83.57**	**14.07**	**10.47**	**7.60**	**10.07**	**6.83**	**16.54**	**23.01**
		[M2]	**12**	**2**	**4**	**14**	**(10,0)**	**28.18**	**0.3930**	**0.2292**	**11.89**	**72.51**	**19.82**	**7.93**	**15.89**	**15.82**	**13.85**	**19.75**	**23.68**
							(0,10)	16.14	0.6471	0.3113	**3.60**	**83.57**	**14.07**	**10.47**	**7.60**	**10.07**	**6.83**	**16.54**	**23.01**
							(2,8)	16.58	0.6223	0.2989	3.84	82.41	**14.06**	**10.22**	**7.84**	**10.06**	**6.95**	**16.29**	**22.51**
							(7,3)	19.68	0.5139	0.2538	5.34	78.19	14.48	9.14	9.34	10.48	7.91	**15.62**	**20.76**
							(8,2)	21.13	0.4810	0.2432	6.15	76.87	14.96	8.81	10.15	10.96	8.56	15.77	**20.58**
	3	[M1/M2]	**8**	**2**	**6**	**18**	**(6,0)**	**35.86**	**0.3394**	**0.1501**	**14.85**	**139.03**	**24.24**	**9.39**	**20.85**	**18.24**	**16.54**	**21.64**	**25.03**
							(0,6)	25.23	0.4445	**0.1246**	**6.77**	**155.72**	**17.22**	**10.44**	**12.77**	**11.22**	**8.99**	**15.66**	**20.11**
2-out-of-n	3	[M1]	**10**	**2**	**6**	**13**	**(6,0)**	**24.78**	**0.4787**	**0.2929**	**6.81**	**54.21**	**15.60**	**8.79**	**10.81**	**11.60**	**9.20**	**16.39**	**21.17**

							(0,6)	15.66	0.8093	0.4781	2.67	60.81	14.76	12.09	6.67	10.76	6.71	18.85	26.95
		[M2]	10	2	6	13	(6,0)	24.78	0.4787	0.2929	6.81	54.21	15.60	8.79	10.81	11.60	9.20	16.39	21.17
							(0,6)	15.66	0.8093	0.4781	2.67	60.81	14.76	12.09	6.67	10.76	6.71	18.85	26.95
							(4,2)	18.53	0.6432	0.3756	3.72	57.65	14.16	10.43	7.72	10.16	6.94	16.59	23.02
							(1,5)	16.10	0.7765	0.4565	2.81	60.24	14.58	11.76	6.81	10.58	6.69	18.34	26.10
							(5,1)	20.54	0.5724	0.3371	4.55	56.28	14.28	9.72	8.55	10.28	7.42	16.00	21.73
Series-parallel	3	[M1]	8	2	4 2/3	18	(6,0)	12.73	0.6738	0.4688	5.51	33.34	16.92	11.41	10.18	12.25	8.88	18.99	25.73
							(0,6)	6.93	1.3139	0.9259	1.62	36.33	19.42	17.81	6.28	14.76	8.19	27.89	41.03
		[M2]	8	2	4 2/3	18	(6,0)	12.73	0.6738	0.4688	5.51	33.34	16.92	11.41	10.18	12.25	8.88	18.99	25.73
							(0,6)	6.93	1.3139	0.9259	1.62	36.33	19.42	17.81	6.28	14.76	8.19	27.89	41.03
							(5,1)	10.11	0.8283	0.5637	3.35	34.48	16.30	12.95	8.02	11.63	7.49	19.92	28.20
							(4,2)	8.78	0.9656	0.6593	2.57	35.13	16.89	14.32	7.24	12.23	7.40	21.88	31.54
Parallel-series	3	[M1]	8	2	3 1/3	10	(6,0)	22.11	0.4650	0.2784	9.56	71.24	17.55	7.98	12.90	14.21	11.89	18.86	23.51
							(0,6)	14.70	0.6976	0.3678	3.40	77.58	13.71	10.31	6.73	10.38	6.89	17.35	24.33
		[M2]	8	2	3 1/3	10	(6,0)	22.11	0.4650	0.2784	9.56	71.24	17.55	7.98	12.90	14.21	11.89	18.86	23.51
							(0,6)	14.70	0.6976	0.3678	3.40	77.58	13.71	10.31	6.73	10.38	6.89	17.35	24.33
							(2,4)	15.56	0.6456	0.3402	3.86	76.14	13.65	9.79	7.19	10.32	7.09	16.77	23.23
							(4,2)	17.21	0.5740	0.3078	4.86	74.15	13.93	9.07	8.19	10.60	7.73	16.34	22.08
							(5,1)	18.86	0.5252	0.2908	6.06	72.82	14.64	8.58	9.39	11.31	8.68	16.56	21.81

[a] For criterion **[C6]**, the objective function is $\psi(\mathbf{R}) = c_1 E(R) + c_2 E(TTE) + c_3(Var(\hat{\mu}) + Var(\hat{\sigma}))$.

TABLE 6 Optimal allocations for n-component systems with Weibull distributed components when $N = 18$, by fixing $R_{max} = 8$ with $n = 1$ and 2 under setting [M1] and [M2]. The corresponding optimal value for a criterion is highlighted in bold.

								[C1]	[C2]	[C3]	[C4]	[C5]	[C6] with (c_1, c_2, c_3)						
System type	n	Setting	m	r	$E(R)$	R_{max}	R	$\|I\|$	$V\mu + V\sigma$	V_μ	$\overline{TTE}$	$\overline{TTT}$	(1,1,10)	(1,0,10)	(1,1,0)	(0,1,10)	(0,1,5)	(0,1,20)	(0,1,30)
Series	1	[M1]	18	8	8	8	**(0,10,0,0,0,0,0,0)**	**150.55**	**0.1783**	**0.1251**	**18.95**	**59.11**	**28.73**	**9.78**	**26.95**	**20.73**	**19.84**	**22.51**	**24.30**
							(0,0,0,0,0,0,0,10)	79.52	0.2664	0.1658	**4.18**	**59.11**	**14.85**	**10.66**	**12.18**	**6.85**	**5.52**	**9.51**	**12.17**
		[M2]	18	8	8	8	**(0,10,0,0,0,0,0,0)**	**150.55**	**0.1783**	**0.1251**	**18.95**	**59.11**	**28.73**	**9.78**	**26.95**	**20.73**	**19.84**	**22.51**	**24.30**
							(4,4,1,0,1,0,0,0)	141.87	0.1814	**0.1250**	**18.16**	**59.11**	**27.97**	**9.81**	**26.16**	**19.97**	**19.07**	**21.79**	**23.60**
							(0,0,0,0,0,0,0,10)	79.52	0.2664	0.1658	**4.18**	**59.11**	**14.85**	**10.66**	**12.18**	**6.85**	**5.52**	**9.51**	**12.17**
	2	[M1]/[M2]	9	8	8	8	**(0,1,0,0,0,0,0,0)**	**110.90**	**0.2050**	**0.1329**	**9.92**	**59.11**	**19.97**	**10.05**	**17.92**	**11.97**	**10.95**	**14.02**	**16.07**
							(1,0,0,0,0,0,0,0)	110.03	**0.2045**	**0.1318**	**9.99**	**59.11**	**20.03**	**10.04**	**17.99**	**12.03**	**11.01**	**14.08**	**16.12**
							(0,0,0,0,0,0,0,1)	95.99	0.2261	0.1427	**6.76**	**59.11**	**17.02**	**10.26**	**14.76**	**9.02**	**7.89**	**11.28**	**13.54**
2-out-of-n	3	[M1]	6	2	4	8	**(4,0)**	**21.87**	**0.4807**	**0.2766**	7.01	49.67	**19.82**	**12.81**	**15.01**	**11.82**	**9.42**	**16.63**	**21.43**
							(0,4)	15.14	0.7112	0.3857	**3.25**	**54.42**	**18.36**	**15.11**	**11.25**	**10.36**	**6.80**	**17.47**	**24.58**
		[M2]	6	2	4	8	**(4,0)**	**21.87**	**0.4807**	**0.2766**	7.01	**49.67**	**19.82**	**12.81**	**15.01**	**11.82**	**9.42**	**16.63**	**21.43**
							(0,4)	15.14	0.7112	0.3857	**3.25**	**54.42**	**18.36**	**15.11**	**11.25**	**10.36**	**6.80**	**17.47**	**24.58**
							(2,2)	16.60	0.6294	0.3407	3.95	52.77	**18.25**	**14.29**	**11.95**	**10.25**	**7.10**	**16.54**	**22.84**
							(3,1)	18.24	0.5671	0.3105	4.77	51.56	18.44	13.67	12.77	10.44	7.61	**16.11**	**21.78**
Series-parallel	3	[M1]	6	2	3 1/3	8	**(4,0)**	**11.31**	**0.6708**	**0.4379**	**5.70**	**32.60**	**20.41**	**14.71**	**13.70**	**12.41**	**9.06**	**19.12**	**25.83**
							(0,4)	6.94	1.1000	0.7067	**2.13**	**35.03**	**21.13**	**19.00**	**10.13**	**13.13**	**7.63**	**24.13**	**35.13**
		[M2]	6	2	3 1/3	8	**(4,0)**	**11.31**	**0.6708**	**0.4379**	**5.70**	**32.60**	**20.41**	**14.71**	**13.70**	**12.41**	**9.06**	**19.12**	**25.83**
							(0,4)	6.94	1.1000	0.7067	**2.13**	**35.03**	**21.13**	**19.00**	**10.13**	**13.13**	**7.63**	**24.13**	**35.13**
							(3,1)	8.99	0.8145	0.5141	3.55	33.70	**19.70**	**16.15**	**11.55**	**11.70**	**7.62**	**19.84**	**27.99**
							(2,2)	7.91	0.9328	0.5890	2.78	34.31	20.11	17.33	10.78	12.11	**7.44**	**21.44**	**30.76**

[a] For criterion **[C6]**, the objective function is $\psi(\mathbf{R}) = c_1 R_{max} + c_2 E(TTE) + c_3(Var(\hat{\mu}) + Var(\hat{\sigma}))$.

systems, the values of $E(R)$ are slightly larger than those for other systems, which might bring some disadvantages in comparison.

Suppose one considers the optimality criterion **[C3]** (V-optimality) that minimizes the asymptotic variance of the MLE, the optimal values of the objective function (i.e., V_{μ}) for **[C3]** based on series, parallel, 2-out-of-n, series-parallel, and parallel-series systems are 0.1330, 0.0980, 0.1181, 0.1381, and 0.1144, respectively. From these results, we suggest performing the progressively Type-II censored experiment by putting the 30 components into ten 3-component parallel systems and censoring eight systems at the second failure. However, the trade-off of using 3-component parallel systems for the progressively Type-II censored experiment is that the expected total time on test and the expected total experimental time based on 3-component parallel systems are the longest compared to the other systems considered here.

From Tables 4 and 5, we observe a similar pattern when the underlying component lifetimes follow Weibull distribution. For example, suppose there are 24 components available for the experiment, and we plan to fail four components. From Table 5, when $N = 24$, $E(R) = 4$ and the component lifetimes follow a Weibull distribution, for optimality criterion **[C3]**, the optimal experimental plan based on 3-component and 2-component parallel systems give the optimal values 0.1246 and 0.2292, respectively, while the optimal experimental plan based on individual components gives 0.2553. In other words, using 3-component and 2-component parallel systems for the progressively Type-II censored experiment provides 51.19 % and 10.22 % improvements, respectively, in terms of the criterion **[C3]** compared to using individual components for the experiment. However, the value of $E(R)$ based on 3-component parallel systems is 6, which is larger than the target value of $E(R) = 4$.

Instead of fixing the value of $E(R)$, in Tables 3 and 6, we compare the optimal progressive censoring schemes based on different coherent systems by fixing the value of $R_{\max}$. From Tables 3 and 6, we observe that for a fixed value of $R_{\max}$, although using 2-out-of-3 or series-parallel systems in a progressively Type-II censored experiment do not always give an advantage over using series systems; they always give shorter expected *TTE* and *TTT* for both lognormal and Weibull distributed components.

The numerical results show the advantages and disadvantages of using systems in an experiment with progressively Type-II censored samples that do not show a specific pattern. The performance depends on the type of systems, the value of $E(R)$ (or $R_{\max}$), and the underlying component distributions. Therefore, it is crucial to evaluate the performance of the optimal experimental plans under different types of systems and different values of n to determine the optimal experimental planning for a specific situation, and the methodologies proposed here are helpful for this purpose.

8 Practical example

In this section, we use a practical example of insulating fluids testing presented in Viveros and Balakrishnan (1994) and Ng et al. (2004) to illustrate the methodologies developed in this chapter for planning a future progressively Type-II censored experiment. Ng et al. (2004) compared the efficiency of different progressively Type-II censoring plans with $N = 19$ components effective sample size $r = 8$ (i.e., the number of observed failures). For a future test, suppose the total number of insulating fluids available is $N = 19$, and we consider putting the insulating fluids into a n-component series system with $n = 1$ (i.e., component-wise experiment), and $n = 2$ (i.e., nine series systems).

Assuming the lifetimes of the insulating fluids follow Weibull distribution and lognormal distribution and considering the optimality criteria and settings described in Section 5, we determine the optimal experimental plans for different optimality criteria under different settings. The optimal progressive censoring schemes for different criteria under the assumption that the component lifetimes follow Weibull and lognormal distributions are presented in Tables 7 and 8, respectively.

From the results in Table 7, the progressively Type-II censored experimental plans based on $n = 1$ are preferred over $n = 2$ for different optimality criteria. For example, for optimality criterion **[C2]**, the optimal value of $Var(\hat{\mu}) + Var(\hat{\sigma})$ based on setting **[M2]** for $n = 1$ and $n = 2$ are 0.1770 and 0.2045, respectively. For various combinations of costs c_1, c_2, and c_3, the performance of optimal experimental plans for $n = 1$ still outperforms those for $n = 2$. In practice, engineers might need to do all censoring at one time in certain circumstances. Therefore, the setting **[M1]** (i.e., censoring all of the $m - r$ systems or components at a certain failure) can be considered instead of setting **[M2]** in a practical situation. If setting **[M1]** is considered, based on the results presented in Table 7, the optimal experimental plans based on $n = 1$ are still preferred over $n = 2$.

We have similar observations in Table 8. If we consider setting **[M2]**, the progressively Type-II censored experimental plans with a single component ($n = 1$) consistently outperform the plans with two components ($n = 2$) across multiple criteria. For instance, for optimality criterion **[C3]**, the optimal value of $Var(\hat{\mu})$ based on setting **[M2]** for $n = 1$ and $n = 2$ are 0.0958 and 0.0963, respectively. For setting **[M1]**, if we use criteria **[C3]** and **[C5]**, then using two-component systems in progressively Type-II censored experiments would yield better performance than those of using components.

TABLE 7 Optimal progressive censoring schemes for n-component series systems with Weibull distributed components when $N = 19$, by fixing $r = 8$ with $n = 1$ and 2 under settings [M1] and [M2]. The corresponding optimal value for a criterion is highlighted in bold.

							[C1]	[C2]	[C3]	[C4]	[C5]	[C6] with (c_1, c_2, c_3)						
n	Setting	m	r	$E(R)$	R_{max}	R	$\|I\|$	$Var(\hat{\mu}) + Var(\hat{\sigma})$	$Var(\hat{\mu})$	$\overline{TTE}$	$\overline{TTT}$	(1,1,10)	(1,0,10)	(1,1,0)	(0,1,10)	(0,1,5)	(0,1,20)	(0,1,30)
1	[M1]	19	8	8	8	**(11,0,0,0,0,0,0,0)**	144.42	**0.1826**	**0.1272**	**19.55**	**59.11**	**29.37**	**9.83**	**27.55**	**21.37**	**20.46**	**23.2**	**25.02**
						(0,11,0,0,0,0,0,0)	**154.11**	**0.1770**	**0.1250**	**18.90**	**59.11**	**28.67**	**9.77**	**26.90**	**20.67**	**19.79**	**22.44**	**24.21**
						(0,0,0,0,0,0,0,11)	78.98	0.2756	0.1743	**3.90**	**59.11**	**14.66**	**10.76**	**11.90**	**6.66**	**5.28**	**9.41**	**12.17**
	[M2]	19	8	8	8	**(0,11,0,0,0,0,0,0)**	**154.11**	**0.1770**	**0.1250**	**18.90**	**59.11**	**28.67**	**9.77**	**26.90**	**20.67**	**19.79**	**22.44**	**24.21**
						(8,0,1,1,1,0,0,0)	140.40	0.1820	**0.1250**	**18.03**	**59.11**	**27.85**	**9.82**	**26.03**	**19.85**	**18.94**	**21.67**	**23.49**
						(0,0,0,0,0,0,0,11)	78.98	0.2756	0.1743	**3.90**	**59.11**	**14.66**	**10.76**	**11.90**	**6.66**	**5.28**	**9.41**	**12.17**
						(1,0,0,0,0,0,0,10)	81.24	0.2656	0.1671	4.16	59.11	14.82	10.66	12.16	6.82	5.49	9.47	**12.13**
2	[M1]	9	8	8	8	**(0,1,0,0,0,0,0,0)**	**110.90**	**0.2050**	**0.1329**	**9.92**	**59.11**	**19.97**	**10.05**	**17.92**	**11.97**	**10.95**	**14.02**	**16.07**
	and [M2]					**(1,0,0,0,0,0,0,0)**	**110.03**	**0.2045**	**0.1318**	**9.99**	**59.11**	**20.03**	**10.04**	**17.99**	**12.03**	**11.01**	**14.08**	**16.12**
						(0,0,0,0,0,0,0,1)	95.99	0.2261	0.1427	**6.76**	**59.11**	**17.02**	**10.26**	**14.76**	**9.02**	**7.89**	**11.28**	**13.54**

TABLE 8 Optimal progressive censoring schemes for n-component series systems with lognormal distributed components when $N = 19$, by fixing $r = 8$ with $n = 1$ and 2 under settings [M1] and [M2]. The corresponding optimal value for a criterion is highlighted in bold.

							[C1]	[C2]	[C3]	[C4]	[C5]	[C6] with (c_1, c_2, c_3)						
n	Setting	m	r	$E(R)$	R_{max}	R	$\|I\|$	$Var(\hat{\mu}) + Var(\hat{\sigma})$	$Var(\hat{\mu})$	$\overline{TTE}$	$\overline{TTT}$	(1,1,10)	(1,0,10)	(1,1,0)	(0,1,10)	(0,1,5)	(0,1,20)	(0,1,30)
1	[M1]	19	8	8	8	**(0,0,11,0,0,0,0,0)**	**225.27**	**0.1543**	**0.1020**	**32.82**	**105.26**	**42.36**	**9.54**	**40.82**	**34.36**	**33.59**	**35.91**	**37.45**
						(0,11,0,0,0,0,0,0)	221.84	**0.1540**	**0.1047**	**34.70**	**105.99**	**44.24**	**9.54**	**42.70**	**36.24**	**35.47**	**37.78**	**39.32**
						(0,0,0,0,0,0,0,11)	186.53	0.1741	**0.0975**	**5.92**	**93.61**	**15.66**	**9.74**	**13.92**	**7.66**	**6.79**	**9.40**	**11.14**
	[M2]	19	8	8	8	**(0,0,11,0,0,0,0,0)**	**225.27**	**0.1543**	**0.1020**	**32.82**	**105.26**	**42.36**	**9.54**	**40.82**	**34.36**	**33.59**	**35.91**	**37.45**
						(0,10,1,0,0,0,0,0)	222.06	**0.1540**	**0.1041**	**34.28**	**105.72**	**43.82**	**9.54**	**42.28**	**35.82**	**35.05**	**37.36**	**38.90**
						(4,0,0,0,0,0,0,7)	191.72	0.1639	**0.0958**	**7.64**	**93.02**	**17.28**	**9.64**	**15.64**	**9.28**	**8.46**	**10.92**	**12.56**
						(0,0,0,0,0,0,0,11)	186.53	0.1741	0.0975	**5.92**	**93.61**	**15.66**	**9.74**	**13.92**	**7.66**	**6.79**	**9.40**	**11.14**
2	[M1]	9	8	8	8	**(0,1,0,0,0,0,0,0)**	**200.68**	**0.1590**	**0.0976**	**14.76**	**95.39**	**24.35**	**9.59**	**22.76**	**16.35**	**15.56**	**17.94**	**19.53**
	and [M2]					**(1,0,0,0,0,0,0,0)**	**198.19**	**0.1589**	**0.0976**	**14.85**	**95.11**	**24.44**	**9.59**	**22.85**	**16.44**	**15.65**	**18.03**	**19.62**
						(0,0,0,0,0,0,0,1)	192.95	0.1633	**0.0963**	**9.71**	**93.42**	**19.34**	**9.63**	**17.71**	**11.34**	**10.52**	**12.97**	**14.61**

9 Concluding remarks and future research directions

In this chapter, we explored optimal experimental design for progressively Type-II censored experiments, focusing on cases where experimental units are arranged into coherent systems for life testing. The component lifetimes were modeled using distributions from the log-location-scale family, a versatile class that includes commonly used distributions such as the Weibull and lognormal. The formulations necessary to derive optimal experimental designs were presented using system signatures, addressing various optimality criteria. Motivated by insulating fluid testing studies in Balakrishnan and Aggarwala (2000) and Ng et al. (2004), we illustrated our approach using series systems and provided optimal progressive Type-II censoring schemes for different system types and component counts. Our results demonstrated that using systems of test units in such experiments can offer greater flexibility and, in some cases, enhance efficiency.

In conclusion, the performance of using systems versus individual components in progressively Type-II censored experiments is context-dependent, varying with the system structure and chosen criteria. While system-based designs do not always outperform designs based on individual components, they can significantly improve estimation efficiency and reduce the overall experimental time when they do.

This chapter primarily addresses optimal experimental designs for a limited number of components and system structures, a constraint imposed by the available number of components. For example, with 20 components, one could only construct 10, 5, 4, or 2 systems if the systems contain 2, 4, 5, or 10 components, respectively. Future research could extend this work by investigating experiments with larger numbers of components-potentially in the hundreds or thousands-and exploring a wider variety of system configurations. Such an expansion would provide a more comprehensive understanding of optimal experimental design across diverse scenarios.

Additionally, our analysis is based on log-location-scale family distributions, chosen for their flexibility in modeling lifetime data. In future studies, it would be valuable to explore other distribution families, such as scale-shape distributions like the gamma and Birnbaum-Saunders. However, these distributions present additional challenges, such as the absence of standardized forms, making careful execution essential to obtaining reliable results.

Appendix A. Supporting information

Supplementary data associated with this article can be found in the online version at https://doi.org/10.1016/bs.host.2025.04.006.

References

Aslett, L.J.M., Coolen, F.P.A., Wilson, S.P., 2015. Bayesian inference for reliability of systems and networks using the survival signature. Risk Anal. 35 (9), 1640–1651.

Balakrishnan, N., Aggarwala, R., 2000. Progressive Censoring: Theory, Methods, and Applications. Springer, New York.

Balakrishnan, N., Burkschat, M., Cramer, E., Hofmann, G., 2008. Fisher information based progressive censoring plans. Comput. Statist. Data Anal. 53, 366–380.

Balakrishnan, N., Cramer, E., 2014. The Art of Progressive Censoring - Applications to Reliability and Quality. Springer, New York.

Balakrishnan, N., Cramer, E., 2023. Progressive censoring methodology: a review. In: Pham, H. (Ed.), Springer Handbook of Engineering Statistics, 2nd ed. Springer, New York, pp. 153–183.

Balakrishnan, N., Han, D., 2007. Optimal progressive type-II censoring schemes for nonparametric confidence intervals of quantiles. Commun. Stat. - Simul. Comput. 46, 1247–1262.

Balakrishnan, N., Ng, H.K.T., Navarro, J., 2011a. Exact nonparametric inference for component lifetime distribution based on lifetime data from systems with known signatures. J. Nonparametr. Stat. 23 (3), 741–752.

Balakrishnan, N., Ng, H.K.T., Navarro, J., 2011b. Linear inference for type-II censored lifetime data of reliability systems with known signatures. IEEE Trans. Reliab. 60 (2), 426–440.

Bhattacharya, D., Samaniego, F.J., 2010. Estimating component characteristics from system failure-time data. Nav. Res. Logist. (NRL) 57 (4), 380–389.

Bhattacharya, R., Balakrishnan, N., 2024. A MCMC-type simple probabilistic approach for determining optimal progressive censoring schemes. Commun. Stat. - Simul. Comput. 53, 1758–1767.

Burkschat, M., Cramer, E., Kamps, U., 2006. On optimal schemes in progressive censoring. Stat. & Probability Lett. 76, 1032–1036.

Burkschat, M., Cramer, E., Kamps, U., 2007. Optimality criteria and optimal schemes in progressive censoring. Commun. Stat. - Theory Methods 36, 1419–1431.

Coolen, F.P.A., Al-nefaiee, A.H., 2012. Nonparametric predictive inference for failure times of systems with exchangeable components. Proc. Inst. Mech. Engineers, Part. O: J. Risk Reliab. 226 (3), 262–273.

Cramer, E., Davies, K., 2018. Restricted optimal progressive censoring. Commun. Stat. - Simul. Comput. 47, 1216–1239.

Cramer, E., Navarro, J., 2015. Progressive type-II censoring and coherent systems. Nav. Res. Logist. 62, 512–530.

Cramer, E., Navarro, J., 2016. The progressive censoring signature of coherent systems. Appl. Stoch. Model. Bus. Ind. 32 (5), 697–710.

Dahmen, K., Burkschat, M., Cramer, E., 2012. A- and D-optimal progressive type-II censoring designs based on fisher information. J. Stat. Comput. Simul. 82, 879–905.

Fallah, A., Asgharzadeh, A., Ng, H.K.T., 2021. Statistical inference for component lifetime distribution from coherent system lifetimes under a proportional reversed hazard model. Commun. Stat. - Theory Methods 50 (16), 3809–3833.

Fan, T., Wang, W., 2011. Accelerated life tests for Weibull series systems with masked data. IEEE Trans. Reliab. 60 (3), 557–569.

Hall, P., Jin, Y., Samaniego, F.J., 2015. Nonparametric estimation of component reliability based on lifetime data from systems of varying design. Statistica Sin. 25 (4), 1313–1335.

Hermanns, M., Cramer, E., 2017. Likelihood inference for the component lifetime distribution based on progressively censored parallel systems data. J. Stat. Comput. Simul. 87 (3), 607–630.

Hermanns, M., Cramer, E., 2018. Inference with progressively censored k-out-of-n system lifetime data. TEST 27, 787–810.

Hermanns, M., Cramer, E., Ng, H.K.T., 2020. EM algorithms for ordered and censored system lifetime data under a proportional hazard rate model. J. Stat. Comput. Simul. 90 (18), 3301–3337.

Jin, Y., Hall, P.G., Jiang, J., Samaniego, F.J., 2017. Estimating component reliability based on failure time data from a system of unknown design. Statistica Sin. 27 (2), 479–499.

Klos, R., Fabisiak, B., Ng, H.K.T., 2018. Comparative reliability analysis of selected joints for case furniture. BioResources 13 (3), 5111–5123.

Navarro, J., 2022. Introduction to System Reliability Theory. Springer Nature, Switzerland AG.

Navarro, J., Rubio, R., 2009. Computations of signatures of coherent systems with five components. Commun. Stat. - Simul. Comput. 39 (1), 68–84.

Navarro, J., Ruiz, J., Sandoval, C., 2007. Properties of coherent systems with dependent components. Commun. Stat. - Theory Methods 36 (1), 175–191.

Ng, H.K.T., 2010. Censoring methodology. In: Lovric, M. (Ed.), International Encyclopedia of Statistical Science. Springer, pp. 221–224.

Ng, H.K.T., Chan, P.S., Balakrishnan, N., 2004. Optimal progressive censoring plans for the weibull distribution. Technometrics 46, 470–481.

Ng, H.K.T., Knac, I., Kus, C., Chan, P.S., 2017. Optimal experimental plan for multi-level stress testing with Weibull regression under progressive type-II extremal censoring'. Commun. Stat. - Simul. Comput. 46 (4), 2611–2637.

Proschan, F., Barlow, R.E., 1965. Mathematical Theory of Reliability. John Wiley & Sons, New York, NY.

Qu, J., Ng, H.K.T., 2024. Tests for homogeneity of distributions of component lifetimes from type-II censored system lifetime data with known system structures. Nav. Res. Logist. (NRL) 71 (6), 831–853.

Qu, J., Ng, H.K.T., Moon, C., 2024. Empirical likelihood ratio tests for homogeneity of component lifetime distributions based on system lifetime data. Comput. Stat 39 (6), 3007–3029.

Rausand, M., Barros, A., Hoyland, A., 2020. System Reliability Theory - Models, Statistical Methods, and Applications. John Wiley & Sons, Hoboken, NJ.

Rausand, M., Hoyland, A., 2004. System Reliability Theory - Models, Statistical Methods, and Applications. Wiley, Hoboken, New Jersey.

Salemi, U.H., Rezaei, S., Nadarajah, S., 2019. A-optimal and D-optimal censoring plans in progressively type-II right censored order statistics. J. Stat. Comput. Simul 60, 1349–1367.

Samaniego, F.J., 1985. On closure of the IFR class under formation of coherent systems. IEEE Trans. Reliab. 34 (1), 69–72.

Samaniego, F.J., 2007. System Signatures and their Applications in Engineering Reliability. Springer, New York.

Shaked, M., Suarez-Llorens, A., 2003. On the comparison of reliability experiments based on the convolution order. J. Am. Stat. Assoc. 98 (463), 693–702.

Tavangar, M., Asadi, M., 2020. Component reliability estimation based on system failure-time data. J. Stat. Comput. Simul. 90 (17), 3232–3249.

Viveros, R., Balakrishnan, N., 1994. Interval estimation of parameters of life from progressively censored data. Technometrics 36 (1), 84–91.

Wu, S.-J., Kus, C., 2009. On estimation based on progressive first-failure-censored sampling. Comput. Stat. & Data Anal. 53 (10), 3659–3670.

Yang, Y., Ng, H.K.T., Balakrishnan, N., 2016. A stochastic expectation-maximization algorithm for the analysis of system lifetime data with known signature. Comput. Stat. 31 (2), 609–641.

Yang, Y., Ng, H.K.T., Balakrishnan, N., 2019. Expectation–maximization algorithm for system-based lifetime data with unknown system structure. AStA Adv. Stat. Ana.l 103 (1), 69–98.

Zhang, J., Ng, H.K.T., Balakrishnan, N., 2015. Statistical inference of component lifetimes with location-scale distributions from censored system failure data with known signature. IEEE Trans. Reliab. 64 (2), 613–626.

Chapter 4

Active learning of computer experiment with both quantitative and qualitative inputs

Anita Shahrokhian[a], Xinwei Deng[b], and C. Devon Lin[a,*]
[a]*Department of Mathematics and Statistics, Queen's University, Kingston, ON, Canada,*
[b]*Department of Statistics, Virginia Tech, Blacksburg, VA, United States of America*
[*]*Corresponding author. e-mail address: devon.lin@queensu.ca*

Chapter Outline

Abstract

Computer experiments refer to the study of real systems using complex simulation models. They have been widely used as alternatives to physical experiments. Design and analysis of computer experiments have attracted great attention in past three decades. The bulk of the work, however, often focus on experiments with only quantitative inputs. In recent years, research on design and analysis for computer experiments have gain momentum. Statistical methodology for design, modeling and inference of such experiments have been developed. In this chapter, we review some of those key developments, and propose active learning approaches for modeling, optimization, contour estimation of computer experiments with both types of inputs. Numerical studies are conducted to evaluate the performance of the proposed methods in comparison with other existing methods.

Handbook of Statistics, Vol. 53. https://doi.org/10.1016/bs.host.2025.04.003

1 Introduction

Computer experiments are widely used to analyze computer simulators, which are mathematical models implemented in code to study complex systems (Sacks et al., 1989). These experiments provide a cost-effective and practical alternative to physical experiments. They have been successfully applied across various scientific and engineering domains, including rocket booster simulations (Gramacy, 2020) and high-performance computing systems analysis (Cameron et al., 2019). One challenge in using computer experiments is that they can be computationally expensive in certain applications, with some simulations requiring hours or even days to generate a single output. To address this challenge, statistical emulators, such as Gaussian process (GP) models, are employed to approximate the behavior of computer simulators. These emulators significantly support the follow-up analysis and inference goals such as optimization (Jones et al., 1998), contour estimation (Ranjan et al., 2008), and prediction (Yang et al., 2020), among others.

Computer experiments often involve diverse data inputs, including scenarios where both quantitative and qualitative inputs are involved. As emulators, such as GP models, typically assume quantitative inputs, there is a growing need for advanced modeling approaches to accommodate mixed inputs. Recent developments have addressed this challenge following early work such as Qian et al. (2008), Han et al. (2009), Zhou et al. (2011). For example, Deng et al. (2017) introduced additive Gaussian process (AGP) models, incorporating an additive structure for qualitative inputs while assuming a multiplicative correlation between qualitative and quantitative inputs. Building on this foundation, Xiao et al. (2021) proposed the easy-to-interpret Gaussian process (EzGP) model, which numerically outperforms AGP. Their approach eliminates the need to explicitly construct a correlation function for qualitative inputs, instead accounting for the effects of different level combinations through an additive GP structure. Zhang et al. (2021) focuses on the case when there is a large number of qualitative inputs. They proposed a sparse covariance estimation approach to estimate the covariance matrix for the mixed-input Gaussian process emulator. In addition, Zhang et al. (2020b) developed the latent variable Gaussian process (LvGP), a novel method that maps qualitative inputs to underlying numerical latent variables to define the correlation function. More recently, Lin et al. (2024) advanced GP models for mixed inputs by introducing a tree structure to partition the levels of qualitative inputs, resulting in enhanced model performance. These innovations have significantly expanded the capabilities of GP models to handle mixed input scenarios in computer experiments.

Another fundamental issue in computer experiments is determining how to select inputs for running computer codes. This is known as *experimental design*. Significant progress has been made in developing experimental designs for computer experiments involving mixed inputs. Notable examples include sliced Latin hypercube designs (Qian, 2012; Hwang et al., 2016; Yang et al., 2013) and marginally and doubly coupled designs (Deng et al., 2015; He et al., 2017a, 2017b; Yang et al., 2023). These designs are space-filling and do not rely on specific models. However, such model-independent designs can be inefficient because they do not explicitly take into account the study's objectives. To address this limitation, adaptive designs provide a more effective alternative. Adaptive designs select inputs based on the specific analysis goals, thereby reducing unnecessary computational time and costs. They typically begin with an initial design, fit a model to the available data, and iteratively update the model after each selection until the desired objective is achieved.

A critical step in adaptive designs is determining next input, which is selected based on a specific criterion aligned with the study's objectives. Here we focus on commonly used adaptive design criteria for optimization, contour estimation, and prediction, although there are other objectives such as the estimation of a probability of failure (Bect et al., 2012). For optimization, widely discussed criteria include expected improvement (EI) (Jones et al., 1998), upper confidence bound (UCB) (Srinivas et al., 2010), quantile-based (Picheny et al., 2013) and entropy-based approaches (Hernández-Lobato et al., 2014). For contour estimation, Ranjan et al. (2008) extended the EI criterion originally proposed for optimization to contour estimation, while Cole et al. (2023) applied entropy-based designs, introducing a failure region to avoid complex integral evaluations. For prediction, Yang et al. (2020) use the EI criterion for multiple contours estimation for adaptive designs for prediction. In addition, Mohammadi et al. (2022) proposed an adaptive design approach using leave-one-out cross-validation to enhance Gaussian process (GP) models for prediction. Most of these criteria have been applied primarily to models with quantitative inputs. Notable exceptions include Cai et al. (2024), Zhang et al. (2020a), and Luo et al. (2024), who considered adaptive designs for optimization with mixed inputs. Furthermore, Shahrokhian et al. (2024) proposed a comprehensive framework for adaptive designs for contour estimation in computer experiments involving mixed inputs.

The goal of this chapter is to survey recent developments in adaptive designs for contour estimation and optimization in computer experiments with mixed inputs and to provide a detailed comparison of these methods. In addition, we evaluate the performance of adaptive design criteria, originally proposed for prediction in computer experiments with only quantitative inputs (Yang et al., 2020), when applied to computer experiments with mixed inputs.

The rest of the chapter is organized as follows. Section 2 provides the necessary background on GP models for computer experiments with both types of inputs, and reviews different adaptive designs used in models with quantitative inputs for optimization, contour estimation and prediction. Section 3 presents different adaptive designs for optimization and contour estimation for computer experiments with mixed inputs. We illustrate the comparison of these methods through the simulation studies in Section 4. Section 5 concludes the paper.

2 Background

This section provides the foundational background for adaptive designs in computer experiments involving both quantitative and qualitative inputs, with applications in contour estimation, optimization, and prediction. Section 2.1 reviews the easy-to-interpret Gaussian process (EzGP) model, a framework for emulating computer simulators with mixed inputs. Section 2.2 summarizes adaptive design strategies tailored to optimization, contour estimation, and prediction in experiments with quantitative inputs.

2.1 Modeling

To review the EzGP, we first review Gaussian process models. Consider a computer experiment with p quantitative input variables. Suppose $i = 1, ..., n$, design points $\boldsymbol{x}_i = (x_{i1}, ..., x_{ip})^{\mathrm{T}}$ are used and the corresponding responses $y_1, ..., y_n$ are collected where $y_i = y(\boldsymbol{x}_i)$ and $\boldsymbol{x}_i \in (0,1)^p$. A Gaussian process model assumes that,

$$Y(\boldsymbol{x}_i) = F\beta + Z(\boldsymbol{x}_i) \quad i = 1, ..., n,$$

where F is a vector of the known basis function of the input, β is the vector of the regression coefficients, $Z(\boldsymbol{x}_i)$ is a GP with $E[Z(\boldsymbol{x}_i)] = 0$, $\mathrm{Var}(Z(\boldsymbol{x}_i)) = \sigma^2$ and $\mathrm{Cov}(Z(\boldsymbol{x}_i), Z(\boldsymbol{x}_j)) = \sigma^2 R_{ij}$. When the mean part $F\beta$ equals to a constant μ, it is known as ordinary kriging. There are many choices for the correlation functions. One of the most commonly used correlation functions is the Gaussian correlation that is,

$$R_{ij} = \mathrm{Corr}(Z(\boldsymbol{x}_i), Z(\boldsymbol{x}_j)) = \prod_{k=1}^{p} \exp\{-\theta_k (x_{ik} - x_{jk})^2\}, \tag{1}$$

where the parameters $\boldsymbol{\theta} = (\theta_1, ..., \theta_p)^T$ are the correlation parameters. A GP assumes $(Y_1, ..., Y_n)^T$ follows a multivariate normal (MVN) distribution $N_n(\mu \mathbb{1}_n, \Sigma)$ where $\mathbb{1}_n$ is an n-vector of ones and $\Sigma = \sigma^2 R_{ij}$. To estimate the parameters, we can use the maximum likelihood approach. In this model we have $p + 2$ parameters, μ, σ^2 and $\boldsymbol{\theta} = (\theta_1, ..., \theta_p)^T$. Let $\mathbf{y} = (y_1, ..., y_n)^T$ denote the vector of n responses, $\boldsymbol{R}$ denote the $n \times n$ matrix whose (i, j) entry is $\mathrm{Corr}(Y(x_i), Y(x_j))$. Then the likelihood function is,

$$L(\theta, \mu, \sigma^2) = \frac{1}{(2\pi)^{(n/2)} |\Sigma|^{1/2}} \exp\left\{-\frac{1}{2\sigma^2} (\mathbf{y} - \mathbb{1}_n \mu)^T \boldsymbol{R}^{-1} (\mathbf{y} - \mathbb{1}_n \mu)\right\}. \tag{2}$$

Note that the dependence of $\boldsymbol{\theta}$ is through the correlation function in (1). Given the correlation parameters $\boldsymbol{\theta} = (\theta_1, ..., \theta_p)^T$, the maximum likelihood estimates of μ and σ^2 are given by,

$$\hat{\mu} = (\mathbb{1}_n^T \boldsymbol{R}^{-1} \mathbb{1}_n)^{-1} (\mathbb{1}_n^T \boldsymbol{R}^{-1} \mathbf{y}) \quad \text{and} \quad \hat{\sigma}^2 = \frac{(\mathbf{y} - \mathbb{1}_n \hat{\mu})^T \boldsymbol{R}^{-1} (\mathbf{y} - \mathbb{1}_n \hat{\mu})}{n}. \tag{3}$$

By substituting the estimates in (3) into the likelihood function (2), we get the likelihood function that only depends on $\boldsymbol{\theta}$ and can be maximized for $\boldsymbol{\theta}$. This model can now be used to make a prediction at any unsampled point $\boldsymbol{x}^*$. The predictive mean and the predictive variance are given by,

$$\hat{\mu}(\boldsymbol{x}^*) = \hat{\mu} + \boldsymbol{r}^T\boldsymbol{R}^{-1}(\mathbf{y} - \mathbb{1}_n\hat{\mu}),$$
$$\hat{\sigma}^2(\boldsymbol{x}^*) = \sigma^2\left(1 - \mathbf{r}^T\boldsymbol{R}^{-1}\boldsymbol{r} + \frac{(1 - \mathbb{1}_n^T\boldsymbol{R}^{-1}\boldsymbol{r})^2}{\mathbb{1}_n^T\boldsymbol{R}^{-1}\mathbb{1}_n}\right), \tag{4}$$

where $\boldsymbol{r} = (r_1(\boldsymbol{x}^*), ..., r_n(\boldsymbol{x}^*))^T$, and $r_i(\boldsymbol{x}^*) = \mathrm{Corr}(Y(\boldsymbol{x}^*), Y(\boldsymbol{x}_i))$.

For computer experiments with quantitative and qualitative inputs, Xiao et al. (2021) proposed the EzGP model. Consider an n-run computer experiment with p quantitative variables and q qualitative factors. Let $x^{(k)}$ be the kth quantitative variable and $z^{(h)}$ be the hth qualitative factor having m_h levels, for $k = 1, ..., p$, $h = 1, ..., q$. For $i = 1, ..., n$, denote $\boldsymbol{x}_i = (x_{i1}, ..., x_{ip})^T$ and $z_i = (z_{i1}, ..., z_{iq})^T$, and let $\boldsymbol{w}_i = (\boldsymbol{x}_i^T, z_i^T)^T$ be the ith input and y_i be the corresponding output. To model the relationship between output Y and an input $\boldsymbol{w} = (\boldsymbol{x}^T, z^T)^T$, Xiao et al. (2021) proposed the model,

$$Y(\boldsymbol{x}, z) = \mu + G_0(\boldsymbol{x}) + G_{z^{(1)}}(\boldsymbol{x}) + ... + G_{z^{(q)}}(\boldsymbol{x}), \tag{5}$$

where μ is a constant mean, G_0 and $G_{z^{(h)}}$'s are independent GPs with mean zero and covariance function ϕ_0 and ϕ_h, $h = 1, ..., q$. Xiao et al. (2021) assumes that G_0 is a standard GP with only quantitative inputs $\boldsymbol{x}$ with ϕ_0 given by,

$$\phi_0(\boldsymbol{x}_i, \boldsymbol{x}_j|\Theta_0) = \sigma_0^2 \exp\left\{-\sum_{k=1}^{p} \theta_k^{(0)}(x_{ik} - x_{jk})^2\right\}, \tag{6}$$

where the correlation parameter $\Theta_{\mathbf{0}} = (\theta_1^{(0)}, ..., \theta_p^{(0)})^T$ are all positive. They proposed the covariance function of $G_{z^{(h)}}$ such that,

$$\phi_h((\boldsymbol{x}_i^T, z_{ih})^T, (\boldsymbol{x}_j^T, z_{jh})^T|\Theta^{(h)})$$
$$= \sigma_h^2 \exp\left\{-\sum_{k=1}^{p} \theta_{kl_h}^{(h)}(x_{ik} - x_{jk})^2\right\}\mathbb{1}(z_{ih} = z_{jh} \equiv l_h), \tag{7}$$

where z_{ih} and z_{jh} are the ith and jth value of the input variable $z^{(h)}$, $l_h = 1, ..., m_h$, m_h is the number of the levels of the qualitative input variable $z^{(h)}$, $\Theta^{(h)} = (\theta_{kl_h}^{(h)})_{p\times m_h}$ is the matrix for the correlation parameters, and the indicator function is 1 when $z_{ih} = z_{jh} \equiv l_h$ and 0 otherwise.

Therefore, from (6) and (7), the covariance function in the model in (5) is given by,

$$\begin{aligned}\phi(\boldsymbol{w}_i, \boldsymbol{w}_j) &= \mathrm{Cov}(Y(\boldsymbol{w}_i), Y(\boldsymbol{w}_j)) \\ &= \phi_0(\boldsymbol{x}_i, \boldsymbol{x}_j|\theta_0) + \sum_{h=1}^{q} \phi_h(\boldsymbol{x}_i, \boldsymbol{x}_j|\Theta^{(h)}) \\ &= \sigma_0^2 \exp\left\{-\sum_{k=1}^{p} \theta_k^{(0)}(x_{ik} - x_{jk})^2\right\} \\ &\quad + \sum_{h=1}^{q} \sum_{l_h=1}^{m_h} \mathbb{1}(z_{ih} = z_{jh} \equiv l_h)\sigma_h^2 \exp\left\{-\sum_{k=1}^{p} \theta_{kl_h}^{(h)}\left(x_{ik} - x_{jk}\right)^2\right\}.\end{aligned} \tag{8}$$

Given the covariance function, there are $2 + p + q + p\sum_{h=1}^{q} m_h$ parameters $\mu, \sigma_0^2, \theta_k^{(0)}, \sigma_h^2$, and $\theta_{kl_h}^{(h)}$ for $h = 1, ..., q$, $k = 1, ..., p$ and $l_h = 1, ..., m_h$. They can be estimated using the maximum likelihood estimation (MLE) method. After dropping the constant terms of the log-likelihood function, under the GP model in (5), the MLE method aims to minimize,

$$\log|\Phi| + (\mathbf{y} - \mu\mathbb{1}_n)^T\Phi^{-1}(\mathbf{y} - \mu\mathbb{1}_n), \tag{9}$$

with $\mathbb{1}_n$ being a column of n ones, $\sigma^2 = (\sigma_0^2, ..., \sigma_q^2)^T$ and $\Theta = (\theta^{(0)}, \Theta^{(1)}, ..., \Theta^{(q)})$ where $\theta^{(0)} = (\theta_k^{(0)})_{p\times 1}, \Theta^{(h)} = (\theta_{kl_h}^{(h)})_{p\times m_h}, \mathbf{y} = (y_1, ..., y_n)^T$ and $\Phi = (\phi(\boldsymbol{w}_i, \boldsymbol{w}_j))_{n\times n}$ with the covariance given in (8). For a given σ^2 and Θ, the maximum likelihood estimate of μ is,

$$\hat{\mu} = (\mathbb{1}_n^T\Phi^{-1}\mathbb{1}_n)^{-1}\mathbb{1}_n^T\Phi^{-1}\mathbf{y}, \tag{10}$$

and σ^2 and Θ can be obtained from,

$$\{\sigma^2, \Theta\} = \underset{\sigma^2, \Theta}{\mathrm{argmin}}\{\log|\Phi| + \mathbf{y}^T\Phi^{-1}\mathbf{y} - (\mathbb{1}_n^T\Phi^{-1}\mathbb{1}_n)^{-1}(\mathbb{1}_n\Phi^{-1}\mathbf{y})^2\},$$

using a minimization algorithm, such as rgenoud in R (Mebane et al., 2011). Denote $Y_* = Y(\boldsymbol{w}^*)$ as the prediction of Y at a new input $\boldsymbol{w}^*$. Given $\hat{\mu}$, σ^2 and Θ, the predictive mean and the predictive variance at a new location $\boldsymbol{w}^*$ are given by,

$$\begin{aligned}E(Y_*|\mathbf{y}) &= \mu_* = \hat{\mu} + \boldsymbol{r}_0^T\Phi^{-1}(\mathbf{y} - \hat{\mu}\mathbb{1}_n), \\ \mathrm{Var}(Y_*|\mathbf{y}) &= \sigma_*^2 = \sum_{i=0}^{q} \sigma_i^2 - \boldsymbol{r}_0^T\Phi^{-1}\boldsymbol{r}_0 + \frac{(1 - \mathbb{1}_n^T\Phi^{-1}\boldsymbol{r}_0)^2}{\mathbb{1}_n^T\Phi^{-1}\mathbb{1}_n},\end{aligned} \tag{11}$$

where $\boldsymbol{r}_0$ is the covariance vector of $\phi(\boldsymbol{w}^*, \boldsymbol{w}_i)_{n\times 1}$ for $i = 1, ..., n$.

2.2 Adaptive designs in computer experiments with quantitative inputs

This section examines the criteria for adaptive designs in computer experiments with quantitative inputs, focusing on their applications in optimization, contour estimation, and prediction. Before delving into the specifics of these criteria, we first outline the general steps of adaptive design, as summarized in Algorithm 1.

Adaptive designs start with an initial design with n_0 runs. Then based on a fitted model we search for the next input by maximizing a particular criterion $C(\boldsymbol{x})$ depending on the goal of the study that will be discussed in this section. The stopping criterion can be the budget sample size or cost, or when there is little improvement in the prediction accuracy.

Algorithm 1. Adaptive Designs/Active Learning.

Require: a function $f(\cdot)$ providing output $y = f(\boldsymbol{x})$ for input $\boldsymbol{x}$, the initial design of size n_0, the budget sample size N, and a criterion $C(\boldsymbol{x})$ to search for the next point.
Run a small initial design $\boldsymbol{X}_{n_0} = (\boldsymbol{x}_1^\top, \ldots, \boldsymbol{x}_{n_0}^\top)$. Obtain $\boldsymbol{D}_{n_0} = (\boldsymbol{X}_{n_0}, \boldsymbol{Y}_{n_0})$ where $\boldsymbol{Y}_{n_0} = (f(\boldsymbol{x}_1), \ldots, f(\boldsymbol{x}_{n_0}))$. Let $n = n_0$.
while $n < N$ **do**
  Obtain the surrogate model using $\boldsymbol{D}_n$.
  Solve the criterion $C(\boldsymbol{x})$ based on the fitted model for the choice of $\boldsymbol{x}_{n+1}|\boldsymbol{D}_n$: $\boldsymbol{x}_{n+1} = \operatorname{argmax}_{\boldsymbol{x}\in\mathcal{X}} C(\boldsymbol{x})|\boldsymbol{D}_n$.
  Observe the response at the chosen location by $y_{n+1} = f(\boldsymbol{x}_{n+1})$.
  Update $\boldsymbol{D}_{n+1} = \boldsymbol{D}_n \bigcup (\boldsymbol{x}_{n+1}, y_{n+1})$, and set $n \leftarrow n + 1$.
end while
Return the chosen design and the fitted model based on $\boldsymbol{D}_n$.

Note that Algorithm 1 is general in the sense that it works for models with only quantitative inputs, models with both types of inputs, and for the different goals of analysis including optimization, contour estimation and prediction and so on.

2.2.1 Optimization

The optimization problem in adaptive designs involves identifying the input values that maximize or minimize the computer simulator. Commonly used criteria in optimization include the expected improvement (EI) criterion proposed by Jones et al. (1998) and the upper confidence bound (UCB) or lower confidence bound (LCB) criteria introduced by Srinivas et al. (2010).

The EI criterion, originally introduced by Jones et al. (1998), is designed for minimizing black-box functions. The criterion measures the potential improvement of a solution over the current best result by accounting for both the predicted mean and the uncertainty provided by a surrogate model, such as a GP. Let $f_{\min}^n = \min\{y_1, ..., y_n\}$, then improvement over $f_{\min}^n$ at an input location $\boldsymbol{x}$ can be defined as,

$$I_o(\boldsymbol{x}) = \max\{0, f_{\min}^n - y(\boldsymbol{x})\}, \tag{12}$$

where $y(\boldsymbol{x}) \sim N(\hat{\mu}(\boldsymbol{x}), \hat{\sigma}^2(\boldsymbol{x}))$. Note that the improvement function in Jones et al. (1998) is denoted by $I(\boldsymbol{x})$ but we use $I_o(\boldsymbol{x})$ instead to distinguish the improvement function in later sections. The expected value of the improvement function in (12) is,

$$E[I_o(\boldsymbol{x})] = \left(f_{\min}^n - \hat{\mu}(\boldsymbol{x})\right)\Phi\left(\frac{f_{\min}^n - \hat{\mu}(\boldsymbol{x})}{\hat{\sigma}(\boldsymbol{x})}\right) + \hat{\sigma}(\boldsymbol{x})\phi\left(\frac{f_{\min}^n - \hat{\mu}(\boldsymbol{x})}{\hat{\sigma}(\boldsymbol{x})}\right), \tag{13}$$

where $\hat{\mu}(\boldsymbol{x})$ and $\hat{\sigma}(\boldsymbol{x})$ are the predictive mean, the predictive standard deviation given in (4), Φ and ϕ are the cumulative distribution function (CDF) and the probability density function (PDF) respectively. Let χ denote the design space throughout, the goal is to maximize (13) such that,

$$\boldsymbol{x}_{n+1} = \underset{x \in \chi}{\operatorname{argmax}}\, E\left[I_o(\boldsymbol{x})\right]. \tag{14}$$

On the other hand, to balance the exploration and exploitation, Srinivas et al (2010) discussed two criteria, one for maximization where the UCB criterion selects the next candidate point to be,

$$\boldsymbol{x}_{n+1} = \underset{x \in \chi}{\operatorname{argmax}} \{\hat{\mu}(\boldsymbol{x}) + \rho\hat{\sigma}(\boldsymbol{x})\},$$

and the other for minimization that the LCB criterion selects the next point to be,

$$\boldsymbol{x}_{n+1} = \underset{x \in \chi}{\operatorname{argmin}} \{\hat{\mu}(\boldsymbol{x}) - \rho\hat{\sigma}(\boldsymbol{x})\}. \tag{15}$$

where $\rho \geq 0$ is tuning parameter, $\hat{\mu}(\boldsymbol{x})$ and $\hat{\sigma}(\boldsymbol{x})$ are the predictive mean, the predictive standard deviation given in (4).

2.2.2 Contour estimation

The contour estimation problem involves identifying the inputs that correspond to a specified contour level of interest. Suppose a is the contour level of interest and the contour is defined to be $S(a) = \{\boldsymbol{x} \in \chi: y(\boldsymbol{x}) = a\}$. For computer experiments with quantitative inputs, Ranjan et al. (2008) introduced the EI criterion with the improvement function defined as,

$$I(\boldsymbol{x}) = \epsilon^2 - \min\{(y(\boldsymbol{x}) - a)^2, \epsilon^2\}, \tag{16}$$

where $\epsilon = \alpha\hat{\sigma}(\boldsymbol{x})$ for $\alpha > 0$, and $y(\boldsymbol{x}) \sim N(\hat{\mu}(\boldsymbol{x}), \hat{\sigma}^2(\boldsymbol{x}))$ with the predictive mean $\hat{\mu}(\boldsymbol{x})$ and the predictive standard deviation $\hat{\sigma}(\boldsymbol{x})$ given in (4). Let $u_1 = (a - \hat{\mu}(\boldsymbol{x}) - \epsilon)/\hat{\sigma}(\boldsymbol{x})$ and $u_2 = (a - \hat{\mu}(\boldsymbol{x}) + \epsilon)/\hat{\sigma}(\boldsymbol{x})$. Then expected value of the improvement function in (16) is,

$$\begin{aligned} E[I(\boldsymbol{x})] = {} & [\epsilon^2 - (\hat{\mu}(\boldsymbol{x}) - a)^2 - \hat{\sigma}^2(\boldsymbol{x})](\Phi(u_2) - \Phi(u_1)) + \\ & \hat{\sigma}^2(\boldsymbol{x})(u_2\phi(u_2) - u_1\phi(u_1)) + 2(\hat{\mu}(\boldsymbol{x}) - a)\hat{\sigma}(\boldsymbol{x})(\phi(u_2) - \phi(u_1)), \end{aligned} \tag{17}$$

where Φ and ϕ are the CDF and the PDF of the standard normal distribution, respectively. The EI criterion in (17) chooses the next input,

$$\boldsymbol{x}_{n+1} = \underset{x \in \chi}{\operatorname{argmax}}\, E[I(\boldsymbol{x})].$$

Cole et al. (2023) studied an entropy-based adaptive design for contour estimation. In their study, the problem depends on the definition of the failure region which is characterized as an output Y exceeding a particular contour level a. That is, for a limit state function $g(Y) = Y - a$, such that,

$$\begin{aligned} \mathbb{G} &= \{\boldsymbol{x} \in \chi: g(Y(\boldsymbol{x})) > 0\} \text{ implying contour} \\ \mathbb{C} &= \{\boldsymbol{x} \in \chi: g(Y(\boldsymbol{x})) = 0\}, \end{aligned} \tag{18}$$

where $\mathbb{G}$ is the failure region and $\mathbb{C}$ is the contour of interest. They introduced the entropy-based contour locator (ECL) criterion,

$$\begin{aligned} \text{ECL}(\boldsymbol{x}|g) &= -P(g(Y(\boldsymbol{x})) > 0) \log P(g(Y(\boldsymbol{x})) > 0) \\ &\quad - P(g(Y(\boldsymbol{x})) \leq 0) \log P(g(Y(\boldsymbol{x})) \leq 0) \\ &= -\left(1 - \Phi\left(\frac{\hat{\mu}(\boldsymbol{x}) - a}{\hat{\sigma}(\boldsymbol{x})}\right)\right) \log\left(1 - \Phi\left(\frac{\hat{\mu}(\boldsymbol{x}) - a}{\hat{\sigma}(\boldsymbol{x})}\right)\right) \\ &\quad - \Phi\left(\frac{\hat{\mu}(\boldsymbol{x}) - a}{\hat{\sigma}(\boldsymbol{x})}\right) \log\left(\Phi\left(\frac{\hat{\mu}(\boldsymbol{x}) - a}{\hat{\sigma}(\boldsymbol{x})}\right)\right), \end{aligned} \tag{19}$$

where Φ is the CDF of the standard normal distribution, for any limit state function $g(Y) = b(Y - a)$ where $b \in \{-1, 1\}$. It shall be noted because of symmetry in (19), the criterion works identically for $g(Y) = b(Y - a)$ where $b \in \{-1, 1\}$, meaning that the criterion is unchanged for $b = -1$ or $b = 1$. The ECL criterion chooses the next input as,

$$\boldsymbol{x}_{n+1} = \underset{x \in \chi}{\operatorname{argmax}} \, \text{ECL}(\boldsymbol{x}|g).$$

2.2.3 Prediction

Prediction aims to accurately estimate the model output and ensure reliable prediction. While several adaptive design approaches exist for enhancing prediction, this section focuses on the methods employed in our numerical studies. We review the EI multiple contours (EI-MC) criterion and the EI single contour (EI-SC) criterion proposed by Yang et al. (2020), which have been developed as adaptive design strategies for prediction.

Yang et al. (2020) consider the improvement function in (16) for a given integer $c > 0$ and a set of scalar $a_1, ..., a_c \in [y_{\min}, y_{\max}]$. Suppose we are interested in estimating $S(a_1), ..., S(a_c)$, where $a_1, ..., a_c \in [y_{\min}, y_{\max}]$. The values of $y_{\min}$ and $y_{\max}$ are unknown contours in general but it can be estimated by a large input from the design domain. They suggest, $1000p$ design points. Note that $a_1 < ... < a_c$ then the improvement function in (16) is adjusted such that,

$$I_c(\boldsymbol{x}) = \epsilon^2 - \min\{(y(\boldsymbol{x}) - a_1)^2, ..., (y(\boldsymbol{x}) - a_c)^2, \epsilon^2\}, \tag{20}$$

where $y(\boldsymbol{x}) \sim N(\hat{\mu}(\boldsymbol{x}), \hat{\sigma}^2(\boldsymbol{x}))$, $\epsilon = \alpha\hat{\sigma}(\boldsymbol{x})$ for some positive constant α. The improvement function will be nonzero if $(y(\boldsymbol{x}) - a_j)^2 < \epsilon^2$ for some j.

Therefore the expectation of the improvement function in (20) is sum of the individual contour estimation EI criterion in (17) over c cases given by,

$$E[I_c(\boldsymbol{x})] = \sum_{j=1}^{c} [\epsilon^2 - (\hat{\mu}(\boldsymbol{x}) - a_j)^2 - \hat{\sigma}^2(\boldsymbol{x})](\Phi(u_2) - \Phi(u_1)) + \\ \hat{\sigma}^2(\boldsymbol{x})(u_2\phi(u_2) - u_1\phi(u_1)) + 2(\hat{\mu}(\boldsymbol{x}) - a_j)\hat{\sigma}(\boldsymbol{x})(\phi(u_2) - \phi(u_1)), \tag{21}$$

where ϕ and Φ are the PDF and the CDF of the standard normal distribution, respectively. The EI-MC criterion chooses the next input,

$$\boldsymbol{x}_{n+1} = \underset{x \in \chi}{\operatorname{argmax}} E[I_c(\boldsymbol{x})].$$

Yang et al. (2020) considered an other way for EI-SC, that they use the EI function in (16). They consider an automatic approach to select a_j in each adaptive design. The contour level is selected from the candidate points to select the next input. Let the candidate points for the next input be $\boldsymbol{x}_1^*, \ldots, \boldsymbol{x}_m^*$ and,

$$\boldsymbol{x}_{opt}^* = \underset{1<i<m}{\operatorname{argmax}} \sigma^2(\boldsymbol{x}_i^*).$$

They set $a_j = \hat{y}(\boldsymbol{x}_{opt}^*)$ and use the EI criterion in (17) given a_j in each adaptive design.

3 Active designs for computer experiments with quantitative and qualitative inputs

This section examines adaptive designs for computer experiments involving both quantitative and qualitative inputs, focusing on the goals of optimization and contour estimation. Incorporating qualitative inputs introduces several unique challenges. First, while criteria in experiments with quantitative inputs often depend solely on the predictive mean and variance, the influence of qualitative inputs on these criteria becomes difficult to quantify. Second, optimizing the criteria is more complex because the search space encompasses both continuous and discrete regions. These challenges necessitate innovative methodologies for adaptive design in experiments with mixed inputs. The following subsections review recent advancements in adaptive designs for optimization and contour estimation in such contexts.

3.1 Optimization

Cai et al. (2024) and Luo et al. (2024) have explored optimization methods for computer experiments with mixed inputs. In this subsection, we review their approaches and provide a numerical example to illustrate their methodologies.

Cai et al. (2024) introduced the *adaptive-region sequential design* (ARSD) approach for addressing optimization problems in computer experiments with mixed inputs. The ARSD method selects the next candidate point to be

$$\boldsymbol{w}_{n+1} = \underset{\boldsymbol{w} \in \mathbb{A}_n}{\operatorname{argmin}} \{\hat{\mu}(\boldsymbol{w}) - \rho\hat{\sigma}(\boldsymbol{w})\}, \tag{22}$$

where $\mathbb{A}_n \subset \mathbb{A}$ is the adaptive design region as,

$$\mathbb{A}_n = \{w \in \mathbb{A}: \hat{\mu}(w) - \sqrt{\beta_{0|n}}\hat{\sigma}(w) \leq \min [\hat{\mu}(w) + \sqrt{\beta_{0|n}}\hat{\sigma}(w)]\}, \tag{23}$$

$\rho \geq 0$ is a tuning parameter, $\mathbb{A} = \{(x, z)|x \in \chi, z \in Z\}$ is the whole design space and Z is all level combinations of the qualitative inputs, $\hat{\mu}(w)$ and $\hat{\sigma}(w)$ are the predicted mean and the predicted standard deviation, $\beta_{0|n} = 2\log\left(\frac{\pi^2 n^2 M}{6\alpha}\right)$, $\alpha \in (0,1)$ and $M = \prod_{j=1}^{q} m_j$, and m_j is the number of level of the jth qualitative input variable. The criterion in (22) is essentially a LCB criterion tailored for minimization problems, combined with an adaptive search region defined in (23). This adaptive region includes inputs whose predictive lower bounds are smaller than the minimum predictive upper bound across the entire design space. By narrowing the search region, this approach significantly reduces computational time. Moreover, the efficiency loss is minimal, as Cai et al. (2024) provides a theoretical guarantee that the solution to the minimization problem lies within the adaptive search region $\mathbb{A}_n$ with a high probability.

Luo et al. (2024) introduces a novel sequential approach for selecting inputs to optimize a black-box function with mixed variables. While not specifically designed for computer experiments with mixed inputs, their method can effectively address optimization problems in such settings. The approach is described as *Hybrid* because it combines Monte Carlo tree search (MCTS) (Kocsis and Szepesvári, 2006) for exploring qualitative inputs with GPs for modeling quantitative inputs. A key contribution of their work is the introduction of a family of candidate kernels for mixed variables, along with a numerically stable and efficient dynamic kernel selection criterion. This criterion identifies the optimal mixed-variable GP surrogate at each stage, which is then used to determine the next set of quantitative inputs.

To outline the methodology, we begin with a discussion of MCTS. MCTS incrementally constructs a search tree through an iterative process that has four key steps: selection, expansion, simulation, and back-propagation.

- Selection: The algorithm navigates the tree using a policy that balances exploration and exploitation, often guided by criteria such as the Upper Confidence Bound for Trees (UCT) introduced by Auer (2002).
- Expansion: New nodes representing previously unexplored states are added to the tree.
- Simulation: The newly expanded nodes are evaluated using a reward function, defined in accordance with the specific objectives of the study.
- Back-propagation: The results from the simulation are propagated back up the tree, updating value estimates for all nodes visited during the iteration.

In their study, MCTS is implemented as a one-layer tree with leaves corresponding to level combinations of qualitative inputs. A path along a tree with L

layers is represented as a string $\mathbf{s} = (s_1, ..., s_L)$ of length L, where each $\mathbf{s}$ denotes a node on the path. During each search, the algorithm conditions on the path $(s_1, ..., s_i)$, C and computes a reward function value. This information is back-propagated through the tree using update strategies to guide subsequent decisions in the adaptive design process. The update strategy follows the UCT principle:

$$\bar{r}(s_1, ..., s_i) + C\sqrt{\frac{\log(n(s_1, ..., s_{i-1}))}{n(s_1, ..., s_i)}}, \tag{24}$$

where $\bar{r}(s_1, ..., s_i)$ is the average reward from all evaluations along the path $(s_1, ..., s_i)$, C is a pre-specified constant controlling the exploration-exploitation trade-off, and $n(\mathbf{s})$ represents the number of times the path $\mathbf{s}$ has been visited. New qualitative variable values are then determined by maximizing the criterion in (24). Once the new qualitative values have been chosen, they follow an update strategy, by updating $n(\mathbf{s})$ and $\bar{r}(\mathbf{s})$ where $\mathbf{s}$ denotes a vector like $(s_1, ..., s_i)$, and the full path $(s_1, ..., s_L)$ corresponds to a qualitative variable value and propagates the updated value along the complete path for each non-leaf node. The updated value is propagated along the entire path for each non-leaf node. Once a new sample location is evaluated, the reward and visit count for all nodes along the tree path leading to the sample are updated. For each node s_i in the path, the visit count $n(s_1, ..., s_i)$ is increased by 1, and the average reward $\bar{r}(s_1, ..., s_i)$ is also updated to reflect the new sample's reward.

After choosing the new qualitative variable values, they dynamically select the mixed-variable GP surrogate. At each adaptive design, a kernel $\boldsymbol{K}$ is selected from a pool of candidates, including additive and multiplicative kernels. The selection is guided by two metrics, one is the log marginal likelihood, which quantifies the kernel's fit to the observed data, and the other is the acquisition function value, which represents the potential improvement. Each kernel is ranked by its log-likelihood $R_P(\boldsymbol{K})$ and acquisition function $R_A(\boldsymbol{K})$, and the combined rank is computed as,

$$R(\boldsymbol{K}) = R_P(\boldsymbol{K}) + \alpha R_A(\boldsymbol{K}), \tag{25}$$

where $\alpha \in [0,1]$ controls the balance between exploitation $R_P(\boldsymbol{K})$ and exploration $R_A(\boldsymbol{K})$. By default, $\alpha = 0.5$, or α can adapt over iterations as, $\alpha = \frac{2 \cdot i}{N}$, where i is the current iteration, and N is the total sampling budget. The kernel with the highest $R(\boldsymbol{k})$ is selected and used to update the GP surrogate. Now given the input and the selected surrogate, they use the EI criterion in (14) to identify the next quantitative variable values. For rest of the work we call this method as *Hybrid*.

To illustrate the methodology of each adaptive design for optimization, we consider a minimization problem in the following computer experiment with one qualitative input variable and one quantitative input variable. We examine the performance of adaptive designs approaches using the EI in (14), LCB in (15), ARSD in (22) criteria and Hybrid method by Luo et al. (2024) respectively.

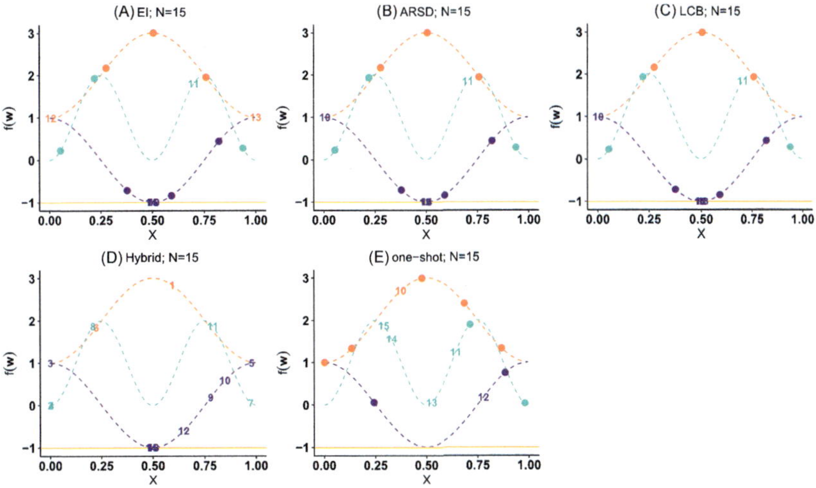

FIG. 1 Adaptive designs versus one-shot designs for Example 1 for (A) EI (B) LCB (C) ARSD (D) Hybrid (E) one-shot for $n_0 = 9$; $N = 15$ for one simulation.

Example 1 *Consider a computer experiment with* $p = 1$ *quantitative input* x *and* $q = 1$ *qualitative input* z *and the computer model is represented by,*

$$f = \begin{cases} 2 - \cos(2\pi x), & \text{if } z = 1 \\ 1 - \cos(4\pi x), & \text{if } z = 2 \\ \cos(2\pi x), & \text{if } z = 3. \end{cases} \tag{26}$$

Here -1 *is the minimum of the response surface and* 3 *is the maximum of the response surface. The solution that minimizes the computer model is* $\mathbf{w} = (0.5, 3)$.

Fig. 1 illustrates the adaptive designs using EI, ARSD, LCB, and Hybrid, along with the one-shot design, for $n_0 = 9$ *and* $N = 15$ *in one simulation to find the minimum of the responses given by the computer model. The dashed lines show the underlying true response of (26) and the solid orange line indicates the minimum of the true responses is −1. The initial points are presented as circled dots in each level, corresponding to their colors, while the added points are labeled numerically (Fig. 1D) starts with a single initial point, with the remaining* $N = 15$ *points adaptively added and labeled numerically using the Hybrid method. As shown in Fig. 1A, B, C and D, the added points in the adaptive designs are mostly selected at level 3 and are located close to the target input* $\mathbf{w} = (0.5, 3)$. *In contrast, as displayed in Fig. 1E, the minimum of the responses given by the one-shot design is given by* $\mathbf{w} = (0.25, 3)$ *so that the corresponding minimum is close to 0.*

3.2 Contour estimation

This subsection discusses adaptive design approaches for contour estimation in computer experiments involving both quantitative and qualitative input variables. To the best of our knowledge, Shahrokhian et al. (2024) is the only work that considers adaptive designs for the contour estimation in computer experiments with both quantitative and qualitative inputs. In this subsection, we review the method proposed by Shahrokhian et al. (2024). Let a denote the contour level of interest; the objective is to identify the contour $\{\boldsymbol{w}: y(\boldsymbol{w}) = a\}$. Let the lower bound (LB) and the upper bound (UB) be given by,

$$\begin{aligned}\mu_h^L(\boldsymbol{w}) &= |\hat{\mu}(\boldsymbol{w}) - a| - \sqrt{\beta_{0|n}}\,\hat{\sigma}(\boldsymbol{w}),\\ \mu_h^U(\boldsymbol{w}) &= |\hat{\mu}(\boldsymbol{w}) - a| + \sqrt{\beta_{0|n}}\,\hat{\sigma}(\boldsymbol{w}),\end{aligned} \tag{27}$$

where $\hat{\mu}(\boldsymbol{w})$ is the predictive mean, $\hat{\sigma}(\boldsymbol{w})$ is the predictive standard deviation defined in (11), $\beta_{0|n} = 2\log\left(\frac{\pi^2 n^2 M}{6\alpha}\right)$, $\alpha \in (0,1)$, $M = \prod_{j=1}^{q} m_j$ and m_j is the number of levels of the jth qualitative input variable. Note that, the choice of $\beta_{0|n}$ is from Cai et al. (2024). Let $\mathbb{A}$ be the whole design space, Shahrokhian et al. (2024) proposed to divide $\mathbb{A}$ into two disjoint groups,

$$\mathbb{A}_1 = \{w \in \mathbb{A}: |\hat{\mu}(w) - a| - \sqrt{\beta_{0|n}}\,\hat{\sigma}(w) > 0\}, \tag{28}$$

$$\mathbb{A}_2 = \left\{w \in \mathbb{A}: |\hat{\mu}(w) - a| - \sqrt{\beta_{0|n}}\,\hat{\sigma}(w) \le 0\right\}. \tag{29}$$

Clearly, $\mathbb{A} = \cup_{i=1}^{2}\mathbb{A}_i$. For $\mathbb{A}_1$ in (28), consider the candidate point given by,

$$w_{n+1} = \underset{w \in \mathbb{A}_{1,\min}}{\operatorname{argmax}}\, \hat{\sigma}(w), \tag{30}$$

where

$$\begin{aligned}\mathbb{A}_{1,\min} = \{w \in \mathbb{A}_1: |\hat{\mu}(w) - a| - \sqrt{\beta_{0|n}}\,\hat{\sigma}(w) \le \\ \min_{w \in \mathbb{A}}[|\hat{\mu}(w) - a| + \sqrt{\beta_{0|n}}\,\hat{\sigma}(w)]\}.\end{aligned} \tag{31}$$

For $\mathbb{A}_2$ in (29), let the candidate point be,

$$w_{n+1} = \underset{w \in \mathbb{A}_2}{\operatorname{argmax}}\, \mathrm{ECL}[w|g], \tag{32}$$

with the ECL criterion in (19).

Among the two candidate points selected from groups $\mathbb{A}_1$ and $\mathbb{A}_2$, Shahrokhian et al. (2024) select the next input with a larger value of $\frac{\hat{\sigma}(w)}{\max(\delta, |\hat{\mu}(w) - a|)}$, where δ is a small positive value. (Shahrokhian et al., 2024) highlighted that the choice of δ plays a critical role in balancing exploration and exploitation. A very small δ strongly favors exploitation, selecting inputs primarily, if not exclusively, from $\mathbb{A}_2$. Conversely, a very large δ shifts the focus entirely to exploration, effectively relying only on the predictive variance and resulting in excessive input selection from $\mathbb{A}_1$. To achieve a balance between exploration and exploitation, δ should be neither too large nor too small, with its optimal value depending on the range of responses. In the numerical examples, the recommended value suggested by Shahrokhian et al. (2024) will be used. In general, it is desirable to select points whose predictive mean is closer to the contour level a while having a larger predictive variance. However, this approach tends to empirically favor selection from a single group after several steps in the sequential procedure. To address this, the denominator $\max(\delta, |\hat{\mu}(w) - a|)$ is introduced to encourage exploration of the design space. This adjustment ensures that when the predictive means of candidate points are close to the contour level, the point with the larger predictive variance is chosen.

Algorithm 2 given in Shahrokhian et al. (2024) describes the adaptive design approach for finding follow-up points for contour estimation. Their algorithm applies the idea of the adaptive search region in the ARSD approach in the design space $\mathbb{A}_1$, and the ECL criterion in the design space $\mathbb{A}_2$. They named their approach *region-based cooperative criterion* (RCC).

Algorithm 2. Region-based Cooperative Contour Estimation.

Run a small initial design $\boldsymbol{W}_{n_0} = (\boldsymbol{w}_1^T, \ldots, \boldsymbol{w}_{n_0}^T)$. Obtain $\boldsymbol{D}_{n_0} = (\boldsymbol{W}_{n_0}, \boldsymbol{Y}_{n_0})$ where $\boldsymbol{Y}_{n_0} = (f(\boldsymbol{w}_1), \ldots, f(\boldsymbol{w}_{n_0}))$. Let $n = n_0$.
while $n < N$ **do**
  Fit the EzGP model using $\boldsymbol{D}_n$.
  Divide the search space into the two groups $\mathbb{A}_1$ and $\mathbb{A}_2$.
  For $\mathbb{A}_1$ defined in (28), find the input that minimizes the criterion in (30).
  For $\mathbb{A}_2$ defined in (29), find the input that maximizes the criterion in (32).
  Choose the input from the two chosen points that $\frac{\hat{\sigma}(\boldsymbol{w})}{\max(\delta, |\hat{\mu}(\boldsymbol{w}) - a|)}$ is larger.
  Run the experiment at the chosen location and let $y_{n+1} = f(\boldsymbol{w}_{n+1})$.
  Update $\boldsymbol{D}_{n+1} = \boldsymbol{D}_n \bigcup (\boldsymbol{w}_{n+1}, y_{n+1})$, and set $n \leftarrow n + 1$.
end while
Return Extract the estimated contour for $\hat{S} = \{\boldsymbol{w} : \hat{y}(\boldsymbol{w}) = a\}$ from the estimated response surface.

Although the LCB or UCB criteria in (15) and the ARSD criterion in (22) are not extended for contour estimation, they can be used. The LCB criterion for contour estimation (LCB-C) chooses the next input as,

$$w_{n+1} = \underset{w \in \mathbb{A}}{\operatorname{argmin}} \{ |\hat{\mu}(w) - a| - \rho \hat{\sigma}(w) \},$$

where ρ is a tuning parameter and $\mathbb{A}$ is the whole design space. Note that the LCB criterion was proposed for the problem of global optimization, but we can apply it to the contour estimation problem by using $|\hat{\mu}(w) - a|$ rather than $\hat{\mu}$.

Similarly, the ARSD approach in (22) for optimization can also be modified for contour estimation. That is, we can choose the next input to be,

$$w_{n+1} = \underset{w \in \mathbb{A}^*}{\operatorname{argmin}} \{ |\hat{\mu}(w) - a| - \rho \hat{\sigma}(w) \},$$

where ρ is a tuning parameter, and

$$\mathbb{A}^* = \{ w \in \mathbb{A} : |\hat{\mu}(w) - a| - \sqrt{\beta_{0In}} \hat{\sigma}(w) \leq \min_{w \in \mathbb{A}} [|\hat{\mu}(w) - a| + \sqrt{\beta_{0In}} \hat{\sigma}(w)] \}.$$

We denote this approach as ARSD for contour estimation (ARSD-C).

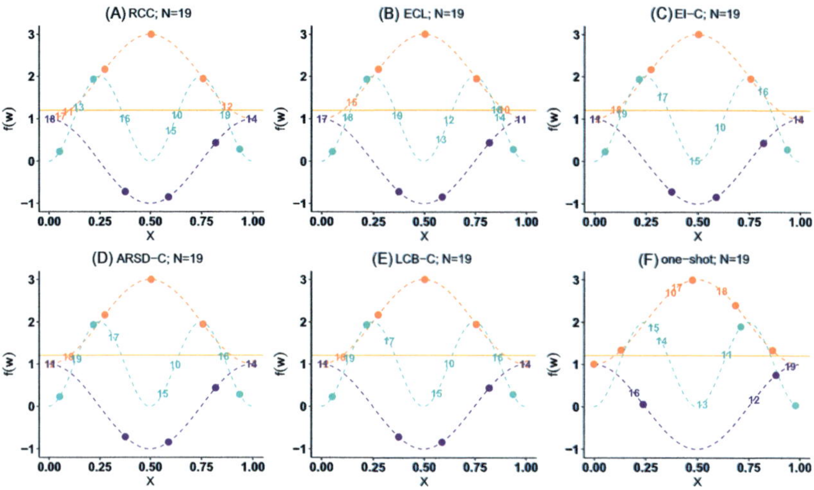

FIG. 2 Adaptive designs versus one-shot designs for Example 1 for (A) RCC, (B) ECL, (C) EI-C, (D) ARSD-C (E) LCB-C, (F) one-shot for $n_0 = 9$; $N = 19$ for one simulation when $a = 1.2$.

Example 1 continued: *Similar to the adaptive designs for optimization discussed in Section 3.1, this section illustrates various adaptive designs for contour estimation and compares them with one-shot designs. Fig. 2 presents the adaptive designs RCC, ECL, EI-C, ARSD-C, and LCB-C, alongside the one-shot designs, for* $n_0 = 9$ *and* $N{=}19$ *in a simulation where* $a = 1.2$*. In the figure, the initial points are shown as circled dots, grouped according to their levels, while the follow-up points are labeled sequentially from 10 to 19. As seen in Fig. 2A, B, C, D and E, the adaptive designs primarily allocate additional points near the orange solid line, effectively targeting the contour corresponding to* $a = 1.2$*. In contrast, the one-shot design in Fig. 2F distributes several points at level 3, many of which do not correspond to the desired contour level* $a = 1.2$*.*

4 Numerical studies

This section presents numerical examples to illustrate various adaptive designs for optimization, contour estimation, and prediction in computer experiments involving both quantitative and qualitative inputs. For all adaptive designs, the process begins with an initial design with n_0 runs. The next input is selected by evaluating a large set of candidate points in the design space. These candidate points include all possible combinations of the qualitative input levels, along with 100 random Latin hypercube designs (LHDs) (McKay et al., 1979) generated for each level combination in the design space. In addition, one-shot designs are constructed by combining random LHDs for the quantitative variables with a (nearly) balanced sample across all level combinations of the qualitative factors. The random LHDs are generated using the randomLHS function from the R package lhs (Carnell, 2022). To model the computer experiment with mixed inputs, we use the EzGP package in R (Li et al., 2023).

Section 4.1 presents the simulation studies for contour estimation, while Section 4.2 focuses on optimization, and Section 4.3 explores the approaches for prediction. For each problem, we examine Example 1 presented in Section 3, Example 2 and Example 3 as detailed below. All examples and implementations may be found on the Github repository https://github.com/devonlin/Handbook-of-Statistics-Chapter.

Example 2.

We consider a computer model with $p = 2$ quantitative inputs $\boldsymbol{x} = (x_1, x_2)$ and $q = 2$ qualitative input $z = (z_1, z_2)$. The computer model is adapted from Xiao et al. (2021) with reduced dimensions and is represented by,

$$i(z_1) = \begin{cases} x_1 + x_2^2, & \text{if } z_1 = 1 \\ x_1^2 + x_2, & \text{if } z_1 = 2 \\ x_1^2 + x_2^2, & \text{if } z_1 = 3, \end{cases} \quad g(z_2)$$

$$= \begin{cases} \cos(x_1) + \cos(2x_2), & \text{if } z_2 = 1 \\ \cos(2x_1) + \cos(x_2), & \text{if } z_2 = 2 \\ \cos(2x_1) + \cos(2x_2), & \text{if } z_2 = 3, \end{cases}$$

$$f = i(z_1) + g(z_2).$$

Here 1 is the minimum of the response surface and 2.7 is the maximum of the response surface.

Example 3.

We consider a computer model with $p = 3$ quantitative inputs $\boldsymbol{x} = (x_1, x_2, x_3)$ and $q = 3$ qualitative input $z = (z_1, z_2, z_3)$ and the computer model Xiao et al. (2021) is represented by,

$$i(z_1) = \begin{cases} x_1 + x_2^2 + x_3, & \text{if } z_1 = 1 \\ x_1^2 + x_2 + x_3, & \text{if } z_1 = 2 \\ x_3 + x_1 + x_2^2, & \text{if } z_1 = 3, \end{cases}$$

$$g(z_2) = \begin{cases} \cos(x_1) + \cos(2x_2) + \cos(x_3), & \text{if } z_2 = 1 \\ \cos(x_1) + \cos(2x_2) + \cos(x_3), & \text{if } z_2 = 2 \\ \cos(2x_1) + \cos(x_2) + \cos(x_3), & \text{if } z_2 = 3, \end{cases}$$

$$h(z_3) = \begin{cases} \sin(x_1) + \sin(2x_2) + \sin(x_3), & \text{if } z_3 = 1 \\ \sin(x_1) + \sin(2x_2) + \sin(x_3), & \text{if } z_3 = 2 \\ \sin(2x_1) + \sin(x_2) + \sin(x_3), & \text{if } z_3 = 3, \end{cases} \quad f = i(z_1) + g(z_2) + h(z_3).$$

Here 3 is the minimum of the response surface and 6.7 is the maximum of the response surface.

4.1 Optimization

We evaluate the performance of adaptive designs for minimization using the EI, LCB, ARSD, and Hybrid methods discussed in Section 2.1 and Section 3.1, alongside one-shot designs. To assess optimization accuracy, the best minimum value of the true surface obtained through the adaptive design is measured. For the ARSD method described in (22), the tuning parameter $\rho = 2$ is set as specified to ensure effective performance.

For Example 1, the initial sample size is set to $n_0 = 9$, with a total number of points $N = 15$. Fig. 3 illustrates the average of the best minimum values obtained for different budget sizes for 50 simulations, demonstrating that the adaptive designs effectively identify the true minimum of the surface for $N = 15$. Table 1 summarizes the average of the best minimum values achieved using both one-shot and adaptive designs across 50 simulations. The results show that, with the addition of 6 points, adaptive designs successfully approach the true minimum value of -1.

For Example 2, the initial sample size is set to $n_0 = 9$, with a total of $N = 18$ points. Fig. 4 shows the average of the best minimum values obtained across 50 simulations for various budget sizes. The results demonstrate that adaptive designs consistently identify a more accurate true minimum of the surface. Among the adaptive methods, EI converges to the minimum faster than the others, while the Hybrid method takes the longest to converge. Table 2 provides the average of the best minimum values achieved using both one-shot and adaptive designs across 50 simulations. The results indicate that ARSD and LCB methods yield minimum values closest to the true minimum for $N = 18$.

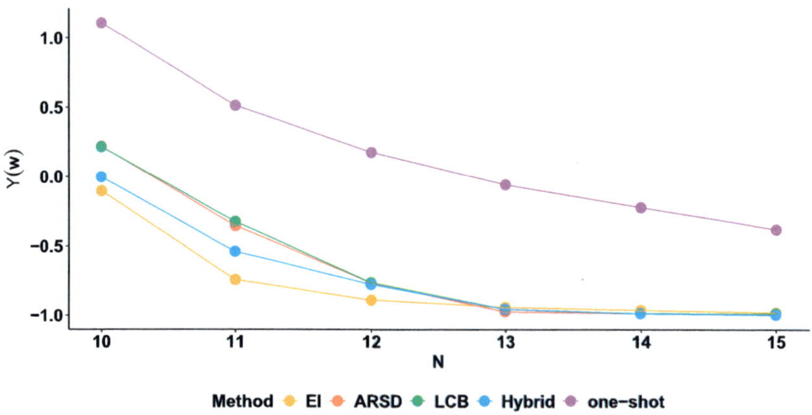

FIG. 3 The average of the best obtained minimum values of the true response surface in Example 1 for the adaptive designs ARSD, EI, LCB, Hybrid and one-shot designs for $n_0 = 9$ and $N = 15$ over 50 simulations.

TABLE 1 The average of the best obtained minimum values of the response in Example 1 for the adaptive designs ARSD, EI, LCB, Hybrid and one-shot designs for $n_0 = 9$ and $N = 15$ over 50 simulations.

N	One-shot	ARSD	EI	LCB	Hybrid
10	1.1092	0.2191	−0.1034	0.2163	−0.0014
11	0.5139	−0.3544	−0.7418	−0.3235	−0.5389
12	0.1751	−0.7654	−0.8898	−0.7639	−0.7766
13	−0.0596	−0.9756	−0.9422	−0.9549	−0.9555
14	−0.2236	−0.9857	−0.9630	−0.9852	−0.9871
15	−0.3805	−0.9910	−0.9794	−0.9863	−0.9962

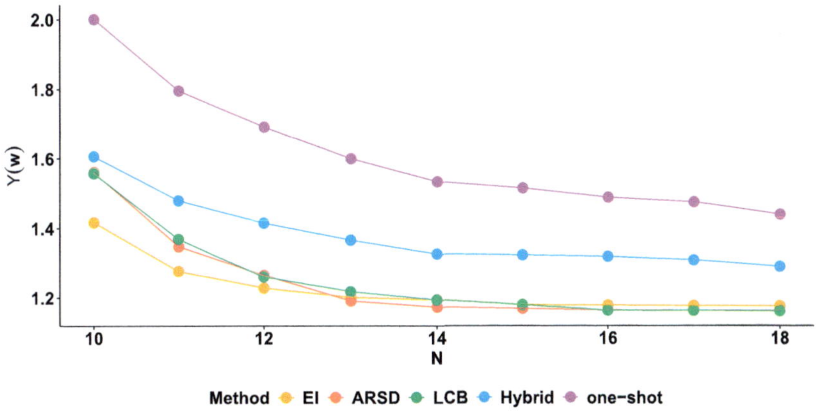

FIG. 4 The boxplot of obtained minimum values of the response in Example 2 for the adaptive designs ARSD, EI, LCB, Hybrid and one-shot designs for $n_0 = 9$ and $N = 18$ over 50 simulations.

For Example 3, the initial sample size is set to $n_0 = 9$, with a total of $N = 18$ points. Fig. 5 illustrates the average of the best minimum values obtained across 50 simulations for various budget sizes. Similar to Example 1 and Example 2, the results show that adaptive designs consistently identify a more accurate true minimum of the surface. In this example, the Hybrid method converges to the minimum faster than the others and outperforms the other adaptive designs. Table 3 provides the average of the best minimum values achieved using both one-shot and adaptive designs across 50 simulations. The results reveal that the Hybrid method successfully achieves a value close to the true minimum for $N = 18$.

TABLE 2 The average of the best obtained minimum values of the response in Example 2 for the adaptive designs ARSD, EI, LCB, Hybrid and one-shot designs for $n_0 = 9$ and $N = 18$ over 50 simulations.

N	One-shot	ARSD	EI	LCB	Hybrid
10	2.0002	1.5607	1.4170	1.5580	1.6067
11	1.7946	1.3474	1.2767	1.3694	1.4797
12	1.6909	1.2652	1.2282	1.2605	1.4153
13	1.5991	1.1905	1.2016	1.2175	1.3663
14	1.5332	1.1730	1.1943	1.1936	1.3262
15	1.5148	1.1685	1.1794	1.1790	1.3229
16	1.4883	1.1637	1.1785	1.1630	1.3184
17	1.4749	1.1622	1.1768	1.1617	1.3079
18	1.4399	1.1621	1.1768	1.1613	1.2897

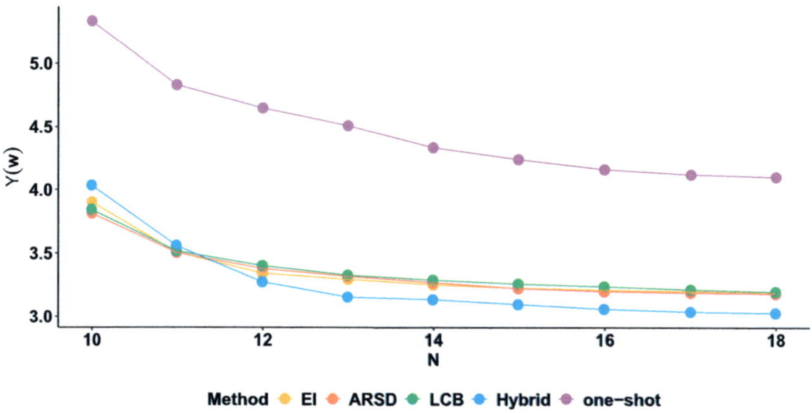

FIG. 5 The boxplot of obtained minimum values of the response in Example 3 for the adaptive designs ARSD, EI, LCB, Hybrid and one-shot designs for $n_0 = 9$ and $N = 18$ over 50 simulations.

4.2 Contour estimation

In contour estimation, we call the EI for contour estimation (EI-C) to make it different from EI for optimization in (14). In our comparison, we consider RCC, EI-C, ECL, ARSD-C, LCB-C and one-shot designs discussed in Section 3.2.

To evaluate and compare the different approaches for contour estimation, we employ the following measurement M_{C_0} used in Shahrokhian et al. (2024), where

$$M_{C_0} = \frac{1}{|C_0|} \sum_{w \in C_0} |Y(w) - \hat{Y}(w)|, \tag{33}$$

TABLE 3 The average of the best obtained minimum values of the response in Example 3 for the adaptive designs ARSD, EI, LCB, Hybrid and one-shot designs for $n_0 = 9$ and $N = 18$ over 50 simulations.

N	One-shot	ARSD	EI	LCB	Hybrid
10	5.3346	3.8132	3.9019	3.8437	4.0355
11	4.8285	3.5031	3.5061	3.5170	3.5603
12	4.6456	3.3773	3.3408	3.4010	3.2724
13	4.5068	3.3159	3.2926	3.3262	3.1529
14	4.3331	3.2652	3.2479	3.2858	3.1320
15	4.2401	3.2190	3.2212	3.2560	3.0945
16	4.1610	3.1957	3.2070	3.2355	3.0583
17	4.1203	3.1847	3.1964	3.2112	3.0359
18	4.1036	3.1800	3.1837	3.1944	3.0273

where $C_0 = \{\boldsymbol{w}: Y(\boldsymbol{w}) = a\}$ is a subset of candidate points that satisfies $a \pm \epsilon$, with ϵ being a small positive value. As remarked by Shahrokhian et al. (2024), the value of ε should be carefully chosen based on the complexity of the response function to ensure that a sufficient number of points are included. To determine the size of the large set of candidate points used in M_{C_0}, we adopt the approach outlined by Shahrokhian et al. (2024). In all examples, to calculate M_{C_0}, we generate 200-run LHDs for the quantitative inputs at each level combination of the qualitative inputs. Note that, we consider all possible combinations of the qualitative input levels. It is important to note that the same dataset is utilized for approximating multiple contours in the EI-MC method described in (21).

For Example 1, the initial sample size is set to $n_0 = 9$, with a total number of points $N = \{11, 13, 15, 17, 19\}$. We examine contour levels $a = \{-0.7, 1.2, 2.2\}$, with $\epsilon = 0.05$ for the M_{C_0} measurement in (33) and $\delta = 0.05$. Table 4 reports the average of the M_{C_0} values, obtained through both one-shot and adaptive designs across 50 simulations, demonstrate that adaptive designs provide more accurate contour estimations compared to one-shot designs. Although no single adaptive design method consistently outperforms the others, the RCC method often proves to be comparable to or superior to alternative approaches. Table 5 presents the average computation time (in seconds) required to add 10 points adaptively. The results reveal that the ARSD-C and RCC methods are more efficient, requiring less computation time compared to the ECL, EI-C, and LCB-C methods.

For Example 2, the contour levels of interest are set to $a = \{1.2, 1.7, 2.1\}$. The initial sample size is $n_0 = 9$, with total points $N = \{27, 36, 45, 54, 63\}$. We use $\epsilon = 0.05$ for the M_{C_0} measurement in (33) and $\delta = 0.02$. Table 6

TABLE 4 The average of the measurements M_{C_0} in Example 1 for the adaptive designs ARSD-C, RCC, ECL, EI-C, LCB-C, and one-shot designs for $n_0 = 9$ and $N = \{11, 13, 15, 17, 19\}$ over 50 simulations for the contour level $a = \{-0.7, 1.2, 2.2\}$. The values in the parentheses are the relative efficiency of the criteria over one-shot designs.

N	One-shot	ARSD-C	RCC	ECL	EI-C	LCB-C
a = −0.7						
11	0.2361	0.0956(2.47)	0.1591(1.48)	0.1959(1.21)	0.2633(0.90)	**0.0893(2.64)**
13	0.1646	0.0360(4.57)	**0.0291(5.66)**	0.0678(2.43)	0.1517(1.09)	0.0440(3.74)
15	0.0959	**0.0022(43.5)**	0.0052(18.4)	0.0050 (19.1)	0.0494(1.94)	0.0075(12.8)
17	0.0823	0.0016(51.4)	0.0002(412)	**0.0001(823)**	0.0039(21.1)	0.0016(51.4)
19	0.0364	**0.0001(364)**	**0.0001(364)**	**0.0001(364)**	0.0004(91.0)	**0.0001(364)**
a = 1.2						
11	0.4156	0.4598(0.90)	0.4512(0.92)	0.4478(0.93)	**0.4262(0.98)**	0.4598(0.90)
13	0.3434	0.3908(0.88)	**0.3625(0.95)**	0.3869(0.89)	0.3763(0.91)	0.3899(0.88)
15	0.2555	0.2514(1.02)	**0.2220(1.15)**	0.2466(1.04)	0.2350 (1.09)	0.2536(1.01)
17	0.1580	0.1247(1.27)	0.1300(1.22)	0.1479(1.07)	**0.0887(1.78)**	0.1251(1.26)
19	0.0901	0.0298(3.02)	0.0316(2.85)	0.0295(3.05)	**0.0216(4.17)**	0.0300(3.00)
a = 2.2						
11	0.2846	0.1470(1.94)	**0.0239(11.9)**	0.0404(7.04)	0.1630(1.75)	0.1469(1.94)
13	0.2275	0.0448(5.08)	0.0061(37.3)	**0.0028(81.2)**	0.1488(1.53)	0.0449(5.07)
15	0.0540	0.0061(8.85)	**0.0008(67.5)**	0.0009(60.0)	0.0500(1.08)	0.0061(8.85)
17	0.0266	0.0024(11.1)	0.0005(53.2)	**0.0004(66.5)**	0.0090(2.96)	0.0024(11.1)
19	0.0096	**0.0001(96.0)**	**0.0001(96.0)**	0.0002(48.0)	0.0008(12.0)	**0.0001(96.0)**

TABLE 5 The average computation time (in seconds) in Example 1 for the adaptive designs ARSD-C, RCC, ECL, EI-C, and LCB-C for $n_0 = 9$ and $N = 19$ over 50 simulations for the contour level $a = \{-0.7, 1.2, 2.2\}$.

a	ARSD-C	RCC	ECL	EI-C	LCB-C
−0.7	0.14	0.13	0.77	0.78	0.79
1.2	0.22	0.49	0.80	0.77	0.75
2.2	0.17	0.15	0.73	0.72	0.73

provides a summary of the average M_{C_0} values obtained using one-shot and adaptive designs across 50 simulations for the specified contour levels and run sizes. The results indicate that the RCC method consistently exhibits superior relative efficiency for different contour levels and most of the run sizes, outperforming the other criteria in the comparison. In addition, Table 7 presents the average computation time (in seconds) required to add 54 points adaptively. The findings demonstrate that the ARSD-C and RCC methods are significantly faster than the ECL, EI-C, and LCB-C methods, further emphasizing their computational efficiency.

For Example 3, the contour levels of interest are set as $a = \{4.5, 5.5, 6.5\}$. We use an initial sample size of $n_0 = 9$, with total points $N = \{27, 36, 45, 54, 63\}$. The measurement M_{C_0} in (33) is evaluated using $\epsilon = 0.1$, and we set $\delta = 0.1$. Table 8 presents the average M_{C_0} values obtained from one-shot designs and adaptive designs over 50 simulations. Table 9 provides the average computation time required to add 54 points adaptively for each adaptive design method. The results are consistent with those observed in Example 1 and Example 2, demonstrating similar trends in performance and efficiency.

4.3 Prediction

This subsection applies the criteria EI-MC and EI-SC (discussed in Section 2.2) that were originally proposed for the problem of prediction for computer experiments with only quantitative inputs to the problem of prediction for computer experiments with both quantitative and qualitative inputs.

For the initial run size and budget size, we follow the recommendations provided by Yang et al. (2020). The initial sample size is set to be $n_0 = 5(p + q)$ the maximum budget is set to be $20(p + q)$. For EI-MC we set c for 10 multiple contours. Similar to optimization and contour estimation the next input is selected by evaluating a large number of candidate points in the design space.

TABLE 6 The average of the measurements M_{C_0} in Example 2 for the adaptive designs ARSD-C, RCC, ECL, EI-C, LCB-C and one-shot designs for $n_0 = 9$ and $N = \{27, 36, 45, 54, 63\}$ over 50 simulations for the contour level $a = \{1.2, 1.7, 2.1\}$. The values in the parentheses are the relative efficiency of the criteria over one-shot designs.

N	One-shot	ARSD-C	RCC	ECL	EI-C	LCB-C
$a = 1.2$						
27	0.0824	0.0139(5.93)	0.0081(10.17)	**0.0070(11.7)**	0.0247(3.34)	0.0134(6.15)
36	0.0335	0.0016(20.9)	0.0010(33.5)	**0.0008(41.8)**	0.0050(6.70)	0.0023(14.5)
45	0.0157	0.0003(52.3)	**0.0002(78.5)**	0.0003(52.3)	0.0005(31.4)	0.0003(52.3)
54	0.0073	**0.0001(73.0)**	**0.0001(73.0)**	**0.0001(73.0)**	0.0002(36.5)	**0.0001(73.0)**
63	0.0045	**0.0001(45.0)**	**0.0001(45.0)**	**0.0001(45.0)**	0.0002(22.5)	**0.0001(45.0)**
$a = 1.7$						
27	0.0643	**0.0500(1.29)**	0.0555(1.16)	0.0577(1.11)	0.0556(1.16)	0.0514(1.25)
36	0.0214	0.0170(1.26)	**0.0147(1.46)**	0.0167(1.28)	0.0175(1.22)	0.0156(1.37)
45	0.0115	0.0034(3.38)	**0.0031(3.71)**	**0.0031(3.71)**	0.0045(2.56)	0.0037(3.11)
54	0.0042	0.0009(4.67)	**0.0009(4.67)**	0.0010(4.20)	0.0011(3.82)	0.0010(4.20)
63	0.0020	**0.0004(5.00)**	**0.0004(5.00)**	0.0005(4.00)	**0.0004(5.00)**	**0.0004(5.00)**
$a = 2.1$						
27	0.0663	0.0528(1.26)	0.0553(1.20)	**0.0503(1.32)**	0.0563(1.18)	0.0540(1.23)
36	0.0239	0.0160(1.49)	0.0197(1.21)	**0.0141(1.70)**	0.0200(1.20)	0.0175(1.37)
45	0.0094	0.0038(2.47)	**0.0034(2.76)**	**0.0034(2.76)**	0.0046(2.04)	0.0039(2.41)
54	0.0040	0.0012(3.33)	**0.0011(3.64)**	0.0013(3.08)	0.0012(3.33)	0.0011(3.64)
63	0.0018	0.0005(3.60)	**0.0004(4.50)**	0.0005(3.60)	0.0006(3.00)	0.0005(3.60)

TABLE 7 The average computation time per second in Example 2 for the adaptive designs ARSD-C, RCC, ECL, EI-C, and LCB-C for $n_0 = 9$ and $N = 63$ over 50 simulations for the contour level $a = \{1.2, 1.7, 2.1\}$.

a	ARSD-C	RCC	ECL	EI-C	LCB-C
1.2	7.14	7.77	34.8	34.3	34.9
1.7	6.49	7.28	34.3	36.2	35.4
2.1	8.14	8.18	34.8	36.3	35.3

To evaluate prediction accuracy, the predicted function is compared with the true function using the root mean square error (RMSE), defined as

$$\text{RMSE} = \sqrt{\frac{1}{n_w} \sum_{i=1}^{n_w} \left(\hat{Y}(\boldsymbol{w}_i^*) - Y(\boldsymbol{w}_i^*)\right)^2}, \tag{34}$$

where n_w is the number of data points in the test set. In the captions of the figures, the values in parentheses represent the total size of the data for each method, and in the figures, we present the logarithm of the RMSE (base e). We record the computation time in minutes for adaptive designs EI-MC and EI-SC. Figs. 6 to 8 each present four panels. Panel (A) displays the logarithm of the RMSE in (34), to evaluate prediction accuracy for each adaptive design. Panel (B) shows the average total time required for the adaptive design to sequentially add points and update the model. Panel (C) provides the time each taken to update the model. Panel (D) illustrates the time needed to select the next input in each adaptive design.

For Example 1 with $n_0 = 10$ and $N = \{15, 21\}$, Fig. 6 demonstrates that adaptive designs achieve significantly better prediction accuracy than one-shot designs. However, the total computational time for EI-MC is higher than EI-SC, as observed in panel (B). This discrepancy stems from the time-intensive calculation of the EI-MC criterion, which requires evaluating EI across 10 contours.

In Example 2 with $n_0 = 20$ and $N = \{30, 40\}$, Fig. 7 similarly shows that adaptive designs outperform one-shot designs. However, the computation time increases substantially due to the higher dimensionality, larger initial design size, and expanded budget. The analysis is capped at $N = 40$, as the advantage of adaptive designs becomes apparent by this point.

For Example 3 with $n_0 = 30$ and $N = \{80, 100\}$, Fig. 8 confirms that adaptive designs consistently deliver more accurate predictions than one-shot designs. The computation time for model updates in this case is notably higher than in the previous examples, reflecting the increased complexity of modeling computer experiments with mixed inputs, which involves estimating a larger number of parameters.

TABLE 8 The average of the measurements M_{C_0} in Example 3 for the adaptive designs ARSD-C, RCC, ECL, EI-C, LCB-C and one-shot designs for $n_0 = 9$ and $N = \{27, 36, 45, 54, 63\}$ over 50 simulations for the contour level $a = \{4.5, 5.5, 6.5\}$. The values in the parentheses are the relative efficiency of the criteria over one-shot designs.

N	One-shot	ARSD-C	RCC	ECL	EI-C	LCB-C
$a = 4.5$						
27	0.1671	**0.1232(1.36)**	0.1412(1.18)	0.1434(1.17)	0.1353(1.24)	0.1248(1.34)
36	0.1256	0.1040(1.21)	0.1028(1.22)	**0.0980(1.28)**	0.1088(1.15)	0.1028(1.22)
45	0.0975	0.0889(1.10)	**0.0714(1.37)**	0.0719(1.36)	0.0947(1.03)	0.0886(1.10)
54	0.0709	0.0619(1.15)	**0.0374(1.90)**	0.0414(1.71)	0.0724(0.98)	0.0642(1.10)
63	0.0384	0.0293(1.31)	0.0227(1.69)	**0.0207(1.86)**	0.0383(1.00)	0.0300(1.28)
$a = 5.5$						
27	0.1453	**0.1215(1.20)**	0.1265(1.15)	0.1296(1.12)	0.1224(1.19)	0.1177(1.23)
36	0.1144	0.1016(1.13)	**0.0967(1.18)**	0.0979(1.17)	0.1036(1.10)	0.1029(1.11)
45	0.0908	0.0874(1.04)	0.0745(1.22)	**0.0713(1.27)**	0.0845(1.07)	0.0814(1.12)
54	0.0643	0.0531(1.21)	**0.0406(1.58)**	0.0421(1.53)	0.0600(1.07)	0.0494(1.30)
63	0.0338	0.0216(1.56)	0.0202(1.67)	**0.0200(1.69)**	0.0251(1.35)	0.0201(1.68)
$a = 6.5$						
27	0.3184	0.2901(1.10)	0.3487(0.91)	0.3521(0.90)	**0.2728(1.17)**	0.2978(1.07)
36	0.2623	**0.1175(2.23)**	0.1606(1.63)	0.1787(1.47)	0.1403(1.87)	0.1624(1.62)
45	0.2142	0.0592(3.62)	**0.0414(5.17)**	0.0523(4.10)	0.0444(4.82)	0.0644(3.33)
54	0.1463	0.0190(7.70)	**0.0088(16.6)**	0.0134(10.9)	0.0138(10.6)	0.0158(9.26)
63	0.0863	0.0117(7.38)	0.0038(22.7)	0.0067(12.8)	0.0040(21.6)	**0.0030(28.7)**

TABLE 9 The average computation time per second in Example 3 for the adaptive designs ARSD-C, RCC, ECL, EI-C, and LCB-C for $n_0 = 9$ and $N = 63$ over 50 simulations for the contour level $a = \{4.5, 5.5, 6.5\}$.

a	ARSD-C	RCC	ECL	EI-C	LCB-C
4.5	57.7	55.6	116.8	116.8	118.5
5.5	69.1	63.5	116.9	107.3	117.7
6.5	26.7	24.3	123.6	120.7	117.9

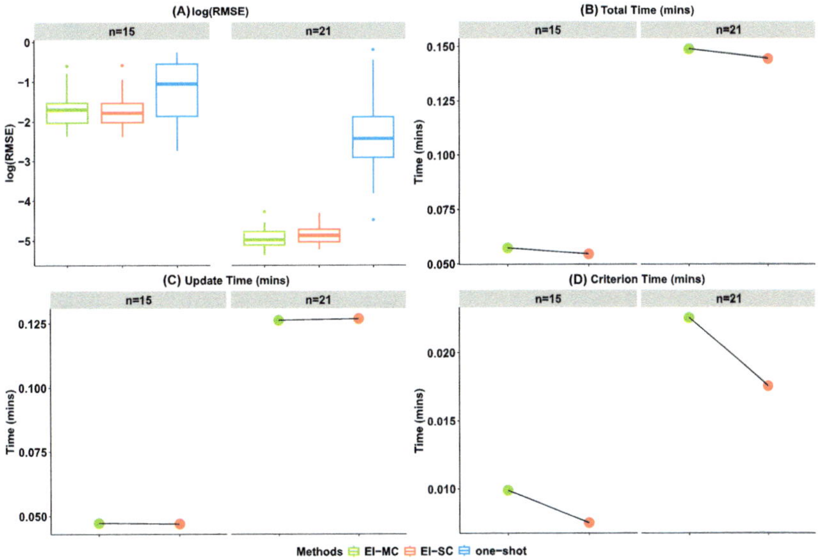

FIG. 6 (A) The boxplot of the log of RMSEs; (B) The average of total time in minutes; (C) The average of update time in minutes; (D) The average of criterion time in minutes; for $n_0 = 10$ and $N = 30$ over 50 simulations in Example 1.

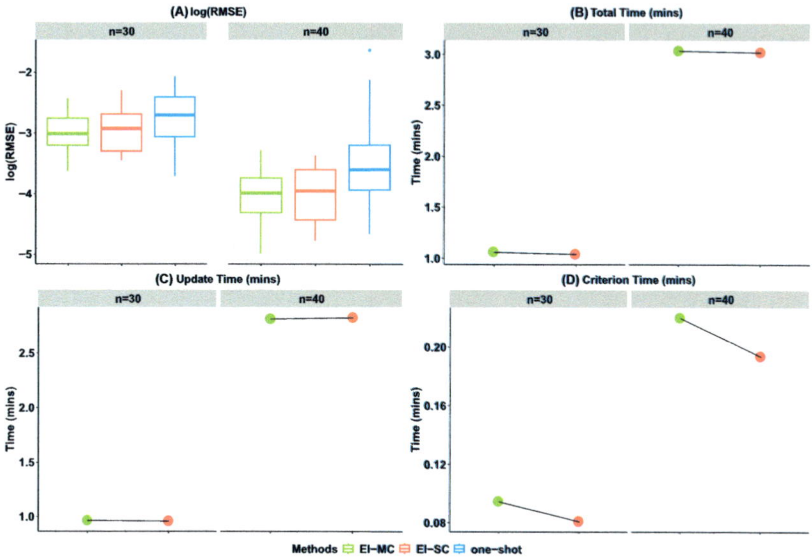

FIG. 7 (A) The boxplot of the log of RMSEs; (B) The average of total time in minutes; (C) The average of update time in minutes; (D) The average of criterion time in minutes; for $n_0 = 20$ and $N = 40$ over 50 simulations in Example 2.

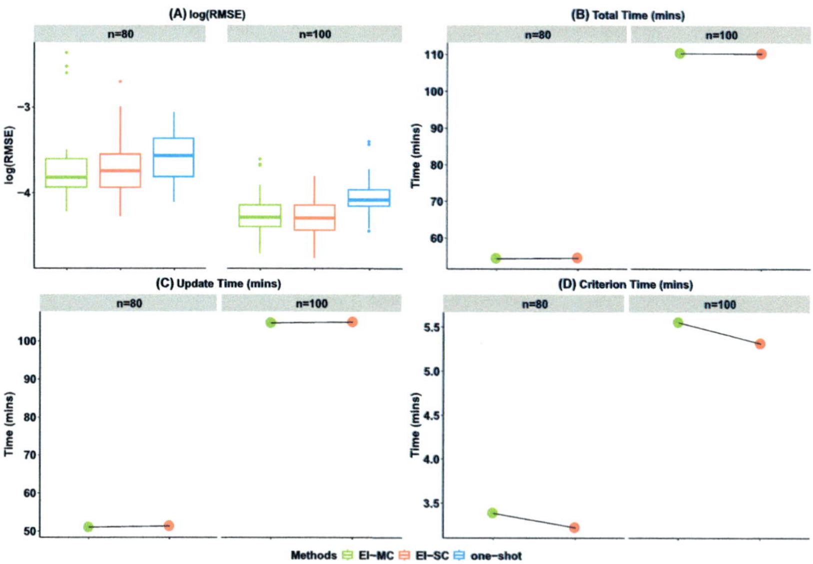

FIG. 8 (A) The boxplot of the log of RMSEs; (B) The average of total time in minutes; (C) The average of update time in minutes; (D) The average of criterion time in minutes; for $n_0 = 30$ and $N = 100$ over 50 simulations in Example 3.

5 Concluding remarks

Adaptive designs are an iterative and goal-oriented approach to selecting inputs for computer experiments. They dynamically update the emulators based on the data at hand, improving efficiency and accuracy. These designs are particularly beneficial in scenarios where computational resources are limited or objectives such as optimization, contour estimation, or prediction need to be achieved with minimal computational cost. They continue to evolve with innovations in modeling and computation, particularly for computer experiments with both quantitative and qualitative input variables. In this work, we review and discuss various adaptive designs for optimization, contour estimation, and prediction in computer experiments involving mixed inputs.

We highlight several key issues that are worth further investigation in computer experiments involving mixed inputs. The efficiency of an adaptive design relies heavily on a well-structured surrogate model and an effective criterion to achieve the desired objectives. While this study utilized the EzGP model as the surrogate, alternative models could also be explored. However, a significant challenge with models for mixed inputs is the computational cost, as they often require estimating a larger number of parameters. This challenge becomes particularly pronounced in high-dimensional settings, calling for cost-effective modeling approaches tailored to mixed inputs.

Another direction for future research would be adaptive designs for prediction when the model involves both quantitative and qualitative inputs. We applied the criteria originally proposed for prediction in computer experiments with only quantitative inputs to computer experiments with mixed inputs. Although the results showed that such adaptive designs outperform one-shot designs. To the best of our knowledge, no specific work has been done on adaptive designs for prediction in computer experiments with both quantitative and qualitative inputs, and this area remains open for exploration.

In addition, for models with mixed inputs, it is not straightforward to directly identify the inputs that optimize an adaptive design criterion. In this work, we addressed this limitation by evaluating a large set of candidate points to approximate the optimal solution. However, this approach is computationally intensive and represents an open and critical challenge for further research. Developing more efficient methods to identify optimal inputs in adaptive designs for mixed inputs remains an important area for future exploration.

References

Auer, P., 2002. Using confidence bounds for exploitation-exploration trade-offs. J. Mach. Learn. Res. 3, 397–422.

Bect, J., Ginsbourger, D., Li, L., Picheny, V., Vazquez, E., 2012. Sequential design of computer experiments for the estimation of a probability of failure. Stat. Comput. 22, 773–793.

Cai, X., Xu, L., Lin, C.D., Hong, Y., Deng, X., 2024. Adaptive-region sequential design with quantitative and qualitative factors in application to HPC configuration. J. Qual. Technol. 56 (1), 5–19.

Cameron, K.W., Anwar, A., Cheng, Y., Xu, L., Li, B., Ananth, U., et al., 2019. Moana: modeling and analyzing I/O variability in parallel system experimental design. IEEE Trans. Parallel Distrib. Syst. 30 (8), 1843–1856.

Carnell, R., 2022. Latin hypercube samples. R. Found. Stat. Comput URL. https://cran.r-project.org/package=lhs.

Cole, D.A., Gramacy, R.B., Warner, J.E., Bomarito, G.F., Leser, P.E., Leser, W.P., 2023. Entropy-based adaptive design for contour finding and estimating reliability. J. Qual. Technol. 55 (1), 43–60.

Deng, X., Lin, C.D., Liu, K.-W., Rowe, R.K., 2017. Additive Gaussian process for computer models with qualitative and quantitative factors. Technometrics 59 (3), 283–292.

Deng, X., Hung, Y., Lin, C.D., 2015. Design for computer experiments with qualitative and quantitative factors. Stat. Sin. 25, 1567–1581.

Gramacy, R.B., 2020. Surrogates: Gaussian Process Modeling, Design, and Optimization for the Applied Sciences. Chapman and Hall/CRC.

Han, G., Santner, T.J., Notz, W.I., Bartel, D.L., 2009. Prediction for computer experiments having quantitative and qualitative input variables. Technometrics 51 (3), 278–288.

He, Y., Lin, C.D., Sun, F., 2017a. On construction of marginality coupled designs. Stat. Sin. 27, 665–683.

He, Y., Lin, C.D., Sun, F., Lv, B., 2017b. Marginally coupled designs for two-level qualitative factors. J. Stat. Plan. Inference 187, 103–108.

Hernández-Lobato, J.M., Hoffman, M.W., Ghahramani, Z., 2014. Predictive entropy search for efficient global optimization of black-box functions. Adv. Neural Inf. Process. Syst. 27.

Hwang, Y., He, X., Qian, P.Z.G., 2016. Sliced orthogonal array-based Latin hypercube designs. Technometrics 58 (1), 50–61.

Jones, D.R., Schonlau, M., Welch, W.J., 1998. Efficient global optimization of expensive black-box functions. J. Glob. Optim. 13 (4), 455–492.

Kocsis, L., Szepesvári, C., 2006. Bandit based Monte-Carlo planning. Eur. Conf. Mach. Learn 282–293.

Li, J., Xiao, Q., Mandal, A., Lin, C.D., Deng, X., 2023. Easy-to-interpret Gaussian process models for computer experiments with both quantitative and qualitative factors. R. Found. Stat. Comput URL. https://cran.r-project.org/package=EzGP.

Lin, W.-A., Sung, C.-L., Chen, R.-B., 2024. Category tree Gaussian process for computer experiments with many-category qualitative factors and application to cooling system design. J. Qual. Technol. 56 (5), 391–408.

Luo, H., Cho, Y., Demmel, J.W., Li, X.S., Liu, Y., 2024. Hybrid parameter search and dynamic model selection for mixed-variable Bayesian optimization. J. Comput. Graph. Stat. 1–14.

McKay, M.D., Beckman, R.J., Conover, W.J., 1979. A comparison of three methods for selecting values of input variables in the analysis of output from a computer code. Technometrics 42 (1), 55–61.

Mebane, W.R., Sekhon, J.S., Saarinen, T., 2011. Genetic optimization using derivatives. R. Found. Stat. Comput URL. https://cran.r-project.org/package=rgenoud.

Mohammadi, H., Challenor, P., Williamson, D., Goodfellow, M., 2022. Cross-validation-based adaptive sampling for Gaussian process models. SIAM/ASA J. Uncertain. Quantification 10 (1), 294–316.

Picheny, V., Ginsbourger, D., Richet, Y., Caplin, G., 2013. Quantile-based optimization of noisy computer experiments with tunable precision. Technometrics 55 (1), 2–13.

Qian, P.Z.G., Wu, H., Wu, C.F.J., 2008. Gaussian process models for computer experiments with qualitative and quantitative factors. Technometrics 50 (3), 383–396.

Qian, P.Z.G., 2012. Sliced Latin hypercube designs. J. Am. Stat. Assoc. 1070 (497), 393–399.

Ranjan, P., Bingham, D., Michailidis, G., 2008. Sequential experiment design for contour estimation from complex computer codes. Technometrics 50 (4), 527–541.

Sacks, J., Welch, W.J., Mitchell, T.J., Wynn, H.P., 1989. Design and analysis of computer experiments. Stat. Sci. 4 (4), 409–423.

Shahrokhian, A., Deng, X., Lin, C.D., Ranjan, P., Xu, L., 2024. Adaptivedesign for contour estimation from computer experiments with quantitative and qualitative inputs. http://arxiv.org/abs/2504.05498.

Srinivas, N., Krause, A., Kakade, S.M., Seeger, M., 2010. Gaussian process optimization in the bandit setting: No regret and experimental design. Int. Conf. Mach. Learn.

Xiao, Q., Mandal, A., Lin, C.D., Deng, X., 2021. EzGP: easy-to-interpret Gaussian process models for computer experiments with both quantitative and qualitative factors. SIAM/ASA J. Uncertain. Quantification 9 (2), 333–353.

Yang, F., Lin, C.D., Ranjan, P., 2020. Global fitting of the response surface via estimating multiple contours of a simulator. J. Stat. Theory Pract. 14, 1–21.

Yang, F., Lin, C.D., Zhou, Y., He, Y., 2023. Doubly coupled designs for computer experiments with both qualitative and quantitative factors. Statistica Sin. 33, 1923–1942.

Yang, J.-F., Lin, C.D., ZG Qian, P., Lin, D.K.J., 2013. Construction of sliced orthogonal Latin hypercube designs. Stat. Sin. 23, 1117–1130.

Zhang, Q., Chien, P., Liu, Q., Xu, L., Hong, Y., 2021. Mixed-input Gaussian process emulators for computer experiments with a large number of categorical levels. J. Qual. Technol. 53 (4), 410–420.

Zhang, Y., Apley, D.W., Chen, W., 2020a. Bayesian optimization for materials design with mixed quantitative and qualitative variables. Sci. Rep. 10 (1), 1–13.

Zhang, Y., Tao, S., Chen, W., Apley, D.W., 2020b. A latent variable approach to Gaussian process modeling with qualitative and quantitative factors. Technometrics 62 (3), 291–302.

Zhou, Q., Qian, P.Z.G., Zhou, S., 2011. A simple approach to emulation for computer models with qualitative and quantitative factors. Technometrics 53 (3), 266–273.

Chapter 5

On the use of machine learning methods for missing data problems

Sixia Chen* **and Chao Xu**

Department of Biostatistics and Epidemiology, The University of Oklahoma Health Sciences, Oklahoma City, OK, United States

**Corresponding author. e-mail address: Sixia-Chen@ouhsc.edu*

Chapter Outline

Abstract

Missing data frequently occurs in government and industrial datasets. Ignoring missing data in statistical analysis may lead to biased results. Commonly used methods for handling missing data include imputation, inverse propensity score weighting, and hybrid methods that combine both. However, traditional methods may not be valid for dealing with high-dimensional, large-scale datasets with highly nonlinear relationships. Modern machine learning methods, such as XGBoost, Support Vector Machines, and Deep Neural Networks, have been shown to be effective in such scenarios. In this paper, we provide a unified framework for the use of machine learning methods in addressing missing data problems. Specifically, we consider imputation methods, inverse propensity score methods, calibration methods, and cross-fit double machine learning methods. We evaluate the performance of several machine learning-based methods using Monte Carlo simulation study and one real data application with Household Pulse Survey.

1 Introduction

Handling missing data in large-scale government and industrial datasets, such as the National Health and Nutrition Examination Survey (NHANES) and the

Handbook of Statistics, Vol. 53. https://doi.org/10.1016/bs.host.2025.05.002

Crash Report Sampling System (CRSS) from the National Highway Traffic Safety Administration (NHTSA), presents unique challenges. These datasets often exhibit complex survey designs, high dimensionality, and varying patterns of missingness, including Missing Completely at Random (MCAR), Missing at Random (MAR), and Missing Not at Random (MNAR). Traditional imputation methods, such as mean substitution or regression-based techniques, may not adequately capture the underlying data structure, leading to biased estimates and reduced statistical power. For instance, in NHANES, missing data can result from nonresponse or skip patterns inherent in the survey design, necessitating sophisticated imputation strategies that account for survey weights and complex sampling designs. Ignoring missing data in statistical analysis can lead to biased results. Commonly used statistical methods for handling missing data include imputation methods, propensity score weighting, doubly robust, and multiple robust methods, see Little and Rubin (2002) and Kim and Shao (2013) for more details. In practice, missing data can be generally classified into two types: unit nonresponse and item nonresponse. Unit nonresponse occurs when an entire observational unit does not participate in the study or survey. Item nonresponse occurs when a respondent participates in a study or survey but fails to provide answers to some specific questions, leading to missing values for certain variables. While imputation methods can be used for dealing with unit nonresponse, the commonly used methods are propensity score methods including parametric logistic regression model as example (Franklin et al., 2017). Imputation methods are mainly used for handling item nonresponse. They include single mean imputation, mode imputation, regression imputation, nearest neighbor imputation, hot-deck imputation, fractional imputation, and multiple imputation, see Chen and Haziza (2019), Little and Rubin (2002), and Kim and Shao (2013) for comprehensive discussion for those imputation methods. To improve the robustness of underlying model assumptions, doubly and multiply robust methods (Bang and Robins, 2005; Cao et al., 2009; Han, 2014; Chen and Haziza, 2017) have been used. Doubly robust methods used both inverse propensity score model and outcome regression model in constructing the final estimators which have the so-called doubly robustness such that if one of the models is correctly specified, then they are consistent. Multiply robust methods extend the idea of doubly robust methods by considering multiple inverse propensity score models and outcome regression models and if one of the models if correctly specified, then the final estimator would be consistent.

Recent advancements in machine learning have introduced more flexible and robust imputation techniques suitable for such complex datasets. Methods like Multiple Imputation by Chained Equations (MICE) have been widely adopted for their ability to handle various types of variables and missing data patterns. Moreover, machine learning algorithms, including k-Nearest Neighbors (k-NN), Random Forests, and deep learning models like autoencoders, have shown promise in capturing nonlinear relationships and interactions among variables, leading to more accurate imputations. For example, a study utilizing NHANES data demonstrated that a deep learning-based imputation model outperformed traditional methods in imputing missing values in actigraphy data. Similarly, in the context of CRSS, where missing data may arise from incomplete crash reports, machine learning approaches can leverage patterns in the available data to predict missing entries, enhancing the quality of traffic safety analyses.

Nowadays, it becomes more cost effective to collect large complex data files such as social media data or internet data. As both the sample size and the number of variables get large in the data file, traditional methods for handling missing data become less effective and even infeasible. Machine learning based methods have been developed to improve such methods under such scenarios. Machine learning-based methods for handling missing data include the following as examples: regression tree and random forest methods (Burgette and Reiter, 2010; Rahman and Islam, 2011; Tang and Ishwaran, 2017; Lee et al., 2010), K nearest neighbor methods (Beretta and Santaniello, 2016; Pujianto et al., 2019), Support Vector Machine methods (Sivapriya et al., 2012), and Deep neural network methods (Khan et al., 2022). Empirical comparison of some commonly used machine learning methods with Monte Carlo simulation and real data examples have been done by Chen and Xu (2023) and Dagdoug et al. (2023).

Recently, Cross-Fit Double Machine Learning (DM) has emerged as a powerful method for handling missing data while accounting for both selection bias and high-dimensional covariates (Chernozhukov et al., 2018). DM is built on the principles of semiparametric efficiency theory, where flexible machine learning models are used to estimate nuisance functions, such as the missingness mechanism (propensity score model) and the outcome model, while reducing overfitting through cross-fitting. The key advantage of DM is its double robustness, meaning that the final estimator remains consistent as long as at least one of the models (propensity score or outcome model) is correctly specified. Moreover, cross-fitting reduces the bias introduced by overfitting in high-dimensional settings, ensuring better finite-sample performance. Unlike traditional imputation or weighting-based methods, DM allows the integration of modern machine learning techniques, such as random forests, neural networks, or gradient boosting, to handle complex, nonlinear missing data patterns. Due to its theoretical guarantees with valid statistical inference and flexibility, DM has gained increasing attention in both theoretical and applied missing data research (Athey and Imbens, 2019; Farrell et al., 2021).

In this paper, we provide unified framework of using machine learning for handling missing data. Specifically, we considered imputation, propensity score weighting, calibration method, and cross-fit double machine learning method. We further evaluate the performance of some methods by using Monte Carlo simulation study. Our paper is organized as follows: Section 2 introduces mathematical notations; discussion of different machine learning based missing data methods is presented in Section 3; Monte Carlo simulation study is contained in Section 4; One real data application by using Household Pulse Survey is discussed in Sections 5 and 6 present the conclusion of this paper.

2 Basic setups

Suppose a random sample $\{(\mathbf{X}_i, Y_i), i = 1, 2, ..., n\}$ with sample size n is generated from some unknown model $F(\mathbf{x}, y)$ which is a joint distribution function of $\mathbf{X}$ and Y, where $\mathbf{X} = (X_1, X_2, \ldots, X_q)$ is a q dimensional vector and Y is a scalar. Note that q can be larger than n. Assume $\mathbf{X}_i$ is fully observed for $i = 1, 2, \ldots, n$ and Y_i might suffer from missingness. Denote R_i as the response indicator of Y_i such that $R_i = 1$ if unit i is observed and 0 otherwise. The response mechanism is assumed to be missing at random (MAR) in the sense of Little and Rubin (2002) with the following

$$\Pr(R = 1|\mathbf{X}, Y) = \Pr(R = 1|\mathbf{X}) \triangleq p(\mathbf{X}), \tag{1}$$

where $p(\mathbf{X})$ is a unknown function of $\mathbf{X}$. In addition, the outcome regression model is assumed to be

$$Y = m(\mathbf{X}) + \epsilon, \tag{2}$$

where $m(\mathbf{X})$ is a unknown function of $\mathbf{X}$ and $E(\epsilon|\mathbf{X}) = 0$. Assume our parameter of interest is defined as the unique solution of the estimating equation $E\{\mathbf{U}(\mathbf{X}, Y; \theta_0)\} = 0$, where $\mathbf{U}(\mathbf{X}, Y; \boldsymbol{\theta}_0)$ satisfies certain regularity conditions in Chen and Kim (2017). Special examples include the population mean Acknowledgmentswith $\mathbf{U}(\mathbf{X}, Y; \boldsymbol{\theta}_0) = Y - \theta_0$, population regression coefficient with $\mathbf{U}(\mathbf{X}, Y; \boldsymbol{\theta}_0) = \mathbf{X}^{\mathsf{T}}(Y - \mathbf{X}\boldsymbol{\theta}_0)$, and population 100τ-th percentile with $\mathbf{U}(\mathbf{X}, Y; \boldsymbol{\theta}_0) = I(Y \leq \theta_0) - \tau$.

3 Missing data methods

3.1 Propensity score weighting

Denote $\hat{p}(\mathbf{X})$ as the estimator of $p(\mathbf{X})$ based on a machine learning method such as random forest or deep neural networks of predicting R by using $\mathbf{X}$ and entire sample, then the propensity score weighing estimator $\hat{\theta}_{PS}$ of $\boldsymbol{\theta}_0$ can be obtained by solving

$$\hat{U}_1(\theta) = \frac{1}{n}\sum_{i=1}^{n} \frac{R_i}{\hat{p}(\mathbf{X})}\mathbf{U}(\mathbf{X}_i, Y_i; \theta) = \mathbf{0}. \tag{3}$$

If $\hat{p}(\mathbf{X})$ is a consistent estimator, one can show that $\hat{\theta}_{PS}$ is consistent under certain regularity conditions. The consistency of random forest methods have been discussed by To improve the robustness against the failure of underlying model assumptions and avoid extreme weights, One can first obtain the nonresponse adjustment cells $C_p = \{C_{p,k}\}, k = 1, 2, ..., K$ based on the percentiles of $\hat{p}(\mathbf{X})$. Then the nonresponse adjustment factor for cell C_k can be written as

$$F_{np,C_k} = \frac{n_{C_k}}{n_{r,C_k}}, \tag{4}$$

where n_{C_k} is the sample size in cell C_k and n_{r,C_k} is the respondents sample size in cell C_k. Then one can obtain the weighting cell based estimator $\hat{\theta}_{WC}$ of $\boldsymbol{\theta}_0$ by solving

$$\hat{U}_2(\theta) = \frac{1}{n} \sum_{i=1}^{n} R_i \sum_{k=1}^{K} D_{k,i} F_{np,C_k} \mathbf{U}(\mathbf{X}_i, Y_i; \theta) = \mathbf{0}, \tag{5}$$

where $D_{k,i} = 1$ if unit i belongs to cell C_k and 0 otherwise. The confidence interval can be obtained by using resampling methods such as bootstrap (Hastie et al., 2009).

3.2 Imputation

Denote $\hat{m}(X)$ as the estimator of $m(X)$ based on a machine learning method of predicting Y by using $\mathbf{X}$ and respondents, then the imputed estimator $\hat{\theta}_I$ of $\boldsymbol{\theta}_0$ can be obtained by solving

$$\hat{U}_3(\theta) = \frac{1}{n} \sum_{i=1}^{n} \{R_i \mathbf{U}(\mathbf{X}_i, Y_i; \theta) + (1 - R_i)\mathbf{U}(\mathbf{X}_i, Y_i^*; \theta)\} = \mathbf{0}. \tag{6}$$

where Y_i^* is the imputed value. For deterministic imputation, it can be written as $Y_i^* = \hat{m}(\mathbf{X}_i)$. It works for estimating the population mean or total. For stochastic imputation, it can be written as $Y_i^* = \hat{m}(\mathbf{X}_i) + \epsilon_i^*$ with ϵ_i^* selected from the pool of estimated residuals $\{\hat{\epsilon}_i, i \in S_r\}$ by using simple random sampling with replacement method, where S_r denotes the set of respondents and $\hat{\epsilon}_i = y_i - \hat{m}(\mathbf{X}_i)$. It works for general parameter of interest, which is defined as the solution of the estimating equation. If the machine learning based estimator $\hat{m}(\mathbf{X})$ is consistent, then it can be shown that the estimator $\hat{\theta}_I$ is consistent for estimating $\boldsymbol{\theta}_0$. To improve the efficiency of stochastic imputation, one can use the semi-parametric fractional imputation procedure proposed by Chen and Kim (2017). One can first obtain the fractional weights $\{\hat{w}_j: j \in S_r\}$ by maximizing the empirical likelihood distance function

$$l = \sum_{j \in S_r} \log(w_j), \tag{7}$$

subject to

$$\sum_{j \in S_r} w_j = 1, \quad \sum_{j \in S_r} w_j \hat{\epsilon}_j = 0, \quad w_j > 0. \tag{8}$$

By using the Lagrange multiplier method, the fractional weight can be written as

$$\hat{w}_j = \frac{1}{n_r} \frac{1}{1 + \hat{\lambda}\hat{\epsilon}_j}, \tag{9}$$

with $\hat{\lambda}$ as the solution of the equation

$$\sum_{j \in S_r} \hat{w}_j \hat{\epsilon}_j = 0. \tag{10}$$

Then, the semi-parametric fractional imputed estimator $\hat{\theta}_{SI}$ of $\boldsymbol{\theta}_0$ can be obtained by solving

$$\hat{U}_4(\boldsymbol{\theta}) = \frac{1}{n} \sum_{i=1}^{n} \left\{ R_i \mathbf{U}(\mathbf{X}_i, Y_i; \boldsymbol{\theta}) + (1 - R_i) \sum_{j \in S_r} w_{ij} \mathbf{U}(\mathbf{X}_i, Y_{ij}^*; \boldsymbol{\theta}) \right\} = \mathbf{0}, \tag{11}$$

where $w_{ij} = \hat{w}_j$ and $Y_{ij}^* = m(\mathbf{X}_i) + \hat{\epsilon}_j$. To improve the robustness of previous machine learning based imputation methods, one can first construct hot-deck cells $C_m = \{C_{m,k}, k = 1, 2, \ldots, K\}$ based on the percentiles of $\hat{m}(\mathbf{X})$. Then the imputed values can be generated by randomly selecting from the respondents in the same cell as the nonrespondents. Statistical inference can be conducted by using resampling methods such as bootstrap.

3.3 Calibration

Assume one is interested in estimating the population mean $\theta_0 = E(Y)$. Suppose we have a class of candidate propensity score models denoted as $C_1 = \{p^{(k)}(\mathbf{X}), k = 1, 2, \ldots, K\}$ and a class of candidate imputation models denoted as $C_2 = \{m^{(l)}(\mathbf{X}), l = 1, 2, \ldots, L\}$ where those models can include parametric model and non-parametric machine learning models such as random forest, support vector machine, and deep neural networks. Then we can first obtain the calibration weights $\hat{w}_i$ by maximizing the empirical likelihood distance function

$$l = \sum_{i \in S_r} \log(w_i), \tag{12}$$

subject to

$$\sum_{i \in S_r} w_i = 1, \quad w_i > 0, \tag{13}$$

$$\sum_{i \in S_r} w_i \hat{p}^{(k)}(\mathbf{X}_i) = n^{-1} \sum_{i \in S} \hat{p}^{(k)}(\mathbf{X}_i), \quad \sum_{i \in S_r} w_i \hat{m}^{(l)}(\mathbf{X}_i) = n^{-1} \sum_{i \in S} \hat{m}^{(l)}(\mathbf{X}_i), \tag{14}$$

for $k = 1, 2, \ldots, K$ and $l = 1, 2, \ldots, L$, where $\hat{p}^{(k)}(\mathbf{X})$ and $\hat{m}^{(l)}(\mathbf{X})$ are the corresponding estimators of $p^{(k)}(\mathbf{X})$ and $m^{(l)}(\mathbf{X})$. Chen and Haziza (2017) discussed the use of other distance functions in the calibration process. Once the calibration weights are obtained, the machine learning based calibration estimator $\hat{\theta}_C$ of θ_0 can be written as

$$\hat{\theta}_C = \sum_{i \in S_r} \hat{w}_i Y_i. \tag{15}$$

It can be shown that $\hat{\theta}_C$ enjoys the multiply robustness property such that it is consistent if one of the candidate models is correctly specified. It can also be regarded as the super-learner method since it aggregates multiple machine learning models. To prevent over-fitting and promote valid statistical inference, one can consider using cross-fitting method with random sample splitting (Chernozhukov et al., 2018; Jacob, 2020; Zivich and Breskin, 2021). The variance estimator can be obtained by using resampling methods such as bootstrap.

3.4 Cross-fit double machine learning

In this section, we describe the cross-fit double machine learning (DML) method (Chernozhukov et al., 2018; Jacob, 2020; Zivich and Breskin, 2021) for handling missing data in the estimation of the parameter of interest $\theta_0 = E(Y)$. The method combines outcome regression modeling $m(\mathbf{X})$ and propensity score model $p(\mathbf{X})$ in a doubly robust manner, while leveraging cross-fitting to improve statistical inference. Cross-fitting helps mitigate overfitting by splitting data into folds and training separate machine learning models in each fold. One can first randomly split original sample into K equal-sized folds: S_k for $k = 1, 2, \ldots, K$. For each fold k, we can train a propensity score model $\hat{p}^{(-k)}(\mathbf{X})$ and an imputation model $m^{(-k)}(\mathbf{X})$ on the remaining data $S^{(-k)}$ by using machine learning methods separately. The Cross-Fit Double Machine Learning estimator (DM) $\hat{\theta}_{DM}$ of θ_0 can be written as

$$\hat{\theta}_{DM} = \frac{1}{K} \sum_{k=1}^{K} \hat{\theta}^{(k)}, \tag{16}$$

where

$$\hat{\theta}^{(k)} = \frac{1}{n_k} \sum_{i \in S_k} \left\{ \frac{R_i Y_i}{\hat{p}^{(-k)}(\mathbf{X}_i)} - \frac{R_i - \hat{p}^{(-k)}(\mathbf{X}_i)}{\hat{p}^{(-k)}(\mathbf{X}_i)} m^{(-k)}(\mathbf{X}_i) \right\}, \tag{17}$$

with n_k as the sample size for S_k. Under some regularity conditions, it can be shown that $\hat{\theta}_{DM}$ is consistent for estimating θ_0. The cross-fit variance estimator of $\hat{\theta}_{DM}$ can be written as

$$\hat{V}(\hat{\theta}_{DM}) = \frac{1}{K} \sum_{k=1}^{K} \frac{1}{n_k} \sum_{i \in S_k} \left[\frac{R_i Y_i}{\hat{p}^{(-k)}(\mathbf{X}_i)} - \left\{ \frac{R_i - \hat{p}^{(-k)}(\mathbf{X}_i)}{\hat{p}^{(-k)}(\mathbf{X}_i)} \right\} m^{(-k)}(\mathbf{X}_i) - \hat{\theta}_{DM} \right]^2. \tag{18}$$

Using asymptotic normality of $\hat{\theta}_{DM}$, the confidence interval can be constructed by using $\hat{V}(\hat{\theta}_{DM})$. Alternatively, bootstrap methods can be used to construct empirical confidence intervals.

4 Simulation study

We generate $B = 100$ Monte Carlo samples with sample size $n = 500$ from the following outcome regression models:

(1) $Y = 1 + 0.6X_1 - 0.3X_2 + 1.1X_3 + 0.65X_4^{(1)} + 2.47X_4^{(2)} + \epsilon$, where $X_1 \sim N(0, 1)$, $X_2 \sim \chi^2(5)$, $X_3 \sim binom(k = 1, 0.4)$, $X_4 \sim multinom(k = 1, 0.2, 0.3, 0.5)$, and $\epsilon \sim N(0, 1)$. $X_4^{(1)}$ and $X_4^{(2)}$ are two dummy variables of X_4, k indicates the number of trials.
(2) $Y = 1 + |X|_1^{1/3} * X_2^{1/5} * \log(|X_1 + X_3|) + sin(X_1^2 + 1.5 * X_3 - 0.2 * X_4^{(1)}) + \epsilon$.

The response model is assumed to be.

(1) $R \sim binom(p)$ with $logit(p) = c + 0.1X_1 + 0.3X_2 + 0.6X_3 + 0.5X_4^{(1)} + 0.9X_4^{(2)}$.
(2) $R \sim binom(p)$ with $logit(p) = c + cos(X_1^3 X_2^2 X_3) + \log(|X_1 X_2|) + X_2^{1/3} X_4^{(1)} + X_3 X_4^{(2)}$.

Where c is selected to have about 50 % response rate. We consider 3 different model scenarios with paired OM and RM models by series id 1–2, which represent linear and non-linear scenarios. The parameter of interest is the population mean of Y, which is $\theta_0 = E(Y)$. The following methods are compared: (1). Simple linear regression model (LM); (2). Generalized additive modeling (GAM); (3). K nearest neighbor method (KNN); (4). Regression tree method (RT); (5). Random forest method (RF); (6). XGboosting method (XGB); (7). Support vector machine (SVM); (8). Deep neural network method (DL). All methods will be applied to imputation estimator (IM) and propensity score weighting estimator (PS). Tuning parameters of GAM, KNN, RT, RF, SVM, and XGB were selected by 10-fold cross validation method implemended using R packages caret, tune, and parsnip. In the GAM model, we optimized the number of knots in the spline function for all continuous variables, evaluating knot counts of 3, 5, 7, 9, and 11. For the KNN method, we considered tuning parameters such as the number of neighbors to consider, a single number for the parameter used in calculating Minkowski distance, and the type of kernel function used to weight distances between samples. The default value ranges provided by the R packages for each tuning parameter were applied to KNN and other ML methods described below. For the RT method, tuning parameters included the maximum depth of the tree, complexity parameter, and the minimum number of data points in a node required for further split. For the RF method, we tuned the number of variables randomly selected to consider at each split when growing a decision tree, the total number of decision trees that are grown to form the Random Forest, the minimum number of data points required to split a node further in a decision tree. For SVM method, we tuned the penalty assigned to misclassified data points, the influence of a data point based on its distance from others, and the desired separation distance between different classes in the decision boundary. In the XGB method, tuning parameters considered were the maximum depth of the tree, the number of trees in the ensemble, learning rate, and the minimum number of data points in a node required for further split. For the DL methods, the considerations involved the 3 layers with the number of nodes (50, 50, and 1 at each layer), learning rates (0.01 and 0.001), and ReLu activation function. According to our previous research experience, those parameters will often lead to great performance. The following Monte Carlo relative bias (RB), relative standard error (RSE), and relative root mean squared (RRMSE) error are calculated and presented in Table 1 for linear model (M1) and Table 2 for non-linear model (M2).

$$RB = \frac{\left| B^{-1}\Sigma_{b=1}^{B}\hat{\theta}^{(b)} - \theta_0 \right|}{|\theta_0|}, \tag{19}$$

$$RSE = \frac{\left\{ (B-1)^{-1}\Sigma_{b=1}^{B}\left(\hat{\theta}^{(b)} - \bar{\hat{\theta}}_B\right)^2 \right\}^{1/2}}{|\theta_0|}, \tag{20}$$

and

TABLE 1 Relative bias (RB) (%), Relative standard error (RSE) (%), and Relative root mean squared error (RRMSE) (%) of different methods under model (M1).

Model	Method	RB	RSE	RRMSE
LM	IM	0.11	7.40	7.40
	PS	0.00	8.16	8.16
GAM	IM	0.02	7.42	7.42
	PS	0.31	8.03	8.04
KNN	IM	4.03	7.70	8.69
	PS	20.43	18.15	27.33
XGB	IM	0.41	7.73	7.74
	PS	5.69	8.84	10.51
RT	IM	2.14	8.71	8.97
	PS	9.59	9.37	13.41
SVM	IM	0.05	7.83	7.83
	PS	2.21	8.98	9.25
RF	IM	0.60	7.92	7.94
	PS	3.40	16.19	16.54
DL (0.01)	IM	0.78	7.50	7.54
	PS	1.09	11.26	11.31
DL (0.001)	IM	1.38	7.60	7.72
	PS	0.97	10.82	10.87

TABLE 2 Relative bias (RB) (%), Relative standard error (RSE) (%), and Relative root mean squared error (RRMSE) (%) of different methods under model (M2).

Model	Method	RB	RSE	RRMSE
LM	IM	14.15	9.06	16.80
	PS	14.28	9.22	17.00
GAM	IM	3.65	9.61	10.28
	PS	7.62	9.66	12.30
KNN	IM	4.85	8.92	10.16
	PS	14.32	16.56	21.89
XGB	IM	2.77	8.38	8.83
	PS	0.13	11.81	11.81
RT	IM	1.95	9.12	9.32
	PS	5.12	11.87	12.93
SVM	IM	9.58	9.06	13.18
	PS	2.37	9.87	10.15
RF	IM	1.90	8.19	8.41
	PS	2.96	15.67	15.95
DL (0.01)	IM	2.59	9.45	9.80
	PS	2.12	9.76	9.99
DL (0.001)	IM	2.43	9.30	9.61
	PS	2.02	10.75	10.93

$$RRMSE = (RB^2 + RSE^2)^{1/2}, \tag{21}$$

where $\hat{\theta}^{(b)}$ is the estimator of θ_0 based on the b-th Monte Carlo sample, θ_0 is the true parameter of interest, and $\bar{\hat{\theta}}_B = B^{-1}\sum_{b=1}^{B}\hat{\theta}^{(b)}$.

According to Table 1, LM and GAM exhibit the best performance in terms of RB, RSE, and RRMSE, as the model is correctly specified. Among the imputation methods, XGB, SVM, RF, and DL perform comparably to LM and GAM, while KNN and RT show slightly worse results. For propensity score weighting methods, KNN has the highest RB, RSE, and RRMSE, whereas DL methods yield lower RB than KNN, XGB, RT, SVM, and RF. The RRMSE values for XGB, SVM, and DL are similar. Overall, imputation methods achieve lower RB, RSE, and RRMSE than propensity score weighting methods. According to Table 2, LM performs the worst in terms of RB and RRMSE due to incorrect model assumptions. Among imputation methods, XGB, RT, RF, and DL achieve comparable RRMSE, with SVM showing a slightly larger RRMSE. For propensity score weighting methods, DL achieves the smallest RRMSE, while KNN has the largest RB, RSE, and RRMSE. In summary, for weak non-linearity, simpler models like LM and GAM are preferable. For complex non-linearity, XGB and DL methods demonstrate the most promising performance. The computational code is available at: https://github.com/xu1912/ImputationMethods.

5 Real data application

We conducted a real data analysis using the Household Pulse Survey (HPS) Phase 4.2 Cycle 09 data collected by United States Census Bureau in 2024 (https://www.census.gov/programs-surveys/household-pulse-survey/data/tables.html). HPS provides a rich resource to study how emergent issues are impacting U.S. households from a social and economic perspective. There are 220 variables and 51,280 records in the downloaded data table. We picked 3 variables as outcome and 16 variables as predictors in the real data analysis. After removing records missing any variable, there are 42,492 records in the complete dataset. Most of the variables are categorical, while there are a few continuous variables. The three outcome variables are Anxious status (43.8 % Yes), Trouble paying energy bill (26.1 % Yes), and Perception of price changes (90.4 % Yes). The predictor variables contain age, gender, race, education, marital status, income, number of adults/kids in the household, employment status, limitations in seeing, hearing, mobility, remembering, selfcare, understanding, and social support. The overall and stratified distribution by outcome variables of predictors were summarized in Real Data Tables.pdf at https://github.com/xu1912/ImputationMethods. Wilcoxon and Chi-square test were used to test the univariate association between predictor and outcome variable. The missing rate in HPS data table is low ($< 10\,\%$) for most of the variables. We manually generated the response index using $logit(p) = -1.14 + Race^{1/3} log_{10}(Gender + Numberofkids) + sin(Gender * Number\ of\ kids + 0.2 * Income) + Race^{1/2}/Income$, which resulted in $\sim 50\,\%$ missing rate.

We applied the previously discussed methods and procedures in estimating the proportion of Yes in the outcome variables, including LM, GAM, KNN, RF, RT, SVM, XGB, and DL. A random sample of 1000 records were drawn from the complete dataset and used in the analysis. The estimates from various methods were evaluated against the value in the complete dataset to calculate the relative bias. The result is shown in Table 3. Overall, the DML approach consistently demonstrates lower relative bias across most models and outcomes. For instance, in the Linear Model (LM) framework, DML achieves RBs of 1.92 %, 0.79 %, and 0.91 % for the respective outcomes, outperforming both IM and PS methods. Similarly, under the Generalized Additive Model (GAM), DML maintains low RBs, indicating its robustness in capturing complex, nonlinear relationships. Notably, while methods like k-Nearest Neighbors (KNN) and Random Forests (RF) exhibit higher biases, especially with the PS

TABLE 3 Relative bias (RB) (%) of different methods for the three outcome variables.

Model	Method	Anxious	Trouble paying energy bill	Perception of price changes
LM	IM	1.38	0.67	1.22
	PS	2.28	2.37	1.45
	DML	1.92	0.79	0.91
GAM	IM	1.38	0.67	1.22
	PS	2.28	2.37	1.45
	DML	2.04	1.15	1.26
KNN	IM	2.21	5.09	2.14
	PS	13.81	13.21	17.15
	DML	19.30	41.71	3.30
XGB	IM	2.95	0.54	21.45
	PS	7.77	11.78	5.69
	DML	0.95	2.84	1.01
RT	IM	2.21	0.46	1.22
	PS	2.55	4.77	0.73
	DML	1.25	2.41	0.73
SVM	IM	0.52	1.81	1.22
	PS	1.29	2.22	3.81
	DML	0.53	1.81	1.11
RF	IM	0.76	2.50	3.92
	PS	23.91	22.91	25.74
	DML	2.79	2.22	1.93
DL (0.01)	IM	5.65	8.26	2.84
	PS	5.78	10.64	6.86
	DML	1.81	4.41	0.85
DL (0.001)	IM	6.79	5.71	2.83
	PS	8.20	13.75	8.49
	DML	1.02	2.49	1.03

approach (e.g., RF-PS shows RBs exceeding 20 % across all outcomes), the DML method mitigates these biases effectively, even when paired with complex models like Extreme Gradient Boosting (XGB) and Deep Learning (DL). These findings suggest that DML, by integrating both outcome modeling and treatment assignment modeling, offers a more reliable framework for handling missing data, particularly in scenarios involving intricate data structures and potential confounding factors. The computational code and data is available at: https://github.com/xu1912/ImputationMethods.

6 Conclusion

Addressing missing data in complex, high-dimensional datasets is increasingly important in government and industrial datasets, as ignoring missingness can lead to serious nonresponse bias. Traditional statistical methods, including parametric and some nonparametric approaches, may not be well-suited for handling such data structures due to computational complexity and restrictive assumptions. Modern machine learning-based methods, such as XGBoost and deep learning, have demonstrated effectiveness in handling missing data in complex settings. However, these methods require careful tuning of hyperparameters, with techniques such as 10-fold cross-validation ensuring optimal model performance. Before applying complex machine learning methods, it is crucial to assess the presence of nonlinearity in the data. If no significant nonlinearity exists, simpler models like LM and GAM should be preferred due to their interpretability and efficiency. However, if strong nonlinear relationships are detected, advanced machine learning methods such as XGBoost and deep learning offer promising solutions. Future research should focus on statistical inference after applying machine learning-based imputation methods, ensuring that valid uncertainty quantification and inferential properties are maintained. More convinient and practical computational tools such as SAS macro and R packages need to be developed.

Acknowledgments

Dr. Chen was supported by the Oklahoma Shared Clinical and Translational Resources (U54GM104938) with an Institutional Development Award from National Institute of General Medical Sciences. The content is solely the responsibility of the authors and does not necessarily represent the official views of the National Institutes of Health. Part of the computing for this project was performed at the OU Supercomputing Center for Education & Research (OSCER) at the University of Oklahoma (OU).

References

Athey, S., Imbens, G.W., 2019. Machine learning methods that economists should know about. Annu. Rev. Econom. 11, 685–725.

Bang, H., Robins, J.M., 2005. Doubly robust estimation in missing data and causal inference models. Biometrics 61, 962–973.

Beretta, L., Santaniello, A., 2016. Nearest neighbor imputation algorithms: a critical evaluation. BMC Med. Inform. Decis. Mak. 16, 197–208.

Burgette, L.F., Reiter, J.P., 2010. Multiple imputation for missing data via sequential regression trees. Am. J. Epidemiol. 172 (9), 1070–1076.

Cao, W., Tsiatis, A.A., Davidian, M., 2009. Improving efficiency and robustness of the doubly robust estimator for a population mean with incomplete data. Biometrika 96, 723–734.

Chen, S., Haziza, D., 2017. Multiply robust imputation procedures for the treatment of item nonresponse in surveys. Biometrika 104 (2), 439–453.

Chen, S., Haziza, D., 2019. Recent developments in dealing with item non-response in surveys: a critical review. Int. Stat. Rev. 87, S192–S218.

Chen, S., Kim, J.K., 2017. Semiparametric fractional imputation using empirical likelihood in survey sampling. Statist. Theor. Relat. Fields 1 (1), 69–81.

Chen, S., Xu, C., 2023. Handling high-dimensional data with missing values by modern machine learning techniques. J. Appl. Stat. 50 (3), 786–804.

Chernozhukov, V., Chetverikov, D., Demirer, M., Duflo, E., Hansen, C., Newey, W., Robins, J., 2018. Double/debiased machine learning for treatment and structural parameters. Econom. J. 21 (1), C1–C68.

Dagdoug, M., Goga, C., Haziza, D., 2023. Imputation procedures in surveys using nonparametric and machine learning methods: an empirical comparison. J. Surv. Stat. Methodol. 11 (1), 141–188.

Farrell, M.H., Liang, T., Misra, S., 2021. Deep neural networks for estimation and inference. Econometrica 89 (1), 181–213.

Franklin, J.M., Eddings, W., Austin, P.C., Stuart, E.A., Schneeweiss, S., 2017. Comparing the performance of propensity score methods in healthcare database studies with rare outcomes. Stat. Med. 36 (12), 1946–1963.

Han, P., 2014. Multiply robust estimation in regression analysis with missing data. J. Am. Stat. Assoc. 109 (507), 1159–1173.

Hastie, T., Tibshirani, R., Friedman, J.H., Friedman, J.H., 2009. vol. 2. The Elements of Statistical Learning: Data Mining, Inference, and Prediction. Springer.

Jacob, D., 2020. Cross-fitting and averaging for machine learning estimation of heterogeneous treatment effects. arXiv Preprint arXiv:2007.02852.

Khan, H., Wang, X., Liu, H., 2022. Handling missing data through deep convolutional neural network. Inf. Sci. 595, 278–293.

Kim, J.K., Shao, J., 2013. Statistical Methods for Handling Incomplete Data. CRC Press.

Lee, B.K., Lessler, J., Stuart, E.A., 2010. Improving propensity score weighting using machine learning. Statistics in medicine 29 (3), 337–346.

Little, R.J., Rubin, D.B., 2002. Statistical Analysis with Missing Data. Wiley, Hoboken.

Pujianto, U., Wibawa, A.P., Akbar, M.I., et al., 2019. K-nearest neighbor (k-nn) based missing data imputation, 2019 5th International Conference on Science in Information Technology (ICSITech), IEEE, pp. 83–88.

Rahman, M.G., Islam, M.Z., 2011. A decision tree-based missing value imputation technique for data pre-processing, The 9th Australasian Data Mining Conference: AusDM 2011, Australian Computer Society Inc, pp. 41–50.

Sivapriya, T., Kamal, A., Thavavel, V., 2012. Imputation and classification of missing data using least square support vector machines–a new approach in dementia diagnosis. Int. J. Adv. Res. J. Adv. Res. 1 (4), 29–33.

Tang, F., Ishwaran, H., 2017. Random forest missing data algorithms. Statistical Analysis and Data Mining: The ASA Data Science Journal 10 (6), 363–377.

Zivich, P.N., Breskin, A., 2021. Machine learning for causal inference: on the use of cross-fit estimators. Epidemiology 32 (3), 393–401.

Chapter 6

Robust Bayesian approaches for leveraging external controls in drug development: Recent advances, statistical and regulatory challenges

Jixian Wang[a,*], Hongtao Zhang[b], and Ram Tiwari[c,d]

[a]*Bristol Myers Squibb, Boudry, Switzerland,* [b]*Merck & Co., Inc., North Wales, PA, United States,* [c]*Bristol Myers Squibb, Berkeley Heights, NJ, United States,* [d]*Global StatSolutions, Reston, VA, United States*

Corresponding author. e-mail address: jixianwang@hotmail.com

Chapter Outline

Handbook of Statistics, Vol. 53. https://doi.org/10.1016/bs.host.2025.06.001

Abstract

To accelerate medical device and drug development, the regulatory agencies such as the U.S. Food and Drug Administration (FDA) and the European Medicines Agency (EMA), have recently issued guidance documents advocating incorporation of real-world data (RWD) or external data, including historical clinical trial data, into current clinical trial/study. We review approaches for leveraging external data in the current study, in particular, Bayesian dynamic borrowing. As this is a wide and rapidly emerging area, our focus is on approaches that are robust to the model and prior misspecification, and confounding biases due to insufficient adjustment of prognostic factors. By robust approaches, we mean a collection of methods that not only adjust for population differences but also mitigate the impact of remaining bias after adjustment. Furthermore, they do not need full specification of the underlying model and parameter distribution and are robust to misspecification. In addition to a critical review, we also propose (1) a calibration based hypothesis testing approach based on the empirical Bayes and a minimum mean squared error (minMSE)-based treatment estimators, and (2) a general approach that modifies some estimators to be "information borrowing consistent (IBC)" while they are originally not, and report a small simulation study to evaluate the type I error and power of these tests with and without the modification for IBC.

1 Introduction

Although randomized clinical trials (RCTs) are considered the gold standard for evaluating causal treatment effects, external data, i.e., data external to the current RCT of interest, including relevant historical trials, are important information for both sponsors and regulators in the pharmaceutical industry. Early work on this topic can be traced back to Pocock (1976) who proposed six criteria for combining historical and RCT data. In summary, the criteria request that external controls have comparable baseline characteristics; were treated by the same treatment in recent clinical studies; and conducted by the same organization, with the same treatment evaluation method. These criteria seemed rather restrictive, which rule out many external data sources under consideration.

The use of external data to supplement an RCT can be considered as an information update, with external data as prior information. Hence, the Bayesian framework is a natural fit for this purpose. Full Bayesian approaches for the use of external data for RCTs, notably power prior-based approaches have been proposed; see, e.g., (Chen and Ibrahim, 2000; Ibrahim and Chen, 2000; Hobbs et al., 2011, 2013; Ibrahim et al., 2015). The power prior (PP) is a convenient way to discount external data to mitigate their impact in the Bayesian framework. How to determine the amount of discounting or the amount of borrowing (AoB) is a challenging task. An intuitive idea is to determine AoB based on similarity of the internal and external data, this adaptive approach is known as dynamic borrowing. We particularly note the empirical Bayes approach (Gravestock et al., 2017) which determines the AoB based on similarity of the internal external controls to be combined, and leads to simple borrowing rules.

External outcome data are often quite different from those of the RCT, usually due to differences in prognostic or predictive factors in the two populations, so preadjustment for population differences before borrowing can make external data more similar to RCT data. Covariate adjustment can also be incorporated in a full Bayesian modeling approach using a regression model. Although Bayesian approaches are robust to prior misspecification, it takes a much larger sample than the small control arm in the RCT to correct a misspecified external data based prior. In addition, the Bayesian model components have an impact on each other, which further decreases the robustness to model misspecification.

Some frequentist approaches rely on less model specification and can be made robust to model misspecification. They also have less interdependence between the model components. For example, an analysis with inverse probability weighting (IPW) for covariate adjustment has no impact on the propensity score. However, these frequentist approaches rely on large sample properties. Other approaches, as a compromise between the Bayesian and the frequentist approaches, have been developed for different purposes. A typical example is Bayesian bootstrap (BB) Rubin (1981), which has a close link to its frequentist counterpart: the weighted likelihood approach Newton and Raftery (1994). Recently, BB approaches based not on a likelihood but a loss function have been proposed as an approximate Bayesian inference tool (Lyddon et al., 2019; Pompe, 2021), and have been used for statistical inference with dynamic borrowing (Wang et al., 2023a; Wang and Tiwari, 2025, 2024). These approaches can use a modular structure consisting of three modules: (1) adjustment for covariates, (2) a borrowing rule determining the AoB based on the similarity after the adjustment by Module 1, and (3) the final analysis that uses the RCT and adjusted external data, and the AoB from Module 2. This sequential approach ensures no impact of the later modules on the previous ones, known as the cut-feedback approach in Bayesian analysis using IPW (McCandless et al., 2009). The overall uncertainty of the three modules can be taken into account with BB approaches without the need to specify the joint likelihood model and its priors (Wang et al., 2023a).

Treatment comparisons using an RCT and external data may include a hypothesis test and estimation. For the latter, there is an additional challenge in determining causal estimands as outlined in the ICH E9(R1) guidance (Food and Drug Administration, 2018). The causal estimand is specified in a target population, which is important for the adjustment of baseline characteristics.

Another regulatory challenge is the control of type I error, which has been a general request for RCTs with regulatory purposes. In general, using the combined RCT data and external data with dynamic borrowing for hypothesis tests, without any prespecified adjustment, cannot control type I error. It is possible to calibrate the critical boundary for restricted type I error control, but there is no guarantee that borrowing leads to power gain if RCT data are tested with the uniformly most powerful (UMP) test (Kopp-Schneider et al., 2019; van Rosmalen et al., 2018).

2 The causal estimand framework for treatment comparison using RCT and external data

First, we state formally our task of borrowing external controls to augment an internal control arm. Let $Y_i(t)$ be the counterfactual outcome of subject i when (s)he received treatment t, $t = 1$ for the test treatment and $t = 0$ for the control treatment. One common causal estimand of treatment effect is the average treatment effect (ATE) (Rubin, 1978):

$$\Delta = \mu(1) - \mu(0) \tag{1}$$

where $\mu(t) = E(Y_i(t))$, $t = 0, 1$; and the expectation is taken over the target population. In an RCT with subjects randomly assigned to the test and control arms, it can easily be estimated by the mean sample difference between the two treatment arms. If the trial is small, its variance may be very large. Therefore, often external control data are valuable to improve the comparison. However, borrowing external data is also subject to confounding biases, since the external controls are not randomized. To eliminate or reduce confounding biases, we can adjust the population difference between the two data sources using observed prognostic factors. Since this adjustment may not be sufficient, a robust approach to mitigate the impact of remaining confounding is also important for using external controls.

Next, we introduce the data structure that we are going to use. For subject i, $i = 1, \ldots, n$, the combined data from the internal trial and the external data source, $(Y_i, \mathbf{X}_i, I_i, T_i)$, consists of the outcome, covariates, indicators $I_i = 0$ for the external control and $I_i = 1$ for the internal trial, and indicators $T_i = 1$ for the test treatment and $T_i = 0$ for controls. Let $S_0 = \{i: I_i = 0\}$ and $S_1 = \{i: I_i = 1\}$ denote the sets of n_0 external controls and n_1 internal controls, respectively. As is often the case, we assume that the test treatment is not included in the external data, therefore $T_i = 0$ if $I_i = 0$. We take the RCT population as the target, then Δ in (1) can be written as

$$\Delta = E(Y_i | T_i = 1, I_i = 1) - E(Y_i | T_i = 0, I_i = 1)$$

due to the randomization of T_i. Hence, $\mu(1) = E(Y_i | T_i = 1, I_i = 1)$ can only be estimated from RCT data and dynamic borrowing involves only control patients. The above suggests that the key to using external controls in estimating Δ is to improve the estimation of $\mu(0)$, which is our main focus. For simplicity of notation, we will drop (t) in $\mu(t)$ and denote $\mu(0)$ by μ when there is no ambiguity and $\mu_j = E(Y_i | T_i = 0, I_i = j), j = 0, 1$ as the mean of controls in population j. Their difference $\delta = \mu_1 - \mu_0$ is a key parameter for dynamic borrowing. Specifically, we want to estimate μ with combined data, but mitigate the bias due to δ.

For some types of outcomes, alternative estimands have been widely used, such as the hazard ratio (HR) for time-to-event, and odds ratio (OR) for binary outcomes. Using them as causal estimands is less straightforward than using ATE or other additive estimands. Here, we also consider these estimands and assume that they are appropriate causal estimands for the RCT data.

3 Using existing external data to supplement RCT as information update: A Bayesian framework

The Bayesian framework is ideal for information update and decision making taking into account information and its uncertainty. We will show that external data borrowing, especially the determination of the AoB, can be considered in this framework. A classical approach to the use of external data is likelihood-based. Let D_0 and D_1 be the external and trial data, respectively, and $L(\theta|D)$ be the likelihood function of the parameters θ, given data D. Given a prior distribution $\pi(\theta)$, in the Bayesian framework; see, e.g., (Chen and Ibrahim, 2000; Ibrahim and Chen, 2000; Ibrahim et al., 2015), we can write the posterior density of θ given data D_0 and D_1, as

$$\pi(\theta|D_0, D_1) \propto L(\theta|D_1)L(\theta|D_0)\pi(\theta) \propto L(\theta|D_1)\pi(\theta|D_0) \tag{2}$$

where $\pi(\theta|D_0) \propto L(\theta|D_0)\pi(\theta)$ is the posterior after observing D_0, which can be considered as a prior density after updating $\pi(\theta)$ with data D_0. However, the formulation (2) treats D_0 and D_1 equally, consequently, if D_0 is quite different from D_1 due to, e.g., population differences, D_0 may have an adverse impact on the inference of θ based on (2). To discount the contribution of $L(\theta|D_0)$ in $\pi(\theta|D_0)$, the PP was proposed:

$$\pi(\theta|D_0, a_0) \propto L(\theta|D_0)^{a_0}\pi(\theta)$$

where $0 \le a_0 \le 1$ is the discounting parameter. We can also specify a prior $\pi_0(a_0)$ independent of θ. The above framework can also be extended to accommodate multiple external data sources with multiple a_0s, one for each source. Instead of independent priors for them, a joint Dirichlet prior can also be used. However, in this chapter, we will focus on the simple situation above.

Because $L(\theta|D_0)^{a_0}$ is not a proper likelihood, an approach to overcome this issue is to normalize it to make it a proper density function, which leads to the modified PP (Neuenschwander et al., 2009):

$$\pi(\theta|D_0) \propto \frac{L(\theta|D_0)^{a_0}\pi(\theta)}{\int L(\theta|D_0)^{a_0}\pi(\theta)d\theta}\pi_0(a_0).$$

When Y_i follows a normal distribution with $\theta = \mu$ as the mean and a known variance σ_e^2, with a fixed a_0, the posterior distribution of μ given D_0 and D_1 is

$$\mu \sim N(\hat{\mu}(a_0), \hat{\sigma}^2)$$

where

$$\hat{\mu}(a_0) = \hat{\sigma}^2\left(a_0\sigma_0^{-2}\bar{y}_0 + \sigma_1^{-2}\bar{y}_1\right) \tag{3}$$

$$\hat{\sigma}^2 = \left(a_0\sigma_0^{-2} + \sigma_1^{-2}\right)^{-1} \tag{4}$$

and $\bar{y}_j$ is the sample mean of Y_i in the external or internal control population $j, j=0, 1$, with variances $\sigma_j^2 = \sigma_e^2/n_j$. Therefore, the posterior mean is a weighted sum of the two sample means with the impact of $\bar{y}_0$ increasing with a_0.

The above prior structures assume that the internal and external controls share the same parameters θ. A variant of the PP is the commensurate PP (CPP) (Hobbs et al., 2011) that accommodates parameter θ_0 for the external control, e.g., by

$$\pi(\theta|\tau, D_0) \propto N(\theta|\theta_0, \sigma_\theta^2)\pi(\theta_0|D_0)$$

where σ_θ^2 is the variance of θ given θ_0, representing commensurability of D_0. With CCP, we can explicitly specify the estimand as μ_1, the mean in the RCT and the conditional distribution $N(\mu_1|\mu_0, \sigma_\theta^2)$ link the two means and provide a way to improve the estimation of μ_1 with D_0. Similarly to the discussion above for the PP, CPP also leads to a posterior mean as a weighted sum of two sample means with σ_θ^2 playing the same role as a_0.

For binary outcomes with unknown probability μ_0 as θ and a beta prior $Beta(\alpha_0, \beta_0)$, the posterior distribution of the probability of response $\pi(\mu_0|D_0)$ also takes the form of a beta distribution. Let $y_{0.} = \Sigma_{i\in S_0} Y_i$ be the sum of responded subjects our of n_0 in the external data, with PP parameter a_0, we have

$$\pi(\mu_0|D_0)^{a_0} = Beta(\alpha_0 + a_0 y_{0.}, \beta_0 + a_0(n_0 - y_{0.})) \tag{5}$$

The posterior distribution gives both D_0 and D_1, $\pi(\mu|D_0, D_1)$, is also a beta distribution (Gravestock et al., 2017).

When there are multiple external data sources, the meta-analytic-predictive (MAP) prior (Neuenschwander et al., 2010) is another way to discount external controls. It synthesizes data from all relevant data sources to form a prior distribution for the parameter of interest, θ, using a meta-analysis approach, and then this prior is updated (using Bayes theorem) by combining data from internal controls. The prior MAP can also be obtained (usually at the end of the current trial) by carrying out the meta-analysis of historical controls and internal control, assuming that they are exchangeable. In most cases, the MAP prior does not follow a known distribution. Therefore, in practice, it is usually approximated by a weighted mixture of distributions that are conjugate to the distribution of the response (Schmidli et al., 2014). The general form can be expressed as

$$\pi_{MAP}(\theta|D_{0k}, k = 1, \ldots, K) \approx \sum_{k=1}^{K} w_k \pi_k(\theta)$$

where $\Sigma_{k=1}^{K} w_k = 1$ and $\pi_k(\theta)$ is a conjugate prior for the parameter of interest θ. For example, if the parameter of interest is the binomial rate, the $\pi_k(\theta)$ components in the approximation would be beta distributions. When obtaining the approximation, only the number of components K needs to be prespecified. An approximation with 3 or 4 components is generally adequate in most cases.

The robust MAP prior (Schmidli et al., 2014; Neuenschwander et al., 2016; Zhang et al., 2023), or rMAP, is constructed as a mixture of the MAP prior and a weakly informative prior, namely a vague prior, to alleviate prior-data conflicts.

$$\pi_{rMAP}(\theta|D_{0k},\ k = 1,...,K) = w\pi_{MAP}(\theta|D_{0k},\ k = 1,...,K) + (1 - w)\pi_0(\theta).$$

The weight w controls the contribution of the data prior and plays a similar role as a_0 in the PP. A similar approach using finite mixture construction is the latent exchangeability prior (LEAP) (Alt et al., 2024), in which each historical data belongs to one of K latent classes with weights $\gamma_1, \ldots, \gamma_k$, one has the same distribution as the RCT. With this prior, a classical Bayesian approach using Markov chain Monte Carlo (MCMC) estimates estimands of interest and also γ_k's.

Both PP (CPP) and MAP (rMAP) have been widely used. MAP is based on a finite mixture distribution, while in the PP, a_0 is introduced as a discount parameter, hence the likelihood principle is not fully observed. Although it can be modified to make proper data prior with the modified PP, some may prefer the straightforward interpretation the MAP has. However, the PP has a close link to approximate Bayesian approaches which we will explore later.

Some other proposed approaches also discount the impact of D_0 without fully following the likelihood principle. For example, the elastic prior (Jiang et al., 2023) discounts the binary data prior by increasing its variance by a factor $g(T)$, where T is a measure of $\delta = \mu_1 - \mu_0$ via its estimate,

$$\pi(\mu_0|D_0, g) \sim Beta(g(T)(\alpha_0 + y_{0.}), g(T)(\beta_0 + (n_0 - y_{0.}))).$$

$g(T)$ has the same role as τ, the precision in the CPP, or as the weight of the external data. Note that $g(T)$ also discounts the original prior $Beta(\alpha_0, \beta_0)$, therefore, the relative importance between it and the external data is not changed.

Bayesian approaches are robust in the sense that if $n_1 \to \infty$ much faster than n_0, then the impact of external data tends to be null for any bounded discount parameter. However, in our context, n_1 is small, which is the main reason to borrow external data, so normal robustness does not apply here. Instead, we want the methods to have good "large" sample properties when $n_0 \to \infty$ and $n_1 \to \infty$ in the same order, i.e., $O(n_1^{-1}) = O(n_0^{-1})$. For example, the "information borrowing consistency" (Jiang et al., 2023) is one of such properties.

4 The amount of borrowing and data adaptive discount

We start with the PP and then extend our discussion to other prior structures. Determining a_0 is a challenging task, since the similarity between D_0 and D_1 is unknown. Multiple approaches have been proposed to estimate the similarity and then use it to determine the AoB.

Selecting the right similarity measure is important for determining a_0. Although (2) is based on the distributions of D_0 and D_1, our analysis often focuses on a key parameter, e.g., Δ given in (1), or specifically on the mean of controls. Therefore, we only need similarity between μ_1 and μ_0, the mean responses of the internal and external controls, respectively; hence $\delta = \mu_1 - \mu_0$ is a key measure for it. For example, if $\mu_0 = \mu_1$, but Y_i in D_0 has a higher variance than in D_1, then, although the distributions of D_0 and D_1 are different, we should still take $a_0 = 1$.

One simple way to determine a_0 is the test-then-pool approach (Li et al., 2020) that tests the difference between D_0 and D_1, then sets $a_0 = 1$ if the test is not significant, otherwise sets $a_0 = 0$, which means no use of D_0. However, a statistically insignificant difference does not mean that $\mu_0 = \mu_1$. A natural extension of this approach is to make a_0 as a function of the p-value of the test. This approach can also be made Bayesian, e.g., by using the prior predictive p-values, or posterior predictive probability (PPP) to determine a_0 (Zhang et al., 2023). Another approach is to weight the external data by the probability of equivalence: $w = P(|\delta| < d|D_0, D_1)$, with the threshold d to be determined (Bennett et al., 2021). Jiang et al. (2023) proposed for their elastic priors that the weight $g(T)$ depends on a test statistic T for $\delta = 0$.

An empirical Bayes (EB) approach estimates a_0 by maximizing its marginal likelihood given D_0 and D_1. For some simple form of $f(D|\theta)$, e.g., normal or binomial densities, the empirical Bayes approach leads to simple analytical forms for a_0 (Gravestock et al., 2017; Gravestock and Held, 2018). Assuming a normal distribution for $\bar{y}_j$, $j = 0, 1$, we can maximize the marginal likelihood $L(a_0|D_0, D_1)$, which leads to a closed form for a_0:

$$\hat{a}_0 = \frac{\hat{\sigma}_1^2}{\max\left[\hat{\delta}^2, \hat{\sigma}_1^2 + \hat{\sigma}_0^2\right] - \hat{\sigma}_0^2}. \tag{6}$$

where, $\hat{\delta} = \bar{y}_1 - \bar{y}_0$, $\bar{y}_1$, and $\bar{y}_0$ are the sample mean response of internal and external controls with sample variances $\hat{\sigma}_1^2$ and $\hat{\sigma}_0^2$, respectively (Gravestock et al., 2017). Then, μ can be estimated by (3) with a_0 replaced by $\hat{a}_0$ above. Generally, some other approaches for estimation of a_0 or equivalent can also be considered as EB, and can also be used on other prior structures. For example, applying it to the rMAP prior also results in a simple borrowing rule for the binary outcome and the beta prior, with a weight determined by the posterior p-value (Zhang et al., 2023).

Without relying on the likelihood framework, we can consider a simple linear combination estimator for μ

$$\hat{\mu}_c(a^*) = \frac{\bar{y}_1 + a^*\bar{y}_0}{1 + a^*}$$

in which a^* is the relative weight for data D_0. As a linear combination of $\bar{y}_1$ and $\bar{y}_0$, this estimator is equivalent to that in (4) with a 1–1 relationship between a^*

and a_0: $a^* = a_0\sigma_1^2/\sigma_0^2 = a_0 n_0/n_1$. We can also choose a^* optimally according to some criteria, e.g., to minimize the mean squared error (MSE) of the combined estimator, which was proposed in some early effort to use external data for clinical trials (Galwey, 2016). One key step to apply this rule is to estimate δ^2 in MSE. One natural way is to use $\hat{\delta}^2$ and correct for the variance included in it. Interestingly, in the situation of normally distributed $\bar{y}_j$, this leads to the same estimator as the empirical Bayes approach (Galwey, 2016); that is, a^* that minimizing the MSE with variance correction in $\hat{\delta}^2$ is equivalent to $\hat{a}_0$ determined by (6). This will be referred to as the cMinMSE rule.

However, with a moderate sample size, this rule leads to full borrowing ($\hat{a}^0 = 1$) until $\hat{\delta}^2$ becomes relatively large compared to the variance in the MSE. Therefore, this rule leads to considerable biases when $\hat{\delta}^2$ is moderate. An alternative is to not adjust for the variances in $\hat{\delta}^2$. This is equivalent to

$$\hat{a}^* = \frac{\hat{\sigma}_1^2}{\hat{\delta}^2 + \hat{\sigma}_0^2}. \tag{7}$$

which will be referred to as minMSE rule. In this way, $\hat{a}^*$ reduces with an increase in $\hat{\delta}^2$ even when it is small. Numerical examples (Wang and Tiwari, 2024) showed that this approach leads to a smaller MSE than cMinMSE, except when the true δ is quite small.

Now, we are in a position to discuss some ideal large sample properties of the AoB such as $\hat{a}_0$, $\hat{a}^*$ or their equivalence, such as $g(T)$ in the elastic prior and w in the rMAP. These may be summarized by "information borrowing consistent" (IBC) (Jiang et al., 2023), that is, as $n_1 \to \infty$ (and $n_0 \to \infty$), $\hat{a}_0 \to_p 1$, where $\to_p$ denotes convergence in probability, when $\delta = 0$; and $\hat{a}_0 \to_p 0$ when $\delta \neq 0$. Some borrowing rules, including the empirical Bayesian rule (Gravestock et al., 2017) and the test statistic-based one rescaled by the sample size (Jiang et al., 2023) satisfy this criterion.

For the minMSE rule, although we also have $\hat{a}^* \to_p 0$ when $\delta \neq 0$; when $\delta = 0$ it does not lead to full borrowing, due to the impact of the variance in $\hat{\delta}^2$, hence it is not IBC. Although this $\hat{a}^*$ has the inconsistency issue in this ideal situation, it may not be a serious practical problem, since the AoB is often capped and full borrowing, which treats external data the same as internal ones, is often not desirable.

In fact, $\hat{a}^*$ can easily be made IBC. For example, we can weight the $\hat{\delta}^2$ term by $\max(n_1, n_0)^{-1/2}$, the same weight as used to rescale the test statistic for $\delta = 0$ by Jiang et al. (2023) so that

$$\hat{a}^*_{ibc} = \frac{\hat{\sigma}_1^2}{\hat{\delta}^2/\max(n_1, n_0)^{1/2} + \hat{\sigma}_0^2}$$

is IBC. To verify this claim, when $\delta = 0$, we have $\hat{\delta}^2 \to_p \sigma_1^2 + \sigma_0^2$ and $\hat{\sigma}_k^2 \to_p \sigma_k^2$, $k = 0, 1$, so $\hat{a}_{ibc}^* \to_p n_0/n_1$ as desired. When $\delta > 0$, since $\sigma_k^2 = \sigma_e^2/n_k$ and $\hat{\delta}^2 \to_p \delta^2 + \sigma_1^2 + \sigma_0^2 \to_p \delta^2$, hence $\hat{a}_{ibc}^* \to_p 0$. Obviously, $\max(n_1, n_0)^{-1/2}$ is not the only IBC weight. For example, $\log(\max(n_1, n_0))^{-1}$ also leads to an IBC, but with very different small sample properties from that with weight $\max(n_1, n_0)^{-1/2}$.

The similar approach can also be used to convert some other non-IBC borrowing rules into an IBC rule. For example, the test-then-pool rule is not IBC, since when $\delta = 0$ the p-value has the uniform distribution within [0,1], hence a threshold at α level leads to $\alpha\%$ chance to have $\hat{a} = 0$. However, with a weight w_n on the test statistic T_n so that $w_n T_n \to_p 0$ under H_0: $\delta = 0$, the weighted test-then-pool rule leads to an IBC rule. A weight $w_n \to_p 0$ fits this purpose, since $T_n \sim_a N(0, 1)$, where $\sim_a$ denotes weak convergence, under H_0, then $w_n T_n \to_p 0$ by Slutsky's lemma (van der Vaart, 1998). In fact, this is the approach of rescaled test statistic in Jiang et al. (2023). The same adaptation can also be applied to, e.g., the PPP based Zhang et al. (2023) and the equivalence probability weighted (Bennett et al., 2021) rules.

As a general approach we can consider the term $\max(n_1, n_0)^{-1/2}$ as a function $r(n_1, n_0)$ and, for $\hat{a}_{ibc}^*$ to be IBC, we only need that when $\delta = 0$,

$$(1/n_1 + 1/n_0) r(n_1, n_0) n_0 \to_p 0,$$

and when $\delta \neq 0$ and is fixed,

$$n_0^{-1}/r(n_1, n_0) \to_p 0.$$

Therefore, $r(n_1, n_0)$ should tend to zero, but at a slower rate than n_0^{-1} (and n_1^{-1}, since we have assumed $O(n_1^{-1}) = O(n_0^{-1})$). As a simple example, we can use

$$r(n_0, n_1) = n_0^{-\rho}$$

with $0 < \rho < 1$. The optimal choice of ρ obviously depends on the true parameter values, and therefore how to determine it is another technical issue.

The empirical Bayes approach (Gravestock et al., 2017) for the estimation of μ_1 based on a normal or binomial distribution is simple to use. It can also be used when similarity is measured by other estimands such as the hazard ratio or odds ratio. For the former, denote the log HR of TTE of internal controls versus external controls by τ, which can be estimated by fitting a Cox regression model with I_i as covariate only. Its variance $\text{var}(\hat{\tau}) = \sigma_\tau^2$ is approximately proportional to $1/d_0 + 1/d_1$, where d_0 and d_1 are the numbers of events of internal and external controls, respectively. Consequently, the relative contributions to $\text{var}(\hat{\tau})$ of the internal and external controls are $1/d_0$ and $1/d_1$, respectively (Wang et al., 2025). Therefore, let $\hat{\sigma}_\tau^2$ be the estimated variance given by a fitted Cox model. We first calculate

$\bar{\sigma}_\tau^2 = \hat{\sigma}_\tau^2/(1/d_0 + 1/d_1)$. Then, we can approximate the portions of the variance from internal and external controls as $\bar{\sigma}_\tau^2/d_0$ and $\bar{\sigma}_\tau^2/d_1$ while keeping the total variance $\hat{\sigma}_\tau^2$ as given in the fitted Cox model unchanged. To determine a_0, $\hat{\delta}$, $\hat{\sigma}_0^2$ and $\hat{\sigma}_1^2$ in Eq. (6), given below in Algorithm 3, can be replaced by $\hat{\tau}$, $\bar{\sigma}_\tau^2/d_0$ and $\bar{\sigma}_\tau^2/d_1$, respectively.

The same approach of Wang et al. (2025) can be adapted when similarity is measured by the odds ratio of the response of the internal and external controls. Let $\hat{p}_1$ and $\hat{p}_0$ be the (adjusted) means of response of internal and external controls, respectively, and β_r be log-OR, which can be estimated by logistic regressions. A well known variance estimator for the estimated log-OR: $\hat{\beta}_r$ is

$$\mathrm{var}(\hat{\beta}_r) \approx \frac{1}{n_1\hat{p}_1(1-\hat{p}_1)} + \frac{1}{n_0\hat{p}_0(1-\hat{p}_0)}$$

in which the first and second terms are the variances of log-Odd of external and internal controls, respectively. Therefore, to determine a_0, $\hat{\delta}$, $\hat{\sigma}_0^2$ and $\hat{\sigma}_1^2$ in (6), below in Algorithm 3, can be replaced by $\hat{\beta}_r$, $\hat{\sigma}_0^2 = 1/(n_0\hat{p}_0(1-\hat{p}_0))$ and $\hat{\sigma}_1^2 = 1/(n_1\hat{p}_1(1-\hat{p}_1))$, respectively.

Finally, since both the HR and log-OR approaches are based on asymptotic normal distributions, both the cMinMSE and minMSE rules can apply. The approach may not be sensitive to differences between the two hazard functions if they are not proportional; therefore, special care should be taken to check in such a situation.

An interesting question is whether or not the borrowings based on log-HR and log-OR are ICB. It is reasonable to assume that the proportional hazard assumption holds with the RCT data, i.e., let $h_0(t)$ be the baseline hazard function of the internal controls and let us consider the borrowing based on log-HR uses a Cox model

$$h_0(t)\exp(\tau I_i).$$

When $\tau \neq 0$, it is obvious that $\hat{a}^* \to 0$. Because $E(\hat{\tau}^2) \approx 1/d_0 + 1/d_1$, when $\tau = 0$, without a correction for variances, it is not IBC. However, the approaches to making (7) IBC also applies here. The borrowing based on log-OR has similar properties when $\beta_r \neq 0$, and can be made IBC in the same way using the approach for the same task for $\hat{a}^*$. However, $\beta_r = 0$ is equivalent to $p_1 = p_0$, so any difference between p_0 and p_1 is reflected in β_r, and there is no such situation of insensitive difference from β_r, as we might have in log-HR based borrowing.

5 Adjustment for population differences

The external outcome data are often rather different from those of the RCT, usually due to differences in prognostic or predictive factors in the two populations, so adjustment for population differences can make external data more similar to RCT data. Specifically, we would like to make μ_0 as close to μ_1 as possible after adjustment for covariates.

The most simple approach is to directly adjust for the difference in covariates $\mathbf{X}_i$ using an outcome regression (OR) model

$$Y_i = g(\mathbf{X}_i, \beta) + \epsilon_i \tag{8}$$

where $g(\mathbf{X}_i, \beta)$ is a mean model (e.g., a linear model or a GLM) with parameters β and ϵ_i is a zero-mean error term. The model can be fitted to data D_0 to obtain the estimate of β: $\hat{\beta}$. Then we can predict the outcome for the population of S_1 and estimate its mean by

$$\hat{\mu}_0 = n_1^{-1} \sum_{i \in S_1} g(\mathbf{X}_i, \hat{\beta}) \tag{9}$$

where we use $\hat{\mu}_0$ to emphasize that it is a D_0 based estimate, although is aimed at estimating μ_1. This approach eliminates the bias due to the difference in $\mathbf{X}_i$ if the model is specified correctly. Obviously, confounding factors not included in $\mathbf{X}_i$ cannot be controlled in this way.

An interesting question is whether we should estimate μ_1 of the whole trial population S_1, as in Eq. (9) or only the internal control patients. The former has the advantage of adjusting the D_0 data to the target population, but when the RCT is not well balanced, this approach may cause problems when determining the AoB based on similarity between external and internal controls, since dissimilarity may be caused due to the difference between internal control and active treated patient.

A commonly used approach for population adjustment is based on propensity score (PS) defined as the probability of subject i belonging to the internal control, given the covariates $\mathbf{X}_i$:

$$e_i = P(I_i = 1|\mathbf{X}_i) = E(I_i|\mathbf{X}_i) \tag{10}$$

which is often modeled by a logistic regression as $e_i = 1/(1 + \exp(-\gamma^T \mathbf{X}_i))$ with parameters γ. The key assumptions for PS-based adjustment include $0 < E(I_i|\mathbf{X}) < 1$ for any $\mathbf{X}$, and that potential outcomes are independent of I_i conditional on $\mathbf{X}_i$. Details can be found in Rosenbaum and Rubin (1983). Under these assumptions, $y_i \perp I_i | e_i$, we can use the IPW estimator

$$\hat{\mu}_{0,ipw} = n_0^{-1} \sum_{i \in S_0} w_i Y_i \tag{11}$$

where the weight $w_i = e_i/(1 - e_i)$ is the odd of being in the RCT population to transport external data to the internal population, to estimate the control mean μ. It is straightforward to show that $E(\hat{\mu}_{0,ipw}) = \mu_1$ when the PS model is specified correctly. This approach is particularly useful when the outcome model such as (8) is more difficult to specify or fit than the PS model (10), e.g., when Y_i is the indicator of a rare event.

In addition to IPW, PS can also be used for matching and stratification to borrow external data; see, e.g., (Lin et al., 2018). The adjustment can be combined with dynamic borrowing in the following steps. First, data are stratified by e_i into K strata. Then, similarity is measured with each stratum, and the AoB is determined by the similarity. Therefore, the contribution of D_0 patients is individually determined for each stratum.

Another powerful but less used approach is covariate balancing (CB), which can be traced back to classical calibration methods in survey sampling (Deville and Särndal, 1992) and can be combined with PS approaches (CBPS) (Imai and Ratkovic, 2014). Let $\bar{\mathbf{X}}_1$ be the sample mean of $\mathbf{X}_i$ of the internal controls. Our goal is to balance $\mathbf{X}_i$ between the trial and external populations by weighting n_0 subjects in D_0 such that

$$\bar{\mathbf{X}}_1 = \sum_{i \in S_0} w_i \mathbf{X}_i. \tag{12}$$

For this purpose, we find the weights w_i's that

$$\begin{aligned} &\min_{w} \sum_{i \in S_0} \rho(w_i) \\ &subj. \sum_{i \in S_0} w_i \mathbf{X}_i = \bar{\mathbf{X}}_1 \\ &\sum_{i \in S_0} w_i = 1, \end{aligned} \tag{13}$$

where $\rho(w_i)$ is a distance between w_i and a constant weight. Commonly used ones are quadratic and entropy functions. The last constraint can be absorbed into the first one by including an intercept in $\mathbf{X}_{ij}$. The weights can be found using the Lagrange method.

With the weights determined above, the CB estimator is

$$\hat{\mu}_{0,cb} = \sum_{i \in S_0} w_i Y_{i1} \tag{14}$$

which depends on the distance function via w_i. It is a consistent estimator only when the CB removes all confounders due to population differences.

For some distance functions, the CB estimator is equivalent to the generalized regression estimator based on a linear regression model $Y_i = \beta^T \mathbf{X}_i + \epsilon_i$ with $E(\epsilon_i) = 0$. To predict μ_1, the model can be fitted to the data D_0 to obtain $\hat{\beta}$, then the mean response is predicted by $\bar{\mathbf{X}}_1 \hat{\beta}$, which is exactly (14) (Deville and Särndal, 1992). Therefore, when the model is correct, we have $E(\hat{\mu}_{0,cb}) \to \mu_1$ as $n_0 \to \infty$. However, this equivalence, that is, $\bar{\mathbf{X}}_1 \hat{\beta} = \hat{\mu}_{0,cb}$, holds even when the linear model is misspecified.

To improve the CB adjustment for more complex relationships than the linear model, one can replace $\mathbf{X}_i$ with a set of prognostic scores as functions of $\mathbf{X}_i$, estimated by fitting response-covariate models. This approach is known as model-assisted CB (Wu and Sitter, 2001). However, we will focus on the simple version.

IPW can also be considered as a CB approach as it balances the distributions of $\mathbf{X}_i$ in D_0 and D_1, since with IPW each weighted patient has the same probability of belonging to any of the populations. However, this is a balance in probability, and some covariates may be quite imbalanced by chance when the sample size is small. In contrast, the above CB approach ensures balancing of all covariates in the weighted data. Nevertheless, IPW is useful when $\sum_{i \in S_0} w_i \mathbf{X}_i = \bar{\mathbf{X}}_1$ cannot be achieved or when some weights become extreme or negative.

6 The final analysis using combined data

The final step of the analysis combines the data D_0, D_1 with the covariate adjustment and the AoB. In contrast to the previous sections in which the response under control treatment is the only concern, this step also involves the active arm in the RCT, since the final statistical inference should take into account all uncertainty in the comparison with the active treatment. Although the mean response for the treated: $\mu(1)$ can be estimated separately using the RCT data only, to estimate Δ, estimating equations (EE) for combined data is more convenient to use. For example, to use combined data to estimate the odds ratio of the active treatment verses the control, with IPW adjustment, we fit a logistic regression by solving the EEs:

$$\begin{aligned} \sum_{i=1}^{n} (1 - T_i) W_i (Y_i - \text{expit}(\beta_0)) &= 0 \\ \sum_{i \in S_1} T_i (Y_i - \text{expit}(\beta_0 + \beta_t T_i)) &= 0 \end{aligned} \tag{15}$$

where $\text{expit}(a) \equiv 1/(1 + \exp(-a))$, $W_i = \hat{a}^0 e_i/(1 - e_i)$ when $I_i = 0$ and $W_i = 1$ otherwise. The weight W_i is a combination of IPW and the AoB for external data. When $\hat{a}^0 = 0$, the external data do not have a contribution and β_t is estimated only with the RCT data. The EE approach applies to other weighting approaches, such as CB weighting.

The same approach as above can be applied to the estimation of HR for TTE endpoint. For weighting-based approaches, we can replace (15) with the partial likelihood score equation for Cox regression. With direct adjustment, we can also augment the partial likelihood score equation with a term with predicted event time of the pseudo-controls, as we did for log-OR in (15). The same approach of augmenting D_1, as we did for estimating OR, can also be used.

Next, we consider the asymptotic properties of the final estimators, particularly the impact of AoB when $O(n_1^{-1}) = O(n_0^{-1})$, hence $O(\hat{a}^0) = O(\hat{a}^*)$. Since the key role of determining a^* is to mitigate the impact of δ when $\delta \neq 0$. Intuitively, we would like $\hat{a}^* \rightarrow_p 0$ sufficiently fast when $n_0 \rightarrow \infty$, so that asymptotically the external data have no impact on $\hat{\mu}_c$. We focus on the simple form of the estimator

$$\hat{\mu}_c = \frac{\hat{\mu}_1 + \hat{a}^*\hat{\mu}_0}{1 + \hat{a}^*}$$

which can be written as

$$\hat{\mu}_c = \hat{\mu}_1 + \frac{\hat{a}^*\hat{\delta}}{1 + \hat{a}^*}.$$

Asymptotically, we have

$$\sqrt{n_1}\,(\hat{\mu}_c - \mu_c) = \sqrt{n_1}\,(\hat{\mu}_1 - \mu_1) + \sqrt{n_1}\frac{\hat{a}^*\hat{\delta}}{1 + \hat{a}^*}$$

When $\delta \neq 0$, $\hat{\delta} \to_p \delta$, hence for $\sqrt{n_1}\,(\hat{\mu}_1 - \mu_1) \sim_a N(0, \sigma_1)$, we need $\sqrt{n_1}\frac{\hat{a}^*\hat{\delta}}{1+\hat{a}^*} \to 0$ in probability. That is, $\hat{a}^* = O(n_1^{-\rho})$ with $\rho > 1/2$. In the last section, we have shown that for IBC we need $r(n_0, n_1) = n_1^{-\rho}$ with $0 < \rho < 1$. The above request to limit the range further to $1/2 < \rho < 1$.

For other estimands such as the log-odds estimated by a logistic model, a similar analysis can be carried out on the EE for the estimation. Taking the estimation of OR with IPW (or any other) weights as an example, for a fixed $\hat{a}^*$, the first equation of (15) can be written

$$S_n(\beta) = \sum_{i\in S_1} (1 - T_i)(Y_i - \text{expit}(\beta_0)) + \hat{a}^* \sum_{i\in S_0} \frac{e_i}{1 - e_i}(Y_i - \text{expit}(\beta_0)) = 0$$

in which the first sum is unbiased, hence

$$n_1^{-1/2} \sum_{i\in S_1} (Y_i - \text{expit}(\beta_0)) \sim_a N(0, \sigma^2).$$

The second term, i.e., the contribution of external data to the EE, may not be unbiased if the preadjustment does not eliminate all the bias. Let $\delta_e \equiv E(e_i(Y_i - \text{expit}(\beta_0))/(1 - e_i))$, for the second term to have asymptotically negligible impact, that is, $n_1^{-1/2} S_n(\boldsymbol{\beta}) \sim_a N(0, \sigma^2)$; it is sufficient to choose an $\hat{a}^*$ so that

$$n_0^{-1/2}\hat{a}^* \sum_{i\in S_0} \frac{1 - e_i}{e_i}(Y_i - \text{expit}(\beta_0)) \to_p 0$$

(van der Vaart, 1998). Since $n^{-1}\sum_{i\in S_0} e_i(Y_i - \text{expit}(\beta_0))/(1 - e_i) \to_p \delta_e$, the above holds for $\hat{a}^* = O(n^{-\rho})$ with $\rho > 1/2$, the same condition as for that in $\hat{\mu}_c$.

7 A modular integrated approach

In this section we propose a modular approach to integrate population adjustment, borrowing rule, and analysis with combined data together as three modules in a sequential way. Depending on specific scenarios, a module can be replaced or adapted as needed. In the following, we present approaches for some typical tasks and show how to adapt them.

7.1 Estimation of ATE estimand with IPW adjustment

We consider dynamic borrowing for a small RCT to compare, to compare the mean difference Δ between treated and control patients in the RCT population. The IPW approach is used for preadjustment. The approach may consist of the following steps:

1. Fit a PS model to determine the weights for IPW
2. Estimate the weighted mean of external controls
3. Use mean difference between internal and weighted external controls to determine the AoB
4. Calculate the mean of combined internal and external controls, the latter are weighted by the product of the IPW weights (as used in Step 2) and the AoB as a percentage, and the difference between it and the mean response of the treated.

Steps 1 and 2 consist of the pre-adjustment module, Step 3 the borrowing rule module, and Step 4 the final analysis module.

The same steps applies to all weight-based adjustments and one can replace the IPW weights by the CB and CBPS weights. Less obvious is that it also applies to the adjusted mean with a linear regression model, since it can also be written in the form of a weighted mean.

7.2 Estimating treatment difference in hazard ratio with IPW adjustment

We consider the same task as above, but the endpoint is time-to-event, e.g. progression-free survival and use the HR as the causal estimand and, under the assumption that proportional hazard holds in the RCT data. A modified approach may consist of the following steps: .

1. Fit a PS model to determine the weights for IPW.
2. Estimate the log-HR with a Cox regression with an indicator for internal or external data with IPW weights for the external controls.
3. Use the log-HR to determine the AoB.
4. Fit a Cox regression with the RCT (treated and controls) and external controls, the latter are weighted by the product of the IPW weights (as used in step 2) and the AoB as a percentage.

The same steps can be used for PS matching and stratification. Taking the latter as an example, we can replace Steps 2 and 4 with a stratified Cox regression. For, e.g., the HR in the RCT population, stratified or matched external data should also be properly weighted to match the RCT population.

The two examples above show that the three modules can be adapted individually to meet their own needs. To state the algorithm in a general form, we use the estimator $\hat{\mu}_c$ as an example. The three modules are denoted by $PopAdj(Y_i, I_i, ...)$, $Borrow(\hat{\sigma}_j, \hat{\mu}_j, ...)$ and $CombEst(\hat{\mu}_j, T_i, \hat{a}^*, ...)$ with the necessary variables. They may also depend on some other variables, denoted as "...", that are not from the upstream modules. The BB approach in the next section also uses the modular structure.

8 Approximate robust Bayesian inference

The above approaches combine three modules: population adjustment, determining the AoB and analyzing the combined data. The final inference of the causal treatment effect such as Δ in (1) should take into account the uncertainty in all modules. For Bayesian inference, the classical Bayesian approaches need the specification of the model and the distribution of model parameters and may not be robust to their misspecification. Recent development on approximate Bayesian approaches, in particular the BB method, provides an alternative to the classical Bayesian approaches. These approaches are robust to model misspecification to some extent and can be easily implemented.

8.1 Approximate Bayesian analysis with loss functions

Let the likelihood given the external data be

$$L(\theta|D_0) = \prod_{i \in S_0} f(Y_i|\theta),$$

where $f(Y|\theta)$ is the density function of Y_i. It is well known that MLE $\hat{\theta}$ has an asymptotically normal distribution: $\sqrt{n_0}(\hat{\theta} - \theta) \sim N(0, \sigma_\theta^2)$, while the asymptotic posterior distribution of θ satisfies $\sqrt{n_0}(\theta - \hat{\theta}) \sim N(0, \sigma_b^2)$, where $\sigma_\theta^2, \sigma_b^2$ are unknown variances. The same property holds when $f(Y|D)$ is replaced by $\exp(-l(Y, \theta))$ where $l(Y, \theta)$ is a loss function, e.g., a normal quasi-log-likelihood with known variance $l(Y, \theta) = (Y - \theta)^2$. A quasi-likelihood can be written as

$$L^q(\theta|D_0) = \prod_{i \in S_0} \exp(-l(Y_i, \theta))$$

and, with some conditions, the maximum quasi-likelihood estimate also follows $\sqrt{n_0}(\hat{\theta} - \theta) \sim N(0, \sigma_q^2)$, but in general $\sigma_q^2 \geq \sigma_\theta^2$.

This loss function has the same structure to the PP, since the data prior $L^q(\theta|D_0)$ can be discounted by a parameter a_0 as $\exp(-a_0 l(Y, \theta))$. Together with the internal controls, a quasi-likelihood can be written as

$$L^q(\theta|D_0, D_1) = \prod_{i \in S_1} \exp(-l(Y_i, \theta)) \prod_{i \in S_0} \exp(-al(Y_i, \theta)),$$

and the contribution of external controls depends on a_0.

The loss function can be generalized to include covariates, e.g., $l(Y|X, \theta) = (Y - X^T\theta)^2$ is the loss function for a regression analysis. More generally, an M-estimator (van der Vaart, 1998) which minimizes a loss function, an estimator that maximizes a composite likelihood, or a Z-estimation which solves estimating equations can also be considered as a maximum quasi-likelihood approach. Furthermore, CB can also be considered as an M-estimation approach (Wang and Tiwari, 2025). Therefore, most pre-adjustment and the final analysis methods can be combined in this framework. The advantage of this approach can be seen when using BB to generate a posterior sample of the estimand of interest, taking into account all uncertainty in these steps, including the estimation of AoB.

8.2 Bayesian bootstrap for approximate Bayesian inference

This section briefly reviews BB for generating posterior samples with non-informative nonparametric priors (Rubin, 1981). Assuming a uniform Dirichlet prior $Dir(1, \ldots, 1)$, a posterior sample of θ, given D_0, can be generated by a weighted MLE:

$$\hat{\theta}^b = \arg\max_{\theta} \prod_{i \in S_0} f(Y_i|\theta)^{\xi_i^b} \tag{16}$$

with weights generated from the uniform Dirichlet $\xi_i^b \sim Dir(1, \ldots, 1)$. Given MLE $\hat{\theta}$, $\sqrt{n_0}(\hat{\theta}^b - \hat{\theta})$ and $\sqrt{n_0}(\theta - \hat{\theta})$ have the same asymptotic posterior distribution (Newton and Raftery, 1994). Similar properties hold when $f(Y_i|\theta)$ in (16) is replaced by $\exp(-l(Y_i, \theta))$ (Lyddon et al., 2019). Although the properties are derived based on minimizing loss function, under some technical conditions, this approach is equivalent to solving a set of estimating equations corresponding to (16) either with $f(Y_i|\theta)$ or with $\exp(-l(Y_i, \theta))$ (van der Vaart, 1998). Furthermore, the weighted likelihood or negative loss function leads to weighted estimating equations. The connection to the CB based adjustment method is less obvious, but we can show, based on the recent development of asymptotic properties of CB estimators, that similar asymptotic properties hold for estimators based on CB adjustment (Wang and Tiwari, 2025).

Although asymptotic distributions of $\hat{\theta}$ are available, the BB sample generally better represents the distribution with moderate n_0 and n_1 and is easier to implement in practice than to calculate the asymptotic variance.

For Bayesian inference based on $\hat{\mu}_c(a)$, our approach is to generate BB-based posterior samples $\hat{\mu}_1^b$ and $\hat{\mu}_0^b$ given D_1 and D_0, respectively, as described above, then obtain the corresponding BB sample $\hat{a}^{*b}$ from (7) using BB posterior samples to replace $\hat{\sigma}_j^2, j = 0, 1$ and $\hat{\delta} = \hat{\mu}_1 - \hat{\mu}_0$. Finally, we put the three components in (3) to obtain a posterior sample based on BB $\hat{\mu}_c^b$, given D_0, D_1. It can be shown that $\hat{\mu}_c^b$ is a posterior sample of $\mu_c(a) = (\mu_1 + a\mu_0)/(1+ a)$ with some desired asymptotic properties and small sample properties demonstrated with simulation studies (Wang and Marion-Gallois, 2023; Wang and Tiwari, 2025).

The following general algorithm for BB uses the notation of the three modules with BB weights when needed, e.g., *PopAdj*(Y_i, I_i, ξ_i, $\mathbf{X}_i$, …): Some remarks are in order:

(i) To generate a BB posterior sample $\hat{\mu}_c^1$, ..., $\hat{\mu}_c^B$, the algorithm requests BB for each of the modules. In the following, we will examine if BB is applicable to some common approaches for these modules or not.

(ii) This approach is sequential, so that a downstream one does not have an impact on its upstream ones. In contrast to Bayesian joint modeling approaches, this approach belongs to the class of "cut feedback" in the Bayesian IPW approach (McCandless et al., 2009). It is more robust to model misspecification and an easy implementation with a bootstrap algorithm without the need to fit joint models.

(iii) We explicitly include the BB weights ξ_i^b in each of the modules, which may not be necessary, depending on the method used for them.

Algorithm 1. Bayesian bootstrap for posterior sample of μ_0, given D_0, D_1.

Require: $Y_i, \mathbf{X}_i, B, n_j$

for Bootstrap run $b = 1$ to B **do**

For each data source, generate weights $\xi_i^b, i = 1, ..., n_j$, from the uniform Dirichlet distribution $Dir(1, ..., 1)$

Population adjustment

$$\hat{\mu}_j^b = PopAdj(Y_i, I_i, \xi_i^b, \mathbf{X}_i), j = 0, 1.$$

Determine AoB

$$\hat{a}^{*b} = Borrow(\hat{\mu}_j^b, T_i, \xi_i^b,)$$

Calculate combined estimate for μ

$$\hat{\mu}_c^b = CombEst(\mu_j^b, \hat{a}^{*b}, \xi_i^b)$$

end for

Output $\hat{\mu}_c^1, ..., \hat{\mu}_c^B$.

We give two examples of applying the above approach based on the methods for pre-adjustment, determining AoB, and final analysis with BB for each step. The algorithm also considers the comparison with the treated patients and produces the BB posterior sample for the causal estimand. The algorithm can be adapted for other preadjustment, by replacing Steps 3 and 4 with, for example, the steps to generate the CB weights as described in Section 5. The AoB rule can also be changed to the maxML approach, as shown in the next example.

Algorithm 2. Bayesian bootstrap for posterior sample of μ_1, given D_0, D_1 with IPW, minMSE for determining AoB.

Require: $Y_i, T_i, I_i, \mathbf{X}_i, B, n_j$

1: **for** Bootstrap run $b = 1$ to B **do**

2: Generate weights $\xi_i^b, i = 1, ..., n$ from the uniform Dirichlet distribution $Dir(1, ..., 1)$

3: (For IPW) Fit a logistic regression model with I_i as the dependent variable and $\mathbf{X}_i$ as covariates using weights ξ_i and obtain estimated parameters $\hat{\gamma}^b$

4: Calculate weights

5:
$$w_i^b = \frac{e_i^b}{1 - e_i^b}, \quad e_i^b = \text{expit}(\mathbf{X}_i \hat{\gamma}^b)$$
$$\hat{\mu}_1^b = \sum_{i \in S_1} \xi_i^b Y_i / \sum_{i \in S_1} \xi_i^b$$
$$\hat{\mu}_0^b = \sum_{i \in S_0} \xi_i^b w_i^b Y_i / \sum_{i \in S_0} \xi_i^b w_i^b.$$

6: Obtain BB sample $\hat{\sigma}_0^{2b}$ and $\hat{\sigma}_1^{2b}$ using the same BB weights.

7: Obtain $\hat{\delta}^b = \hat{\mu}_1^b - \hat{\mu}_0^b$, and
$$\hat{a}^{*b} = \hat{\sigma}_0^{2b} / (\hat{\sigma}_1^{2b} + \hat{\delta}^{2b})$$

8: Calculate
$$\hat{\mu}_c^b = \frac{\hat{\mu}_0^b + \hat{a}^{*b} \hat{\mu}_1^b}{1 + \hat{a}^{*b}}$$

9: Calculate $\hat{\Delta}^b$
$$\hat{\Delta}^b = \hat{\mu}^b(1) - \hat{\mu}_c^b$$
where
$$\hat{\mu}^b(1) = \sum_{i \in S_1} T_i \xi_i^b Y_i / \sum_{i \in S_1} T_i \xi_i^b$$
is the BB mean of Y_i of treated patients.

10: **end for**

11: Output $\hat{\Delta}^1, ..., \hat{\Delta}^B$.

A more complex example is the estimation of HR for TTE data. For this purpose, we redefine Y_i as the TTE with censoring time C_i. The bootstrap posterior sample $\hat{\tau}^1, ..., \hat{\tau}^B$ can be used to infer HR between treated and combined control patients.

Algorithm 3. Bayesian bootstrap for posterior sample of HR with IPW adjustment and maxML as the AoB rule.

Require: $Y_i, T_i, C_i, I_i, i = 1, ..., n$

1: **for** Bootstrap run $b = 1$ to B **do**

2: Generate weights $\xi_i^b, i = 1, ..., n$, from the uniform Dirichlet distribution $Dirichlet(1, ..., 1)$.

3: Fit the propensity score model to external and internal control patients using weights ξ_i^b and calculate weights w_i^b.

4: Fit a Cox model to data Y_i, C_i, I_i with $T_i = 0$ using I_i as the covariate and weights $w_i^b \xi_i^b$ for external patients and ξ_i^b for trial controls and obtain $\hat{\tau}^b$ and its SE $\hat{\sigma}_\tau^b$.

5: Obtain the weighted number of events d_1^b, d_0^b of trial and external controls.

6: Calculate $\bar{\sigma}_\tau^{2b} = \hat{\sigma}_\tau^{2b}/(1/d_0^b + 1/d_1^b), \hat{\sigma}_1^{2b} = \bar{\sigma}_\tau^{2b}/d_1^b$ and $\hat{\sigma}_0^{2b} = \bar{\sigma}_\tau^{2b}/d_0^b$.

7: Obtain

$$\hat{a}_0^b = \frac{\hat{\sigma}_0^{2b}}{\max[\hat{\tau}^{b2}, \hat{\sigma}_0^{2b} + \hat{\sigma}_1^{2b}] + \hat{\sigma}_1^{2b}}. \tag{17}$$

8: Fit a Cox model to the data $Y_i, T_i, C_i, i = 1, ..., n$ with T_i as the covariate and weights ξ_i^b when $I_i = 1$, and $w_i^b \xi_i^b \hat{a}_0^b$ when $I_i = 0$, to obtain the HR estimate $\hat{\tau}^b$ of treatments.

9: **end for**

10: Output $\hat{\tau}^1, ..., \hat{\tau}^B$.

---

9 Regulatory and industry aspects and applications

The propensity score methodology, formulated by Rosenbaum and Rubin (1983), mainly for the mitigation of confounding bias in non-randomized comparative studies, was adopted by FDA/CDRH in 2002 and applied in the evaluation of medical device pre-market confirmatory studies including those with a control group leveraged from a well-designed and executed registry database or historical clinical studies; see, e.g., in (Yue, 2007, 2012; Campbell and Yue, 2016). To address the regulatory concerns with study design integrity and study result interpretability, for carrying out PS based analysis, when both covariate and outcome data are in sight, (Rubin, 2001, 2007) put forward the outcome-free principle; see also Yue et al. (2014) for details on operationally executable two-stage study design framework for regulatory submissions for post-market observational studies, commonly employed for medical device regulatory applications of the PS methodology.The two-stage study design framework for medical device studies, based on PS methodology, proposes the investigator of the study to identify an independent statistician, in the fist design stage, to serve as a firewall, to perform the PS design with no patient-level outcome data provided to him/her. The other elements of the first design stage are the specification of the study endpoints, the study hypothesis, and the "initial" sample sizes for the treated and control groups. The sample sizes are revisited in the second design stage and may be revised later prior to the unblinding of the outcome data.

The second stage of the two-stage design starts when the patient's enrollment in the treated group is complete, and so is the patient leveraging from the registry/historical clinical data, at which point the baseline covariate data are available for all patients. In the second design stage, the independent statistician carries out the PS estimation, performs PS matching, weighting, or stratification, and assesses covariate balance. The power revisited to find if it is adequate, say due to unbalanced "randomization ratio" from stratification.

The main purpose of the PS design is to balance observed covariates. Methods for assessing covariate balance, also called balance diagnostics, fall into two categories: graphical and numerical; with most common numerically assessing method called absolute standardized mean difference. if the balance is adequate for all observed covariates, the PS design is complete. It is possible that adequate balancing cannot be achieved. This is a risk inherent in the application of the PS methodology. After completion of the two-stage design, the outcome data are analyzed.

A number of articles have been published from FDA/CDRH statisticians on leveraging external data to augment a single-arm study, or to augment the control arm of an RCT, where the "nominal number" of subjects to be borrowed from the external data, A, is fixed and specified in the planning stage of the study. Here, the word "nominal" indicates the actual number of subjects identified from the external data, using inclusion/exclusion criteria, that may be much larger than A. How to ensure that the nominal number of external subjects to be leveraged does not exceed A relates to down-weighting or "discounting". There are two alternative methods for discounting that are proposed, one Bayesian and the other frequentist. The Bayesian method is that of PP and the frequentist method is that of composite likelihood. The details of these methods are available in FDA/CDRH statisticians' publications; e.g., (Wang et al., 2020; Chen et al., 2020; Lu et al., 2022).

The propensity-score integrated composite likelihood (PSCL) method with PS stratification (Propensity Bin) presented above, along with two other methods (namely, a two-step Cox model; a fixed Cox model Borrowing) were applied by Sengupta et al. (2023) to emulate the CheckMate-057 RCT study, retrospectively, by comparing hybrid borrowing results with the RCT results for the estimation of the overall survival (OS) hazard ratio (estimand). The two-step Cox model is an outcome-based dynamic borrowing that downweighs the RWD based on the patient- level covariate similarity and outcome similarity. The fixed

Cox model is a static borrowing method similar to the two-step Cox model, except that the weights are based on a prespecified number of events borrowed from RWD. The PSCL method is an outcome-free dynamic borrowing method, based on a single Cox regression model with stratum-specific weights estimated from 5 strata based on propensity scores of RCT patients, following procedures proposed in Chen et al. (2020). All three borrowing methods demonstrated similar results and showed that it is feasible to emulate an RCT by using a fit-for-purpose RWD; see, Viraswami-Appanna et al. (2024).

Another application of a hybrid control, where a frequentist conditional borrowing method developed by Li et al. (2022), integrating the PS matching, also mentioned in Viraswami-Appanna et al. (2024) above, is to evaluate the contribution of components of combination therapies in oncology, where the trial subjects were randomized to three arms: combination therapy, monotherapy, and standard of care (SOC). The primary objective of the study was to evaluate the efficacy and safety of the combination therapy compared to the SOC control. The study was powered for the primary objective. It was not possible to power a comparison between combination therapy and monotherapy possibly due to treatment difference being small. For this comparison, data from (small) historical trials, which were conducted with similar methodologies to those used in the current trial, were used to augment the monotherapy arm, to form a hybrid control arm, and to evaluate the contributions of combination therap and monotherapy (dropping the SOC arm). The method proposed to augment the control arm in RCT, by Li et al. (2022), is a conditional borrowing framework. The method establishes empirical distributions for the hazard ratio in outcomes between the external and trial control subjects. The borrowing decision is then derived from this empirical distribution using a measure of similarity. Once the borrowing decision is established, the borrowing weights for the external subjects, based on the similarity measure, are incorporated into the weighted partial likelihood to evaluate the treatment effect. The details of forming a hybrid control arm, based on the similarity of baseline covariates, are provided in Li et al. (2022).

A novel approach for augmenting a single-arm study with RWD, that can also be easily applied to construct a hybrid control of an RCT is achieved by adopting the "sampling importance resampling (SIR) method" described in Sachdeva et al. (2021). Define $0 \leq a_0 \leq 1$, as the borrowing parameter such that the number of subjects borrowed from RWD is equal to $n_0 a_0$, where n_0 is the number of subjects in RWD that meet the inclusion/exclusion criteria. Let $g_i(e)$; $i = 0, 1$ denote the estimated densities of the propensity scores for current and external data sources. We compute the importance sampling weight for each of the n_0 subjects, $w_k = g_1(e_k)/g_0(e_k)$; $k = 1, \ldots, n_0$ in the external data source, where $e_0 = (e_1, \ldots, e_{n_0})$, denote the PS for the n_0 external subjects. Then, we sample $n_0 a_0$ subjects from n_0 categories $e_0 = (e_1, \ldots, e_{n_0})$ using multinomial distribution with n_0 categories $e_0 = (e_1, \ldots, e_{n_0})$ and their associated probabilities $w_k^* = w_k/(w_1 + \cdots + w_{n_0})$; $k = 1, \ldots, n_0$ from RWD using these weights. For details, see Sachdeva et al. (2021). Here, the borrowing parameter a_0 is either prespecified or determined in an outcome-free manner (see, e.g., Wang and Marion-Gallois, 2023; Wang et al., 2020; Chen et al., 2020) using the overlapping area between the PS densities, $g_i(e)$; $i = 0, 1$.

For some earlier applications of dynamic borrowing using PP methodology in the context of determining the non-inferiority margin, leveraging information from carefully selected well-controlled historical clinical trials, to compare treatment arm with an active comparator, see (Gamalo et al., 2014; Lin et al., 2016, 2018).

10 Statistical challenges

There are multiple statistical challenges, in particular the choices of data prior model and population adjustment methods. Using these approaches for regulatory purposes may face additional challenges, e.g., the control of type I error (referred to as α henceforth) in a frequentist test for treatment effects. We will review the key ones, especially those related to the α control.

10.1 Is power gain possible when exact frequentist type I error control is required?

The classical frequentist approach is to control unconditional α when $\hat{\mu}_1$ and $\hat{\mu}_0$ are random variables with $\mu_1 = 0$ and any arbitrary δ. Obviously, for fixed $\hat{a} = a$, α cannot be controlled at all, since δ can be arbitrarily large. However, with an appropriate borrowing rule, the impact of a large δ can be mitigated. Multiple simulation studies have shown that in multiple settings the power gain is possible with slightly inflated α and with some δ values the power gain is possible without α inflation (Viele et al., 2014; van Rosmalen et al., 2018).

Recently, some research has focused on conditional α control, treating $\hat{\mu}_0$ as fixed. In this setting, the test function or equivalently the test threshold may be calibrated so that α is controlled (Nikolakopoulos et al., 2017; Kopp-Schneider et al., 2019). However, if the internal data D_1 is tested with a uniformly most powerful (UMP) test, the power gain by including the data D_0 is generally impossible if exact α control is required. Without a detailed technical setting, an intuitive explanation is as follows. The threshold of the UMP test is set to control α at a given level. When a fixed $\hat{\mu}_0$ is added to the test statistic, to control α, a recalibration of the threshold has to undo the impact of $\hat{\mu}_0$ so that the values of $\hat{\mu}_1$ leading to the rejection or no rejection of H_0 are the same as in the test using D_1 only. A numerical example can be found in Kopp-Schneider et al. (2019).

Although most previous work for answering this question lead to a negative answer, we examine if this is possible based on the simple linear estimator

$$\hat{\mu}_c = \frac{\hat{\mu}_1 + \hat{a}\hat{\mu}_0}{1 + \hat{a}} \tag{18}$$

where $\hat{a}$ is determined by either the cMinMSE or minMSE approach. For power calculation, we introduce an active treatment arm of size n_a. Let

$\Delta = \mu(1) - \mu(0)$ be the treatment effect, and we estimate $\mu(0)$ by $\hat{\mu}_c$, and $\mu(1)$ by $\hat{\mu}_a$: the mean response of the active treatment arm. Then, Δ can be estimated by

$$\hat{\Delta} \equiv \hat{\mu}_a - \frac{\hat{\mu}_1 + \hat{a}\hat{\mu}_0}{1 + \hat{a}}.$$

We assume that $\hat{\mu}_a$, $\hat{\mu}_1$ and $\hat{\mu}_0$ are all normally distributed but make no assumption about the key parameter: $\delta = \mu_1 - \mu_0$. Our goal is to test the one-sided test H_0: $\Delta = 0 vs.$ H_1: $\Delta > 0$ based on $\hat{\Delta}$. Due to the impact of $\hat{a}$ and the impact of $\delta \neq 0$, a rejection boundary based on the normal distribution does not control α at the nominal level.

To exactly control α, we recalibrate the rejection boundary based on the empirical distribution of $\hat{\Delta}$ in the worst scenario of δ. Let the test function be $\phi(\hat{\Delta}) = I(\hat{\Delta} > C_\alpha)$ with a threshold C_α. The power under H_1: $\Delta > 0$ can be written as $E_\Delta(\phi(\hat{\Delta}))$, and the type I error $\alpha = E_0(\phi(\hat{\mu}_c))$, where $E_0(.)$ is the expectation under $\Delta = 0$. Then, α can be controlled if a finite threshold C_α exists such that $\alpha \leq E_0(\phi(\hat{\Delta}))$. In practice, we can set C_α as the $1 - \alpha$ percentile of the empirical distribution of $\hat{\mu}_c$ in the worst case among all the values of δ. Specifically, for a grid of δ values, we simulate the data m times under H_0; that is, $\Delta = 0$, $\mu_1 = 0$, $\mu_0 = \delta$, each time we calculate (18) and denote it by $\hat{\mu}_m(\delta)$. Let $\hat{F}_m(\hat{\Delta}: \delta)$ be its empirical distribution. Then, the rejection boundary at the α level is $C_\alpha = \max_\delta \hat{F}_m^{-1}(1 - \alpha: \delta)$.

We conducted a small simulation to examine whether the power gain is possible with this calibrated C_α to exactly control α or not. We vary n_1 and n_a, and Δ while fixing $n_0 = 100$ to explore whether the power gain occurs at any combination of their values. The following result compares the type I error and the power of this approach with $\hat{\alpha}$ estimated by the cMinMSE and minMSE rules, compared to those that use only internal data. Figs. 1 and 2 show the type I error (top left panel) and power curves of the cMinMSE and minMSE approaches when $\Delta = 0.5$ and $\Delta = 0.7$, respectively. These curves indicate that the worst-case scenario for the α control is when the δ value is moderate. However, a power gain is possible if δ is small. Also, although the minMSE approach leads to less AoB than the cminMSE approach, it is more powerful after recalibration for the α control.

Although we can only simulate with finite values of δ, with appropriate $\hat{g}(\hat{\delta})$), we can ensure that the worst-case scenario for the α control is not when $\delta \to \infty$. For example, with the minMSE rule and assuming $n = n_a = n_0 = n_1$ for notation simplicity, we have

$$\sqrt{n}\hat{\Delta} = \sqrt{n}(\hat{\mu}_a - \hat{\mu}_1) + \sqrt{n}\frac{\hat{a}^*\hat{\delta}}{1 + \hat{a}^*}.$$

Recall that from (7), $\hat{a}^* \propto O(n^{-1}\delta^{-2})$ for sufficiently large δ and n. The second term $\sqrt{n}\hat{a}^*\hat{\delta}/(1 + \hat{a}^*)$ converges to 0 in probability when $\delta \to \infty$, hence $\sqrt{n}\hat{\Delta} \sim_a \sqrt{n}(\hat{\mu}_a - \hat{\mu}_1)$, so $\hat{\Delta}$ is essentially the mean difference between the two

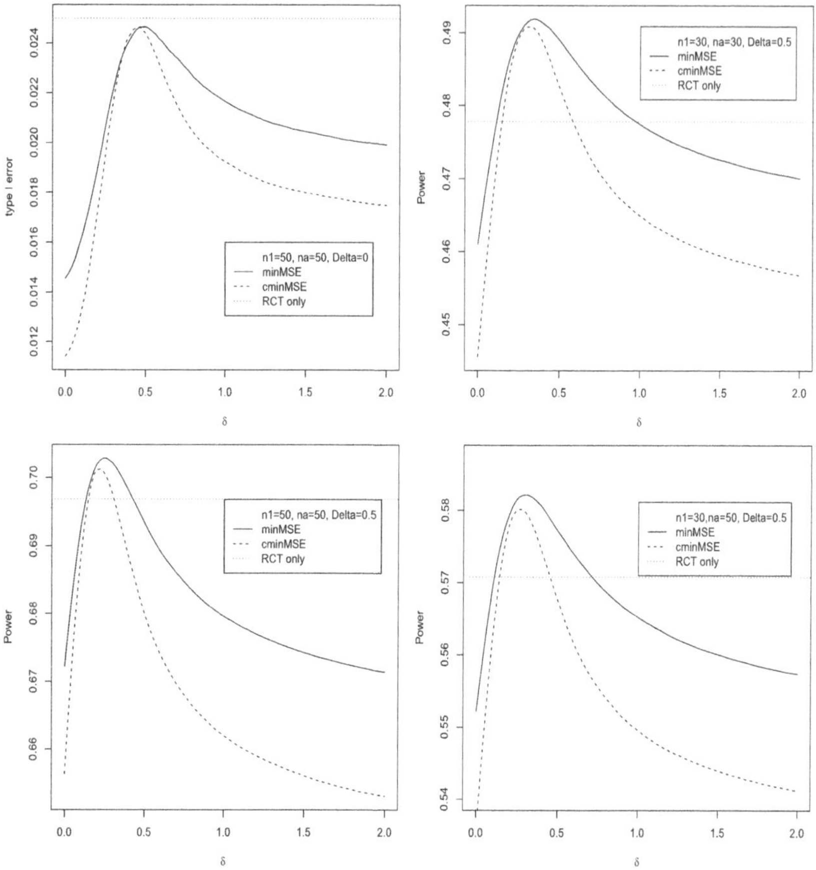

FIG. 1 Power and type I error curves of cMinMSE and minMSE rules as a function of δ with varying sample sizes and treatment effect $\Delta = 0.5$.

arms of the trial. Consequently, a very large δ has no impact on type I error asymptotically. The simulated α curves also suggest that α does not increase with δ after the peak at a relatively small δ.

The simulation results show that, in some settings, power gain is possible when α is controlled exactly and the amount of power gain depends on multiple factors, particularly on n_1. Although most of the power gain occurs at small δ, which coincides with the range named the "sweet spot" in Viele et al. (2014), our test and simulation results are fundamentally different from theirs, since our test is calibrated for exact α control at **any** δ. Since the maximum amount of power gain is rather small, e.g., 1 % in Fig. 2, this result mainly shows the possibility of power gain under exact α control. For practical use, using alternatives to this worst-case scenario recalibration may be a better choice if exact α control is not required.

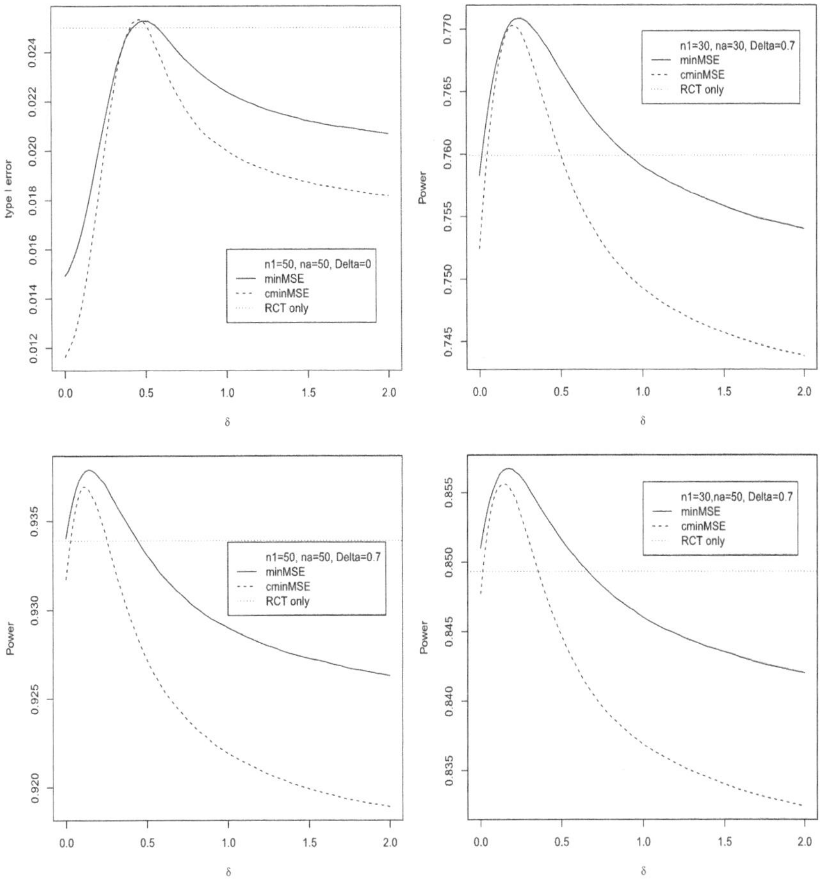

FIG. 2 Power and type I error curves of cMinMSE and minMSE rules as a function of δ with varying sample sizes and treatment effect $\Delta = 0.7$.

10.2 Estimation or hypothesis testing?

The primary analysis for an RCT is often a hypothesis test for, e.g., H_0: $\Delta = 0$ vs H_a: $\Delta = \Delta_a \neq 0$. Recently, the ICH E9 (R1) guidance requires careful consideration of causal estimand in the design and analysis of RCTs. This is a welcoming change, since often causal estimands should be a key consideration for RCTs. However, it also raises multiple challenges from practical and methodology aspects, in particular, for RCTs using external controls.

A key attribute, among four of them, of a causal estimand is the population in which it is defined. In an RCT, often it is the trial population and the estimand is the average (or marginal) treatment effect. The goal of the pre-adjustment is to make D_0 similar to D_1. However, the similarity in the outcome can only be assessed between controls. Therefore, a practical question is: What is the target we adjust (e.g, by IPW or PS matching) D_0 to? One may think of the whole D_1 as the target; however, the similarity of the outcomes cannot be measured in a large part of D_1, especially when the internal control arm is small (the exact reason for borrowing).

One possible solution is to use ANCOVA to adjust the RCT data (which is often beneficial even without external controls), and adjust the preadjustment to the same target. Suppose that a 1:2 randomized trial results in the mean age 40 and 50 years in the control and treated arms, respectively; hence the mean age of the RCT is $(40 \times 2 + 50)/3 = 43.3$ years. Then the preadjustment may also target the same mean age. Note that the adjusted D_0 is comparable to the adjusted RCT controls, and therefore the assessment of similarity is still valid. ANCOVA adjustment is rather robust against model misspecification, but there is no guarantee when the RCT control arm is very small. Sometimes, the RCT is designed with stratification. In this case, a simple borrowing approach is to adjust within the strata. However, this approach does not overcome the problem of a small control arm in the RCT, since we may see a very small sample size in some strata in which the assessment of similarity becomes unreliable. The above suggests that a control arm that is too small in the RCT is not a good option.

Another issue with respect to causal estimation is the selection of the final analysis using both D_1 and D_0. This is particularly important when HR or OR is the estimand. They are defined as the marginal estimands over the whole population, in contrast to subject- or subgroup-specific estimands. However, for hypothesis testing, the difference is less important, since H_0 holds for all subjects or subgroups implicating the same over the population. Hence, if a PS matching or stratification is used for adjustment, to estimate a marginal estimand, a pooled analysis should be used, while a stratified hypothesis test is more powerful.

10.3 Firewall and data privacy protection

When the RCT is for regulatory purposes, it is important to ensure that the selection and analysis of external data for preadjustment are as independent of the RCT data as possible. A common approach is to perform external data selection and key steps for preadjustment (including fitting PS model, calculating IPW weights or performing PS matching or stratification) without accessing internal outcome data. A systematic approach is to set a firewall between the internal outcome data and the personal (known as the firewall statistician/analyst) performing the above tasks. In this way, the key components for determining the AoB and the final analysis (e.g., IPW weights or matched or stratified patient IDs) are determined without the internal outcome data and can be locked in the RCT database lock. Note that direct adjustment with a regression model does not need internal outcome data, so it can be performed without violating the firewall.

However, sometimes accessing individual data, particularly outcome data, may be a sensitive issue due to data privacy protection laws, which may vary from one country to another. A practical advantage of the CB approach is important when D_1 and D_0 are owned by different owners and individual-level data cannot be pooled due to data privacy protection. The CB approach can avoid the pooling of individual data since it only needs $\bar{X}_1$.

Weighting approaches are generally easier to adapt to avoid sharing individual outcome data, as in most cases, such as the estimation of Δ, all parties involved only need to share weighted means. However, even for PS-matched and stratified analyzes, some privacy-protected data approaches are also available without pooling individual outcome data (Wang and Marion-Gallois, 2023).

11 Concluding remarks

This chapter gives a critical review of recent advances in developing and using robust approaches to borrowing external data for clinical trials for drug development. Our focus is mainly on borrowing external controls for RCTs with population adjustment and also dynamic borrowing that adaptively determines AoB. However, a review of regulatory and industry aspects and applications can be found in Section 9. Another important topic that we did not discuss is the selection of external data. Although Pocock (1976) proposed his criteria, they seemed too restrictive in practice. With recent development of statistical methodology, particularly dynamic borrowing, we should be able to borrow external data that are less ideal.

Although we have briefly mentioned borrowing multiple external datasets when discussing rMAP priors, most parts of this chapter focus on borrowing from a single data source. Often times, there could be multiple external data sources relevant to the (current) RCT. Using them properly will not only make inference of the RCT more efficient but also avoid selection bias to pick a single data source. Although borrowing multiple external datasets most commonly uses rMAP priors, PP or CPP can also be adapted to accommodate this task. In fact, robust approaches based on $\mu_c = (\mu_1 + a^* \mu_0)/(1 + a^*)$ can also be easily adapted. Let μ_{0k} be the mean of the kth external data source, the robust borrowing approaches can be adapted by considering

$$\mu_c = \frac{\mu_1 + \sum_{k=1}^{K} a_k^* \mu_{0k}}{1 + \sum_{k=1}^{K} a_k^*}$$

One can also specify a joint prior for a_k^*'s, e.g., a Dirichlet distribution. For the α control, the empirical distribution recalibration approach can also be used here. However, the power gain is probably low because of the nature of the method to protect the worst scenario.

The use of external TTE data is an important topic for drug development, which we have not covered extensively here. A book chapter in Wang et al. (2023b) provides a detailed review of this topic. However, as this is a rapidly developing area, we included some recent developments, particularly in dynamic borrowing.

The use of external data as sole controls for a single-arm trial is possible if all confounders are properly adjusted for; however, this approach is based on assumptions such as that there are no unobserved confounders, which cannot be verified by the data. The role of the control arm in the trial is to provide not only internal control data but also a gold standard to verify the similarity of external data, with or without preadjustment. Therefore, its sample size is a key design parameter for planning a trial.

In summary, leveraging external data for the analysis of RCTs can reduce the costs and duration of drug development. Although external data borrowing may introduce biases, robust borrowing approaches can mitigate the risk by proper preadjustment and the control of AoB. Multiple challenges remain in the regulatory, statistical, and practical aspects. We hope that further development of statistical methodology, including Bayesian approaches for leveraging external data, demonstrating desirable statistical properties such as exact or approximate control of type I error, and their use in practice in drug development and regulatory adaptation will evolve over time and lead to more efficient drug development.

References

Alt, E.M., Chang, X., Jiang, X., Liu, Q., Mo, M., Xia, H.A., Ibrahim, J.G., 2024. Leap: the latent exchangeability prior for borrowing information from historical data. Biometrics 80 (3), ujae083.

Bennett, M., White, S., Best, N., Mander, A., 2021. A novel equivalence probability weighted power prior for using historical control data in an adaptive clinical trial design: a comparison to standard methods. Pharm. Stat. 20 (3), 462–484.

Campbell, G., Yue, L.Q., 2016. Statistical innovations in the medical device world sparked by the fda. J. Biopharm. Stat. 26 (1), 3–16.

Chen, M.-H., Ibrahim, J.G., 2000. Power prior distributions for regression models. Stat. Sci. 15 (1).

Chen, W.-C., Wang, C., Li, H., Lu, N., Tiwari, R., Xu, Y., Yue, L.Q., 2020. Propensity score-integrated composite likelihood approach for augmenting the control arm of a randomized controlled trial by incorporating real-world data. J. Biopharm. Stat. 30 (3), 508–520.

Deville, J.-C., Särndal, C.-E., 1992. Calibration estimators in survey sampling. J. Am. Stat. Assoc. 87 (418), 376–382.

Food and Drug Administration, 2018. Use of real-world evidence to support regulatory decision-making for medical devices. Food and Drug Administration.

Galwey, N.W., 2016. Supplementation of a clinical trial by historical control data: is the prospect of dynamic borrowing an illusion? Stat. Med. 36 (6), 899–916.

Gamalo, M.A., Tiwari, R.C., LaVange, L.M., 2014. Bayesian approach to the design and analysis of non-inferiority trials for anti-infective products. Pharm. Stat. 13 (1), 25–40.

Gravestock, I., Held, L., COMBACTE-Net consortium, 2017. Adaptive power priors with empirical Bayes for clinical trials. Pharm. Stat. 16 (5), 349–360.

Gravestock, I., Held, L., 2018. Power priors based on multiple historical studies for binary outcomes. Biom. J. 61 (5), 1201–1218.

Hobbs, B.P., Carlin, B.P., Mandrekar, S.J., Sargent, D.J., 2011. Hierarchical commensurate and power prior models for adaptive incorporation of historical information in clinical trials. Biometrics 67 (3), 1047–1056.

Hobbs, B.P., Carlin, B.P., Sargent, D.J., 2013. Adaptive adjustment of the randomization ratio using historical control data. Clinical Trials 10 (3), 430–440.

Ibrahim, J.G., Chen, M.-H., 2000. Power prior distributions for regression models. Stat. Sci. 46–60.

Ibrahim, J.G., Chen, M.-H., Gwon, Y., Chen, F., 2015. The power prior: theory and applications. Stat. Med. 34 (28), 3724–3749.

Imai, K., Ratkovic, M., 2014. Covariate balancing propensity score. J. R. Stat. Soc. Series B: Stat. Methodol. 76 (1), 243–263.

Jiang, L., Nie, L., Yuan, Y., 2023. Elastic priors to dynamically borrow information from historical data in clinical trials. Biometrics 79 (1), 49–60.

Kopp-Schneider, A., Calderazzo, S., Wiesenfarth, M., 2019. Power gains by using external information in clinical trials are typically not possible when requiring strict type i error control. Biom. J. 62 (2), 361–374.

Li, H., Tiwari, R., Li, Q.H., 2022. Conditional borrowing external data to establish a hybrid control arm in randomized clinical trials. J. Biopharm. Stat. 32 (6), 954–968.

Li, W., Liu, F., Snavely, D., 2020. Revisit of test-then-pool methods and some practical considerations. Pharm. Stat. 19 (5), 498–517.

Lin, J., Gamalo-Siebers, M., Tiwari, R., 2016. Non-inferiority and networks: inferring efficacy from a web of data. Pharm. Stat. 15 (1), 54–67.

Lin, J., Gamalo-Siebers, M., Tiwari, R., 2018. Propensity score matched augmented controls in randomized clinical trials: a case study. Pharm. Stat. 17 (5), 629–647.

Lu, N., Wang, C., Chen, W.-C., Li, H., Song, C., Tiwari, R., Xu, Y., Yue, L.Q., 2022. Propensity score-integrated power prior approach for augmenting the control arm of a randomized controlled trial by incorporating multiple external data sources. J. Biopharm. Stat. 32 (1), 158–169.

Lyddon, S.P., Holmes, C.C., Walker, S.G., 2019. General Bayesian updating and the loss-likelihood bootstrap. Biometrika 106 (2), 465–478.

McCandless, L.C., Gustafson, P., Austin, P.C., 2009. Bayesian propensity score analysis for observational data. Stat. Med. 28 (1), 94–112.

Neuenschwander, B., Branson, M., Spiegelhalter, D.J., 2009. A note on the power prior. Stat. Med. 28 (28), 3562–3566.

Neuenschwander, B., Capkun-Niggli, G., Branson, M., Spiegelhalter, D.J., 2010. Summarizing historical information on controls in clinical trials. Clinical Trials 7 (1), 5–18.

Neuenschwander, B., Wandel, S., Roychoudhury, S., Bailey, S., 2016. Robust exchangeability designs for early phase clinical trials with multiple strata. Pharm. Stat. 15 (2), 123–134.

Newton, M.A., Raftery, A.E., 1994. Approximate Bayesian inference with the weighted likelihood bootstrap. J. R. Stat. Soc. Series B: Stat. Methodol. 56 (1), 3–26.

Nikolakopoulos, S., van der Tweel, I., Roes, K.C.B., 2017. Dynamic borrowing through empirical power priors that control type I error. Biometrics 74 (3), 874–880.

Pocock, S.J., 1976. The combination of randomized and historical controls in clinical trials. J. Chronic Dis. 29 (3), 175–188.

Pompe, E., 2021. Introducing prior information in weighted likelihood bootstrap with applications to model misspecification. arXiv preprint arXiv:2103.14445.

Rosenbaum, P.R., Rubin, D.B., 1983. The central role of the propensity score in observational studies for causal effects. Biometrika 70 (1), 41–55.

Rubin, D.B., 1978. Bayesian inference for causal effects: the role of randomization. Ann. Stat. 6, 34–58.

Rubin, D.B., 1981. The Bayesian bootstrap. Ann. Stat. 9 (1).

Rubin, D.B., 2001. Using propensity scores to help design observational studies: application to the tobacco litigation. Health Serv. Outcomes Res. Methodol. 2, 169–188.

Rubin, D.B., 2007. The design versus the analysis of observational studies for causal effects: parallels with the design of randomized trials. Stat. Med. 26 (1), 20–36.

Sachdeva, A., Tiwari, R.C., Guha, S., 2021. A novel approach to augment single-arm clinical studies with real-world data. J. Biopharm. Stat. 32 (1), 141–157.

Schmidli, H., Gsteiger, S., Roychoudhury, S., O'Hagan, A., Spiegelhalter, D., Neuenschwander, B., 2014. Robust meta-analytic-predictive priors in clinical trials with historical control information. Biometrics 70 (4), 1023–1032.

Sengupta, S., Ntambwe, I., Tan, K., Liang, Q., Paulucci, D., Castellanos, E., Fiore, J., Lane, S., Micsinai Balan, M., Viraswami-Apanna, K., et al., 2023. Emulating randomized controlled trials with hybrid control arms in oncology: a case study. Clin. Pharmacol. Ther. 113 (4), 867–877.

van der Vaart, A.W., 1998. Asymptotic Statistics. Cambridge University Press.

van Rosmalen, J., Dejardin, D., van Norden, Y., Löwenberg, B., Lesaffre, E., 2018. Including historical data in the analysis of clinical trials: is it worth the effort? Stat. Methods Med. Res. 27 (10), 3167–3182.

Viele, K., Berry, S., Neuenschwander, B., Amzal, B., Chen, F., Enas, N., Hobbs, B., Ibrahim, J.G., Kinnersley, N., Lindborg, S., et al., 2014. Use of historical control data for assessing treatment effects in clinical trials. Pharm. Stat. 13 (1), 41–54.

Viraswami-Appanna, K., Buenconsejo, J., Baidoo, C., Chan, I., Li, D., Micsinai-Balan, M., Tiwari, R., Yang, L., Sethuraman, V., 2024. Accelerating drug development at bristol myers squibb through innovation. Drug Discov. Today, 103952.

Wang, C., Lu, N., Chen, W.-C., Li, H., Tiwari, R., Xu, Y., Yue, L.Q., 2020. Propensity score-integrated composite likelihood approach for incorporating real-world evidence in single-arm clinical studies. J. Biopharm. Stat. 30 (3), 495–507.

Wang, J., Marion-Gallois, R., 2023. Propensity score matching and stratification using multiparty data without pooling. Pharm. Stat. 22 (1), 4–19.

Wang, J., Sengupta, S., Tiwari, R., 2025. Bayesian dynamic power prior borrowing for augmenting a control arm for survival analysis. J. Biopharm. Stat. 1–22.

Wang, J., Tiwari, R., 2024. A Bayesian bootstrap approach for dynamic borrowing for minimizing mean squared error. arXiv Preprint arXiv:2407.21588.

Wang, J., Tiwari, R., 2025. Approximate Bayesian analysis for borrowing external controls for randomized controlled trials with dynamic borrowing and covariate balancing adjustment. Pharm. Stat. 24. https://doi.org/10.1002/pst.2474

Wang, J., Zhang, H., Tiwari, R., 2023a. A propensity-score integrated approach to Bayesian dynamic power prior borrowing. Stat. Biopharm. Res. https://doi.org/10.1080/19466315.2023.2223533

Wang, J., Zhang, H., Tiwari, R., 2023b. Statistical challenges for causal inference using time-to-event real-world data. Real-World Evidence in Medical Product Development. Springer, pp. 233–254.

Wu, C., Sitter, R.R., 2001. A model-calibration approach to using complete auxiliary information from survey data. J. Am. Stat. Assoc. 96 (453), 185–193.

Yue, L.Q., 2007. Statistical and regulatory issues with the application of propensity score analysis to nonrandomized medical device clinical studies. J. Biopharm. Stat. 17 (1), 1–13.

Yue, L.Q., 2012. Regulatory considerations in the design of comparative observational studies using propensity scores. J. Biopharm. Stat. 22 (6), 1272–1279.

Yue, L.Q., Lu, N., Xu, Y., 2014. Designing premarket observational comparative studies using existing data as controls: challenges and opportunities. J. Biopharm. Stat. 24 (5), 994–1010.

Zhang, H., Shen, Y., Li, J., Ye, H., Chiang, A.Y., 2023. Adaptively leveraging external data with robust meta-analytical-predictive prior using empirical bayes. Pharm. Stat. 22 (5), 846–860.

Chapter 7

Statistical innovation: Transforming pharmaceutical research and development

Pandurang M. Kulkarni[a], Demissie Alemayehu[b], Yongming Qu[a], Daniel Backenroth[c], Brian Millen[d], Haoda Fu[e], Wim Van Der Elst[f], Satrajit Roychoudhury[b], Karen Price[a], Tianle Chen[d], Kannan Natarajan[b], and Wei Shen[a,*]

[a]*Eli Lilly and Company, Indianapolis, IN, United States,* [b]*Pfizer Inc., New York, NY, United States,* [c]*Johnson & Johnson Innovative Medicine, Raritan, NJ, United States,* [d]*Biogen, Inc., Cambridge, MA, United States,* [e]*Amgen Inc., Thousand Oaks, CA, United States,* [f]*Johnson & Johnson Innovative Medicine, Beerse, Belgium*

Corresponding author. e-mail address: shen@lilly.com

Chapter Outline

Handbook of Statistics, Vol. 53. https://doi.org/10.1016/bs.host.2025.04.005

Abstract

Breakthroughs in statistical innovation have played paramount roles in enhancing quality, integrity, speed and efficiency in pharmaceutical research and development. This chapter provides an overview of innovative techniques that have had considerable impacts in changing the landscape of the pharmaceutical industry and in transforming the drug development paradigm accelerating the discovery, development, approval and manufacturing of new medicines for patients with unmet medical needs. We highlight how flexible and adaptive designs have helped accelerate drug development timelines and reduce cost, and we provide examples of how this has impacted the advancement of public health in the recent past, including the remarkable development of the COVID-19 messenger RNA vaccine. Advances in modeling and simulation, as well as Bayesian Statistics, have enabled integration of quantitative decision making in drug development in all phases and at all levels - from study design to portfolio prioritization. Novel statistical approaches have also helped with validation of surrogate endpoints and handling of multiplicity issues that support decision making by regulators and other stakeholders, while safeguarding the safety and well-being of patients. Among the emerging innovations, special attention is paid to such novel technologies and methodologies as causal inference and modern analytics. New approaches for causal inference have proved promising in the use of data beyond randomized controlled trials to generate reliable evidence concerning the relative risks and benefits of alternative investigational treatment options. Lastly, thanks to the unfolding digital revolution, there is a growing adoption of artificial intelligence and machine learning tools at every stage of drug development, from discovery through loss of exclusivity.

1 Introduction

Over the last several decades, developing safe and effective drugs has become an increasingly costly and lengthy process. The issue has especially been critical in rare diseases and other illnesses with unmet medical needs. According to recent reports, the average cost of bringing a new medicine to patients is over 1 billion dollars (Wouters et al., 2020). The corresponding duration of time is estimated to be at least 12 years (Van Norman, 2016). While various stakeholders have undertaken many measures to address the issue, statistical innovations in the field of biopharmaceutical industry have contributed significantly to speed the drug development process through creative and innovative designs and analysis methodologies, while maintaining the sanctity of statistical integrity and rigor. These have contributed to the execution of trials and regulatory approvals of potentially lifesaving treatments faster and with higher efficiency.

It is widely known that Statistics plays a critical role throughout the entire drug development continuum, including the screening of molecules with promising therapeutic properties, the design and analysis of trials involving animal and human subjects, supporting health technology assessment, and commercialization of licensed products (Senn, 2021). With the increasing challenges of drug development, the impact and influence of statistics as a discipline has also evolved over time with the introduction of novel trial design and analysis methodologies, which have not only accelerated speed and reduced cost but have also elevated the role of statisticians as indispensable strategic partners enabling quantitative and data-driven decision making.

Advances in statistical methodology have resulted in the traditional fixed trial design framework being supplemented by sequential and flexible trials, which have enabled early decision making and calibration of various aspects of a trial without compromising the study's validity and integrity (Chow and Chang, 2012). Adaptive trial designs permit the incorporation of prespecified changes to study protocols, such as sample size estimation, removal of potentially ineffective or unsafe investigational treatment arms, characterizing the patient groups that are most likely to respond to investigational treatment, and terminating a trial early for overwhelming success or lack of efficacy. Of particular importance are seamless trial designs, which have enabled uninterrupted transition between different phases of development, thereby reducing time and cost.

The frequentist paradigm, which has been the cornerstone of traditional analysis approaches in the past, is now reliably buttressed with Bayesian theories. Notably, the latter are specifically deployed in trials with small sample sizes, clinical trials using external controls, or studies that incorporate flexible learning. With the availability of software to implement them, such techniques are now routinely used in tandem with frequentist approaches.

The Complex Innovative Trial Design (CID) initiative that was recently launched by the United States Food and Drug Administration (FDA) has opened a new door for pharmaceutical statisticians to partner with regulators in fostering innovation in drug development (Food and Drug Administration, 2023a). The CID framework permits use of novel statistical techniques and relies heavily on computer simulation to determine procedure operating characteristics. In addition to optimizing resource use, CID can maximize the success rate of clinical trials and help address ethical concerns that may arise with the assignment of subjects to a placebo arm.

Statistics has also become an indispensable tool in portfolio prioritization, enabling probabilistic decision making through the prediction of trial, regulatory, and commercial successes. Determining the probability of regulatory success for a given asset not only ensures effective budgeting and resource planning through data-driven decision making, but it also allows for the advancement of public health by facilitating the speedy development of medicines for patients with unmet medical needs.

New techniques for causal inference now enable supplementing randomized controlled trials (RCTs) with other data sources including claims databases and electronic health records. Counterfactual models are honed to mitigate bias associated with nonrandomized studies, which are increasingly used in situations where RCTs are not feasible or ethical. Techniques such as the targeted maximum likelihood estimation approach and target trials are now available to generate relatively reliable real-world evidence or implement external controls (Dang and Balzer, 2023).

In comparative effectiveness analysis, novel approaches have been developed to draw conclusions about the relative benefits and risks of alternative investigational treatments that have not been studied in head-to-head trials. Bayesian Network meta-analysis and other indirect comparison procedures now permit decision making about access to potentially lifesaving medicines without the need to conduct costly and time-consuming comparative trials (Kiefer et al., 2015).

The digital revolution—coupled with access to high dimensional data, efficient algorithmic models, and high-speed computers—has also opened a new frontier for statisticians to enhance their traditional inferential tools and advance precision medicine. Promising research is underway to integrate the new tools into the classical inferential framework (Zhou, 2021).

In this chapter, we review major statistical breakthroughs that have been used to accelerate the drug development process in the recent past. Section 2 provides an overview of innovative statistical approaches, with an emphasis on adaptive designs, master protocols, external control and Bayesian inference. Section 3 considers quantification of uncertainty in decision making, while Section 4 addresses advances in multiple testing as a foundational tool in confirmatory clinical trials. Section 5 highlights the importance of leveraging surrogate endpoints for statistical innovation. Emerging trends and techniques are discussed in Sections 6 and 7, with reference to causality and real-world evidence as well as the role of machine learning (ML) and artificial intelligence (AI). Finally, concluding remarks are presented in Section 8.

2 Innovative trial designs

As indicated in the previous section, the clinical trial landscape has undergone a transformative shift, driven by the need for more adaptable, patient-centric, and efficient trials. In this effort, statisticians have played significant roles by advancing innovative designs that encompass a variety of strategies that leverage new technologies, data sources, analytics, and regulatory flexibility to enhance the drug development process through improved decision making and efficiency, reduced costs, and enhanced patient outcomes. Recent examples of innovative approaches that have been deployed with significant impact include, but are not limited to, adaptive designs, master protocols, external controls, and technology-enabled trials.

Innovative designs are flexible and offer specific tailoring to each phase of clinical development, enhancing the overall research process. For example, in phase 1, adaptive designs may allow for dose adjustments based on initial safety data, maximizing participant safety while efficiently identifying optimal dosing. Phase 2 trials can benefit from novel endpoints or enrichment strategies, targeting specific patient populations to better assess efficacy. Seamless designs can integrate different phases, streamlining the transition from efficacy testing to confirmatory trials and significantly reducing the number of patients needed (which is a major part of the cost and time for any trial) and the overall timelines. Throughout all phases, incorporating digital tools, external data sources, and real-time data collection have also enhanced participant engagement and improved data accuracy, ultimately leading to more robust outcomes and quicker pathways to deliver medicines to patients.

2.1 Group sequential and adaptive designs

Limitations of traditional clinical trial designs include their prohibitive length, rigidity, and resource intensiveness. Group sequential and adaptive designs represent innovative approaches that allow for more flexible trial conduct, enabling researchers to make informed decisions based on accumulating information. While the choice between group sequential and adaptive designs often depends on the need for flexibility, both approaches have proved to be compellingly effective in enhancing the drug development process. However, these approaches also require careful planning and statistical rigor to ensure trial integrity.

Group sequential designs are ideal when one wants to conduct preplanned interim analyses to assess efficacy or safety while maintaining strict control over type I error rates. This approach is particularly useful in trials with well-defined objectives that allow early stopping for either promising efficacy or futility. Unlike traditional fixed sample designs, where the sample size is predetermined at the outset, group sequential designs provide several advantages, including the possibility to reduce the number of patients exposed to potentially ineffective investigational treatments, accelerate the development of promising therapies, and save resources. A key component of this design is careful specification of the spending function, which determines how much of the overall type I error (a α spending) or type II error rates (beta spending) can be allocated at each interim analysis. In general, group sequential designs are mostly used for confirmatory trials.

Adaptive designs offer greater flexibility, enabling real-time modifications based on accumulating data, such as changing sample sizes, treatment assignment, or population allocation. This adaptability is beneficial in dynamic research environments. As indicated in a recent FDA guideline (Food and Drug Administration, 2019), there is a wide range of adaptive designs used in clinical studies. The use of adaptive designs is more common in early phase trials (e.g., dose-finding studies), while the number of examples in the confirmatory phase is still limited.

In general, most adaptive trials combine two main stages: a *learning stage*, which would otherwise be achieved in a phase 2 trial; and a *confirmatory stage,* as in a pivotal phase 3 trial. If prespecified, certain elements of the *confirmatory stage* design can be modified based on the outcome from the *learning stage* or interim analyses of the trial. At the end of the trial, data from both stages are combined for final analysis using appropriate statistical methodology. Some statistical considerations for confirmatory adaptive design include controlling the overall type I error rate, establishing criteria for modification in confirmatory stage design, and correcting potential bias in the treatment effect estimates and associated confidence intervals during the reporting of trial results. When interest lies in seamless designs, methodologies based on the conditional invariance principle are powerful tools to combine data from multiple stages (Brannath et al., 2007). This methodology may accommodate a wide range of metrics based on frequentist and Bayesian statistical frameworks in the interim decision making and adaptation. A popular and intuitive metric for adaptations is the so-called *predictive probability of success*, which is defined as the conditional probability of achieving statistical significance at the end of the study given the interim data. Of course, on the flip side, it also offers the evaluation of the flip-flop rate to enable informed decision making. Together with other relevant interim information (e.g., effect estimates and standard errors), predictive probability of success is often used to guide adaptations. Another concern in adaptive designs is the potential bias associated with estimates of treatment effects and associated confidence intervals. Because interim adaptations often select groups with promising estimates, there is the potential to overestimate treatment effects. Although point estimates and confidence intervals are usually reported without an appropriate bias correction, we perform numerous sensitivity analyses with different methods to assess the impact of adaptation on treatment effect estimates.

Finally, a common feature of adaptive designs is the use of simulations to estimate frequentist operating characteristics (e.g., type I error and power) or to evaluate design parameters (e.g., timing of interim analyses and probability of stopping for efficacy or futility at that time). We routinely make these available as computational capabilities have become much more readily available due to the advancement of high-speed computing.

2.2 Master protocols

Master protocols permit investigation of multiple drugs, diseases, or both under a single protocol and infrastructure, thereby avoiding the need to establish new protocols for every trial as is traditionally done. This, of course, results in efficient and accelerated drug development, with a substantial reduction in cost and time.

A master protocol may involve one of several approaches, including umbrella, basket, or platform designs (Woodcock and LaVange, 2017). An umbrella trial can be conducted when there is interest in evaluating multiple investigational drugs administered as single drugs or as drug combinations in a single disease population. On the other hand, a study of a single investigational drug or drug combination in different populations defined by different cancers, disease stages for a specific cancer, histologies, number of prior therapies, genetic or other biomarkers, or demographic characteristics may involve the so-called basket trial. Other master protocol designs may involve features of both basket and umbrella trials (often referred to as platform trials).

Since master protocols are inherently complex, the role of Statistical thinking and hence of the statistician is critical to ensure the attainment of desired regulatory outcome as well as patient safety while maintaining statistical rigor and trial integrity. Notably, there should be adequate justification for the sample size, the selection of control arms, and methods used to control for potential sources of bias.

While many applications of the master protocol framework have been reported in oncology research, the approach has proved to be of tremendous value in the study of COVID-19 treatments, when the pandemic was a major global disruption. Examples of the latter include the RECOVERY (Randomized Evaluation of COVID-19 Therapy), I-SPY COVID, and COMMUNITY (COVID-19 Multiple Agents and Modulators Unified Industry Members) trials (Lu et al., 2021). Master protocols enabled multiple companies and regulators to work together and collaborate to meet the needs of patients in unprecedented speed and efficiency.

2.3 Bayesian approaches

Clinical trials have historically been analyzed using frequentist statistics, which is grounded in the concept of the P value. As such, the major challenges of frequentist analysis include difficulty in communicating results to non-statisticians and the lack of a convenient mechanism for incorporating external information with internal trial data. By contrast, Bayesian Statistics is inherently well-suited to address these challenges. Rather than focusing on P values, Bayesian approach estimates the probability of a clinically meaningful treatment benefit, which is appealing to clinicians. Moreover, as discussed in other sections of this chapter, the Bayesian framework provides a formal mechanism to incorporate external information into the analysis of a clinical trial and can help improve interim decision making about a development program or a single trial. In this context, a major appeal of the Bayesian paradigm is that it allows the calculation of predictive probability, which may include partial follow-up times of patients who have been enrolled but have not yet experienced an event of interest or completed the follow-up period.

The choice of a prior distribution is a controversial topic in Bayesian analysis, as it could lead to inconsistent conclusions about the trial, particularly in research involving rare diseases and small sample sizes. One approach to the choice of a reasonable prior distribution is through elicitation of expert opinions. Recently, the Sheffield Elicitation Framework (SHELF) approach has been increasingly used as a formal prior elicitation by some researchers and may be an especially reasonable choice in the rare disease setting (Gosling, 2018). Another family of methods employs an informative prior distribution based on available historical and external data. Alternatively, when there is little information available to guide the choice of the prior distribution, a natural approach is the use of a noninformative or weakly informative prior distribution. Although noninformative prior distributions may yield similar results to the frequentist approach, the prior distributions may be difficult to interpret because they attempt to assign equal weight to all values of interest. In general, there should be a strong rationale for the choice of a prior distribution, including appropriateness of the external data or expert opinion. To address these, we routinely perform sensitivity analyses to assess the validity of alternative prior distribution assumptions and to ensure that the prior distribution does not have unintended influence on the inference and trial conclusion.

2.4 Use of external and synthetic controls

In rare disease research and other cases where it might not be ethically appropriate to use human subjects, we may consider external and synthetic control arms as viable options (Thorlund et al., 2020). Both external and synthetic control methods can help reduce the number of patients needed in the control group, thereby minimizing cost and accelerating the development of a new treatment. An external control arm typically consists of actual real-world data (RWD) or data from a previous clinical study. In contrast, a synthetic control arm is typically constructed using models, potentially combining different data sources, including RWD and data from previous RCTs, with the aim of simulating the treated group.

There is an intrinsic interest in and need for leveraging all available information, especially in rare disease clinical trials, to ensure optimization of the trial's design and analysis (Kidwell et al., 2022; Williamson et al., 2017). Further, including external information is ethically appealing, as it allows for trials to be conducted with smaller sample sizes or unequal randomization (e.g., more subjects receiving investigational treatment than control). It also ensures that a single underpowered trial can provide sufficient information for decision making, so that a potentially lifesaving medication would not be erroneously abandoned due to lack of adequate data.

There are several considerations that need to be taken into account in using external or synthetic control arms. First, it is important to ensure that the data for the control group are reflective of the population being studied. In addition, there should be similarity between the patients in the study drug and control arm groups with respect to relevant prognostic factors. To minimize the introduction of bias in the use of external control arms, alternative statistical approaches have been developed, most of which are based on a Bayesian framework. The Bayesian methods typically allow incorporation of historical data in the form of a prior distribution and aim at appropriately discounting historical information.

As mentioned earlier, external control information can come from different sources, including previous RCTs, registries, or electronic health records; however, a careful selection of the historical data is necessary to ensure the exchangeability assumption for the control parameters is plausible. Pocock (1976) proposed a set of criteria to assess the similarity between external control and trial control data that requires considering several different aspects, such as inclusion and exclusion criteria, the distribution of baseline characteristics, disease characteristics, treatment, relevant medical practice, endpoint definition, and follow-up. Thorlund et al. (2020) provided a checklist on similarity evaluation when using external controls in clinical trials. The choice should not be biased by the outcome of interest or the actual analysis of candidate external control group data. This may happen, for example, if historical data/trials with superior results are preferentially omitted. As a result, it is important that the entire selection process of the dataset and patient-level data be prespecified, without regard to the outcome data. Nevertheless, one must acknowledge the possible conflict between external control and trial control due to unknown factors.

With the increased interest in external controls, several statistical methods are now available in the literature, such as propensity score, power priors, commensurate priors, and meta-analytic predictive priors (for example, see Hobbs et al. (2012) and Schmidli et al. (2014)). The FDA has also issued draft guidelines on the use of external controls in regulatory decision making (Food and Drug Administration, 2023b). Most notably, there have been several examples of regulatory approvals where the FDA has accepted external control data in their benefit/risk assessment (Jahanshahi et al., 2021). Indisputably, the contributions of statisticians in these efforts, which have transformed the drug development paradigm, saved considerable resources, and have accelerated the delivery of medicines to patients, cannot be overestimated.

2.5 Statistical innovations for bioequivalence studies

Bioequivalence (BE) studies are conducted to determine if a new drug formulation has similar bioavailability to a reference product. Two drugs are considered bioequivalent if their absorption rates and extents are similar when administered in the same dosage form. In practice, BE is assessed using pharmacokinetic parameters such as the area under the curve (AUC) and maximum concentration (Cmax). When designing BE studies, the sample size must be large enough to detect a meaningful difference between test and reference products. The analysis plan should also outline how to handle outliers, adjust for baseline covariates, and transform PK data if needed. For the latter, if log transformation is applied, the ratio of geometric means of the log-transformed AUC and Cmax for the test product to the reference product is calculated. In this case, a 90 % confidence interval for the ratio of AUC or Cmax must fall within 0.8 and 1.25 to confirm BE (see, e.g., Food and Drug Administration, 2022a).

Traditional BE studies use fixed designs, often with a two-period, two-sequence crossover approach. Fixed designs generally lack flexibility and can be inefficient if model assumptions are not satisfied. On the other hand, adaptive designs provide a more flexible and efficient framework, allowing modifications of certain aspects of the trials based on interim data while maintaining study integrity. Approaches that are commonly used adaptively in BE studies include sample size re-estimation (SSR) and group sequential designs (GSDs).

Sample size determination in BE studies is usually performed based on assumptions about intra-subject variability and the expected PK parameter ratio between test and reference products. When the assumptions are violated, the trial will be underpowered and fail to establish BE (Type II Error), even for equivalent products. Adaptive designs can mitigate the issue by performing SSR at an interim stage, ensuring sufficient power without unnecessarily overrecruiting participants. This is especially beneficial for highly variable drugs (HVDs), e.g., with a coefficient of variation $> 30\ \%$ (Xu, et al., 2016). SSR can be done blinded or unblinded, with appropriate safeguards to control type I error and ensure study integrity.

As in other trials, GSDs in BE studies enable interim analyses to stop trials early for success or futility, reducing study time and costs. In a two-stage adaptive BE design, for example, the first stage uses an initial sample size to evaluate results. If equivalence is not established but seems plausible with more participants, the study moves to the second stage with a larger sample. Typically, combination tests, such as the inverse normal method or Fisher's method, are used to minimize increased false positive rates. Group sequential methods, although less common in bioequivalence (BE) studies, can also be implemented to enable early stopping based on efficacy or futility. Other approaches involve two-stage adaptive designs for three-treatment bioequivalence studies (Grayling, et al., 2019).

The scaled average bioequivalence (SABE) method is an important tool for HVDs. The approach adjusts BE limits based on the within-subject variability for the reference product (Tothfalusi et al., 2009). A related approach, the reference-scaled two one-sided tests (RS-TOST) method, modifies the traditional approach by incorporating variability scaling (US FDA 2022).

Bayesian statistical methods are also increasingly used in BE trials, incorporating prior knowledge to refine estimates of variability or treatment differences. Adaptive designs involving Bayesian approaches are particularly useful in situations where sample sizes are limited.

Effective integration of statistical innovation and adaptive design in BE studies requires adherence to regulatory requirements, including prospective planning and detailed documentation within the protocol, pre-specification of adaptation rules and statistical methodologies, control of the overall type I error rate, utilization of validated software and rigorous simulation to evaluate design operating characteristics. With the improvement of computational tools and increased familiarity of regulatory bodies with these innovative approaches, the techniques will likely become important tools in accelerating drug development.

2.6 A noteworthy statistical impact: COVID-19 vaccine development

An illustration of how Bayesian and other innovative methods make remarkable societal and economic impacts at the global scale is the case of the Pfizer and BioNTech trial design for messenger RNA COVID-19 vaccine (BNT162b2) development (Polack et al., 2020). The design was motivated by the high sense of urgency due to the COVID-19 pandemic that called for accelerated but scientifically rigorous vaccine efficacy and safety evaluation. The approach consisted of seamless study phases, moving from dose selection through confirmatory evidence, based on a Bayesian framework that provided flexibility by incorporating early and frequent interim analyses.

The first primary endpoint for the phase 2/3 part was the efficacy of BNT162b2 against confirmed COVID-19 with onset after 7 days since the second dose in participants with no serologic or virologic evidence of SARS-CoV-2 infection up to 7 days after the second dose. Vaccine efficacy (VE) was defined as $100 \times (1 -$ incidence risk ratio), where the incidence risk ratio was the ratio of the confirmed COVID-19 illness rate in the vaccine group to the corresponding illness rate in the placebo group. Sponsors were required to demonstrate a lower bound of the appropriately α-adjusted confidence interval around (or equivalent to) VE $>30\,\%$, with a point estimate of at least 50 %. Due to high public health demand, there was strong interest to bring the vaccine to the market as soon as possible if there was evidence of safety and overwhelming efficacy. Therefore, a frequent interim check for VE was necessary. The phase 2/3 part of the study planned for 4 interim analyses (after the accrual of at least 32, 62, 92, and 120 cases of COVID) and one final analysis after 164 confirmed cases were accrued. Futility and efficacy boundaries for interim and final analyses were appropriately calibrated to ensure that the overall type I error was protected at the desired (one-sided 2.5 %) level.

The assessment for the primary analysis was based on posterior probabilities using a Bayesian model. A weakly informative prior distribution centered around VE of 30 % was used. Planning for several interim analyses was especially useful in case accrual was slower than expected and/or if VE was higher than expected.

There were 43,548 adults in the trial; these subjects were healthy or had stable chronic medical conditions. For operational reasons, the first two planned interim analyses were not performed. The first interim analysis took place after the accrual of 94 cases. The BNT162b2 vaccine met the primary efficacy endpoint with more than 99.5 % probability (efficacy threshold) that the true VE was greater than 30 %. The interim analysis outcome was released by Pfizer on November 9, 2020. Moreover, the final analysis of the primary efficacy endpoint occurred soon after the interim analysis due to the fast accrual of COVID-19 cases caused by the second wave. The final analysis also met the prespecified success criteria, i.e., to establish a probability above 98.6 % (observed value $>99.99\,\%$) that the true VE was greater than 30 %; this greatly exceeded the minimum FDA criteria for authorization. The Emergency Use Authorization was based on the final analysis, which took place only 1 week after the interim analysis. For final approval, the FDA asked for longer-term safety data.

The Bayesian study design provided flexibility without compromising the quality of evidence. The Bayesian design directly quantified the probability that a study hypothesis (VE ≥30 %) was true given the trial data, which provided a straightforward way to set up a meaningful interim analysis plan, and the analysis used for the primary analysis provided strong evidence of VE.

The development of the vaccine in a record time is a testament to the role of statistics in addressing urgent societal needs. Indeed, the speed and precision of the development program—undoubtedly the most ambitious vaccine development program in history—was made possible, in part, thanks to the novel design and analytical strategies that were central to the pivotal trials.

3 Quantitative decision making in drug development

Advances in modeling and simulation and the use of Bayesian statistics have enabled wide use of quantitative decision making in drug development, from the study design to the interpretation of the clinical trial results. During the study design, quantitative analyses are performed to predict the potential efficacy and safety profile, as well as to establish critical success factors (CSFs) that ensure clear decision making will be possible after the study results become available. After the study is completed, quantitative analyses are used to assess the competitiveness of the efficacy and safety profile to inform the design and decision for the next stage of clinical development or commercial strategy. We briefly discuss four areas of advances in the last two decades:

- CSFs and probability of study success
- Forming the prior distribution information before designing a clinical trial
- Assessing the study results using the predefined CSFs and adjusting for the selection bias if moving to the next development stage
- Understanding the competitiveness of the new treatment versus those on the market or in development by competitors

3.1 Probability of study success

Traditionally, a study is designed to have a certain statistical power, say 80 % or 90 %. A fixed mean treatment difference is assumed in the power calculation. This is almost an academic exercise as it does not account for the uncertainty in the assumed value of the mean. Therefore, development of methodologies that account for this variation is important in understanding the probability that the study will succeed. The true treatment effect is always unknown, and our understanding of the treatment effect is based on previous data (analyses). In the last two decades, the variability in the mean treatment effect has been increasingly considered in the "statistical power" assessment, resulting in the terminology coined by industry researchers Wang et al. (2013) of the *probability of study success* (PrSS). Since then, the PrSS has been broadly used across the pharmaceutical industry. This measurement is also known as "assurance" (see O'Hagan et al., 2005). The PrSS is defined as

$$\text{PrSS} = \int \text{Power}(\mu; \sigma)\phi(\mu)d\mu,$$

where μ is the mean treatment difference parameter, $\text{Power}(\mu; \sigma)$ is the power function with respect to the mean μ and standard deviation σ, and $\phi(\mu)$ is the density function for the true treatment effect, which is often called the *prior* distribution despite there being no Bayesian component in the calculation of the PrSS. The PrSS, which incorporates the variability in the true treatment effect, provides a more realistic probability for the study success and has been widely used in clinical trial design. Generally, a fixed value for the standard deviation σ (in a continuous variable) is assumed in the PrSS, but one can certainly incorporate a prior distribution for σ in the PrSS calculation.

While most phase 3 studies often specify a significance level for the primary objective to meet regulatory requirements and provide scientific evidence, a different CSF may be used in phase 1 or 2 studies. A Bayesian CSF, which incorporates the distribution of the posterior distribution rather than the mean effect only, is often utilized. The Bayesian CSF is typically given by.

$$\text{Pr}(\text{treatment effect} > \delta \mid \text{Data}) > p,$$

where δ is generally the treatment effect that is not viable (corresponding to the null scenario) and p (which is generally no less than 0.50) is the minimum required probability. The PrSS for the above CSF is.

$$\text{PrSS} = E\left[\text{Pr}(\text{treatment effect} > \delta \mid \text{Data}) > p\right].$$

3.2 Forming the prior distribution

The most straightforward way to form a prior distribution is to use the estimation from the previous study. For example, a phase 2 study is completed, and the treatment difference between the new experimental treatment (for a specific dose) versus a control is estimated as $\hat{\mu}_{ph2}$ with a standard error of $se(\hat{\mu}_{ph2})$ If the study population and comparator in the phase 3 study are the same as that in the phase 2 study, a normal distribution $N(\hat{\mu}_{ph2}, [se(\hat{\mu}_{ph2})]^2)$ can be used as the prior distribution for the phase 3 study design.

In some cases, the comparator in the phase 3 study might be different from that in the phase 2 study, so direct use of the phase 2 to serve as the prior distribution in phase 3 may not be possible. In this case, an indirect comparison via a (Bayesian) network meta-analysis may be performed to estimate the treatment difference for the experimental treatment and the new control(s) in the phase 3 study, and the results can be used as a prior distribution for planning the phase 3 study.

In some less common cases, the sponsor may not have relevant prior phase 2 data for an indication because either the phase 3 study has a completely different endpoint compared to the phase 2 study or the indication has not been previously studied. In this case, using a prediction model becomes critical in the decision making for pursuing this new indication and in study design. A recent successful example is in the evaluation of tirzepatide as a potential treatment for obstructive sleep apnea (OSA). A meta-regression analysis based on data from a systematic literature search identified a strong relationship between percent weight loss and percent change in the apnea-hypopnea index (Malhotra et al., 2024a). This research was utilized to design a phase 3 study evaluating tirzepatide in treating OSA disease without a prior phase 2 study. Subsequently, the phase 3 study was recently completed with positive results (Malhotra et al., 2024b). Furthermore, sometimes, especially in diseases where large, long, expensive phase 3 studies are needed to show the outcomes, predictive models based on biomarkers are essential before planning such a phase 3 study. Such was the case for Lipoprotein(a) (Lp[*a*]), which has been identified as a new atherosclerotic cardiovascular disease risk factor. We conducted a phase 2 studies which showed an impressive reduction in Lp(a) ($>70\%$). However, we needed to design phase 3 study with cardiovascular endpoint, In order to provide the probability/confidence of success before investing this large phase 3 study, we utilized the predictive models from Madsen et al. (2020) to predict the risk reduction in major adverse cardiovascular events; corresponding phase 3 studies were designed (ClinicalTrials.gov identifiers: NCT04023552, NCT05581303, and NCT06292013). In cases where one wants to test a molecule in a new indication, one may utilize the prior success of that molecule along with external information. For such cases, Proper et al. (2024) has developed a novel Bayesian network meta-analysis approach to predict the efficacy of an approved therapy in a new indication and has illustrated via a case example in psoriasis and psoriatic arthritis.

3.3 Adjusting for selection bias in clinical development

As sponsors often select the most promising compounds for an early proof-of-concept (PoC) or phase 2 study for further development, such a decision process generates selection bias, which is also called regression to the mean. Although, on average, the estimator from a small PoC or phase 2 study is unbiased, there is a selection bias if only the compounds with an impressive observed treatment effect are selected to move to the next development stage. The larger the mean and variance of the estimated treatment effect in a PoC or phase 2 study, the more selection bias it has. A meta-analysis of treatments for rheumatoid arthritis showed that phase 2 studies generally have a better treatment effect than phase 3 studies (Qu et al., 2020).

A similar selection bias or regression to mean phenomenon also occurs in the subgroup analyses. For a clinical trial, in addition to the primary and secondary analyses for the entire population, the treatment effects for some important subgroups are also estimated. The treatment effect in the most promising subgroups is also subject to selection bias. Wallach et al. (2017) conducted a meta-analysis to evaluate the subgroup claims from RCTs: 46 of 117 subgroup claims had evidence of statistically significant heterogeneity from a test for interaction; five of these 46 subgroup findings had led to a subsequent RCT, but none of these RCTs produced significant results. The Bayesian method is widely used in "shrinking" the subgroup treatment effect (Wang et al., 2024). Du et al. (2024) provides a unified approach using Bayesian model averaging to adjust for the selection bias for studies with multiple endpoints, multiple treatment arms, and multiple preidentified subgroups. Du et al. (2024) also provides a decision-making framework that suggests adjustment for the selection biases due to multiple comparisons in the same study and due to selecting promising compounds at the portfolio level.

3.4 Understanding the efficacy and safety of different treatments via indirect comparisons without head-to-head studies

During drug development, especially after the PoC or phase 2 study results, sponsors are interested in understanding the competitiveness (in both efficacy and safety) of the experimental treatment compared to currently marketed drugs or those in development by other pharmaceutical companies. After the phase 3 study is completed and/or the new treatment is approved by regulators, payors have an interest to understand the efficacy of the new treatment versus those on the market. Conducting head-to-head studies is costly, may take a long time, and is often not feasible. Therefore, drawing inferences without conducting additional head-to-head studies has become increasingly important. The network meta-analysis plays an important role in these cases. Frequentist and Bayesian network meta-analysis methods (Ohlssen et al., 2014) have been developed and widely applied to address these questions. Here we use recent research evaluating the efficacy and tolerability of tirzepatide and semaglutide in treating type 2 diabetes (Karagiannis et al., 2024) to illustrate the concept and results of network meta-analyses. Fig. 1A shows the networks for a few treatments for type 2 diabetes. Fig. 1B, C, and D show the network meta-analysis results for the change in glycated hemoglobin, change in body weight, and gastrointestinal tolerability incidence between various treatments, respectively.

Fig. 2 summarizes the utilization of the quantitative approaches in clinical drug development from study design, data analysis, and result interpretation. From the examples mentioned in this section, industry statisticians have contributed tremendously to the statistical advances in quantitative decision making, including the concept of PrSS, the example of using meta-regression for making decision for directly testing tirzepatide in treating OSA disease with a prior phase 2 study, and the selection bias adjustment.

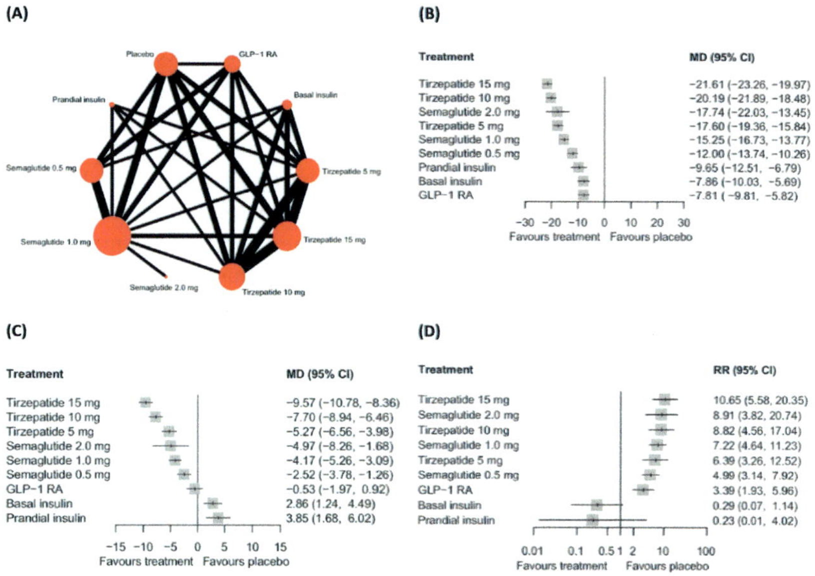

FIG. 1 Network meta-analysis to compare the efficacy and tolerability of tirzepatide and semaglutide in treating type 2 diabetes. (A) Network plot. Each circle indicates a treatment node. Lines connecting two nodes represent direct comparisons between two treatments. The size of the nodes is proportional to the number of trials evaluating each treatment; the thickness of the lines is proportional to the number of trials directly comparing the two connected treatments. (B) Network meta-analysis results for the change in glycated hemoglobin (mmol/mol) compared with placebo. (C) Network meta-analysis results for the change in body weight (kg) compared with placebo. (D) Network meta-analysis results for the discontinuation of treatment due to gastrointestinal adverse events compared with placebo. Abbreviations: CI, confidence interval; GLP-1 RA, glucagon-like peptide-1 receptor agonist; MD, mean difference; RR, risk ratio. *(D) Figure is reprinted from Karagiannis et al. (2024).*

4 Multiple testing: A foundational tool for inference in confirmatory clinical trials

Multiple testing is an area of significant and broad application in clinical drug development. The regulatory importance of this area of application has been reflected in FDA guidance documents (for example, see Food and Drug Administration, 2022a), a European Medicines Agency guidance document (European Medicines Agency, 2016), two European Medicines Agency expert workshops on the topic, and numerous presentations and publications from leaders of major regulatory agencies. Multiple testing procedures are statistical procedures that control the probability of false-positive claims regarding a drug. As such, they provide a safeguard for patients and all those who rely on statistical inference from drug clinical trials. Common sources of multiplicity in clinical trials include interim analyses or multiple looks (e.g., group sequential trials), assessment of multiple endpoints that measure performance

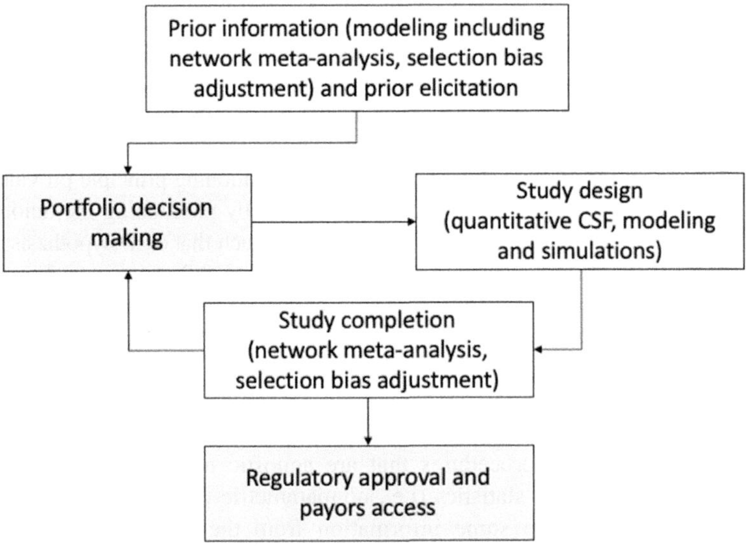

FIG. 2 Quantitative analyses in decision making in clinical drug development. Abbreviation: CSF, critical success factor.

characteristics of treatments, assessment of multiple treatments or dosing regimens, or assessment of treatment effects in multiple populations. Modern clinical trials often include several of these sources of multiplicity simultaneously, resulting in complex multiple testing problems. Over the past 25 years, statisticians from the pharmaceutical industry and regulatory bodies have conducted methods research and proposed new testing procedures to specifically address these complex multiple testing problems.

4.1 Foundations

When testing multiple hypotheses, the probability of obtaining at least one false-positive outcome increases as the number of hypotheses being tested increases. Therefore, it is desirable to control the probability of obtaining at least one false positive at a reasonable level to ensure fidelity of inference. In confirmatory clinical trials, the principal criterion to be met for any multiple testing procedure applied for inference is the strong control of the type I error (i.e., false-positive rate referred to as α) rate. That is, when testing k null hypotheses, H_{01}, H_{02}, …, H_{0k}, the probability of incorrectly rejecting at least one true null hypothesis is required to be no greater than α for *all possible* configurations of true and false null hypotheses. There are two primary principles that provide the foundation for establishing multiple testing procedures, which provide strong control of the type I error rate: the closed testing principle or closure principle (Marcus et al., 1976) and the partitioning principle (Finner and Strassburger, 2002; Steffansson et al., 1988). The closure principle provides a strategy for constructing multiple testing procedures by considering appropriate α -level tests of all intersection hypotheses formed from the k hypotheses H_{01}, H_{02}, ..., H_{0k}. Any hypothesis H_{0i}, $i = 1, 2, ..., k$, is rejected under the multiple testing procedure if its local alpha-level test and the α -level test of every intersection hypothesis that includes it is also rejected. The majority of multiple testing procedures have been constructed/proven based on the closed testing principle. The partitioning principle provides a strategy for constructing multiple testing procedures by partitioning the union of all hypotheses under consideration into disjoint sets, such that each hypothesis can be written as the disjoint union of some of the sets. While employed less frequently than the closed testing principle, this approach to construction of multiple testing procedures has also been used to establish multiple testing approaches for important, complex multiple testing problems in drug development.

By the mid-1990s, there were numerous procedures that had been developed for testing a group of k hypotheses. A summary of these is shown in Table 1. They include procedures that are agnostic to the underlying joint distribution of the k test statistics (i.e., nonparametric procedures), as well as procedures that leverage some information from the underlying joint distribution of the k test statistics (i.e., parametric or semiparametric procedures). While all the procedures provide strong control of the type I error rate when their respective assumptions are met, they differ in their efficiency based on the distributional information used and the details of their closed testing representation. These details impact the performance of the testing approaches and would prompt extensions and adaptations in subsequent methods development.

4.2 Modern bespoke multiple testing procedures

Beginning in the late-1990s, additional multiple testing approaches, often extensions of the methods noted in Table 1, were developed with specific drug development applications in mind. These new methodologies were the product of research conducted by pharmaceutical or regulatory statisticians, often in partnership with academic statisticians, and provided rigorous approaches for evaluating multiple primary and/or secondary hypotheses concerning treatment effect claims across multiple populations, endpoints, and/or dosing regimens. The early 2000s saw the introduction of a class of multiple testing procedures (derived via the closure principle) known as *gatekeeping* procedures (Westfall and Krishen, 2001; Dmitrienko et al., 2003; others). These procedures sought to formally incorporate in the testing paradigm logical constraints reflecting a hierarchy of clinical trial objectives and their associated hypotheses to be tested. As the name suggests, these procedures established a subset of hypotheses as a "gatekeeper" to other subset(s) of hypotheses. The simplest manifestation of gatekeeping came in fixed sequence testing of k hypotheses, in which each hypothesis test served as a "gatekeeper" for the subsequent test(s). That is, testing of k hypotheses commenced sequentially, testing each hypothesis at the α level; testing ended when a hypothesis test failed to reject the null hypothesis. The concept of sequential

TABLE 1 Traditional multiple testing procedures.

Multiple testing procedure	Distributional information	Operational execution
Bonferroni	Nonparametric	Single step
Holm	Nonparametric	Data-driven testing ordering
Hochberg	Semiparametric	Data-driven testing ordering
Hommel	Semiparametric	Data-driven testing ordering
Simes	Semiparametric	Single step
Dunnett	Parametric	Single step
Step-down Dunnett	Parametric	Data-driven testing ordering
Step-up Dunnett	Parametric	Data-driven testing ordering

testing according to hierarchy was extended from single hypotheses to *families* of hypotheses, aligning to a hierarchy of clinical trial objectives. For example, a primary set (family) of hypotheses may be tested first, followed by testing of hypotheses in a secondary family, based on the outcomes from testing the primary family and associated logical constraints. For applications in which testing of a secondary family should commence only if all null hypotheses in the primary family are rejected, serial gatekeeping procedures were defined. For applications in which testing of a secondary family should commence when any null hypothesis in the primary family is rejected, parallel gatekeeping procedures were defined. Example 1 in Section 4.4 provides an illustration of an application of parallel gatekeeping. The general gatekeeping framework of multiple testing across families represented an important foundational shift in multiple testing research. Multiple testing procedures could be developed with a focus on how to mathematically combine relevant component (family-level) multiple testing procedures. Some multiple testing procedures may be directly used as the component gatekeepers in the parallel gatekeeping framework. These procedures are known as *separable* (Dmitrienko and Tamhane, 2011) procedures. Other procedures require adaptation to be used as gatekeepers in the parallel gatekeeping framework. An adjustment, known as *truncation* (Dmitrienko and Tamhane, 2011), was proposed to create separable procedures from these, enabling broad opportunities for constructing application-specific parallel gatekeeping procedures. Throughout the first decade of the 2000s, there was considerable research and development of powerful gatekeeping procedures with application to drug development, including the application of gatekeeping procedures to group sequential and other adaptive designs. The influence of this body of work is evident in the FDA guidance document (Food and Drug Administration, 2022b), which heavily cites principles based on gatekeeping applications.

At the end of the decade, multiple authors independently proposed a multiple testing framework that did not explicitly incorporate notions of families of hypotheses. Instead, this framework focused on the allocation of α across the set of k null hypotheses and the iterative propagation of α across those null hypotheses in a sequential testing approach. These multiple testing procedures are known as graphical procedures or chain procedures (Burman et al., 2009; Bretz et al., 2009, 2011; Millen and Dmitrienko, 2011). They have had a significant impact on how practitioners think about α propagation (i.e., efficient inference through α exhaustive approaches) and how they communicate multiple testing procedures to non-statistician colleagues. The iterative graph introduced by Bretz et al. (2009) provides an intuitive view of what can be complex underlying closed testing procedures. When restricting to the set of nonparametric graphical/chain procedures, the framework provides a shortcut to developing closed tests. Other forms of these procedures are more complex. Nonetheless, the graphical display can be powerful in communicating testing relationships and implications. These graphical/chain procedures are quite popular today, along with gatekeeping procedures. Like gatekeeping procedures, graphical procedures are referenced significantly in the FDA guidance document (Food and Drug Administration, 2022a). Example 2 in Section 4.4 offers an illustration of a (Bonferroni-based) graphical procedure.

4.3 Application

Given the vast recent research to develop new and relevant multiple testing procedures, today's clinical trialists and pharmaceutical statisticians have a broad armamentarium of procedures available to enable the complex inference demanded of modern clinical trials. Selection of an appropriate, powerful multiple testing approach for a given application requires thoughtful consideration and involves evaluation of procedure relevance (i.e., alignment to logical constraints of the clinical trial context) and procedure performance via appropriate application-specific metrics (Dmitrienko et al., 2011). Metrics for evaluation of procedure performance have been discussed in multiple publications and include conjunctive power, disjunctive power, and extensions (for example, see Westfall and Krishen, 2001; Millen and Dmitrienko, 2011). Monte Carlo simulations are generally employed to evaluate procedure performance according to chosen metrics.

The execution of the multiple testing procedures is often carried out somewhat manually, according to the procedure's decision rules, after computation of local P values for the k hypotheses. However, manual execution is challenging in some applications, particularly of complex graphical/chain procedures. Currently, there are open-source software packages to enable execution of these procedures.

The contributions of pharmaceutical and regulatory statisticians have led to a mature and growing armamentarium of multiple testing procedures and supporting frameworks that have had significant, direct impact on inference from clinical trials. Importantly, these advancements have inspired additional statistical research in multiple testing methodologies that may provide improvements to the available methods or address new multiple testing problems.

4.4 Examples

Example 1. ***A parallel gatekeeping application***

Consider a confirmatory clinical trial that tests two doses (low and high) of treatment X versus placebo. Inference is desired for the primary endpoint and an important secondary endpoint whenever statistical significance is achieved on the primary endpoint. The trial is considered successful if at least one dose demonstrates statistical significance on the primary endpoint. A parallel gatekeeping procedure may be constructed using a Dunnett procedure for the primary family of hypotheses and a Dunnett procedure for the secondary family of hypotheses, subject to logical constraints, such that testing of the secondary hypothesis for each dose only occurs if statistical significance for that dose occurred based on testing of the primary family of hypotheses.

Example 2. ***A graphical testing application***

As in Example 1, consider a confirmatory clinical trial that tests two doses (low and high) of treatment X versus placebo. Inference is desired for the primary endpoint and an important secondary endpoint whenever statistical significance is achieved on the primary endpoint. The trial is considered successful if at least one dose demonstrates statistical significance on the primary endpoint. A Bonferroni-based graphical procedure is constructed as shown in Fig. 3. Because of α recycling present in this procedure, it has the potential to reject more hypotheses than the parallel gatekeeping procedure of Example 1, despite its use of less powerful component tests (Bonferroni vs Dunnett).

Let H_{p1} and H_{p2} denote the null hypotheses in the primary family and H_{s1} and H_{s2} denote the null hypotheses in the secondary family, where subscript 1 reflects the low dose and subscript 2 reflects the high dose. Testing proceeds as follows:

- Apply local tests at level $\alpha/2$ for the primary endpoint for each of the two doses. If neither H_{p1} nor H_{p2} is rejected, testing stops.
- If H_{p1} is rejected, then H_{s1} is tested at level $\alpha/2$.
- Similarly, if H_{p2} is rejected, then H_{s2} is tested at level $\alpha/2$.
- If H_{p1} and H_{s1} are rejected but either H_{p2} or H_{s2} is not rejected, these are retested in sequence at the full level α.
- Similarly, if H_{p2} and H_{s2} are rejected but either H_{p1} or H_{s1} is not rejected, these are retested in sequence at the full level α.

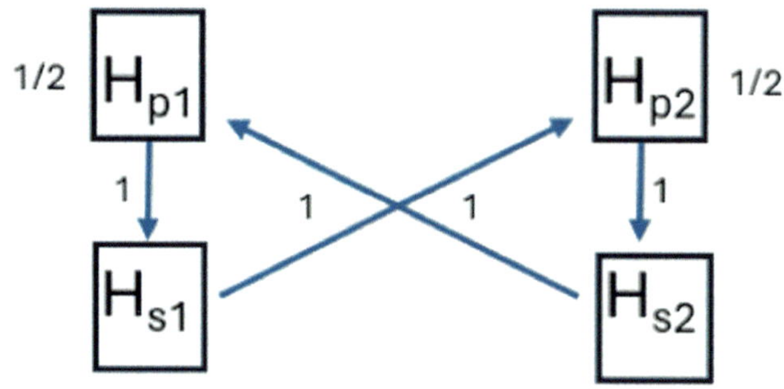

FIG. 3 Graphical multiple testing procedure for Example 2.

5 Leveraging surrogate endpoints for statistical innovation

A surrogate endpoint is an endpoint that is used as a substitute for a direct clinical outcome measure of how a patient feels, functions, or survives. Changes that are induced by a therapy on a surrogate endpoint should reflect changes in a clinically meaningful endpoint (Temple, 1995). The FDA defines two different categories of surrogate endpoints according to their level of clinical validation. A fully validated surrogate endpoint has a clear mechanistic rationale and is known to predict clinical benefit. The FDA considers it as having enough clinical evidence to enable the surrogate to replace the clinical endpoint and may grant traditional full approval based on a validated surrogate endpoint without request for additional evidence of clinical efficacy. A reasonably likely surrogate endpoint requires the same features, that is, biological plausibility and predictive capability for clinical benefit, but there is not enough clinical evidence yet to fully support the latter. The FDA instituted the accelerated approval program to allow for earlier approval of drugs treating serious conditions and filling an unmet medical need based on a reasonably likely surrogate endpoint.

5.1 Surrogate endpoints and their relevance in terms of statistical innovation

An important factor that impacts the duration, complexity, and cost of a clinical trial is the endpoint that is used to study the efficacy of the investigational treatment (Alonso et al., 2016). In some situations, it is difficult to use the clinical endpoint. For example:

- The clinical endpoint may require a long follow-up time (e.g., overall survival in oncology), such that the evaluation of the new therapy using this endpoint would be delayed and potentially confounded by other therapies; or,
- The clinical endpoint may have a low incidence (e.g., pregnancy in severe luteinizing hormone deficiency), such that the evaluation of the new therapy using this endpoint would result in a very large sample size.

In such situations, it can be an attractive strategy to substitute the difficult-to-measure clinical endpoint by a surrogate endpoint that can be measured earlier (e.g., progression-free survival in oncology) or that has a higher prevalence (e.g., follicular development in severe luteinizing hormone deficiency).

The use of surrogate endpoints can have major benefits, including the acceleration of drug development, increased feasibility of clinical trial research, decreased wait time for patients to access a new drug, and a more precise estimation of treatment effects. In fact, since 1992 the FDA has granted about 200 approvals based on surrogate endpoints for cancer drugs alone. Surrogate endpoints are also used in many other diseases. For example, surrogate endpoints were instrumental in the rapid evaluation of new therapies for HIV/AIDS, and surrogate endpoints are routinely used to evaluate VE (called Correlates of Protection in this context). The FDA maintains a list of validated and reasonably likely surrogate endpoints that have been used as the basis of drug approval. The list covers a wide range of diseases and conditions, and approximately 200 surrogate endpoints supporting either traditional or accelerated approvals are listed.

5.2 Statistical methods to validate surrogate endpoints

It is a common misconception that surrogacy automatically follows from a strong association between the surrogate and the clinical endpoint. This fallacy has had some severe consequences in the past. For example, the FDA has approved three drugs (i.e., encainide, flecainide, and moricizine) based on their capacity to suppress arrhythmias. As arrhythmia is known to be associated with a significant increase in death rates due to cardiac complications, it was assumed that these drugs would also reduce the death rate. However, the postmarketing Cardiac Arrhythmia Suppression Trial found that the death rate with the active treatment was actually *higher* than that with placebo (Cardiac Arrhythmia Suppression Trial [CAST] Investigators, 1989).

The key lesson that was learned from the Cardiac Arrhythmia Suppression Trial was that a correlate is not a surrogate (Baker and Kramer, 2003). Relying solely on the biological plausibility of a putative surrogate endpoint, or the simple correlation between both endpoints, can lead to misguided conclusions and even to the approval of treatments that turn out to be *harmful* to patients. Instead, establishing surrogacy requires the use of robust statistical validation methods that demonstrate that the *treatment effect* on the putative surrogate endpoint reliably predicts the *treatment effect* on the clinical endpoint. Innovative statistical methods that are currently used to validate surrogate endpoints are based on mixed-effects/hierarchical models, Bayesian approaches, information theory, and/or causal inference ideas.

The formal evaluation of a putative surrogate endpoint is indeed not a trivial endeavor, and early attempts that were based on classical statistical methodology (such as regression analysis or correlational analysis) turned out to be flawed (Alonso et al., 2016). For example, Prentice (1989) provided the first formal definition of a surrogate endpoint (i.e., "a response variable for which a test of the null hypothesis of no relationship to the investigational treatment groups under comparison is also a valid test of the corresponding null hypothesis based on the clinical endpoint") and formulated four operational criteria that should be fulfilled to validate a candidate surrogate endpoint. In essence, these criteria require that the effect of the clinical endpoint be fully mediated via the surrogate endpoint. This happens when the surrogate endpoint is an integral part of the mechanism of action of the disease (or, in other words, when the surrogate is in the causal chain leading to disease). Unfortunately, there are several nontrivial issues with these criteria, and a surrogate endpoint that passes the proposed criteria is not guaranteed to fulfill the formal definition of surrogacy (Burzykowski et al., 2005).

A key issue with all the initially proposed methodology is that the surrogate endpoint evaluation exercise was carried out based on the data of *a single clinical trial* (Prentice, 1989; Freedman et al., 1992; Buyse and Molenberghs, 1998). Recall that a key property of a good surrogate endpoint is that the treatment effect on the surrogate should be predictive for the treatment effect on the clinical endpoint. When data are available for a single clinical trial, there is only one estimated treatment effect on the surrogate and one estimated treatment effect on the clinical endpoint. Evidently, proper examination of whether the treatment effect on the putative surrogate reliably predicts the treatment effect on the clinical endpoint can never be done based on this information alone (a single "data point"). Indeed, *multiple* clinical trials are needed to quantify the strength of the association between the treatment effects on the endpoints. This key insight, learnt from the drug development paradigm, led to the development of the meta-analytic approach through industry and academic collaborative research (Buyse et al., 2000). This approach assumes that information regarding the surrogate and the clinical endpoints is available from multiple clinical trials. In this way, there is not only replication at the level of the patients (i.e., the surrogate and clinical endpoints are measured across multiple patients) but also at the level of the clinical trials (i.e., the treatment effects on the clinical and surrogate endpoints are estimated across multiple clinical trials).

In the meta-analytic approach of Buyse et al. (2000), a mixed-effects model is fitted to the data of the N randomized clinical trials. The model essentially describes the relationship between the surrogate endpoint and the clinical endpoint while accounting for treatment effects and the variability across the different clinical trials and patients. The fixed effects in the model represent the overall treatment effects on the surrogate and clinical endpoints, whereas the random effects capture the trial-specific variability of the treatment effects. In the meta-analytic framework, surrogacy is quantified by two metrics. The *trial-level coefficient of determination* assesses the strength of the association between the treatment effects on both endpoints across the different clinical trials. This metric quantifies the extent to which the treatment effect on the clinical endpoint can be predicted based on the treatment effect on the surrogate endpoint. The *individual-level coefficient of determination* quantifies the strength of the (treatment- and trial-corrected) association between both endpoints at the level of individual patients. To conclude that a surrogate is valid, both the individual- and trial-level coefficients should be high. The meta-analytic approach is often used in practice. For example, Belin et al. (2020) reported that 83 of 91 studies (91 %) that examined the validity of progression-free survival as a surrogate for overall survival in oncology used this approach. The main challenge with using the meta-analytic approach is that it requires data from multiple clinical trials. At the earliest stages of drug development (where surrogate endpoints are arguably the most useful), this information is often not available. In such cases, alternative methods that require less data (but substantially more assumptions that are often even *un*testable) can be used. For an overview, the reader is referred to Elliott (2023).

6 Causal inference and real-world evidence

Although RCTs are the gold standard for generating reliable evidence concerning the relative risks and benefits of alternative treatment options, they may lack external validity or be impractical for operational and ethical reasons. In this regard, recent developments in statistical innovation have made the use of RWD or routinely collected data relating to health status or healthcare delivery, standard in drug development for use cases ranging from post-marketing studies of safety and effectiveness to external control arms for single-arm trials. For example, after COVID-19 vaccines were approved based on large RCTs, numerous studies were carried out to assess the effectiveness of these vaccines in a real-world setting. Reasons for using RWD to carry out such studies include assessing safety and effectiveness of a therapy in a broader population, not restricted by the inclusion and exclusion criteria of a trial, where patients are treated outside the setting of a protocol-driven study.

Fundamental to the appropriate use of RWD is causal inference, a set of statistical tools that can be used to quantify the effect of treatment on a clinical outcome in the presence of potential confounding factors. When carrying out studies of comparative effectiveness using RWD, several assumptions are required for the analysis to have a causal interpretation. One of the most important assumptions is that of conditional exchangeability. Namely, given a set of covariates or confounders, the assignment of treatment is independent of the outcomes that would result under any treatment (these are called potential outcomes, as they are not all observed). All relevant confounders would need to be measured to draw the causal inference.

The first step in a well-designed observational study should be to design the target randomized trial that the investigators would use to answer the question of interest, including the population of interest, treatment strategies, start and end of follow-up periods, outcomes, and contrasts to be computed. Then, the investigators should describe how they will emulate that target trial using observational data. Once a covariate set considered sufficient for conditional exchangeability is identified, an intuitive way to emulate a target trial that would randomize individuals to a study drug or control is via matching with respect to the covariates. Next, outcomes for the matched sets of patients are compared to one another. If the required assumptions are satisfied, the resulting estimates have a causal interpretation, that is, they represent the difference in outcomes that would be seen between a situation in which all patients could be treated with the study drug versus the control.

Another estimation approach often used in causal inference is inverse probability of treatment weighting. Here, the propensity score, which is the probability of treatment conditional on observed baseline characteristics, is calculated and then used to construct weights that, when applied to the study, create a sample in which the treatment groups are balanced on potentially confounding baseline variables. Various weights are possible, depending on the target population of interest, including average treatment effect for the treated weights, which yield the average effect of treatment on those who would receive the treatment, and average treatment effect weights, which yield the average effect of treatment on the entire population. Overlap weights, which estimate the treatment effect in a population reflective primarily of those in the sample who had a reasonable probability of receiving both treatments, are often used when standard weights would result in some patients receiving extreme propensity scores. These extreme scores result in high variability, as just a few patients get very large weights. Additionally, regression can also be used for causal inference as long as all potential confounding variables are included as covariates.

The "g-methods" introduced by Robins (2000) are also used in causal inference involving observational data. Notably, estimation based on marginal structural models, which permit estimation of marginal quantities without conditioning on other covariates, can be performed using either the g-computation algorithm or inverse probability of treatment weighting, with the former having more desirable properties in terms of efficiency and stability of estimates as well as handling of time-varying covariates (Daniel et al., 2013).

Various methods and guidelines for the selection of covariates to be used for matching or calculation of weights have been developed, with the aim of increasing the likelihood that the conditional exchangeability assumption will be satisfied along with improving precision. For example, it is not advisable to include variables that impact only the exposure but not the outcome. In addition, so-called doubly robust estimation methods have been developed that allow for unbiased causal inference either if exposure is correctly modeled or if the outcome is correctly modeled using regression. Some advanced methods allow for flexible machine learning (ML) models to be used to estimate these exposure and outcome models, further increasing the likelihood that the conditional exchangeability assumption will be satisfied.

Care must be taken when designing comparative effectiveness or safety studies using RWD to avoid several different types of potential bias. One example of such bias is immortal time bias, where patients are incorrectly classified as exposed to a treatment before their exposure begins. For example, one study compared hospital readmission rates among patients with chronic obstructive pulmonary disease who did and did not receive inhaled corticosteroids within 90 days of discharge. In such an analysis, anyone who might have received corticosteroids 20 days after discharge but was readmitted 10 days after discharge would be classified as a corticosteroid nonuser, whereas anyone who received corticosteroids 20 days after discharge but was readmitted later than that would be classified as a corticosteroid user. This introduces bias in favor of the corticosteroid intervention. To avoid this bias, one could use a model that appropriately classifies exposed and unexposed times, such as a time-dependent Cox model. A simpler alternative is landmark analysis. To implement a landmark analysis here, readmission starting 90 days *after* discharge, for patients not readmitted in the first 90 days after discharge, could be compared between patients who did and did not receive corticosteroids in the 90 days after discharge (Gleiss et al., 2018).

Given the strong assumptions required for causal inference using observational data, methods have been developed to assess how sensitive results are to deviations from those assumptions. Another promising set of methods uses negative control outcomes, which are auxiliary outcomes different from the outcome of interest and not thought to be affected by the exposure of interest but thought to be affected by confounders of the relationship between the exposure and outcome of interest. For example, some observational studies in the elderly showed a surprisingly large reduction in the risk of mortality after influenza vaccination, after adjustment for measured covariates. Other investigators then analyzed the effect of influenza vaccination on a negative control outcome that should be unaffected by influenza vaccination (i.e., hospitalization due to accidental injury). The implausible finding that influenza vaccination was protective against hospitalization due to accidental injury suggested that the finding showing protection against mortality was likely not credible.

Another important recent area of application of causal inference in drug development is the use of an external control arm for a single-arm trial, discussed in Section 2.4. Often used in rare diseases or oncology, where it may be infeasible or unethical to carry out an RCT, an external control arm involves the contextualization of a single-arm trial using a control arm comprised of historically or concurrently treated patients, treated either in the real world or prospectively followed using a protocol. This design poses many of the same challenges as an observational study, with the added complexity that data quality may be different between the single-arm trial, carried out pursuant to a protocol, and the external control arm.

In view of the growing importance of RWD in drug development, along with the challenges associated with causal inference, several frameworks and guidelines for the use of RWD have been developed, including the target trial framework (Hernán and Robins, 2016) and the causal roadmap framework

(Dang et al., 2023). Central to these frameworks is the role that statisticians have to play in addressing underlying methodological and conceptual issues to ensure the successful use of RWD to advance public health.

7 Artificial intelligence (AI) and machine learning (ML) in drug discovery and development

This section reviews key contributions of AI and ML methods in medical research and drug development. AI encompasses systems that perform tasks traditionally associated with human intelligence, while ML is a subset of AI focused on data-driven learning. Deep learning, a further subset of ML involves stacking multiple layers of artificial neurons, which in essence are nonlinear transformations with ability to learn higher-order features of the input to process data. Breakthroughs in ML have been facilitated by increased computational power and access to large, high-quality datasets.

While AI/ML methods are deeply rooted in statistical methods, they possess certain distinct features. Traditional statistics and biostatistics research often assumes a true data generation model, leading to concepts such as *P* values, convergence, and efficiency, while the focus of most data analytics fields is on prediction and classification (Breiman, 2001). In addition, AI/ML methods have also demonstrated superior performance in handling unstructured data, including images, videos, and audio. Recent achievements in areas like protein structure (Jumper et al., 2021) prediction have garnered significant attention, culminating in a Nobel Prize in 2024.

In pharmaceutical research and development, there is a growing adoption of AI and ML methods, particularly during critical stages of drug discovery, including target identification and molecular design, as well as late-stage research. The high failure rates of clinical trials, along with their significant costs, drive the need for more efficient, cost-effective approaches. Target identification involves exploring associations between outcomes and various features, such as gene expression and genomic pathways related to diseases. Scalable methods like XGBoost (Extreme Gradient Boosting) (Chen and Guestrin, 2016) and deep neural networks have emerged as vital tools for predicting drug–target interactions, which are crucial for advancing drug discovery. The molecular design phase benefits from high-throughput screening, a technique that quickly identifies active compounds. The integration of ML enables the mining of extensive chemical databases to design drugs with specific desired properties. Selecting appropriate hits from high-throughput screening is critical, as it influences clinical trial success and necessitates consideration of multiple criteria such as potency, toxicity, and pharmacokinetics, leading to complex multi-objective optimization problems. Recent studies have benchmarked various AI/ML methods against traditional techniques for molecular design, providing insights into their comparative effectiveness (Schneider et al., 2020).

In the clinical space, digital health technologies and digital biomarkers have demonstrated promising results across various medical conditions, particularly in areas such as Parkinson's disease, sleep apnea, and itch measurement. For instance, digital biomarkers can offer real-time monitoring and assessment of symptoms, enabling more personalized treatment approaches and improving patient outcomes. In Parkinson's disease, wearable devices and mobile applications track movement patterns, providing clinicians with valuable insights into disease progression and treatment effectiveness. Similarly, advancements in sleep apnea monitoring through digital health tools allow for more accurate diagnoses and tailored therapeutic interventions. AI-based digital technology also expands diagnostic capabilities, potentially aiding clinical trials by reaching a broader patient population through easy-to-access digital diagnostic solutions (Shajari et al., 2023).

Furthermore, large language models have been increasingly integrated into research and development processes to enhance efficiency and productivity. These AI-driven tools facilitate numerous tasks, including drafting research documents, monitoring clinical trial progress, detecting anomalies, copiloting clinical design, and engaging in interactive chat sessions with researchers and clinicians. By streamlining administrative tasks, large language models enable researchers to focus on more complex analytical and creative challenges, ultimately accelerating the development of new treatments and improving collaboration across multidisciplinary teams. Their ability to process and analyze vast amounts of medical literature and data supports informed decision making, potentially leading to breakthroughs in clinical research and patient care (Ghim and Ahn, 2023).

Another key AI tool that is gaining importance among researchers is reinforcement learning—an algorithmic approach where an agent tries to maximize the total amount of reward it receives when interacting with a complex and uncertain environment. The individualized treatment recommendation framework, rooted in reinforcement learning, aims to optimize patient outcomes by treating subgroup identification as an outcome optimization problem. This approach accommodates both randomized trials and observational studies, enhancing performance by removing intercept and covariate effects. Individualized treatment recommendation has been connected to ML through weighted classification problems, offering exciting possibilities for personalized medicine (Zhao et al., 2012). Later, Fu et al. (2016) extended the method for subgroup identification, which was later extended to multiple treatments (Qi et al., 2020) and handling both efficacy and safety endpoints (Wang et al., 2018).

In recent years, ML-based prediction models have also emerged, which allow enhanced understanding of disease progression and prediction of treatment benefit from biomarkers observed in early phases. For example, gradient boosting ML was used to understand the relationship between glucose variability and hypoglycemia events (Qu et al., 2012). Further, multiple cardiovascular risk prediction models have been developed using ML (see Cai et al., 2024 for a systematic review).

The reinforcement learning framework extends personalized medicine from single decision points to multi-stage interventions, particularly relevant for chronic diseases. Dynamic treatment regimens and methods like Q-learning facilitate this transition, with recent studies expanding these concepts into continuous horizons for mobile health applications (Liu et al., 2024).

In conclusion, the integration of AI and ML into medical research and drug development is not just a trend but a transformative shift that promises to enhance patient outcomes and streamline the research and development process. As the power of advanced computational methods is harnessed, personalized medicine will continue to evolve, making healthcare more precise and effective. The ongoing collaboration between clinical research and technology will pave the way for innovative solutions, ultimately reshaping the landscape of modern medicine and fostering a future where tailored therapies are accessible to all patients.

8 Concluding remarks

Over the past few decades, clinical trials that are conducted to establish the safety and efficacy of medicines have become increasingly complex and resource intensive. While traditional frequentist statistical methods have served and continue to serve as the mainstay in the design and analysis of trials, new approaches are constantly needed to address the ever-increasing cost and time of bringing new medicines to patients. As elucidated in this chapter, there have been significant advances in this regard, with statisticians from academics, industry, and FDA playing impactful and collaborative roles in bringing innovative methodologies that have enhanced the design, analyses and interpretation of trials, and have reduced the cost and improved efficiency of drug development. Recent laws and regulations, such as the US 21st Century Cures Act and the US Prescription Drug User Fee Act VI, have also fostered innovations, of which Statistical leadership is a central component. A case in point is the use of advanced innovative trial designs, as previously discussed in this chapter. In this framework, a common theme is the reliance on Bayesian methods and simulations to characterize operating characteristics of inferential procedures. Bayesian methods are especially attractive because they offer a viable framework for decision making under situations that benefit from the use of prior information, updating current belief with accumulating information or borrowing information from other sources to directly address questions that matter. Indeed, in cases such as the use of external controls that are constructed from historical or real-world data, most tools in the design and analysis of trials involve Bayesian techniques.

The US 21st Century Cures Act has specifically underscored the importance of real-world evidence, especially in relation to claims for a secondary indication of a product that has already been approved, as well as for post-approval safety evaluation. This is a fertile area for statistical research and application, in view of the inherent limitations arising from the absence of randomization in observational research with RWD.

Personalized medicine is another horizon of opportunity for statisticians. Modern ML tools have the potential to advance precision medicine, generating valuable information from the massive amounts of data collected through wearable devices and other digital technologies. In addition to promoting process efficiency, these tools can complement traditional methods to analyze high-dimensional and multimodal data, thereby expanding the span of influence and impact of statisticians.

Despite the considerable contributions of statistical innovations in addressing complex challenges in drug development, there is still more that can be done to enhance the delivery of medicines for unmet medical conditions. The pace at which novel approaches are receiving regulatory approval is slow, and the successful submissions involving CID and related approaches are few and far between. Although the digital revolution has transformed many industries, the adoption and implementation of ML and AI leaves much to be desired. This, of course, creates even more opportunities for biopharmaceutical statisticians to innovate further, strengthen their collaboration with regulators and academia, and boost their visibility in decision making and opinion leadership in the pharmaceutical industry.

Acknowledgements

We extend our sincere appreciation to the members of the "Biopharmaceutical Statistics Leaders Consortium" for their collaboration in assembling the team of authors and offering valuable guidance and input.

References

Alonso, A., Bigirumurame, T., Burzykowski, T., Buyse, M., Molenberghs, G., Muchene, L., et al., 2016. Applied Surrogate Endpoint Evaluation Methods with SAS and R, first ed. CRC Press, New York.

Baker, S.G., Kramer, B.S., 2003. A perfect correlate does not a surrogate make. BMC Med. Res. Methodol 3, 1–5.

Belin, L., Tan, A., De Rycke, Y., Dechartres, A., 2020. Progression-free survival as a surrogate for overall survival in oncology trials: a methodological systematic review. Br. J. Cancer 122, 1707–1714.

Brannath, W., Koenig, F., Bauer, P., 2007. Multiplicity and flexibility in clinical trials. Pharm. Stat. 6, 205–216.

Breiman, L., 2001. Statistical modeling: the two cultures. Stat. Sci. 16, 199–231.

Bretz, F., Maurer, W., Brannath, W., Posch, M., 2009. A graphical approach to sequentially rejective multiple test procedures. Stat. Med. 28, 586–604.

Bretz, F., Posch, M., Glimm, E., Klinglmueller, F., Maurer, W., Rohmeyer, K., 2011. Graphical approaches for multiple comparison procedures using weighted Bonferroni, Simes, or parametric tests. Biom. J. 53, 894–913.

Burman, C.F., Sonesson, C., Guilbaud, O., 2009. A recycling framework for the construction of bonferroni-based multiple tests. Stat. Med. 28, 739–761.

Burzykowski, T., Molenberghs, G., Buyse, M., 2005. The Evaluation of Surrogate Endpoints. Springer-Verlag, New York.

Buyse, M., Molenberghs, G., 1998. Criteria for the validation of surrogate endpoints in randomized experiments. Biometrics 54, 1014–1029.

Buyse, M., Molenberghs, G., Burzykowski, T., Renard, D., Geys, H., 2000. The validation of surrogate endpoints in meta-analyses of randomized experiments. Biostatistics 1, 49–67.

Cai, Y., Cai, Y.Q., Tang, L.Y., Wang, Y.H., Gong, M., Jing, T.C., et al., 2024. Artificial intelligence in the risk prediction models of cardiovascular disease and development of an independent validation screening tool: a systematic review. BMC Med. 22, 56.

Cardiac Arrhythmia Suppression Trial (CAST) Investigators, 1989. Preliminary report: effect of encainide and flecainide on mortality in a randomized trial of arrhythmia suppression after myocardial infarction. N. Engl. J. Med. 321, 406–412.

Chen, T., Guestrin, C., 2016. XGBoost:a scalable tree boosting system. Proceedingsof the 22nd ACM SIGKDD International Conference on Knowledge Discovery and Data Mining. 13–17 August 2016, San Francisco CA USA, pp. 785–794.pp. 785–794.

Chow, S.C., Chang, M., 2012. Adaptive Design Methods in Clinical Trials, second ed.,. CRC Press, Boca Raton.

Dang, L.E., Balzer, L.B., 2023. Start with the target trial protocol, then follow the roadmap for causal inference. Epidemiology 34, 619–623.

Dang, L.E., Gruber, S., Lee, H., Dahabreh, I.J., Stuart, E.A., Williamson, B.D., et al., 2023. A causal roadmap for generating high-quality real-world evidence. J. Clin. Transl. Sci. 22, e212.

Daniel, R.M., Cousens, S.N., De Stavola, B.L., Kenward, M.G., Sterne, J.A.C., 2013. Methods for dealing with time-dependent confounding. Stat. Med. 32, 1584–1618.

Dmitrienko, A., Offen, W.W., Westfall, P.H., 2003. Gatekeeping strategies for clinical trials that do not require all primary effects to be significant. Stat. Med. 22, 2387–2400.

Dmitrienko, A., Millen, B.A., Brechenmacher, T., Paux, G., 2011. Development of gatekeeping strategies in confirmatory clinical trials. Biom. J. 53, 875–893.

Dmitrienko, A., Tamhane, A.C., 2011. Mixtures of multiple testing procedures for gatekeeping applications in clinical trials. Stat. Med. 30, 1473–1488.

Du, Y., Li, J., Raha, S., Qu, Y., 2024. A unified Bayesian framework for bias adjustment in multiple comparisons from clinical trials. Stat. Med. 43, 2928–2943.

Elliott, M.R., 2023. Surrogate endpoints in clinical trials. Annu. Rev. Stat. Appl 10, 75–96.

European Medicines Agency, 2016. Guideline on multiplicity issues in clinical trials. Available at: https://www.ema.europa.eu/en/documents/scientific-guideline/draft-guideline-multiplicity-issues-clinical-trials_en.pdf.

Finner, H., Strassburger, K., 2002. The partitioning principle: a powerful tool in multiple decision theory. Ann. Stat 30, 1194–1213.

Food and Drug Administration, 2019. Guidance for industry. Adaptive designs for clinical trials of drugs and biologics. Available at: https://www.fda.gov/media/78495/download.

Food and Drug Administration, 2022a. Guidance for industry. Statistical approaches to establishing bioequivalence. Available at: https://www.fda.gov/media/163638/download.

Food and Drug Administration, 2022b. Guidance for industry. Multiple endpoints in clinical trials. Available at: https://www.fda.gov/media/162416/download.

Food and Drug Administration, 2023a. Complex innovative trial design meeting program. Available at: https://www.fda.gov/drugs/development-resources/complex-innovative-trial-design-meeting-program.

Food and Drug Administration, 2023b. Guidance for industry. Considerations for the design and conduct of externally controlled trials for drug and biological products. Available at: https://www.fda.gov/media/164960/download.

Freedman, L.S., Graubard, B.I., Schatzkin, A., 1992. Statistical validation of intermediate endpoints for chronic diseases. Stat. Med. 11, 167–178.

Fu, H., Zhou, J., Faries, D.E., 2016. Estimating optimal treatment regimes via subgroup identification in randomized control trials and observational studies. Stat. Med. 35, 3285–3302.

Ghim, J.L., Ahn, S., 2023. Transforming clinical trials: the emerging roles of large language models. Transl. Clin. Pharmacol. 31, 131–138.

Gleiss, A., Oberbauer, R., Heinze, G., 2018. An unjustified benefit: immortal time bias in the analysis of time-dependent events. Transpl. Int. 31, 125–130.

Gosling, J.P., 2018. SHELF: the sheffield elicitation framework. In: Dias, L.C., Morton, A., Quigley, J. (Eds.), Elicitation: The Science and Art of Structuring Judgement, vol. 261 Springer, Cham.

Grayling, M.J., Mander, A.P., Wason, J.M.S., 2019. Two-stage adaptive designs for three-treatment bioequivalence studies. Stat. Biopharmaceutical Res. 11 (4), 360–374. https://doi.org/10.1080/19466315.2019.1654911

Hernán, M.A., Robins, J.M., 2016. Using big data to emulate a target trial when a randomized trial is not available. Am. J. Epidemiol. 183, 758–764.

Hobbs, B.P., Sargent, D.J., Carlin, B.P., 2012. Commensurate priors for incorporating historical information in clinical trials using general and generalized linear models. Bayesian Anal. 7, 639–674.

Jahanshahi, M., Gregg, K., Davis, G., Ndu, A., Miller, V., Vockley, J., et al., 2021. The use of external controls in FDA regulatory decision making. Ther. Innov. Regul. Sci. 55, 1019–1035.

Jumper, J., Evans, R., Pritzel, A., Green, T., Figurnov, M., Ronneberger, O., et al., 2021. Highly accurate protein structure prediction with Alpha. Fold. Nat. 596, 583–589.

Karagiannis, T., Malandris, K., Avgerinos, I., Stamati, A., Kakotrichi, P., Liakos, A., et al., 2024. Subcutaneously administered tirzepatide vs semaglutide for adults with type 2 diabetes: a systematic review and network meta-analysis of randomised controlled trials. Diabetologia 67, 1206–1222.

Kidwell, K.M., Roychoudhury, S., Wendelberger, B., Scott, J., Moroz, T., Yin, S., et al., 2022. Application of Bayesian methods to accelerate rare disease drug development: scopes and hurdles. Orphanet J. Rare Dis. 17, 186.

Kiefer, C., Sturtz, S., Bender, R., 2015. Indirect comparisons and network meta-analyses. Dtsch. Arztebl. Int. 112, 803–808.

Liu, M., Wang, Y., Fu, H., Zeng, D., 2024. Learning optimal dynamic treatment regimens subject to stagewise risk controls. J. Mach. Learn. Res. 25, 1–64.

Lu, C.C., Li, X.N., Broglio, K., Bycott, P., Jiang, Q., Li, X., et al., 2021. Practical considerations and recommendations for master protocol framework: basket, umbrella and platform trials. Ther. Innov. Regul. Sci. 55, 1145–1154.

Madsen, C.M., Kamstrup, P.R., Langsted, A., Varbo, A., Nordestgaard, B.G., 2020. Lipoprotein (a)-lowering by 50 mg/dL (105 nmol/L) may be needed to reduce cardiovascular disease 20% in secondary prevention: a population-based study. Arterioscler. Thromb. Vasc. Biol. 40, 255–266.

Malhotra, A., Heilmann, C.R., Banerjee, K.K., Dunn, J.P., Bunck, M.C., Bednarik, J., 2024a. Weight reduction and the impact on apnea-hypopnea index: a systematic meta-analysis. Sleep. Med. 121, 26–31.

Malhotra, A., Grunstein, R.R., Fietze, I., Weaver, T.E., Redline, S., Azarbarzin, A., et al., 2024b. Tirzepatide for the treatment of obstructive sleep apnea and obesity. N. Engl. J. Med. https://doi.org/10.1056/NEJMoa2404881. (Epub ahead of print).

Marcus, R., Peritz, E., Gabriel, K.R., 1976. On closed testing procedures with special reference to ordered analysis of variance. Biometrika 63, 655–660.

Millen, B.A., Dmitrienko, A., 2011. Chain procedures: a class of flexible closed testing procedures with clinical trial applications. Stat. Biopharm. Res. 3, 14–30.

O'Hagan, A., Stevens, J.W., Campbell, M.J., 2005. Assurance in clinical trial design. Pharm. Stat 4, 187–201.

Ohlssen, D., Price, K.L., Xia, H.A., Hong, H., Kerman, J., Fu, H., et al., 2014. Guidance on the implementation and reporting of a drug safety Bayesian network meta-analysis. Pharm. Stat 13, 55–70.

Pocock, S.J., 1976. The combination of randomized and historical controls in clinical trials. J. Chronic Dis. 29, 175–188.

Polack, F.P., Thomas, S.J., Kitchin, N., Absalon, J., Gurtman, A., Lockhart, S., et al., 2020. Safety and efficacy of the BNT162b2 mRNA Covid-19 vaccine. N. Engl. J. Med. 383, 2603–2615.

Prentice, R.L., 1989. Surrogate endpoints in clinical trials: definitions and operational criteria. Stat. Med. 8, 431–440.

Proper, J.L., Chu, H., Prajapati, P., Sonksen, M.D., Murray, T.A., 2024. Network meta analysis to predict the efficacy of an approved treatment in a new indication. Res. Synth. Methods 15, 242–256.

Qi, Z., Liu, D., Fu, H., Liu, Y., 2020. Multi-armed angle-based direct learning for estimating optimal individualized treatment rules with various outcomes. J. Am. Stat. Assoc 115, 678–691.

Qu, Y., Du, Y., Zhang, Y., Shen, L., 2020. Understanding and adjusting for the selection bias from a proof-of-concept study to a more confirmatory study. Stat. Med. 39, 4593–4604.

Qu, Y., Jacober, S.J., Zhang, Q., Wolka, L.L., DeVries, J.H., 2012. Rate of hypoglycemia in insulin-treated patients with type 2 diabetes can be predicted from glycemic variability data. Diabetes Technol. Ther. 14, 1008–1012.

Robins, J.M., 2000. Marginal structural models versus structural nested models as tools for causal inference. In: Halloran, M.E., Berry, D. (Eds.), Statistical Models in Epidemiology, the Environment, and Clinical Trials 16 Springer, New York.

Schmidli, H., Gsteiger, S., Roychoudhury, S., O'Hagan, A., Spiegelhalter, D., Neuenschwander, B., 2014. Robust meta-analytic-predictive priors in clinical trials with historical control information. Biometrics 70, 1023–1032.

Schneider, P., Walters, W.P., Plowright, A.T., Sieroka, N., Listgarten, J., Goodnow Jr., R.A., et al., 2020. Rethinking drug design in the artificial intelligence era. Nat. Rev. Drug. Discov 19, 353–364.

Senn, S., 2021. Statistical Issues in Drug Development, third ed. John Wiley and Sons, Hoboken.

Shajari, S., Kuruvinashetti, K., Komeili, A., Sundararaj, U., 2023. The emergence of AI-based wearable sensors for digital health technology: a review. Sens. (Basel) 23, 9498.

Steffansson, G., Kim, W., Hsu, J., 1988. On confidences sets in multiple comparisons. In: Gupta, S.S., Berger, J.O. (Eds.), Statistical Decision Theory and Related Topics IV, vol. 2 Springer-Verlag, New York.

Temple, R.J., 1995. A regulatory authority's opinion about surrogate endpoints. In: Nimmo, W.S., Tucker, G.T. (Eds.), Clinical Measurement in Drug Evaluation. Wiley, New York.

Thorlund, K., Dron, L., Park, J.J.H., Mills, E.J., 2020. Synthetic and external controls in clinical trials - a primer for researchers. Clin. Epidemiol. 12, 457–467.

Tothfalusi, L., Endrenyi, L., Arieta, A.G., 2009. Evaluation of bioequivalence for highly variable drugs with scaled average bioequivalence. Clin. Pharmacokinet 48 (11), 725–743. https://doi.org/10.2165/11318040-000000000-00000. PMID: 19817502.

Van Norman, G.A., 2016. Drugs, devices, and the FDA: part 1: an overview of approval processes for drugs. JACC Basic. Transl. Sci. 1, 170–179.

Wallach, J.D., Sullivan, P.G., Trepanowski, J.F., Sainani, K.L., Steyerberg, E.W., Ioannidis, J.P., 2017. Evaluation of evidence of statistical support and corroboration of subgroup claims in randomized clinical trials. JAMA Intern. Med. 177, 554–560.

Wang, Y., Fu, H., Kulkarni, P., Kaiser, C., 2013. Evaluating and utilizing probability of study success in clinical development. Clin. Trials 10, 407–413.

Wang, Y., Fu, H., Zeng, D., 2018. Learning optimal personalized treatment rules in consideration of benefit and risk: with an application to treating type 2 diabetes patients with insulin therapies. J. Am. Stat. Assoc 113, 1–13.

Wang, Y., Tu, W., Koh, W., Travis, J., Abugov, R., Hamilton, K., et al., 2024. Bayesian hierarchical models for subgroup analysis. Pharm. Stat. https://doi.org/10.1002/pst.2424. (Epub ahead of print).

Westfall, P.H., Krishen, A., 2001. Optimally weighted, fixed sequence, and gatekeeping multiple testing procedures. J. Stat. Plan. Inference 99, 25–40.

Williamson, S.F., Jacko, P., Villar, S.S., Jaki, T., 2017. A Bayesian adaptive design for clinical trials in rare diseases. Comput. Stat. Data Anal. 113, 136–153.

Woodcock, J., LaVange, L.M., 2017. Master protocols to study multiple therapies, multiple diseases, or both. N. Engl. J. Med. 377, 62–70.

Wouters, O.J., McKee, M., Luyten, J., 2020. Estimated research and development investment needed to bring a new medicine to market, 2009-2018. JAMA 323, 844–853.

Xu, J., Audet, C., DiLiberti, C.E., Hauck, W.W., Montague, T.H., Parr, A.F., et al., 2016. Optimal adaptive sequential designs for crossover bioequivalence studies. Pharm. Stat. 15 (1), 15–27. https://doi.org/10.1002/pst.1721. Epub 2015 Nov 5. PMID: 26538182.

Zhao, Y., Zeng, D., Rush, A.J., Kosorok, M.R., 2012. Estimating individualized treatment rules using outcome weighted learning. J. Am. Stat. Assoc. 107, 1106–1118.

Zhou, Z., 2021. Statistical inference for machine learning: feature importance, uncertainty quantification and interpretation stability. Dissertations & Theses, Cornell University ProQuest Available at: https://www.proquest.com/openview/c2b5598ff59880cf61546dea2bcad5e3/1?pq-origsite=gscholar&cbl=18750&diss=y.

Chapter 8

Analysis of official time series with Ecce Signum, an R package for multivariate signal extraction and forecasting

Tucker McElroy* and James Livsey
US Census Bureau, Washington, DC, United States
**Corresponding author. e-mail address: tucker.s.mcelroy@census.gov*

Chapter Outline

Abstract

The chapter discusses methodology for the analysis of multivariate time series encountered in official statistics, and develops the methods through the R package **Ecce Signum**. These methods involve multivariate time series models with stuctural components, thereby allowing one to extract latent signals such as trends or seasonality. Models are fitted using maximum likelihood estimation, allowing for non-stationarity, fixed regression effects, and ragged-edge missing values. Simple types of extreme values can be corrected using the device of entropy maximization. Model adequacy is assessed through residual diagnostics, and model-based signal extraction filters can be assessed in time domain and frequency domain. Extracted signals are produced with uncertainty measures that account for sample edge effects and missing values, and the signals (as well as the original time series) can be forecasted.

Handbook of Statistics, Vol. 53. https://doi.org/10.1016/bs.host.2025.04.001

1 Introduction

For data published by official statistics agencies there are many applications of time series analysis, including forecasting, nowcasting, privatization, disaggregation, benchmarking/reconciliation, adjusting for survey error, linking, and seasonal adjustment. Many of these topics are discussed in Bell et al. (2024), the preface to a special issue of the Journal of Official Statistics on time series analysis; in the present chapter we primarily focus upon forecasting and seasonal adjustment – or, more broadly, signal extraction. Both applications are classically founded upon time series models, and are more effectively applied when the dataset is multivariate. Therefore, this chapter focuses upon methodology for multivariate time series modeling, developed through the R programming language. The choice of R is natural in the modern milieu of data science; many researchers and practitioners in official statistics agencies have some proficiency in R, which is rapidly becoming ubiquitous in both software development and production. Readers should have some familiarity with time series analysis, and can refer to McElroy and Politis (2020) for background.

The R package **Ecce Signum**,[1] or **sigex** for short, developed organically from R code for multivariate signal extraction required for the empirical applications of McElroy and Trimbur (2015). This methodology employed direct matrix formulas for signal extraction, which has certain advantages over the use of a state space smoother when the sample size is small. For further applications it became necessary to extend the models to handle diverse types of co-integration effects, such as common trends and common seasonality, and the resulting methods were summarized in McElroy (2017). However, at the same time it was recognized that extending the non-state space methodology to longer samples would be advantageous, given that general forms of integrated processes can be handled, including non-Markovian processes and long memory processes not amenable to a state space embedding. Algorithmic work to make feasible these extensions is discussed in McElroy (2018) and McElroy (2022).

The result of these developments is a flexible set of tools for the analysis of multivariate time series. Time series data with some or all of the following attributes can be analyzed with **Ecce Signum**:

- Multivariate time series, or univariate time series with embedded multivariate structure
- Fixed effects, such as holiday patterns
- Aberrancies, such as additive outliers
- Complex dynamics, including trend, seasonality, and business cycles
- Seasonal cycles of diverse period (e.g., weekly and annual effects)
- Linked dynamics, such as common trends and common seasonality

[1] Latin for "Behold the Proof"; cf. Falstaff's speech in Henry IV, Part 1 (Act 2, Scene 4).

- Ragged edge missing values
- Changes in sampling frequency

Highly exotic time series cannot be handled appropriately with **Ecce Signum**. Data that can be easily linearized (i.e., rendered as a linear time series, possibly with Gaussian marginals) through transformations, regression, and differencing is appropriate for analysis. We have in mind several possible objectives of analysis, which the methodological tools can address:

- Modeling of the data via reduction to a maximum entropy residual
- Identification and estimation of fixed effects
- Linearization of the data via maximum entropy transformations
- Extraction of underlying signals, which may be common, such as trends, seasonals, and cycles
- Imputation of missing values
- Forecasting and aftcasting
- Quantification of uncertainty in extractions

The modeling philosophy is informed by the concept of entropy, and is discussed in McElroy and Penny (2019). Tools are available to compare fitted models and evaluate their performance, and two methods of fitting are presented: maximization of the Gaussian likelihood utilizing a proper factorization, and method-of-moments – which is faster and more appropriate for high-dimensional parameter spaces.

The framework of **Ecce Signum** consists of an N-dimensional multivariate time series, which may be observed with missing values at various times for various components. After the possible application of a log transform, the model is written additively in terms of fixed effects (given through user-specified time series regressors) and latent stochastic processes (which the user specifies during model declaration). The class of processes is fairly broad, including ARIMA and SARIMA models (Box et al., 2015), as well as popular structural models discussed in Harvey (1989) and Harvey and Trimbur (2003). In broad strokes, the analyst proceeds from exploratory tools to a putative model, which can then be fitted using Gaussian maximum likelihood estimation – possibly accounting for additive outliers by entropy methods; if there are latent stochastic processes present, these are implicitly fitted as a result. If the modeling is satisfactory, the analyst can then proceed to applications such as forecasting or signal extraction.

The chapter is organized into the following sections: Section 2 proceeds through the various stages of analysis that a user of **Ecce Signum** would utilize sequentially. Then Section 3 provides more statistical detail about model parameterizations, estimation methods, and vector embeddings. Several detailed examples are explored in Section 4; these are vignettes in the package. Section 5 concludes. An expanded version of this paper is available online (arXiv preprint 2201.02148), with appendices providing additional details: Appendix A treats cyclical processes, Appendix B discusses issues related to weekly time series, and Appendix C gives more granular detail on the functionality, with a topology of the function call structure given through look-up tables.

2 Stages of analysis

In the analysis of multivariate time series in **Ecce Signum**, there are several stages that are interdependent. We discuss these in turn, in the sequence that they would typically be utilized in an application.

2.1 Loading, transforming, and exploring

We begin with entering the data into an R variable *data*, which is a $T \times N$-dimensional matrix involving T time points and N series. If there are missing values, these can be encoded with "NA," in which case the full sample size T is defined by taking the size of the union of all temporal indices for each component time series. A first stage of analysis is to call `sigex.load` with metadata such as the start date, period (i.e., number of seasons per year, if applicable), and epithets (i.e., shortened names or mnemonics) for each component series.

Secondly, a call to `sigex.prep` allows the analyst to apply a log transformation, aggregate component series (say if one needs to analyze the sum of all the series), restrict the temporal range, select a subset of component series for analysis, and plot. The output we shall denote as *data.ts*, and for convenience the cross-sectional dimension will be denoted N (even if a subset of series was selected). Note that this function call may also be utilized after some exploratory analysis to identify transformation, range, and subcollection; see Chapter 1 of McElroy and Politis (2020) for background.

Optionally, spectral plots can be viewed with a call by `sigex.specar`, which is just a wrapper for the native `spec.ar` of R. To the trained analyst, such plots provide insight into the salient dynamics of each component time series, and can assist one in positing unit root model structures in subsequent analyses; see Chapter 9 of McElroy and Politis (2020), or the discussion in McElroy and Roy (2021). For instance, a sharp peak in the spectral density at a frequency $\lambda \in [0, \pi]$ indicates that the polynomial of the autoregressive representation of the process may have a root close to the unit circle of the complex plane, of the form $e^{i\lambda}$.

2.2 Model declaration

Models in **Ecce Signum** have two facets: stochastic effects and fixed effects. Fixed effects determine the mean of the time series through specified regressors, one for each component series. Just as with RegARIMA models (Findley et al. (1998)), once these fixed effects are subtracted out what remains is a mean zero stochastic process, which can be modeled with one or more latent processes.[2] Let $\{X_t\}$ denote the transformed multivariate time series, for which our sample corresponds to *data.ts*. Our modeling framework is given by

$$X_t = z_t' \beta + Y_t. \tag{1}$$

Here $\{X_t\}$ is the data process, an N-dimensional time series with time-varying mean $\mathbb{E}[X_t] = z_t' \beta$ by our modeling assumption. The stochastic process $\{Y_t\}$ has mean zero. For each t in the sample period $\{1, 2, \ldots, T\}$, the regressors can be placed in a matrix z_t of dimension $r \times N$, where r represents the union of all fixed effects for all component series. Because each component series has separate regression effects (even if an identical regressor is used for several component series, the parameters will be distinct by assumption), the structure of z_t is block diagonal. In particular, if $z_t(j)$ is an r_j-dimensional vector of regressors for component series j (for $1 \leq j \leq N$), then

$$z_t = \begin{bmatrix} z_t(1) & 0 & \ldots & 0 \\ 0 & z_t(2) & \ldots & 0 \\ \vdots & \vdots & \vdots & \vdots \\ 0 & \ldots & 0 & z_t(N) \end{bmatrix}.$$

Then with $r = \sum_{j=1}^{N} r_j$, we have $\beta' = [\beta(1)', \beta(2)', \ldots, \beta(N)']$ and the fixed effect for the jth component series is $z_t(j)' \beta(j)$. More details on adding fixed effects are given below, and in Section 3.1.3.

Our first step in modeling $\{X_t\}$ is to build the stochastic model for $\{Y_t\}$. Here we use *mdl* as the variable (a list object) storing all the model information. We initially declare *mdl* to be NULL, and then use repeated calls of `sigex.add` to insert latent processes into the model. The key inputs to these function calls are the scalar differencing operator $\delta(B)$ (where the backshift operator B acts on a time series via $B X_t = X_{t-1}$), the rank configuration for the driving white noise, the time series model type (e.g., VARMA or SARMA, etc.) with ancillary information such as the model order, and an epithet (e.g., trend, irregular, weekly seasonal, etc.). If a component is stationary, simply input $\delta(B) = 1$; otherwise, input various polynomials as coefficient strings, taking care that each non-stationary latent process has a unique polynomial.[3] In particular, suppose there are K vector latent processes $\{S_t^{(k)}\}$ (for $1 \leq k \leq K$) that are related to the stochastic portion of $\{X_t\}$ via

$$Y_t = \sum_{k=1}^{K} S_t^{(k)}. \tag{2}$$

[2] Called "unobserved components" in the econometrics literature. We use the term latent process to avoid confusion with the components of a vector time series.

[3] Identification problems in signal extraction occur if diverse latent components share common unit roots; see Bell (1984) and McElroy (2008) for discussion.

For instance, $\{S_t^{(1)}\}$ might correspond to a trend, and $\{S_t^{(2)}\}$ to a transient stationary structure. Each latent component has an associated (scalar) differencing operator $\delta^{(k)}(B)$ such that

$$\delta^{(k)}(B)S_t^{(k)} = \underline{S}_t^{(k)}, \tag{3}$$

where $\{\underline{S}_t^{(k)}\}$ is a mean zero, covariance stationary vector stochastic process. The differencing in (3) is applied in the same way to each component of the vector process. The stationary model structure can be specified through ARMA/SARMA (the same model is used for each component), or a VARMA/SVARMA, or a cyclical model. Generally, all the models implemented in **Ecce Signum** are special cases of the causal moving average form

$$\underline{S}_t^{(k)} = \Psi^{(k)}(B)\epsilon_t^{(k)}, \tag{4}$$

where $\Psi^{(k)}(z)$ is a matrix power series. The coefficient matrices are multiples of the identity in the case of ARMA/SARMA models, or a cycle model, but for VARMA/SVARMA these matrices can be non-trivial; further details are in Section 3.1.2. The vector white noise process $\{\epsilon_t^{(k)}\}$ is independent and identically distributed (i.i.d.) with unspecified marginal distribution, and covariance matrix $\Sigma^{(k)}$ that can be reduced rank; further details are in Section 3.1.1.

The specification of these components takes some care and experience. Note that any unit roots deemed to be present in the autoregressive representation of the data process must be featured among one of the latent processes, since the product of the differencing polynomials for the latent processes equals the differencing polynomial for the data process (Hillmer and Tiao, 1982). In particular, setting $\delta^{(-k)}(z) = \prod_{j \neq k} \delta^{(j)}(z)$, it follows from (2) and (3) that

$$\delta(B)Y_t = \sum_{k=1}^{K} \delta^{(-k)}(B)\underline{S}_t^{(k)}, \tag{5}$$

where $\delta(z) = \prod_{k=1}^{K} \delta^{(k)}(z)$. The right hand side of (5) is a sum of stationary processes, and hence $\delta(B)Y_t$ is stationary too, and yet if there are any common unit roots among any of the $\delta^{(-k)}(z)$ then $\{Y_t\}$ has been over-differenced – this will generate estimation problems, because the resulting process will be non-invertible (Chapter 6 of McElroy and Politis (2020)). Therefore it is important to specify the various $\delta^{(k)}(z)$ such that they are relatively prime. Henceforth $\underline{Y}_t = \delta(B)Y_t$ defines the mean zero stationary process given by (5).

The second stage in model declaration is building in fixed effects; at a minimum, one must call `sigex.meaninit`, which ensures that a polynomial trend is inserted with order corresponding to the number of specified trend differences in the first stage. This is important, because applying differencing to the data process will typically leave a non-zero constant mean, which is automatically associated with the leading coefficient of a polynomial trend effect. Next, if additional fixed effects are desired (e.g., calendar effects, level shifts, etc.) then these can be incorporated through calls to `sigex.reg`. This function requires that the regressors have already been constructed for the full sample span $\{1, 2, \ldots, T\}$, so if a reduced range has been considered any regressors inputted to the program must be appropriately (and manually) modified. Because each component series can have different regressors, one must set up a loop over the N components to allocate the specified regressors.

For moving holiday effects such as Easter, the utility `gethol` for daily time series can be used. One inputs the desired calendar dates, and a window of days before and after each occurrence of the holiday when it is deemed that economic activity will be increased relative to the rest of the year; this mimics the Genhol routine (Census Bureau, U. S., 2020) of X-13ARIMA-SEATS for constructing moving holiday regressors. In particular, if t is the index corresponding to a holiday, and f and a indicate the number of days before and after the holiday to be considered, then the window is $t-f, t-f+1, \ldots, t, \ldots, t+a-1, t+a$; the length of the window is $a+f+1$. A corresponding sequence of regressors, computed using the start and end dates of the time series sample, is generated.

2.3 Model fitting and diagnostics

By the fitting of a model, we mean that parameter estimates are obtained from the data. **Ecce Signum** has two methods for doing this: Method of Moments (MOM) and Maximum Likelihood (ML). Before these can be used, initial values of the parameters need to be set up, and the practitioner makes choices about whether a parameter is fixed or free. A detailed discussion of parameterization and fitting is given in Section 3.1, and MOM is expounded in Section 3.2.

Typically, the ML method is preferred. This is because the MOM method is limited to fairly simple latent process structures; essentially, the causal moving average portion $\Psi^{(k)}$ of the kth latent component model (4) is held fixed at an initial value specification, and only $\Sigma^{(k)}$ is estimated. Moreover, the MOM method is not equipped to handle missing values, and the estimation of fixed effects is through OLS. In contrast, ML allows for ragged edge[4] missing values and the most general types of model structures encoded in **Ecce Signum**; moreover, the statistical estimates are efficient when the model is correctly specified and any non-Gaussian features are absent – see the fourth appendix of Holan et al. (2017) for details. The parameters for fixed effects correspond to Generalized Least Squares (GLS) estimates based upon the time series model parameters; furthermore, we can impose parameter constraints (Section 3.1.4) and generate test statistics for parameter significance.

[4] Ragged edge missing values refer to a situation where some variables have data extending further into the most recent time periods than others. This creates an uneven or "ragged" structure at the edge (end) of the dataset, where not all variables are observed for the same time points.

The demerits of ML are the larger computational resources required, as well as the usual pitfalls of nonlinear optimization. Each likelihood evaluation has an expense relative to the size of T and N, as well as the model complexity (McElroy, 2018). For models with many parameters, e.g., 100 or more, the parameter manifold is large and requires a time-consuming search. As usual, there are no guarantees that the algorithm's termination corresponds to a global maximum. Our approach in **Ecce Signum** is to re-parameterize constraints into an unconstrained (so-called) pre-parameter space homeomorphic to Euclidean space, as advocated by Pinheiro and Bates (1996). Different initializations can be specified by the user, as well as what type of optimizer is desired; the pre-parameter corresponding to the last successful likelihood evaluation is saved to the global variables in the R workspace, so that if *optim* crashes the user can restart with a slight perturbation of this last value.

Upon termination of the optimization routine, the user can examine the parameter values and assess the model fit. This is possible by computing the time series residuals and measuring their entropy via a Shapiro-Wilks normality test (Shapiro and Wilk, 1965) as well as Ljung-Box statistics (Ljung and Box, 1978). More generally, through modeling we are attempting to map the data vector to a maximum entropy residual; see Chapter 8 of McElroy and Politis (2020) for background on this modeling philosophy. A serially uncorrelated Gaussian time series has maximum entropy, and therefore the user can assess departures of the model residuals from this ideal target.

More generally, the transformed time series may have extreme values present – see McElroy (2016b) for an overview of extremal time series, and the impact on signal extraction problems. As discussed in McElroy and Penny (2019), non-Gaussian processes can be handled through non-linear forecasting and signal extraction algorithms, or by first applying an extreme-value adjustment procedure followed by a linear algorithm. **Ecce Signum** follows the latter path, chiefly for simplicity and speed (non-linear methods typically require more nuanced modeling, and more expensive algorithms). This approach is in the tradition of intervention analysis: Chang et al. (1988) use an additive outlier (AO) regressor (Ljung, 1993), and Tsay (1986) considers more general types of outlier effects (e.g., level shifts).

We view the AO as an unknown stochastic additive effect, not a fixed effect; as shown in McElroy and Penny (2019), entropy is maximized by deleting the corresponding observation. Hence, the treatment of AOs in **Ecce Signum** is by inserting an NA for any putative additive extreme, and comparing the likelihood. The difference in log likelihoods has an interpretation as a test statistic for the significance of any collection of AOs. The identification of extremes is an important facet of model building, and can be accomplished in tandem with fitting a specified model; as missing values are introduced for possible extremes, the remaining sample attains to more typically Gaussian characteristics, and the ragged edge Durbin-Levinson algorithms allow for quick computation.

If multiple satisfactory models have been obtained, we can perform nested comparisons using a difference of likelihoods (McElroy, 2016a). If particular pre-parameters are not significantly different from zero, they can be replaced by a zero (this is entered as a constraint) and the model re-fitted (as a nested model). Another possible restriction is through the rank configuration of a latent process' white noise covariance matrix; see Section 3.1.1, although in this case a statistical test arising from the difference of likelihoods is not valid (without a non-standard asymptotic theory) because the restriction lies on the boundary of the parameter space.

Through use of these tools, we suppose the analyst has arrived at a final model, which is bundled using `sigex.bundle`: this includes *data.ts* (where any detected AOs have been marked by inserting NAs), the model *mdl*, the fitted pre-parameters ψ, and the transformation used.

2.4 Applications of the fitted model

At this point in the analysis we can generate output from the fitted model for applications. We focus on casting (forecasting, aftcasting, and midcasting) and signal extraction. The function `sigex.midcast` generates midcasts for all missing values in the data, which are marked as usual with the NA symbol; a number of forecasts and aftcasts to be generated as well. `sigex.midcast` is a "deluxe" version of the casting methodology, also generating the large covariance matrix between all casting errors; since this can be expensive to compute and store if there are hundreds of casts, a faster version is available when uncertainty is not needed – `sigex.cast` generates only forecasts and aftcasts, and presumes there are no missing values.

Neither of these two function calls is complete without accounting for regression effects: both `sigex.cast` and `sigex.midcast` presume that fixed effects have already been removed from *data.ts*, and these regression effects need to be extended and added back to the series. The reason this is not done automatically, is that there is no way to know future values of general regressors encoded in *mdl*; clearly, polynomial and sinusoidal functions can be extrapolated/interpolated, but for holiday regressors or calendar effects a human is required to do the modifications. One approach to analysis is to load the data with extended regressors entered into *mdl*, and *data.ts* padded fore and aft with NA, up to the horizons that are anticipated to be needed.

For signal extraction, there are two main options: Wiener-Kolmogorov (WK) filtering or ad hoc (AC) filtering, as discussed in McElroy (2022). In the first case, we have in mind a model that contains at least two latent processes, and it is of interest to extract one of them (or possibly a sum of a subset of these latent processes). For the specified latencies, **Ecce Signum** computes the minimum mean square linear estimate conditional on the (ragged edge) data, along with time-varying uncertainty. This uncertainty can be viewed as being composed of two portions: the so-called WK error and the casting error. In contrast, the AC case can be applied when there is only a single latency specified, but the analyst must define and supply the desired (non-model based) filter. Like the WK case, the optimal estimate is computed from the ragged edge data, but the time-varying uncertainty only depends on the casting error, as the WK error component is not defined. In either case, the extracted signal $\{U_t\}$ takes the form

$$U_t = \Upsilon(B) Y_t = \sum_{h=-\infty}^{\infty} \Upsilon_h Y_{t-h},$$

where $\Upsilon(B)$ is either a WK or AC filter. In the WK case, $\{U_t\}$ is viewed as an estimate of a stochastic process, such as a latent process $\{S_t^{(k)}\}$ (or aggregation of such). But for AC filtering, $\{U_t\}$ is viewed as the actual target.

We begin with the WK case, and there are two methods available: `sigex.signal` provides the exact filter matrix that yields the WK signal extraction at all time points (derived in McElroy and Trimbur (2015)), but requires there be no missing values. (Although we can apply `sigex.signal` to ragged edge data that has been "patched" with midcasts, we will not be able to account for midcasting uncertainty in the signal extraction uncertainty measure.) A secondary call to `sigex.extract` furnishes the actual extractions (along with point-wise confidence intervals defined as ± 2 standard deviations, based on the extraction error covariance matrix), but the signal extraction error covariances (assuming no missing values) as well as the filter weights are produced by `sigex.signal`. This method is not recommended for longer samples ($T > 500$), because a non-Toeplitz matrix of dimension NT needs to be inverted.

The broader method `sigex.wkextract` is generally preferred, except in cases where the time-varying filter weights are needed. The technique is described in McElroy (2022), and involves first patching the series with midcasts – and a set number of aftcasts and forecasts to handle the sample boundary – followed by filtering with the so-called WK filter. This method also obtains both components of the signal extraction uncertainty, accounting for missing values and the sample boundary. However, unlike `sigex.signal` it assumes a truncation of the WK filter. The truncation error can be reduced to any desired level by taking lag cutoff to be larger, but this induces a computational cost because there is a corresponding increase to the number of aftcasts and forecasts (with uncertainty) that are needed.

An additional feature is the ability to generate casts of signals; automatically the signal is extracted for time points $\{1, 2, \ldots, T\}$, but for a specified horizon H we can obtain forecasts for times $\{T+1, \ldots, T+H\}$ and aftcasts for times $\{1-H, ..., 0\}$. A second auxiliary feature is the extraction of linear combinations of a signal: suppose that our extraction is

$$\Xi(B)U_t = \sum_{j=0}^{p} \Xi_j U_{t-j}$$

for coefficient matrices Ξ_j. The matrix polynomial $\Xi(z)$ is passed into `sigex.wkextract`, and the extraction with appropriate covariance is returned. Note that such extended extractions can also be obtained through use of `sigex.signal`, by applying appropriate linear combinations to both the filter matrix and the extraction error covariance matrix, but in `sigex.wkextract` the error calculations are done using frequency domain instead of time domain. Mathematical details of these matters can be found in McElroy (2022).

A second option is to generate AC extractions. The desired filter $\Upsilon(z)$ is entered as an array object, in a truncated format:

$$\Upsilon(z) \approx \sum_{h=-c}^{m-1-c} \Upsilon_h z^h,$$

where m is the total number of nonzero coefficient matrices, and c is a shift parameter that indicates the proper centering. For instance, setting $c = 0$ indicates that the filter is concurrent, i.e., the output depends only on present and past data. Then `sigex.adhocextract` computes the filtered estimates, first patching and extending the series as needed; this routine functions much like `sigex.wkextract`, only the AC filter is used instead of the WK filter. Also, if we really desired a signal of the form $\Xi(B)U_t$, this amounts to

$$\Xi(B)U_t = \Xi(B)\Upsilon(B)Y_t,$$

so our AC filter would just be redefined as $\Xi(B)\Upsilon(B)$.

Plots of casts and extractions, with uncertainty indicated, is a final step in the analysis. Note that regression effects should be added back to casts, and may be added to certain extractions. For example, an Easter moving holiday regressor would be added to the final extracted seasonal component. To pair estimated parameters with the appropriate regressors, `sigex.fixed` can be used. For plotting, `sigex.graph` can be used to overlay extracted signals if the data has already been plotted; this function presumes the format of the output of `sigex.extract`. The second argument of `sigex.graph` accepts fixed effects that are added onto the extraction in the first argument. The starting date and period are also entered, and it is possible to vertically displace plotted signals (if this is visually desirable). For example, the seasonal component will be plotted around zero, and hence may be too far below the data and trend extraction to be easily visible – it can be shifted up in such a case.

The extracted signals can also be assessed as to whether they contain only the proper dynamics. One can examine the spectral density through a call to `sigex.specar`, and of course the autocorrelations can be checked as usual;

for background on seasonal adjustment diagnostics, see McElroy and Roy (2021), McElroy (2021), and McElroy and Roy (2022b).

It may be of interest to examine the WK filter coefficients, as well as the frequency response function. `sigex.wk` can be used to compute and plot the WK coefficients Υ_h. In particular, there are multiple panels and the jkth panel (for $1 \leq j, k \leq N$) plots the jkth entries of Υ_h for $h = -m, \ldots, m$. For example, the weights on the second input series used to extract the first output series are given by $j = 1$ and $k = 2$. Similarly, `sigex.frf` computes (and can be plotted in multiple panels by calling `sigex.getfrf`) the frequency response function $\Upsilon(e^{-i\lambda})$. The output is an array, with elements corresponding to a mesh of frequencies λ taken in the interval $[0, \pi]$. Note that for purposes of inverting the frequency response function into coefficients, it is always sufficient to restrict to the interval $[0, \pi]$, because

$$\Upsilon_h = \frac{1}{2\pi} \int_{-\pi}^{\pi} e^{i\lambda h} \Upsilon(e^{-i\lambda})\, d\lambda = \frac{1}{\pi} \int_0^{\pi} \Re[e^{i\lambda h} \Upsilon(e^{-i\lambda})]\, d\lambda$$

by change of variable. (This integration is discretized and calculated in `sigex.wk`.).

3 Detailed aspects

3.1 Model parameterization

Ecce Signum makes use of unconstrained parameterization, which means there exists a bijection between all model parameters and Euclidean space. Specifically, all model parameters are stored in a list object *param* (whose format depends on the specified model), and there exists a bijection that maps a real vector ψ to *param*. The vector ψ is further decomposed into three portions: ξ (which maps to covariance matrices for driving white noise of the latent processes), ζ (which maps to time series model parameters, such as ARMA, for the latent processes), and β (which maps to the regression parameters for all component series). The functions `sigex.par2psi` and `sigex.psi2par` pass *param* to ψ and back.

Why do we do this? The format of *param* involves the Generalized Cholesky Decomposition (GCD) of covariance matrices as well as things such as AR coefficients, organized in a list that corresponds to component series; hence a user can more easily interpret the quantities in *param*. But ψ consists of the unconstrained parameterization of covariance matrices and stable AR polynomials and so forth; it is not easy to understand what a particular component entry of ψ corresponds to. Typically the user will print *param*, which has a readable structure; the standard user will not need to understand the mapping of ψ to *param*.

3.1.1 Covariance matrix parameterization

First, the parameterization of a non-negative definite matrix Σ proceeds along the lines described in McElroy (2017). Specifically, the GCD (described in Golub and Van Loan (2012)) decomposes a possibly singular Σ into unit lower triangular L and diagonal D such that

$$\Sigma = L\, D\, L'. \tag{6}$$

For an N-dimensional matrix, there are $\binom{N}{2}$ lower triangular entries in L that can be any real number. The matrix D has N non-negative entries, which are all strictly positive if Σ is positive definite (invertible). Hence the total number of parameters required is $\binom{N+1}{2}$. If the rank of Σ is less than N, then at least one entry of D is zero, and the corresponding column of L becomes obsolete in the decomposition. To determine the requisite number of parameters, one must further specify which entries of D are zero; we call this a *rank configuration.* Clearly, if d_j (the jth diagonal entry of D) is zero, then $N-j$ free parameters in L become obsolete and can be removed from the needed parameterization. (In the case $j=N$, no reduction occurs.).

The complement of the rank configuration corresponds to entries of D that are positive, and the variable *vrank* captures this. If the rank is reduced, the number of parameters needs to be computed, and `sigex.param2gcd` constructs an L matrix from a vector of real numbers, where any columns corresponding to the rank configuration are omitted (i.e., L is a rectangular matrix with potentially more rows than columns). The positive entries of D are parameterized by taking the exponential of a real number. In this way ξ is mapped into the various covariance matrices of *param*, one for each latent process.

3.1.2 Serial dependence parameterization

Second, the parameterization of the serial dependence in the stochastic process is accomplished through `sigex.par2zeta` and `sigex.zeta2par`. The exact mapping is dictated by the specified model structures. For an ARMA model, both the AR and MA polynomials are stably parameterized through the pacf discussed in McElroy and Politis (2020). In particular, for $j=1, 2, \ldots, p$ for a degree p stable polynomial (with a minus convention, so that $\phi(x) = 1 - \sum_{j=1}^{p} \phi_j z^j$) we consider recursively parameterizing all polynomials of degree up to p, denoting the coefficients at the jth stage via $\phi_1^{(j)}, \ldots, \phi_j^{(j)}$. For $p=1$, we have $\phi_1^{(1)} \in (-1, 1)$, which is accomplished by taking $\phi_1^{(1)} = (\exp\{\zeta_1\} - 1)/(\exp\{\zeta_1\} + 1)$. Then for $2 \leq j \leq p$, parameterize $\phi_j^{(j)}$ in the same way and update $\phi_1^{(j)}, \ldots, \phi_{j-1}^{(j)}$ via

$$\begin{bmatrix} \phi_1^{(j)} \\ \vdots \\ \phi_{j-1}^{(j)} \end{bmatrix} = \begin{bmatrix} \phi_1^{(j-1)} \\ \vdots \\ \phi_{j-1}^{(j-1)} \end{bmatrix} - \phi_j^{(j)} \begin{bmatrix} \phi_{j-1}^{(j-1)} \\ \vdots \\ \phi_1^{(1)} \end{bmatrix}.$$

This update can also be expressed as follows: multiply the old coefficient vector by the matrix $A(j-1)$ that is given by the sum of a $(j-1)$-dimensional identity matrix and another matrix with $-\phi_j^{(j)}$ on the transverse-diagonal. In this way we map real numbers $\zeta_1, \ldots, \zeta_p$ to $\phi_1^{(p)}, \ldots, \phi_p^{(p)}$. By applying the inverse of $A(j-1)$ recursively, we can also recover the original pre-parameters from a given stable polynomial. For a SARMA model (with seasonal period s),

$$\phi(B)\,\Phi(B^s)\,Y_t = \theta(B)\,\Theta(B^s)\epsilon_t,$$

and each polynomial is stably parameterized. There are four such polynomials – a regular AR polynomial ϕ, a seasonal AR polynomial Φ, a regular MA polynomial θ, and a seasonal MA polynomial Θ, all of which are defined using the minus convention of Box et al. (2015).

For a VAR model, the parameterization of Roy et al. (2019) is used, whereby real numbers are mapped into the space of stable matrix polynomials. The functions `var2.par2pre` and `var2.pre2par` provide the mapping from $\phi(B)$ to the real numbers, and back again. For VARMA and SVARMA, there are additional polynomials for the moving average and seasonal aspects.

Stochastic cycles come in four varieties. The basic nth order Butterworth cycle is described in Harvey and Trimbur (2003), and takes the form

$$(1 - 2\rho\cos(\omega)B + \rho^2B^2)^n\,C_t = (1 - \rho\cos(\omega)B)^n\,\epsilon_t$$

for a cyclical latent process $\{C_t\}$, where $\{\epsilon_t\}$ is white noise (this can be multivariate). Here ρ corresponds to the persistency of the cycle, and is restricted to (0, 1), while ω is the frequency (or 2π divided by the period of the cyclical effect). A higher order n ensures that the high frequency aspects of the cycle are damped, so that the resulting stochastic process more purely represents only cyclical phenomena. In the model specification, the user specifies n, and can also input range parameters for ρ and ω. This can be used to impose hard boundaries on the parameter values, although **Ecce Signum** considers an open interval dictated by the specified bounds and maps ζ to this interval using a logistic transform.

A second type of process called the Balanced cycle can also be used, and the spectral shape of the resulting process is slightly tilted – as compared to the Butterworth cycle – in such a way that the high frequency content is lowered. Harvey and Trimbur (2003) argue for the more favorable properties of the Balanced cycle, which can be specified in the same way as the Butterworth.

Finally, both the cycles and the ARMA and SARMA models can be "stabilized," which refers to a modification of the model that is non-invertible. Bell and Pugh (1989) first experimented with such models, describing them as canonical component models. Any stationary stochastic process has spectral density f that is non-negative, and if it is strictly positive we can consider subtracting off the minimum value (a positive number), thereby rendering f shifted downwards such that it now takes on a zero value. A process having such a spectral density is said to be non-invertible (Chapter 6 of McElroy and Politis (2020)); while slightly problematic from the standpoint of fitting and forecasting, such a process has the interpretative advantage that it is maximally stable in the sense that as much white noise as possible has been removed. The intuition here is that we can decompose a given stochastic process into the sum of a non-invertible process and an independent white noise process of variance equal to the minimum value of the spectral density – an operation sometimes called "canonization," as it is at the core of the canonical decomposition routine of Hillmer and Tiao (1982).

The last model available is the "damped trend." This is essentially an AR(1) process where the user can place bounds on the range of the ϕ_1 parameter, in the same manner as with the persistency parameter in the cycles. Typically this specification would be used in conjunction with a trend differencing operator such as $\delta(B) = 1 - B$. The resulting model for a trend latent process $\{S_t\}$ would then be

$$(1 - B)(1 - \phi_1 B)S_t = \epsilon_t.$$

This type of trend interpolates continuously between the random walk and integrated random walk trends, allowing one to more finely model and extract a trend with a varying degree of smoothness (higher values of ϕ_1 correspond to smoother extracted trends).

3.1.3 Regression effects parameterization

Third, the regression parameters for each series are given by β, the final portion of ψ. This portion is straightforward; note that a mean effect for the differenced time series is always added through `sigex.meaninit`, so there will always be at least N parameters in β. When a user decides to add a regression effect to a component series, the program checks to see whether this is in the null space of $\delta(B)$, the full differencing operator for $\{Y_t\}$ described in (5). Consider applying $\delta(B)$ to the jth component (for $1 \leq j \leq N$) of (1). We obtain

$$\delta(B)X_{tj} = \delta(B)z_t(j)'\beta(j) + \underline{Y}_{tj},$$

where $\underline{Y}_{tj}$ is the jth component of $\underline{Y}_t = \delta(B)Y_t$. As discussed below (5), the process $\underline{Y}_t$ is invertible so long as the various $\delta^{(k)}(z)$ are relatively prime. If $\delta(B)$ completely annihilates $z_t(j)$, then there are no fixed effects at all; because the fitting of non-stationary processes relies upon a factorization of the Gaussian likelihood that features the likelihood of the differenced data $\{\underline{Y}_t\}$ (see McElroy (2022)), it follows that any covariate with corresponding regressor in the null space of $\delta(B)$ cannot be identified – therefore in **Ecce Signum** such a regressor is removed from the model. This is automatically done in `sigex.reg`, and the removal is done without further warning to the user. (The desired regression effect that the user enters is never incorporated into the model.) For regressors not in the null space of the full differencing operator, $\delta(B)z_t(j)$ will be non-zero and hence generates the time-varying mean for $\delta(B)X_{tj}$. This is accounted for in likelihood and casting calculations.

There is a type of weak converse to this discussion. If we have included differencing in our model, without loss of generality we can envision the inclusion of all regression effects that are in the null space of this $\delta(B)$, although such effects can never be identified. For instance, if $\delta(B) = 1 - B^4$ then a basis of functions composing the null space consists of 1, i^t, $(-1)^t$, and $(-i)^t$ with $i = \sqrt{-1}$. This follows from the theory of ordinary difference equations, discussed in Chapter 5 of McElroy and Politis (2020). An individual latent process may contribute some (real) polynomial factors to $\delta(B)$. In the course of modeling a time series, we may find that the innovation variance of one latent process is very close to zero (relative to the overall scale of the data). In such a case, the stochastic process has limiting form as the innovation variance tends to zero given by only the fixed time-varying mean function in the null space; in practice, we can replace such a latent process with the corresponding fixed effect.

For example, consider a quarterly seasonal process with differencing operator $1 + B + B^2 + B^3$. If this seasonal process were found to have innovation variance close to zero, we could replace that stochastic process with a fixed effect corresponding to the null space of $1 + B + B^2 + B^3$, i.e., i^t, $(-1)^t$, and $(-i)^t$. Because it is awkward to do regression with complex numbers – and using the fact that complex roots will always come in conjugate pairs if we are considering a real-coefficient polynomial – we can instead work with cosines and sines. In particular, if $e^{i\omega}$ (for $0 < \omega < \pi$) is a root of a differencing polynomial that is being eliminated from the model due to low innovation variance, then $e^{-i\omega}$ is also a root and a regression effect of the form

$$c\, e^{i\omega t} + \bar{c}\, e^{-i\omega t}$$

is in the null space, for a complex regression parameter c. However, this can be rewritten as

$$\beta_1 \cos(\omega t) + \beta_2 \sin(\omega t)$$

with $\beta_1 = c + \bar{c}$ and $\beta_2 = i(c - \bar{c})$, which are both real parameters. As a result, we can simply introduce the regressors $\cos(\omega t)$ and $\sin(\omega t)$ into the model. If instead the root of the differencing polynomial is -1, we introduce the regressor $(-1)^t$.

We make a parenthetical remark here about levels and rates. All regression effects are specified for levels (the original, undifferenced data process), and when moving to rates (or a series where $1 - B$ has been applied) the regressors are also differenced in a like manner, so the interpretation of the regression coefficients is unaltered. Sometimes economists enter fixed effects into data that has already been differenced, but don't difference the regressors; then the resulting parameter estimates no longer correspond to that particular fixed effect as it is manifested in levels. Since this is confusing (one can find transformations for the regression parameters), we avoid this and always specify the model for the time series (in levels) before any differencing has been applied.

So far, we have discussed the incorporation of fixed effects that a user thinks may be appropriate. Because a time series after applying full differencing will rarely have mean zero, even if it is stationary, it is prudent to include a constant regressor for each component in the equations for $\delta(B)X_{tj}$. But what regression effect does this imply for the original series $\{X_{tj}\}$? Again, any regressors in the null space of $\delta(B)$ cannot be identified. But we associate the constant regressor for the differenced series with a polynomial trend effect in the original series, which is explained as follows. We say a unit root of order d is present in $\delta(B)$ if $(1-B)^d$ is a factor, where $d \geq 0$. The action of $(1-B)^d$ on a polynomial of order d reduces it to a constant (Chapter 3 of McElroy and Politis (2020)). Hence we can insert the regressor t^d in the model, and its coefficient is proportional to the mean effect. In particular, if $\delta(B) = (1-B)^d\, \delta_S(B)$ is the factorization (for some polynomial $\delta_S(z)$ without trend factors), then

$$\delta(B)\,[\beta t^d] = \delta_S(B)\, d!\beta = \delta_S(1)\, d!\beta.$$

The last equality follows from the fact that passing a constant through a filter yields the sum of the filter coefficients; $\delta_S(1) \neq 0$ because 1 is not a root of $\delta_S(B)$, as all the unit roots were already accounted for in the factorization of $\delta(B)$. Hence the mean is $\delta_S(1)\, d!\beta$, and the coefficient of the trend polynomial equals the mean effect divided by $\delta_S(1)\, d!$. In the case that $d = 0$, `sigex.meaninit` allows the user to input a trend polynomial of any desired order, but if $d > 0$ then this input is ignored and only the regressor t^d is incorporated.

3.1.4 Fixed and free parameters

It is possible to impose general linear constraints on the pre-parameters ψ. For example, one can fix specific elements of ψ to a number, such as zero, or one can impose that a linear combination of components equals a constant. We can enforce that two components of ψ take the same value by imposing a linear constraint consisting of 1 and -1 as the combination, and 0 as the value that the linear combination is enforced to equal. In general these constraints are encoded with the equation

$$C\,\psi = b, \tag{7}$$

where b is a vector with length given by the number of constraints. So each row of C corresponds to one linear constraint, providing the linear combinations of ψ. The row dimension of C must be less than the column dimension, and the difference between these is the number of free variables.

A limitation of this framework is that constraints are enforced on pre-parameters instead of the original, interpretable parameters; in applied work a researcher might be interested in enforcing a constraint on the parameters, but apart from the regression parameters it is difficult to see how elements of ψ

should be bounded to yield such a constraint. Because the mapping of regression parameters to the pre-parameter is the identity (they are mapped to the subvector β of ψ), this case of regression parameters is easily handled. Otherwise, constraints for ξ or ζ could arise from a prior fit of the model where it was subsequently determined that certain pre-parameters were not significantly different from zero.

Here we determine a real vector of free variables, called η, and a real vector of constrained variables ν, such that together η and ν make up the vector ψ. Then likelihood evaluation is in terms of η, and this is the variable that is optimized over for maximum likelihood estimation. To determine η and ν from a given ψ, and vice versa, consider the QR decomposition (Golub and Van Loan, 2012) of the matrix C:

$$C = Q\,R\,\Pi,$$

where Q is an orthogonal matrix, R is unit upper triangular (with dimensions the same as those of C), and Π is a permutation matrix. Let $R = [R_1|R_2]$ be partitioned such that R_1 is square; it is invertible because it is upper triangular with ones on the diagonal. Then we define η to be the bottom portion of $\Pi\,\psi$, and we solve for ν in terms of η as follows:

$$\nu = R_1^{-1}\,Q^{-1}\,b - R_1^{-1}\,R_2\,\eta. \tag{8}$$

Finally, ψ is constructed from η and ν via

$$\psi = \Pi^{-1}\begin{bmatrix}\nu\\ \eta\end{bmatrix}. \tag{9}$$

It follows that (using 1 and 0 to denote either an identity matrix or a zero matrix of appropriate dimension)

$$\psi = \Pi^{-1}\left(\begin{bmatrix}0\\1\end{bmatrix} - \begin{bmatrix}1\\0\end{bmatrix} R_1^{-1}\,R_2\right)\eta + \Pi^{-1}\begin{bmatrix}1\\0\end{bmatrix} R_1^{-1}\,Q^{-1}\,b, \tag{10}$$

which expresses ψ as $A\,\eta$ plus a constant vector. These calculations are performed by `sigex.eta2psi`, and `sigex.psi2eta` produces η and ν from a given ψ. Default values of *param* are determined through `sigex.default`, which occurs by setting η equal to the zero vector and computing ψ via (8) and (9). This requires the matrix *constraint* to be defined, which has the format $[b|C]$. This can be set to NULL if there are no constraints.

The model can then be fitted using `sigex.mlefit` and the output stored to a variable *fit*. Setting the last argument to "bfgs" tells the computer to use the BFGS (Golub and Van Loan, 2012) algorithm for nonlinear optimization of the Gaussian likelihood, based on initial values dictated by *param*, but fixing parameters according to *constraint*, and an overall model structure given by *mdl*. However, we begin by obtaining ψ from *param*, and ψ should satisfy (7); the code checks this condition. At the end of fitting a model, we will typically have the MLE for η (denoted *eta.mle*). Then a call to `sigex.eta2psi` will yield *psi.mle*, and we can apply `sigex.psi2par` to obtain *par.mle*.

One can also extract the Hessian matrix if BFGS was used for model fitting. Note that uncertainty for the model parameters only pertains to η, so the dimension of *hess* is given accordingly. The inverse of *hess* is the asymptotic covariance matrix V of the MLEs for η; to get the asymptotic covariance matrix for ψ, we compute $A\,VA'$, where A is the matrix described below (10).

3.2 Method of moments

The goal of the MOM procedure is to obtain quick, rough estimates that might be used as initial values in the ML routine, or alternatively as final estimates for higher-dimensional time series data where ML is not computationally feasible. Background for MOM for latent process models is given in McElroy and Roy (2022a), although some more specific implementation details are discussed here. If $\Psi^{(k)}(z)$ in (4) is known, then by applying (5) we can express the autocovariances of $\{\underline{Y}_t\}$ in terms of certain known matrix polynomials and the covariance matrices $\Sigma^{(k)}$; by substituting sample autocovariances, we can solve for estimators of the covariance matrices. These are called the MOM estimates.

In practice several provisos mitigate the usefulness of this device. First, the presence of fixed effects interferes with the population equations. Second, the solutions to the equations yield symmetric matrices, but these are not guaranteed to be positive definite (pd). Third, there may be model parameters involved in the $\Psi^{(k)}(z)$. Fourth, missing values in the data generate obstacles to constructing the sample autocovariances.

To handle the first obstacle, we begin with a simple OLS in `sigex.momfit`, which is fast to compute and consistent notwithstanding serial correlation present in $\{Y_t\}$. Subtracting off the fitted regression effects, we proceed and now presume the de-meaned series has no fixed effects present at all. We can then apply the MOM equations so long as there are no missing values (the routine will not work if there are NAs); if parameters ζ are present and enter into the $\Psi^{(k)}(z)$, these are fixed at the values dictated by the default settings.

The second obstacle is important to rectify. In **Ecce Signum** a covariance matrix can be assessed in terms of condition numbers discussed in McElroy (2017), which are direct functions of the diagonal entries of D in (6). In fact, it can be shown that the condition numbers equal D_{jj}/Σ_{jj}, and they are typically viewed in log scale. After taking the logarithm, a value of $-\infty$ corresponds to a singularity in Σ that occurs at a particular index in the rank configuration (certain partial correlations of Σ are equal to ± 1 in this case). Moreover, negative values of large magnitude (in log scale) indicate the matrix is close to singularity; it is possible to alter the condition numbers to positive quantities such that the entries of D are altered and the whole matrix becomes better conditioned. This can be done without altering the various partial correlations embedded in L, and is implemented in `sigex.renderpd`, as shown below.

Let a j-dimensional covariance matrix be denoted Σ_j, which can be decomposed as $\Sigma_j = L_j \, D_j \, L_j'$ for a unit lower-triangular L_j and a diagonal matrix D_j. Letting the lower row of L_j be denoted $[\ell'_j, 1]$, it is known (McElroy, 2017) that

$$g_{j+1} = d_{j+1} + \ell'_{j+1} \, D_j \, \ell_{j+1},$$

where g_{j+1} denotes the lower right entry of Σ_{j+1}. Moreover, the $(j+1)$th condition number can be expressed as

$$\tau_{j+1} = -\log\left(g_{j+1}/d_{j+1}\right) = -\log\left(1 + \ell'_{j+1} \, D_j \, \ell_{j+1}/d_{j+1}\right).$$

These are recursive relations, so that for a given N-dimensional Σ we can examine the upper left submatrix Σ_j for $1 \leq j \leq N$ and recursively compute the condition numbers τ_j. Hence, if we wish to modify a given Σ such that the condition numbers exceed some minimum threshold, then we must modify either ℓ_{j+1} or D_j. Whereas the former is related to linkages between variables, each d_j is a partial variance (of the jth variable given the preceding variables 1 through $j-1$); choosing to modify these to some new $\widetilde{d}_j$, we can see how the condition number changes accordingly. Thus, if we set τ_{j+1} equal to a desired threshold α, and given that we have already modified in a recursive fashion the preceding D_j to $\widetilde{D}_j$, we have the formula

$$\widetilde{d}_{j+1} = \left(\frac{\exp\{-\alpha\} - 1}{\ell'_{j+1} \, \widetilde{D}_j \, \ell_{j+1}} \right)^{-1}. \tag{11}$$

This suggests the following algorithm to produce a positive definite modification of a given Σ matrix: compute the GCD, and iteratively check whether $1/d_{j+1} < (\exp\{-\alpha\} - 1)/\ell'_{j+1} \, \widetilde{D}_j \, \ell_{j+1}$; if so, we replace d_{j+1} by $\widetilde{d}_{j+1}$ in (11) and increment j. This yields a new

$$\widetilde{\Sigma} = L \, \widetilde{D} \, L',$$

where all the condition numbers are bounded below by α. In practice, the thresholds for α can be set according to the criteria discussed in McElroy (2017). For instance, $\alpha = -1.66$ corresponds to a partial correlation of. 90; this is sufficiently distinct from unity, so as to ensure there is no signal leakage. Calling `sigex.reduce` with a specified value of α (denoted by *thresh* in the code) replaces low logged condition numbers with *thresh*, rendering the matrix pd through `sigex.renderpd`.

However, this discussion pertains to matrices that are non-negative definite. The covariance matrices estimated by MOM can potentially have both positive and negative eigenvalues. In this case a second usage of `sigex.reduce` occurs (setting *modelFlag* equal to *TRUE*) whereby logged condition numbers lower than *thresh* result in the corresponding component index being added to the rank configuration. In other words, a low condition number for component j will result in j being added to the rank configuration. Now the GCD algorithm will still run on a symmetric matrix where the entries of D are possibly zero or negative; in such a case `sigex.reduce` will automatically assign a non-positive value of D to being below the threshold, and hence the rank configuration will be incremented. The outcome for MOM is that matrices that are not estimated as being pd get replaced by reduced rank approximations – essentially any negative values of the diagonal of D get replaced by a zero.

Once a reduced rank pd covariance matrix has been obtained, one can then determine *param* accordingly. However, the process of reducing the rank changes the model structure, and hence `sigex.reduce` updates *mdl*. This is a side-effect of MOM estimation, and may be undesirable in some applications (say, where we wish to keep all the covariance matrices as being full rank initially).

3.3 Casting and fixed effects

The basic model structure (1) yields a $\{Y_t\}$ time series that is mean zero. Optimal extraction means that we should filter $\{Y_t\}$ after having removed fixed effects, and then add back those fixed effects that pertain to the particular signal. However, in practice we do not know β exactly, and we instead subtract $z_t'\hat{\beta}$ from the data to get the de-meaned data:

$$\hat{Y}_t = X_t - z_t'\hat{\beta}.$$

If there are missing values, or extreme-value adjustment is being performed for an AO, then there is already an NA in $\hat{Y}_t$, and this will be estimated with a midcast. Importantly, the casting algorithms are predicated on the time series having mean zero, which is approximately true of $\hat{Y}_t$. Let $\widetilde{Y}_t$ denote the modification arising from casting, i.e., imputing missing values and adjusting extremes. The discrepancy between $\widetilde{Y}_t$ and $\hat{Y}_t$ is

$$\varepsilon_t = \hat{Y}_t - \widetilde{Y}_t,$$

which is also the casting error. It is important to keep track of this quantity, because in many applications of signal extraction we want all the extracted latencies to exactly add back up again to the original data. The form this typically takes is the following:

$$\begin{aligned} X_t &= \hat{Y}_t + z_t'\hat{\beta} \\ &= \widetilde{Y}_t + \varepsilon_t + z_t'\hat{\beta} \\ &= \Upsilon(B)\widetilde{Y}_t + (1 - \Upsilon(B))\widetilde{Y}_t + \varepsilon_t + z_t'\hat{\beta}. \end{aligned}$$

In the first line we break out the estimated fixed effects. In the second line we correct for missing values and extreme values. In the last line we apply two complementary filters – such as seasonal adjustment and seasonal filtering. The casting error must be added to one of these two extractions in order that all sums back to X_t. The fixed effects might be split up between the extractions. For example, if $\Upsilon(B)$ is a seasonal adjustment filter (so that $1 - \Upsilon(B)$ is a seasonal filter), then the casting error for extreme-value adjustment should be added to the seasonal adjustment (because it corresponds to an outlier, and is therefore not seasonal); fixed effects that correspond to moving holidays can be added to the seasonal component, but the trend polynomial should be combined with the seasonal adjustment.

The only case where this breaks down is if the original data is missing, i.e., X_t is NA. In such a case, we replace the left hand side with its imputation – essentially by subtracting the casting error from both sides of the equation, along with the fixed effects. For instance, if X_t is missing but we wish to publish a seasonal adjustment, the sum of the adjustment and seasonal factor equals the imputation of X_t given by $\widetilde{Y}_t + z_t'\hat{\beta}$.

3.4 Embedding

A vector embedding of a univariate process consists of rewriting the process as an N-variate time series of lower sampling frequency, such that the new sampling frequency equals the original sampling frequency divided by N. Sometimes embedding facilitates the modeling of a scalar time series. For example, a univariate daily time series might be embedded as a weekly time series with $N=7$, so that day-of-week specific effects can be modeled in the mean or variance directly. The function `sigex.daily2weekly` provides such an embedding given a specification of the day-of-week corresponding to the first component in the vector, and adds missing values to pad out the possibly incomplete first and last weeks. An inverse is provided by `sigex.weekly2daily`, which can be used on extracted components obtained from the weekly series.

3.4.1 Embedding filters

Scalar filters can also be embedded, turning them into a multivariate filter with Toeplitz structure. We can also apply embedding results to the moving average representation of a scalar series, allowing us to re-express it as a vector moving average. We can embed (Tiao and Grupe, 1980) a given high-frequency scalar process $\{X_t\}$ as an s-variate (where s is a positive integer) low-frequency process $\{\mathbf{X}_n\}$, where the jth component (for $1 \leq j \leq s$) of $\mathbf{X}_n$ is X_{ns+j}. This jth component is denoted as the jth *season's series*, and the vector time series is the *seasonal vector series*, which is denoted with a bold font.

Define the backshift operator B, which acts on the high-frequency time index, via $B X_t = X_{t-1}$. Similarly, the low-frequency lag operator is defined to be $L = B^s$, so that $L\mathbf{X}_n = \mathbf{X}_{n-1}$. The $s \times s$ identity matrix is denoted I_s. Given a scalar Laurent series $v(z)$, its embedded matrix Laurent series (cf. equation

(2.7) of Tiao and Grupe (1980)) is defined to be the $s \times s$-dimensional matrix Laurent series $\Upsilon(L) = \Sigma_h \Upsilon_h L^h$, where

$$\text{the } j, k \text{ coefficients of } \Upsilon_h \text{ are given by } \upsilon_{j-k+sh}. \tag{12}$$

It follows that if $\upsilon(B)$ is a polynomial of degree q, then the degree of the matrix polynomial $\Upsilon(L)$ is $\lfloor q/s \rfloor + 1$. Consider the following relations of filter input to filter output (expressed as operating on the de-meaned process):

$$U_t = \upsilon(B)\, Y_t$$
$$\mathbf{U}_n = \Upsilon(L)\, \mathbf{Y}_n.$$

We verify that these two equations are equivalent, if we utilize the rule (12). The jth component (for $1 \leq j \leq s$) of the second equation (with e_j the jth unit vector) indicates

$$\begin{aligned} \mathbf{U}_{n,j} &= e_j' \sum_{h \in \mathbb{Z}} \Upsilon_h \mathbf{Y}_{n-h} = \sum_{h \in \mathbb{Z}} \sum_{k=1}^{s} \upsilon_{j-k+sh}\, Y_{(n-h)s+k} \\ &= \sum_{\ell \in \mathbb{Z}} \upsilon_\ell\, Y_{ns+j-\ell} = \upsilon(B) Y_t, \end{aligned}$$

which uses the change of variable $\ell = j - k + s * h$.

Next, consider the case of a finite filter $\upsilon(B)$ that has a total of m coefficients, and such that υ_0 is the $(c+1)$th coefficient, where $0 \leq c < m$ is the integer shift parameter. (In the documentation for `sigex.hi2low` and `sigex.adhocextract`, L is used for the filter length.) This means that the non-zero coefficients are $\upsilon_{-c}, \upsilon_{1-c}, \ldots, \upsilon_0, \ldots, \upsilon_{m-1-c}$. So $c = 0$ corresponds to a causal filter, where only present and past observations get weighted; if m is odd and $c = (m-1)/2$, then the filter weights an equal number of past and future observations. The function `sigex.hi2low` calculates the corresponding non-zero filter matrices Υ_h, and determines the appropriate shift parameter for the low-frequency multivariate filter. From (12), we have

$$\ldots, \Upsilon_{-1} = \begin{bmatrix} \upsilon_{-s} & \ldots & \upsilon_{1-2s} \\ \vdots & \ddots & \vdots \\ \upsilon_{-1} & \ldots & \upsilon_{-s} \end{bmatrix}, \; \Upsilon_0 = \begin{bmatrix} \upsilon_0 & \ldots & \upsilon_{1-s} \\ \vdots & \ddots & \vdots \\ \upsilon_{s-1} & \ldots & \upsilon_0 \end{bmatrix}, \; \Upsilon_1 = \begin{bmatrix} \upsilon_s & \ldots & \upsilon_1 \\ \vdots & \ddots & \vdots \\ \upsilon_{2s-1} & \ldots & \upsilon_s \end{bmatrix}, \ldots$$

We need to determine the length M of the multivariate filter, and its shift parameter C. Clearly, if any Υ_h has at least one non-zero entry then it must be included in the array object for the multivariate filter. We begin by padding $\upsilon_{-c}, \ldots, \upsilon_{m-1-c}$ with ℓ zeroes on the left and r zeroes on the right, such that: (i) $\ell + c \equiv 0 \bmod s$ and $0 \leq \ell < s$, (ii) $m + \ell + r \equiv 0 \bmod s$, and such that r is the smallest integer $\geq s$ with this property.

First, (i) guarantees that the number of coefficients υ_k with $k < 0$ is divisible by s, and is the smallest such number. Examining the bottom (sth) row of the Υ_h matrix, we have the values $\upsilon_{sh+s-1}, \ldots, \upsilon_{sh}$, so if all $\upsilon_k = 0$ for $k < -(\ell + c)$, then $\Upsilon_h = 0$ if $h < -(\ell + c)/s$. This means that the non-zero matrices are $\Upsilon_{-(\ell+c)/s}, \ldots, \Upsilon_{M-1-(\ell+c)/s}$, with M to be determined. Hence the low-frequency shift parameter is $C = (\ell + c)/s$.

Second, (ii) ensures that the entire length of the padded scalar filter is divisible by s, given that with the padding from (i) the filter is currently of length $\ell + m$. We need to add at least s zeroes on in this stage so that the bottom row of Υ_h with highest index is zero except possibly for the bottom right entry υ_{sh}; if υ_{sh+1} were non-zero, then it would appear in Υ_{h-1}, indicating that the matrix is non-zero. It follows that $M = (m + \ell + r)/s$.

3.4.2 Embedding from daily to weekly

We discuss some of the details on the weekly embedding. Dates are stored in a month-day-year format as a three-component vector, and the utility `date2day` computes the day index (between 1 and 366) corresponding to any such date. The inverse of this routine is `day2date`, and `day2week` yields the day-of-week (1 for Sunday, 2 for Saturday, etc.) corresponding to a given calendar date. To compute the week index corresponding to a given date, `sigex.daily2weekly` uses the following procedure: the day index is first computed from the date, and we determine the day index corresponding to the first day of the first week. Specifically, between 0 and 6 NAs are pre-pended to form the first week, with this number of NAs determined by what day-of-week corresponds to the calendar date, and what type of day corresponds to the first component (this is called *first.day*, and is chosen by the user). Let ℓ denote the day index for the first day of the first week in the embedded series; note that ℓ can be negative (e.g., the calendar date is very early in the year, and we pad with enough NAs). Then the day indices for that first week are $\{\ell, \ell + 1, \ldots, \ell + 6\}$. Our rule is that these indices corresponds to week w of that year if and only if $7(w - 1) + 1 \in \{\ell, \ell + 1, \ldots, \ell + 6\}$. The logic is that if $1 \in \{\ell, \ell + 1, \ldots, \ell + 6\}$ then this must be the first week (as it includes the first day index), and hence $w = 1$; thus the formula follows. An equivalent condition is that

$$w = \lceil (\ell - 1)/7 \rceil + 1.$$

The lowest value that ℓ can take is -5, in the case that 6 NAs were padded on and the calendar date is the first day of the year; in that case $w = 1$, so we always get $1 \leq w \leq 53$. For example, January 1, 2020 was a Wednesday. Suppose we wish to have Saturday as our first series, but our time series has a start date of February 28, 2020. Hence, there will be 6 NAs added onto the first week; the day index for the start date is 59, so $\ell = 53$ and $w = 9$. The first "custom" week covers December 28, 2019 through January 3, 2020, and the ninth custom week includes February 22, 2020 through February 28, 2020. Note that week one will typically have a few days from the prior year; here the start year for the weekly series is 2020, even though four days of the first week are from 2019.

When passing back to a daily series via `sigex.weekly2daily`, the original start date can be recovered as follows. Compute the difference between the day-of-week of January 1 of the start year (for the weekly series) and the given *first.day*, and add seven if this difference is negative; the result, called *day.lead*, gives the number of days prior to January 1 that are included in the first week, and hence

$$\ell = 7\,(w-1) - day.lead + 1.$$

From this day index ℓ, we can determine the corresponding start date via `day2date`; however, if $\ell \leq 0$ (this can happen if $w = 1$) then we need to decrement the year by one and add to ℓ the day index of December 31 for the prior year (i.e., either 365 or 366). If we started by embedding a daily time series of a certain start date, use `sigex.daily2weekly`, and then invert with `sigex.weekly2daily`, we will not recover the exact start date unless *first.day* corresponds to the day-of-week of that calendar date – because otherwise, additional NAs are added and the daily series gets lengthened (so that the total sample size is a multiple of 7).

4 Illustrations

We consider several numerical illustrations of the capabilities. These are meant to be consecutive, in the sense that commands documented in a prior example are not explicitly re-visited in subsequent examples. The R package **Ecce Signum** can be installed from https://github.com/tuckermcelroy/sigex, (which also has R Markdown notebooks for this chapter's illustrations), or by using the command

```
devtools::install_github("tuckermcelroy/sigex").
```

4.1 Non-defense capital goods

We first examine data for monthly shipments and new orders from the Manufacturing, Shipments, and Inventories survey,[5] or *ndc* for short. The data for New Orders is not available at January 1992, since this series starts at February 1992 – so this value is entered as an NA. This is an example of ragged edge data.

We perform a bivariate analysis by fitting a VAR model. After loading the data into a variable named *ndc*, preliminary metadata is assigned as follows:

```
dataALL.ts <- sigex.load(data       = ndc,
                         start.date = c(1992, 1),
                         period     = 12,
                         epithets   = c("Shipments", "NewOrders"),
                         plot       = TRUE )
```

[5] Seasonally adjusted monthly data covering the sample period is January 1992 to April 2020, downloaded July 21, 2020 (4:45 PM), U.S. Census Bureau, obtained from https://www.census.gov/mtis/index.html, by selecting Non-Defense Capital Goods, and either Value of Shipments or New Orders.

These commands attach a starting date and frequency, as well as names for the series. The last argument of `sigex.load` generates a time series plot. On the basis of this plot, we decide (below) to utilize no transformation for the data.

Next, data transformations and subsets of series can be selected. Here, we will examine both series, so *subseries* is set equal to `c(1,2)`. The Boolean *aggregate* can be used to define a cross-sectional sum corresponding to *subseries*, but in this example that option is not utilized. The *range* argument can be used to select a window of times, but is set to `NULL` if the full span of the data is to be used.

```
data.ts <- sigex.prep(data.ts.  = dataALL.ts,
                      transform = "none",
                      aggregate = FALSE,
                      subseries = c(1,2),
                      range     = NULL,
                      plot      = TRUE )
```

Next, we move to model declaration, first recording the dimensions via.

```
N <- dim(data.ts)[2]
T <- dim(data.ts)[1]
```

We will fit a VAR model to the differenced data, and we use Yule-Walker estimation to identify the model order p and get preliminary estimates.

```
ar.fit    <- ar.yw(diff(ts(ndc[2:T, ])))
p.order   <- ar.fit$order
par.yw    <- aperm(ar.fit$ar, c(2, 3, 1))
covmat.yw <- getGCD(ar.fit$var.pred, 2)
var.out   <- var.par2pre(par.yw)
psi.init  <- as.vector(c(covmat.yw[[1]][2, 1], log(covmat.yw[[2]]),
                     var.out, colMeans(diff(ts(ndc[2:T, ])))) )
```

The first line fits a VAR model using the native R function `ar.yw`. Note that we exclude the first observation because it is missing for the NewOrders series. We also apply differencing, as there seems to be some change in level for the series. The second line extracts the identified model order ($p = 7$ in this case), and the third line stores the fitted parameter in an array of dimension $2 \times 2 \times 7$. The fourth line takes the estimated innovation covariance matrix, and produces the GCD via `getGCD`. The output is a list object, where the lower triangular entry of L and the log of the diagonal entries of D are entered into the appropriate positions of *psi.init* in line seven. Line five maps the array of VAR parameters via *var.par2pre* to a pre-parameter vector *var.out* of length 28. Lines six and seven assemble *psi.init* from the Yule-Walker fit, including the sample means for the last two pre-parameters.

Next, we declare the model. In this example there are no regressors specified, and there is only one latent process:

```
mdl <- NULL
mdl <- sigex.add(mdl     = mdl,
                 vrank   = seq(1,N),
                 class   = "varma",
                 order   = c(p.order,0),
                 bounds  = NULL,
                 name    = "process",
                 delta   = c(1,-1) )
# regressors:
mdl <- sigex.meaninit(mdl = mdl, data.ts = data.ts, d = 0)
```

In the second line the main process is defined through a call to `sigex.add`. The list object *mdl* stores all our specifications about the model, but none of the parameter estimates. The second argument indicates the rank configuration for the trend; setting *vrank* equal to `seq(1,N)` ensures a full rank covariance matrix. The third argument gives the model type as *varma*; the order $p=7$, $q=0$ of the VARMA is given in the fourth argument. The fifth argument provides parameter bounds, and does not apply to this type of model, and hence is set to NULL. The sixth argument associates a label, and the last argument is for the scalar differencing operator $\delta^{(1)}(z)$, expressed as a vector of coefficients `c(1,-1)`.

Now we move on to the next stage: model fitting. Here we demonstrate MLE using BFGS. We begin by setting up parameter constraints; there are none in this case.

```
constraint <- NULL
psi.mle    <- psi.init
par.mle    <- sigex.psi2par(psi = psi.mle, mdl = mdl, data.ts = data.ts)
```

The second line assigns *psi.mle* to the initial values of the pre-parameters – this will speed up optimization. The third line converts this pre-parameter to the corresponding value of *param* via `sigex.psi2par`. Next, we run the MLE routine as follows:

```
fit.mle <- sigex.mlefit(data.ts    = data.ts,
                        param      = par.mle,
                        constraint = constraint,
                        mdl        = mdl,
                        method     = "bfgs",
                        debug      = TRUE )
```

With the debug option set to TRUE, the values of the divergence are printed to the console; this is a way to track progress and understand what parts of the likelihood function are being searched. Setting debug to FALSE turns off this screen output. When the routine finally terminates (this can take a few minutes, depending on the machine), we manage the output using

```
psi.mle <- sigex.eta2psi(eta = fit.mle[[1]]$par, constraint = constraint)
hess    <- fit.mle[[1]]$hessian
par.mle <- fit.mle[[2]]
```

The first line ensures that if constraints have been used, the final η is mapped to the corresponding ψ via `sigex.eta2psi`. In this case $\eta=\psi$ because there are no constraints. We also extract the Hessian matrix and the *param* corresponding to ψ in the second and third lines. Next, we can check our model fit by examining the residuals.

```
resid.mle <- sigex.resid(psi = psi.mle, mdl = mdl, data.ts = data.ts)[[1]]
resid.mle <- sigex.load(data       = t(resid.mle),
                        start.date = start(data.ts),
                        period     = frequency(data.ts),
                        epithets   = colnames(data.ts),
                        plot       = TRUE )
resid.acf <- acf(x = resid.mle, lag.max = 4*period, plot = TRUE)$acf
```

The first line extracts the residuals corresponding to the MLE. We want the first item in the list output of `sigex.resid`; the second item is an array of the fitted model's autocovariances. The second call to `sigex.load` does formatting, and the last line computes sample autocorrelations of the residuals. Next, we can examine the condition numbers via

```
log(sigex.conditions(data.ts = data.ts, psi = psi.mle, mdl = mdl))
```

```
     [,1]       [,2]
eta2    0 -0.1349298
```

The results indicate that the innovation covariance matrix is well-conditioned. The model goodness-of-fit can be checked with

```
sigex.portmanteau(resids = resid.mle, lag = 4*period, nump = length(psi.mle))
```

```
[1] 165.1150040   0.3534992
```

This code calls `sigex.portmanteau`, which yields the Portmanteau statistic for residual serial correlation discussed in Lütkepohl (2005); the second argument gives the number of lags to be used (48 in this case), and the third argument is the number of estimated parameters. The output indicates there is little residual autocorrelation, which is confirmed by inspection of the sample autocorrelations. The call to `sigex.gausscheck` yields p-values for both series from the Shapiro-Wilks normality test; although there is some non-normality indicated, this does not invalidate the model.

```
sigex.gausscheck(resids = resid.mle)
```

```
[1] 1.211720e-07 2.548542e-20
```

We can examine the Hessian matrix, and compute the t statistics for pre-parameters via

```
print(eigen(hess)$values)
tstats <- sigex.tstats(mdl,psi.mle,hess,constraint)
print(tstats)
```

If the Hessian is not positive definite (this can happen if nonlinear optimization terminates wrongfully at a saddlepoint), then the standard error is set to zero, and `sigex.tstats` returns $\pm\infty$ for each t statistic. In this case the Hessian is positive definite, and some of the pre-parameters are not significantly different from zero – we could refine the model by imposing zero constraints. Finally, we bundle the analysis with the call

```
analysis.mle <- sigex.bundle(data.ts   = data.ts,
                             transform = "none",
                             mdl       = mdl,
                             psi       = psi.mle )
```

As an application, we generate forecasts and aftcasts with uncertainty; also a midcast at time $t = 1$ is generated for the New Orders series, since this value is missing. Typically we unload the bundled model via.

```
data.ts <- analysis.mle[[1]]
mdl     <- analysis.mle[[3]]
psi     <- analysis.mle[[4]]
param   <- sigex.psi2par(psi = psi, mdl = mdl, data.ts = data.ts)
```

This amounts to a relabeling. Fifty forecasts and aftcasts are generated via

```
window <- 50
data.casts <- sigex.midcast(psi      = psi,
                            mdl      = mdl,
                            data.ts  = data.ts,
                            castspan = window )
```

```
extract.casts <- sigex.castextract(data.ts    = data.ts,
                                   data.casts = data.casts,
                                   mdl        = mdl,
                                   castspan   = window,
                                   param      = param )
```

The output of `sigex.midcast` is a list with two items: the matrix of casts, and a 4-array of casting error covariances. The tth column of the cast matrix contains a vector of casts for the tth index in the time series that has any missing values; this takes into account the specified number of aftcasts and forecasts with the `castspan` variable. Also note that these casts are for the series with all fixed effects removed. This output can be formatted for use via `sigex.castextract`, which inserts the casts in the correct places in the time series and adds an extended trend regression effect (it is extended forward and backward by the `castspan` value). The output of `sigex.castextract` is a list with three items, in the same format as for `sigex.extract` output; therefore the output can be used by `sigex.graph`:

```
dataPad.ts <- rbind(matrix(NA, nrow = window, ncol = N),
                    data.ts,
                    matrix(NA, nrow = window, ncol = N))
for(i in 1:N){
  plot(x    = ts(dataPad.ts[, i], start = c(1992, 1), frequency = 12),
       xlab = "Year",
       ylab = "",
       lwd  = 1,
       col  = 1)
  sigex.graph(extract    = extract.casts,
              reg        = NULL,
              start.date = c(1992,1),
              period     = 12,
              series     = i,
              displace   = 0,
              castcol    = "black",
              fade       = 60 )
}
```

This code sets up grey shading for cast uncertainty in *castcol* and *fade*. For each component time series, the raw data is plotted after first padding with NA before and after. The first argument of `sigex.graph` takes the extracted list object *extract.casts*; other fixed effects could be added in the second argument (these are typically not known at the times covered by the fore and aft windows, unless the regressors are simple to extrapolate). The sixth argument allows for a vertical displacement in case one does not wish to overlay the extraction on top of the data – setting to zero indicates no displacement. The last argument determines the darkness of shading corresponding to the signal extraction uncertainty. This concludes the example.

4.2 Petrol

We next examine two monthly petroleum time series: "Industrial Petroleum Consumption and OPEC Oil Imports," or *petrol* for short.[6] We perform a bivariate analysis by fitting two simple models, both of which are based on a trend-irregular specification referred to as the Local Level Model (LLM) in Harvey (1989). Here there are no regressors specified, and $K = 2$, with $\{S_t^{(1)}\}$ corresponding to a trend process and $\{S_t^{(2)}\}$ a so-called irregular process, which is a bivariate white noise. The trend is a bivariate random walk, i.e., $\delta^{(1)}(z) = (1 - z)$ with $\{\underline{S}_t^{(1)}\}$ a white noise process.

After loading the data into a variable named *petrol*, the preliminary metadata is assigned as discussed in the prior example; a log transformation is used. We perform exploratory analysis by generating spectral density estimates; the function `sigex.specar` generates vertical lines at frequencies that are multiples of 2π divided by the *period* argument. The second argument of `sigex.specar` is a Boolean variable, indicating whether the data should be first differenced, i.e., apply $1 - B$ to each series. Here we set *period* to 12 in `sigex.specar` because we are interested in the monthly frequencies:

```
par(mfrow = c(2, 1))
for(i in subseries){
    sigex.specar(data.ts   = data.ts,
                 diff      = FALSE,
                 subseries = i,
                 period    = period)
}
```

[6] Source: Department of Energy, Energy Information Administration. Both series cover the period January 1973 through December 2016, and are seasonally adjusted (by DOE). The first series, "Industrial Petroleum Consumption," is U.S. Product Supplied of Crude Oil and Petroleum Products, Monthly, in Thousand Barrels per Day. The second series, "OPEC Oil Imports," is Petroleum Imports from OPEC, in Thousand Barrels per Day.

The basic LLM is referred to as a "related trends model," in contrast to a "common trends model" considered subsequently, where it is enforced that $\Sigma^{(1)}$ has rank one. Consider the code

```
mdl <- NULL
mdl <- sigex.add(mdl    = mdl,
                 vrank  = seq(1, N),
                 class  = "arma",
                 order  = c(0, 0),
                 bounds = NULL,
                 name   = "trend",
                 delta  = c(1, -1)
                )
mdl <- sigex.add(mdl    = mdl,
                 vrank  = seq(1, N),
                 class  = "arma",
                 order  = c(0, 0),
                 bounds = NULL,
                 name   = "irregular",
                 delta  = 1
                )
mdl <- sigex.meaninit(mdl = mdl, data.ts = data.ts, d = 0)
```

The first call to `sigex.add` defines the first latent process, labeled as "trend," and the second such call gives the "irregular" latent process. Both latent processes have full rank. Next, the "common trends model" is specified via

```
mdl2 <- NULL
mdl2 <- sigex.add(mdl    = mdl2,
                  vrank  = 1,
                  class  = "arma",
                  order  = c(0, 0),
                  bounds = NULL,
                  name   = "trend",
                  delta  = c(1,-1) )
mdl2 <- sigex.add(mdl    = mdl2,
                  vrank  = seq(1, N),
                  class  = "arma",
                  order  = c(0, 0),
                  bounds = NULL,
                  name   = "irregular",
                  delta  = 1 )
mdl2 <- sigex.meaninit(mdl = mdl2, data.ts = data.ts, d = 0)
```

where the second argument of `sigex.add` in the second line has *vrank* set to 1 instead of `seq(1,N)`, as in *mdl*.

Next, we fit the models, beginning with the related trends specification. The default parameter setting involves setting the pre-parameter vector to zero, via:

```
constraint <- NULL
par.mle <- sigex.default(mdl = mdl, data.ts = data.ts, constraint = constraint)
psi.mle <- sigex.par2psi(param = par.mle, mdl = mdl)
```

The call to `sigex.default` creates a *param* object *par.mle* corresponding to a pre-parameter *psi.mle* of all zeroes – there are 8 components in this case.

The collation and examination of results can be done in the manner described for the *ndc* example in Section 4.1; for the related trends model, results are bundled into *analysis.mle*. In the case of the common trends model, we initialize at a given ψ value obtained from the first fitted model as follows: the pre-parameters associated with the trend innovation covariance matrix are altered to correspond to a full-correlation matrix (but with the same variances). In this case

$$\Sigma = \begin{bmatrix} \sigma_1^2 & \rho\sigma_1\sigma_2 \\ \rho\sigma_1\sigma_2 & \sigma_2^2 \end{bmatrix} \mapsto \begin{bmatrix} \sigma_1^2 & \sigma_1\sigma_2 \\ \sigma_1\sigma_2 & \sigma_2^2 \end{bmatrix},$$

and this mapping is accomplished by setting the lower entry of L in the GCD equal to σ_2/σ_1, and d_1 is set equal to σ_1^2. This is encoded with

```
cov.mat <- par.mle[[1]][[1]] %*% diag(exp(par.mle[[2]][[1]])) %*%
 t(par.mle[[1]][[1]])
l.trend <- sqrt(cov.mat[2,2]/cov.mat[1,1])
d.trend <- cov.mat[1,1]
```

Then the initialization to the common trends model is

```
constraint <- NULL
psi.mle2 <- c(l.trend, log(d.trend), psi.mle[4:8])
par.mle2 <- sigex.psi2par(psi = psi.mle2, mdl = mdl2, data.ts = data.ts)
```

After fitting and managing output, we see that the common trends model is quite a bit worse (not surprising, given that the correlation ρ is estimated at .26 in the related trends model), having a much higher divergence. We can compare the fits via

```
test.glr <- sigex.glr(data.ts      = data.ts,
                      psi.nested   = psi.mle2,
                      psi.nesting  = psi.mle,
                      mdl.nested   = mdl2,
                      mdl.nesting  = mdl)
print(c(test.glr[1], 1-pchisq(test.glr[1], df = test.glr[2])))
```

```
 [1] 445.10      0.00
```

A likelihood comparison statistic (McElroy, 2016a) for nested models is computed using `sigex.glr` by taking the difference of the divergences, and the p-value is reported in the second line. (However, this p-value is not correct when the nested model arises at the boundary of the parameter space, as in this case.) Finally, we consider trend extraction using either model. For example, to load the second model we use

```
data.ts <- analysis.mle2[[1]]
mdl     <- analysis.mle2[[3]]
psi     <- analysis.mle2[[4]]
param   <- sigex.psi2par(psi = psi, mdl = mdl, data.ts = data.ts)
```

Here we use the matrix formula approach to signal extraction. The first step is to get the matrix filters:

```
signal.trend <- sigex.signal(data.ts  = data.ts,
                             param    = param,
                             mdl      = mdl,
                             sigcomps = 1 )
signal.irr   <- sigex.signal(data.ts  = data.ts,
                             param    = param,
                             mdl      = mdl,
                             sigcomps = 2 )
```

There are two latent processes (trend and irregular), and the filters for these are obtained through the two calls to `sigex.signal`; the last argument *sigcomps* of `sigex.signal` is a vector of indices denoting the composition of the desired signal. In this case, *signal.trend* is composed of just the first latent process, so the value of *sigcomps* is 1. The second step is to compute the extractions:

```
extract.trend <- sigex.extract(data.ts = data.ts,
                               filter  = signal.trend,
                               mdl     = mdl,
                               param   = param )
extract.irr   <- sigex.extract(data.ts = data.ts,
                               filter  = signal.irr,
                               mdl     = mdl,
                               param   = param )
```

In the first line, *extract.trend* is defined as a list object, with first item corresponding to a $T \times N$ matrix of trend extractions. The second and third items of the list give upper and lower bounds of confidence intervals, based on $\pm$ 2 square root signal extraction MSE. Next, it is important to re-integrate fixed regression effects, and this is done via

```
reg.trend <- NULL
for(i in 1:N) {
  reg.trend <- cbind(reg.trend, sigex.fixed(data.ts = data.ts,
                                            mdl     = mdl,
                                            series  = i ,
                                            param   = param,
                                            type    = "Trend" ) )
}
```

The function `sigex.fixed` finds a specified regressor (named in the last argument) and multiplies by the corresponding parameter estimate, for the component specified in the third *series* argument. Here, the default linear trend line regressor is passed into the variable *reg.trend*. The results can be displayed using calls to `sigex.graph`, as in the following code:

```
trendcol <- "tomato"
fade     <- 40
par(mfrow=c(2,1),mar=c(5,4,4,5)+.1)
for (i in 1:N){
    plot(x    = data.ts[,i],
         xlab = "Year",
         ylab = colnames(data.ts)[i],
         ylim = c(min(data.ts[, i]), max(data.ts[, i])),
         lwd  = 2,
         yaxt = "n",
         xaxt = "n")
sigex.graph(extract    = extract.trend,
            reg        = reg.trend,
            start.date = begin.date,
            period     = period,
            series     = i,
            displace   = 0,
            color      = trendcol,
            fade       = fade)
axis(1, cex.axis = 1)
axis(2, cex.axis = 1)
}
```

This code sets up a red color for the trend in *trendcol*, and a shading percentage *fade*. For each component time series, the raw data is plotted (in logarithms) and overlaid with the trend; the first argument of `sigex.graph` takes the extracted list object *extract.trend*, and any fixed effects that should be added must be included in the second argument. The sixth argument allows for a vertical displacement in case one does not wish to overlay the extraction on top of the data – setting to zero indicates no displacement. The last argument determines the darkness of shading corresponding to the signal extraction uncertainty. Finally, one can examine the frequency response functions for WK filters through `sigex.getfrf`, which can also plot the real portions.

```
grid <- 1000
frf.trend <- sigex.getfrf(data.ts  = data.ts,
                          param    = param,
                          mdl      = mdl,
                          sigcomps = 1,
                          plotit   = TRUE,
                          grid     = grid )
```

The *grid* argument corresponds to a mesh of frequencies in $[0, \pi]$, and the fourth argument of `sigex.getfrf` indicates the combination of components desired. One can also examine the WK filter coefficients:

```
len      <- 50
target   <- array(diag(N), c(N, N, 1))
wk.trend <- sigex.wk(data.ts  = data.ts,
                     param    = param,
                     mdl      = mdl,
                     sigcomps = 1,
                     target   = target,
                     plotit   = TRUE,
                     grid     = grid,
                     len      = len)
```

The *target* argument indicates that no linear combination of the trend is being considered here; *len* is set to 50, so that the indices of the filter run from −50 up to 50. This finishes the example.

4.3 Housing starts

Here we study "New Residential Construction (1964–2012), Housing Units Started, Single Family Units," or *starts* for short,[7] corresponding to the four U.S. regions of South, West, NorthEast, and MidWest. This example illustrates modeling a multivariate series with 8 components, estimated with both MOM and MLE methods, and the extraction of signals with two methods (direct matrix and truncated WK). We also show how functions of a signal, such as growth rates, can be obtained.

The initial loading scripts are similar to the previous example. Exploratory analysis indicates we should consider a model with trend and seasonal effects; here we consider a structural model with $K = 8$ latent processes:

```
mdl <- NULL
mdl <- sigex.add(mdl,seq(1,N),"arma",c(0,0),NULL,"trend",c(1,-2,1))
mdl <- sigex.add(mdl,seq(1,N),"arma",c(0,0),NULL,"first seasonal",c(1,-sqrt(3),1))
mdl <- sigex.add(mdl,seq(1,N),"arma",c(0,0),NULL,"second seasonal",c(1,-1,1))
mdl <- sigex.add(mdl,seq(1,N),"arma",c(0,0),NULL,"third seasonal",c(1,0,1))
mdl <- sigex.add(mdl,seq(1,N),"arma",c(0,0),NULL,"fourth seasonal",c(1,1,1))
mdl <- sigex.add(mdl,seq(1,N),"arma",c(0,0),NULL,"fifth seasonal",c(1,sqrt(3),1))
mdl <- sigex.add(mdl,seq(1,N),"arma",c(0,0),NULL,"sixth seasonal",c(1,1))
mdl <- sigex.add(mdl,seq(1,N),"arma",c(0,0),NULL,"irregular",1)
mdl <- sigex.meaninit(mdl,data.ts,0)
```

Each of the eight latent processes is labeled, and has a distinct differencing polynomial, whose application produces a full-rank white noise; details of this

[7] The four series are from the Survey of Construction of the U.S. Census Bureau, available at http://www.census.gov/construction/nrc/how_the_data_are_collected/soc.html.

model are discussed in McElroy (2017). Next, we illustrate the MOM to obtain initial estimates.

```
mdl.mom     <- mdl
constraint  <- NULL
par.default <- sigex.default(mdl = mdl.mom, data.ts = data.ts, constraint = constraint)
par.mom     <- sigex.momfit(data.ts = data.ts, param = par.default, mdl = mdl.mom)
psi.mom     <- sigex.par2psi(param = par.mom, mdl = mdl.mom)
sigex.lik(psi     = psi.mom,
          mdl     = mdl.mom,
          data.ts = data.ts,
          debug   = FALSE )
[1] 6329.107
```

First, we define a new model object *mdl.mom*, and the subsequent initialization is standard in the second and third lines. The MOM estimates are obtained through `sigex.momfit` in the fourth line. In the last call we evaluate the divergence, obtaining 6329.107.

Examination of the residuals indicates that a substantial degree of serial dependence remains. We can also examine the condition numbers, and observe that all the six seasonal latent processes seem to have reduced rank covariance matrices, indicating various types of seasonal co-integration. We can modify the initial model to one that has reduced rank components, by using `sigex.reduce`:

```
thresh      <- -6.22
reduced.mom <- sigex.reduce(data.ts   = data.ts,
                            param     = par.mom,
                            mdl       = mdl.mom,
                            thresh    = thresh,
                            modelflag = TRUE )
mdl.mom     <- reduced.mom[[1]]
par.mom     <- reduced.mom[[2]]
psi.mom     <- sigex.par2psi(param = par.mom, mdl = mdl.mom)
sigex.lik(psi     = psi.mom,
          mdl     = mdl.mom,
          data.ts = data.ts,
          debug   = FALSE )
[1] 6292.841
```

Here the threshold $\alpha = -6.22$, and the updated model is stored in *reduced.mom*, a list object. The first item in the list contains the updated information for the model – in particular, the new rank configurations. These are entered into *mdl.mom* by extraction from *reduced.mom*. Then the parameters are updated, and the last call just returns the divergence for the MOM estimates: the value is 6292.841. Curiously, this is less than the value obtained previously for a nesting model; however, the previous divergence of 6329.107 corresponds to a modification of the initial MOM fit where the covariance matrices have been rendered non-negative definite. The reduced rank specification utilizes the same cholesky factors L, only with columns omitted corresponding to the rank configuration; hence it is possible to obtain a lower value of the divergence.

We move to the MLE fitting at this point; initial values from the MOM fit are loaded through

```
data.ts <- analysis.mom[[1]]
mdl     <- analysis.mom[[3]]
psi.mom <- analysis.mom[[4]]
par.mom <- sigex.psi2par(psi = psi.mom, mdl = mdl, data.ts = data.ts)
```

Fitting this reduced rank model can take a substantial amount of time, because there are 67 pre-parameters to be estimated. Optimization terminates at the divergence of 6152.607, which is substantially better than that of the MOM estimate. Now when we examine the condition numbers, we see values of $-\infty$(or very large negative values) corresponding to the complement of the rank configuration.

The next stage of analysis is signal extraction, and both the matrix method and the WK method are illustrated in the script. Here we focus on three latent processes: trend, seasonal, and seasonal adjustment. Since the seasonal process is composed of the six atomic seasonals that occur at indices two through seven, *sigcomps* equals `seq(2,7)`. Similarly, the setting is `c(1,8)` for the seasonal adjustment. In the display we illustrate how the seasonal component can be plotted with a vertical displacement. In order to use `sigex.wkextract`, one sets the *grid* parameter to determine the number of mesh points in the interval $[-\pi, \pi]$ used as a Riemann approximation of the integral; the *horizon* parameter corresponds to H (the number of desired aftcasts and forecasts), whereas *target* is the array storing the coefficients of $\Xi(B)$. Here we set up the WK signal extraction via

```
grid    <- 7000
window  <- 50
horizon <- 0
target  <- array(diag(N), c(N, N, 1))
```

which specifies that the filter will be truncated to length 101 (equal to twice *window* plus one) at the sample boundaries – only 50 forecasts and aftcasts will be generated. Setting *horizon* to zero means that we will not forecast the signal ahead. Finally, in the fourth line the *target* variable is set to the identity (as an array). The calls to `sigex.wkextract` are then given by.

```
extract.trend2 <- sigex.wkextract(psi      = psi,
                                  mdl      = mdl,
                                  data.ts  = data.ts,
                                  sigcomps = 1,
                                  target   = target,
                                  grid     = grid,
                                  window   = window,
                                  horizon  = horizon,
                                  needMSE  = TRUE )
extract.seas2  <- sigex.wkextract(psi      = psi,
                                  mdl      = mdl,
                                  data.ts  = data.ts,
                                  sigcomps = seq(2,7),
                                  target   = target,
                                  grid     = grid,
                                  window   = window,
                                  horizon  = horizon,
                                  needMSE  = TRUE )
extract.sa2    <- sigex.wkextract(psi      = psi,
                                  mdl      = mdl,
                                  data.ts  = data.ts,
                                  sigcomps = c(1,8),
                                  target   = target,
                                  grid     = grid,
                                  window   = window,
                                  horizon  = horizon,
                                  needMSE  = TRUE )
```

The fourth argument in these function calls is *sigcomps*, and parallels the prior calls of `sigex.signal`. Note that whereas in the matrix approach we used two calls in succession (`sigex.signal` followed by `sigex.extract`), here there is a single call with output in the same list format as that provided by `sigex.extract`. The following plot in the script just compares the square root MSE for the trend components arising from both signal extraction methods; with *window* size of 50 there is no discrepancy. (The user can re-run the script with *window* set to 10, to observe some discrepancy between the exact matrix results and the approximate WK results.).

The final stage of this script is extraction of the trend growth rate, i.e., we extract $(1 - B)S_t^{(1)}$. Note that this is an I(2) trend, so the growth rate is a random walk – however, the dynamics of an extracted component in general do not match those of the specified process (see discussion in McElroy (2012)). The first code block extracts the trend growth rate and MSE directly from the output of `sigex.signal` stored in *signal.trend*, and we do not cover this in detail, instead focusing on the second code block which produces approximately the same results via the WK filter. Noting that the growth rate polynomial *gr.poly* has already been set to `c(1,-1)`, we proceed to

```
grid      <- 70000  # need large grid value to get accuracy
window    <- 100
horizon   <- 0
gr.array <- array(t(gr.poly) \%x\% diag(N), c(N, N, p+1))
reg.gr    <- array(0, c(T, N))
for (k in 1:N) {
    reg.gr[, k] <- stats::filter(x = reg.trend[, k],
                                  filter = gr.poly,
                                  method = "convolution",
                                  sides = 1)
}
```

In the first line we take a large *grid* value to get decent accuracy, and take a larger truncation *window* of 100. In the fourth line, the target is defined by *gr.array* by writing *gr.poly* in an array. Finally, we need to modify the trend regressors so that they correctly correspond to the trend growth rate; this is accomplished by filtering the polynomial trend with *gr.poly*. (Recall that *reg.trend* was defined above, and already incorporates the regression parameters.) Next, we call

```
extract.trendgr2 <- sigex.wkextract(psi      = psi,
                                    mdl      = mdl,
                                    data.ts  = data.ts,
                                    sigcomps = 1,
                                    target   = gr.array,
                                    grid     = grid,
                                    window   = window,
                                    horizon  = horizon,
                                    needMSE  = TRUE )
```

which can take some time because of the high value of *grid*. Note that *gr.array* is now in the *target* argument. The output is plotted in the usual way, displaying the trend growth rate for the four series, with appropriate uncertainty. The script repeats the exercise for the seasonal adjustment, and the output can be compared to the direct matrix output. This concludes the example.

4.4 Business formation statistics

Next, we study a weekly univariate series: the business applications component of the "Business Formation Statistics,"[8] or *bfs* for short. We consider the nonseasonally adjusted data at the national level. This component is the main business applications series, which captures the weekly flow of IRS applications for Employer Identification Numbers, mainly for business purposes. For the metadata specifications we enter

[8] Weekly data as of week 27 of 2020, downloaded on July 14, 2020 (11:27 AM), U.S. Census Bureau, obtained from https://www.census.gov/econ/bfs/index.html,.

```
begin  <- c(2006, 1)
end    <- c(2020, 27)
period <- 52
```

which uses the fallacious *period* of 52. The actual period should be approximately $365.25/7 \approx 52.17857$ (which *Ecce Signum* can allow for), but we use an integer value to simplify exposition. Note that a 53rd week is present every five to seven years; we present some ways below to circumvent this difficulty.

In loading the series, we select one of four national series corresponding to business applications, and choose a log transformation. In order to specify the holiday regressors, we need to know the calendar date (in month-day-year format) for the beginning and end of the series, i.e., the first day of the first week and the last day of the last week. The function `weekly2date` provides this:

```
first.day  <- 1
all.date   <- weekly2date(first.day = first.day, begin = begin, T = T)
start.date <- all.date[[1]]
end.date   <- all.date[[2]]
```

Weekly data typically corresponds to measurements over a seven-day period beginning with Sunday, in which case *first.day* is set equal to 1, but nonstandard week configurations can be accommodated as well. Having extracted the output in lines three and four, we can load the holiday regressors. One supplies an ASCII file whose every row contains a calendar date for the relevant holiday, covering a broad range of years. (Because we later remove a long-term mean of the holiday regressor to avoid confounding with seasonal effects, we recommend providing these dates for several centuries.) These date files can be constructed from the Internet, or the `datefinder` program can be used. For Easter we utilize the following code; note `easter500`[9] comes preloaded with the package.

```
easter.path  <- system.file('extdata','easter500.txt',package='sigex')
easter.dates <- read.table(easter.path)
easter.reg   <- gethol(hol.dates  = easter500,
                       hol.fore   = 7,
hol.aft    = 0,
start.date = start.date,
end.date   = end.date )
```

Regressors can simply be created as a daily time series by `gethol`, and then later converted to a weekly regressor. Here we consider a window of 7 days before and 0 days after, so that the holiday effect is presumed to constitute eight days, including the entire prior week. Note that in the call to `gethol`, both *start.date* and *end.date* are required.

[9] This is a conversion of the U.S.Census Bureau's list of Easter dates for a 500 year period (1600–2099): `https://www.census.gov/srd/www/genhol/easter500.html`.

For *bfs* we consider several holiday effects: all the ten federal holidays as well as Easter and Black Friday. In each case we take the actual date of the holiday rather than its observed date (the observation dates are more difficult to determine, but this could be done). Note that holidays that do not have moving dates (such as Independence Day, Veteran's Day, and Christmas) constitute a deterministic annual seasonal effect; because `gethol` removes a long-term mean over many years (so as to avoid confounding with seasonal effects), such regressors become zero. (We still include these three holidays in our analysis to demonstrate below that they are automatically removed from the model.) Next, the daily regressors are converted to weekly flow regressors by first embedding with `sigex.daily2weekly` followed by averaging over each week:

```
easter.reg <- sigex.daily2weekly(data.ts    = easter.reg,
                                 first.day  = first.day,
                                 start.date = start.date)
easter.reg <- rowSums(easter.reg)/7
```

The averaging in the third line corresponds to the construction of monthly and quarterly regressors in X-12-ARIMA described in Findley et al. (1998), whereby the proportion of the effect (over the entire window) that falls in any given week is declared to be the value of the regressor for that week.

After these specifications, we fit a baseline model that involves no fixed effects, but models the serial dependence with a SARMA specification with order one for the AR, MA, seasonal AR, and seasonal MA polynomials:

```
mdl <- NULL
mdl <- sigex.add(mdl    = mdl,
                 vrank  = seq(1,N),
                 class  = "sarma",
                 order  = c(1,1,1,1,52),
                 bounds = NULL,
                 name   = "process",
                 delta  = 1)
mdl <- sigex.meaninit(mdl = mdl, data.ts = data.ts, d = 0)
```

In particular, we do not specify any differencing even though the level of the series seems to change – gradual (but transitory) level changes are not inconsistent with a stationary specification. The divergence is −2076.881, and the fit seems reasonably good. Our next model incorporates the ten federal holidays along with Easter and Black Friday, although Independence Day, Veteran's Day, and Christmas are not included by the call to `sigex.reg` (which checks for regressors that are numerically zero after differencing). For instance,

```
mdl <- sigex.reg(mdl = mdl,
                 series = 1,
                 reg = ts(data      = as.matrix(xmas.reg),
                          start     = start(xmas.reg),
                          frequency = period,
                          names     = "Xmas") )
mdl <- sigex.reg(mdl = mdl,
                 series = 1,
                 reg = ts(data      = as.matrix(black.reg),
                          start     = start(black.reg),
                          frequency = period,
                          names     = "BlackFriday") )
```

will make no modification to *mdl* in the first call, whereas the second call adds the Black Friday regressor. So this specification has seven federal holidays, plus two other holidays and the trend constant, for a total of 10 regressors. This improved model is fitted, producing a divergence of −2095.162. We refine the model by restricting ourselves to the New Year's (NY) and Martin Luther King (MLK) holidays. At this point, we observe in the fitted residuals that there is a substantial outlier in the first week of 2012, which is not explained by the New Year's regressor. We can treat this as an additive outlier, which is handled via

```
AO.times  <- 314
dataNA.ts <- data.ts
dataNA.ts[AO.times] <- NA
```

This defines a new time series object *dataNA.ts* that is identical to *data.ts* but has NA inserted wherever an additive outlier occurs. Our third refined model is therefore

```
mdl <- NULL
mdl <- sigex.add(mdl = mdl,
                 vrank = seq(1,N),
                 class = "sarma",
                 order = c(1, 1, 1, 1, 52),
                 bounds = NULL,
                 name = "process",
                 delta = 1)
mdl <- sigex.meaninit(mdl = mdl, data.ts = dataNA.ts, d= 0)
mdl <- sigex.reg(mdl = mdl,
                 series = 1,
                 reg = ts(data      = as.matrix(nyd.reg),
                          start     = start(nyd.reg),
                          frequency = period,
                          names     = "NewYearDay") )
mdl <- sigex.reg(mdl = mdl,
                 series = 1,
                 reg = ts(data      = as.matrix(mlk.reg),
                          start     = start(mlk.reg),
                          frequency = period,
                          names     = "MLK") )
```

This model is fitted and the divergence is −2013.311, indicating an inferior fit due to having less regressors. For this final model, we get a positive definite Hessian and hence t statistics for the parameters, confirming the significance of all the variables. Residual analysis indicates the model seems to be a good fit. We record the midcast for time $t = 314$ via

```
data.casts <- sigex.midcast(psi      = psi.mle,
                            mdl      = mdl,
                            data.ts  = dataNA.ts,
                            castspan = 0 )
```

and bundle the analysis using *dataNA.ts*. The final stage of analysis is to filter the data. The relevant fixed effects are loaded into variables via

```
reg.trend  <- sigex.fixed(data.ts = data.ts,
                          mdl     = mdl,
                          series  = 1,
                          param   = param,
                          type    = "Trend" )
reg.nyd    <- sigex.fixed(data.ts = data.ts,
                          mdl     = mdl,
                          series  = 1,
                          param   = param,
                          type    = "NewYearDay" )
reg.mlk    <- sigex.fixed(data.ts = data.ts,
                          mdl     = mdl,
                          series  = 1,
                          param   = param,
                          type    = "MLK" )
dataLIN.ts <- data.ts - ts(data      = reg.trend + reg.nyd + reg.mlk,
start     = start(data.ts),
frequency = period)
```

The last instruction creates *dataLIN.ts*, consisting of the original time series with the fixed effects removed, i.e., $\hat{Y}_t$. This de-meaned series shall be adjusted for the AO, and then filtered. We utilize a set of nonparametric filters to extract trend, seasonal, and non-seasonal (or seasonally adjusted) latent processes, implemented in `x11filters`.

```
week.period  <- 365.25 / 7
half.len     <- floor(week.period / 2)
x11.filters  <- x11filters(period = week.period, p.seas = 1)
trend.filter <- x11.filters[[1]]
seas.filter  <- x11.filters[[2]]
sa.filter    <- x11.filters[[3]]
shift        <- (dim(sa.filter)[3] - 1) / 2
```

The first line defines the non-integer period, which `x11filters` is designed to accommodate. The call in the third line returns a three-item list, containing the three filters, each of which is a symmetric sequence of numbers. Next, we call `sigex.adhocextract`:

```
trend.comp <- sigex.adhocextract(psi     = psi,
                                 mdl     = mdl,
                                 data.ts = dataNA.ts,
                                 adhoc   = trend.filter,
                                 shift   = half.len,
                                 horizon = 0,
                                 needMSE = TRUE )
sa.comp    <- sigex.adhocextract(psi     = psi,
                                 mdl     = mdl,
                                 data.ts = dataNA.ts,
                                 adhoc   = sa.filter,
                                 shift   = shift,
                                 horizon = 0,
                                 needMSE = TRUE )
AO.errs <- dataLIN.ts[AO.times] - data.casts[[1]]
sa.comp[[1]][AO.times] <- sa.comp[[1]][AO.times] + AO.errs
```

In the second call, note that we apply the method to *dataNA.ts*, which was defined previously as the de-meaned series with an NA inserted for the AO. The method will automatically generate the appropriate cast for the NA, filter the resulting casted series, and return the output along with appropriate uncertainty. The AC filter *trend.filter* is inputted, along with the *shift* parameter set to the middle position of the filter vector – this corresponds to a symmetric filter. The penultimate line computes the casting error ε_t at the AO time $t = 314$, and this is added onto the seasonal adjustment in the last line – because an AO effect belongs with the irregular in the seasonal adjustment extraction.

The final parts of the example display the extractions. The illustration is complete.

4.5 New Zealand arrivals

Our last illustration considers a daily immigration series (*nz* for short): New Zealand residents arriving in New Zealand after an absence of less than 12 months, covering the period January 1, 2008 through July 31, 2012.[10] The most obvious dynamics are the annual movements due to seasonality; more subtle are the weekly dynamics. We find that the sample autocorrelation functions for the series is somewhat different for each of the seven weekly series corresponding to each day-of-the-week. This phenomenon indicates that the daily series – analogously to the daily retail series analyzed in McElroy (2022) – can be modeled by a vector embedding as a 7-variate weekly series. Some initial processing of the daily data is needed:

[10] The series consists of New Zealand residents arriving in New Zealand after an absence of less than 12 months. These are public use data produced by Stats New Zealand via a customized extract, and correspond to a portion of the "daily border crossings - arrivals" tab (Total) of the Travel category in the Covid-19 portal: https://www.stats.govt.nz/experimental/covid-19-data-portal,.

```
n.months <- dim(imm)[1]/32
imm <- imm[-seq(1,n.months)*32,]     # strip out every 32nd row (totals)
imm <- matrix(imm[imm != 0],ncol=6) # strip out 31st False days
```

The original data is grouped into "months" of 31 days, with extra days of value zero added for months of shorter length; our code removes these features, as well as a monthly summation. Next, a set of customized regressors, *NZregs* comes with the package. The first column is a regressor that captures a moving holiday effect for Easter, and the next six columns correspond to school holidays for public schools in New Zealand. An effect is estimated for the start and end of the holiday for the first three school vacations in each year.[11] The last three columns correspond to Y2K effects, and will not be used. For metadata we enter

```
start.date <- c(9, 1, 1997)
end.date   <- day2date(day = dim(imm)[1] - 1, start.date = start.date)
period     <- 365
```

The full data begins on September 1, 1997 and ends July 31, 2012, though we will be interested in a restricted range. Following are some calendrical calculations, and a call to `sigex.load`:

```
dataALL.ts <- sigex.load(data       = imm,
                         start.date = begin,
                         period     = period,
                         epithets   = c("NZArr", "NZDep", "VisArr",
                                        "VisDep", "PLTArr", "PLTDep"),
                         plot       = TRUE )
```

There are six time series in the data set, though we will be focused on just the first one "NZArr," described above. We restrict the data span and focus on the first series via

```
transform  <- "log"
aggregate  <- FALSE
subseries  <- 1
range      <- list(c(2008, 1), end)
dataONE.ts <- sigex.prep(data.ts   = dataALL.ts,
                         transform = transform,
                         aggregate = aggregate,
                         subseries = subseries,
                         range     = range,
                         plot      = TRUE )
```

We can examine spectral plots to get an idea of the time series dynamics. Also we set the first day to be Sunday, and embed as a 7-variate weekly time series via a call to `sigex.daily2weekly`:

[11] The school holiday regressor for the first day of the vacation period is set to one for that day and one for the day after, zero otherwise. The school holiday regressor for the last day of the vacation period is set to one for that day and one for the day before, zero otherwise. The school holiday regressors are centered by removing the daily mean of each of the regressors over the entire series.

```
first.day <- 1
data.ts   <- sigex.daily2weekly(data.ts = dataONE.ts,
                                first.day = first.day,
                                start.date = start.date)
plot(data.ts)
```

As a result $N = 7$ and $T = 240$, which yields an initial NA (because the first data value is on Monday, but the week begins on Sunday). Also the data ends on a Sunday, so the last week has six NA values. This is a classic example of a ragged edge data pattern. Next, we define appropriate subspans of the holiday regressors via

```
s <- (range[[1]][1] - begin[1]) * period + (range[[1]][2] - begin[2]) + 1
e <- (range[[2]][1] - begin[1]) * period + (range[[2]][2] - begin[2]) + 1
times <- s:e
easter.reg   <- NZregs[times,1]
school1.reg  <- NZregs[times,2]
school1e.reg <- NZregs[times,3]
school2.reg  <- NZregs[times,4]
school2e.reg <- NZregs[times,5]
school3.reg  <- NZregs[times,6]
school3e.reg <- NZregs[times,7]
```

We have labeled the regressors corresponding to Easter and the three school holiday effects (both the beginning and end of each holiday period). These are the regressors for daily data, and need to be embedded as weekly series:

```
easter.reg <- sigex.daily2weekly(data.ts    = easter.reg,
                                 first.day  = first.day,
                                 start.date = start.date )
easter.reg[is.na(easter.reg)] <- 0
```

The last line replaces any ragged edge NA values (arising from embedding) with zero, which is appropriate because the corresponding data values are also missing. This code is repeated for the other 6 regressors.

Now we come to the model specification. The data has a slight upwards trend drift, and there is strong annual correlation present. Although a stationary model can be entertained, we found a fairly good model fit given by a SVARMA for data differenced to stationarity by applying annual differencing $1-L^{52}$. (Recall L is the weekly lag operator, and there are approximately 52 weeks per year.) After trying different possibilities, we consider non-seasonal $q = 1$ and seasonal $P = 1$, which is specified by the code

```
mdl <- NULL
mdl <- sigex.add(mdl    = mdl,
                 vrank  = seq(1,N),
                 class  = "svarma",
                 order  = c(0, 1, 1, 0, 52),
                 bounds = NULL,
                 name   = "process",
                 delta  = c(1, rep(0, 51), -1) )
mdl <- sigex.meaninit(mdl = mdl, data.ts = data.ts, d = 0)
```

The specification of the regressors takes more coding, and we show the code for Easter (the index i loops over the days of the week):

```
mdl <- sigex.reg(mdl     = mdl,
                 series = i,
                 reg    = ts(data      = as.matrix(easter.reg[,i]),
                             start     = start(easter.reg),
                             frequency = frequency(easter.reg),
                             names     = "Easter-day") )
```

Next we consider model fitting. Here we make use of parameter constraints for the regressors: because the same holiday effect is present for each day of the week (recall that the weekly regressors are derived from a single daily regressor), the parameter estimates should be constrained to be identical. We use the `sigex.constrainreg` function to do this: the *regindex* argument specifies indices J_1, J_2, ..., J_N that delineate which regressors are being considered for each series. For instance, $J_1 = \{1, 3, 4\}$ would correspond to the trend, school1, and school1e regressors for the first (Sunday) series. The *combos* argument provides the linear combinations of these regressors, but when set to NULL instead it is enforced that all the regression parameters are the same.

```
constraint <- rbind(constraint,
    sigex.constrainreg(mdl,data.ts,list(2,2,2,2,2,2,2),NULL))
constraint <- rbind(constraint,
    sigex.constrainreg(mdl,data.ts,list(3,3,3,3,3,3,3),NULL))
constraint <- rbind(constraint,
    sigex.constrainreg(mdl,data.ts,list(4,4,4,4,4,4,4),NULL))
constraint <- rbind(constraint,
    sigex.constrainreg(mdl,data.ts,list(5,5,5,5,5,5,5),NULL))
constraint <- rbind(constraint,
    sigex.constrainreg(mdl,data.ts,list(6,6,6,6,6,6,6),NULL))
constraint <- rbind(constraint,
    sigex.constrainreg(mdl,data.ts,list(7,7,7,7,7,7,7),NULL))
```

So the first call here identifies in *regindex* the Easter regressor for each of the seven series, and *combos* set to NULL indicates that these parameters will be equal. The next call does the same for the school1 regressor, and so forth; the first regressor is the trend, and this remains unconstrained. The output of `sigex.constrainreg` is a *constraint* matrix, which can be then passed into the MLE fitting routine. In this case the nonlinear optimization can take a long time – several days – and we just report the output in the script. The remaining code lines review diagnostics, as discussed in previous illustrations.

Moving on to signal extraction, we consider applying a daily seasonal adjustment filter that we embed as a 7-variate weekly filter. Consider the code

```
sa.hifilter <- c(1, rep(2, 365), 1) / (2 * 365)
len       <- 183
hi.freq   <- 7
low.freq <- 1
shift.hi <- len
out       <- sigex.hi2low(filter.hi = sa.hifilter,
                          hi.freq   = hi.freq,
                          low.freq  = low.freq,
                          shift.hi  = shift.hi )
sa.lowfilter <- out[[1]]
shift.low    <- out[[2]]
```

The daily seasonal adjustment filter *sa.hifilter* is an extension of the classic crude trend filter (known as the 2×12 filter) of X11 (discussed in McElroy and Politis (2020)) from monthly to daily frequency; it should suppress all daily frequencies of the form $2\pi j/365$ for $1 \leq j \leq 183$, as well as linear trends. This is embedded with a call to `sigex.hi2low`, where the high and low frequencies are indicated and the *shift.hi* parameter is $c = 183$, corresponding to a symmetric filter. The output is the 7-variate weekly filter *sa.lowfilter*, with shift parameter $C = 27$ given by *shift.low*.

Next, this filter is applied using a standard call of `sigex.adhocextract`. Recall that this automatically supplies missing value imputations for the ragged edges, and does forecast and aftcast extension for application of the filter. However, this should be transformed back to a daily series, so that we can view the daily series with its seasonal adjustment:

```
daily.ts <- ts(data      = sa.low[[1]],
               start     = start(dataONE.ts),
               frequency = frequency(dataONE.ts) )
sa.hi.daily[[1]] <- sigex.weekly2daily(data.ts   = daily.ts,
                                       first.dat = first.day )
```

The function `sigex.weekly2daily` transforms the series, using the indicated value of *first.day*. We repeat this for the lower and upper values of the confidence intervals (not shown). This can now be displayed, but we first generate another signal for comparison – we remove only the weekly dynamics. This is done by considering a simple filter annihilating weekly effects from the daily data, namely $(1 + B + \ldots + B^6)/7$, which we call the Trading Day (TD) filter. The relevant code is

```
td.hifilter <- rep(1, 7) / 7
len       <- 3
hi.freq   <- 7
low.freq <- 1
shift.hi <- len
out <- sigex.hi2low(filter.hi  = td.hifilter,
                    hi.freq    = hi.freq,
                    low.freq   = low.freq,
                    shift.hi   = shift.hi )
td.lowfilter <- out[[1]]
shift.low    <- out[[2]]
```

Now $c = 3$ and $C = 1$ for the TD filter. Before plotting, the trend must be computed. A subtlety is that we have 7 trends, one for each day of week, which therefore combine long-term trend effects and day-of-week effects. The mean of the 7 parameter values corresponds to the overall mean of each week, and the residual is the day-of-week effect in excess of the mean. So we define the daily trend to be the output of applying *td.hifilter* to the regression mean, i.e.,

```
reg.td <- NULL
for (i in 1:N){
  reg.td <- cbind(reg.td, sigex.fixed(data.ts = data.ts,
                                      mdl     = mdl,
                                      series  = i,
                                      param   = param,
                                      type    = "Trend") )
}
reg.td <- rbind(rep(0, N), reg.td)
reg.td <- ts(data      = sigex.weekly2daily(data.ts = reg.td, first.day = first.day),
             start     = start(dataONE.ts),
             frequency = period)
reg.trend <- stats::filter(x      = reg.td,
                           filter = td.hifilter,
                           method = "convolution",
                           sides  = 1)
reg.trend <- as.matrix(reg.trend[8:length(reg.td)])
```

First *reg.td* is computed, which is the trend means for each day-of-week expressed as a daily time series. We augment by one prior week's values, which would be all zero, and then convert to a daily series. Next, *reg.trend* is obtained by filtering with $(1 + B + \ldots + B^6)/7$, and finally removing the NA values (resulting from using *stats::filter*, which does not extend the input series). The illustration is completed with a display: the seasonal adjustment contains some mild weekly effects (because the filter is designed to remove daily, and not weekly frequencies), whereas the TD-adjusted component has annual movements but no weekly effects.

5 Conclusion

This chapter provides a tutorial on the analysis of multivariate time series using **Ecce Signum**, with a view towards applications that are of interest to official statistics, including modeling, forecasting, nowcasting, extreme-value adjustment, imputation, and seasonal adjustment. These methodological tools are appropriate for time series data that can – possibly after transformations and corrections for aberrancies – be plausibly construed as a Gaussian sample; this includes the bulk of economic and demographic time series published at national statistical offices. But the range of possible applications is broader than what has been indicated by this chapter's illustrations; the framework of **Ecce Signum** has been used in other work to analyze business cycles (Chen et al. (2024)), accommodate changes to sampling frequency (e.g., transitions of a survey from quarterly to monthly sampling, as in McElroy and Monsell (2015)), and modify meager time series (i.e., data with a high prevalence of zeroes or small values, due to low economic activity in a narrow time span (McElroy and Penny, 2019)).

This wide breadth of uses demonstrates the paradigm's flexibility. The demerit, however, of this wide scope is a higher burden placed upon the user: whereas a tightly focused application – intended for only certain kinds of data structures – can be written with sensible default settings so that data scientists with little pertinent background can still be successful, our stated endeavor is to build an encompassing methodological scaffold. This demands an educated analyst, and the material of this chapter supplies that pedagogical need by explicating the nuts and bolts of **Ecce Signum**.

This tool arose organically, originally with a focused view upon model-based signal extraction, and gradually expanded to include other capabilities that were important for research and production. It is our intention to continue this development, so as to address the applications of disaggregation, index construction, benchmarking, and privatization, as well as survey-related topics such as linking (i.e., joining two time series divided by a survey discontinuity) and survey errors. Moreover, future development goals include automatic model selection (perhaps through the use of artificial intelligence), improvements to parameter estimation (e.g., see Livsey and McElroy (2024)), the broadening of extreme-value adjustment's scope (e.g., to include level shifts and more general effects, as in McElroy (2025)), and the incorporation of accounting constraints (encountered in hierarchical time series commonly published by national statistical offices) into multivariate systems. These desiderata will be realized within the open source culture of R (or Python) and GitHub, with the new engines of analysis made freely available to the global public.

Acknowledgments

This report is released to inform interested parties of research and to encourage discussion. The views expressed on statistical issues are those of the authors and not those of the U.S. Census Bureau.

References

Bell, W., 1984. Signal extraction for nonstationary time series. Ann. Stat. 12, 646–664.

Bell, W.R., DiFonzo, T., Ladiray, D., McElroy, T.S., Smith, P.A., 2024. Preface to time series special issue of the journal of official statistics. J. Off. Stat. 40 (4), 499–507.

Bell, W.R., Pugh, M.G., 1989. Alternative approaches to the analysis of time series components. In: Proceedings of the Statistics Canada Symposium on Analysis of Data in Time, Ottawa.

Box, G.E., Jenkins, G.M., Reinsel, G.C., Ljung, G.M., 2015. Time Series Analysis: Forecasting and Control. John Wiley & Sons, New York.

Census Bureau, US, 2020. Genhol – A Utility that Generates User-Defined Moving Holiday Regressors for X-13ARIMA-SEATS. U.S. Census Bureau, Washington, DC.

Chang, I., Tiao, G.C., Chen, C., 1988. Estimation of time series parameters in the presence of outliers. Technometrics 30 (2), 193–204.

Chen, B., Hood, K., McElroy, T., Trimbur, T., 2024. Cycle fluctuations in the us real gdp and national income and product accounts aggregates. J. Off. Stat. 40 (4), 554–600.

Findley, D.F., Monsell, B.C., Bell, W.R., Otto, M.C., Chen, B.C., 1998. New capabilities of the x-12-arima seasonal adjustment program (with discussion). J. Bus. Econ. Stat. 16.

Golub, G.H., Van Loan, C.F., 2012. Matrix Computations. vol. 3 JHU Press.

Harvey, A.C., 1989. Forecasting, Structural Time Series Models and the Kalman Filter. Cambridge University Press, Cambridge.

Harvey, A.C., Trimbur, T.M., 2003. General model-based filters for extracting cycles and trends in economic time series. Rev. Econ. Stat. 85 (2), 244–255.

Hillmer, S.C., Tiao, G.C., 1982. An arima-model-based approach to seasonal adjustment. J. Am. Stat. Assoc. 77, 63–70.

Holan, S., McElroy, T.S., Wu, G., 2017. The cepstral model for multivariate time series: the vector exponential model. Stat. Sin. 27, 23–42.

Livsey, J., McElroy, T., 2024. Applying the expectation-maximization algorithm to multivariate signal extraction. J.Off. Stat. 40 (4), 660–684.

Ljung, G., 1993. On outlier detection in time series. J. R. Stat. Soc. Series B 55, 559–567.

Ljung, G.M., Box, G.E., 1978. On a measure of lack of ?t in time series models. Biometrika 65, 297–303.

Lütkepohl, H., 2005. New Introduction to Multiple Time Series Analysis. Springer, Science& Business Media.

McElroy, T., 2025. Analysis of crisis effects via maximum entropy adjustment. J. Time Ser. Anal. 1–15.

McElroy, T.S., 2008. Matrix formulas for nonstationary arima signal extraction. Econ. Theory 24, 1–22.

McElroy, T.S., 2012. An alternative model-based seasonal adjustment that reduces residual seasonal autocorrelation. Taiwan Econ. Forecast Policy 43 (1), 33.

McElroy, T.S., 2016a. Nonnested model comparisons for time series. Biometrika 103, 905–914.

McElroy, T.S., 2016b. On the measurement and treatment of extremes in time series. Extremes 19, 467–490.

McElroy, T.S., 2017. Multivariate seasonal adjustment, economic identities, and seasonal taxonomy. J. Bus. Econ. Stat. 35 (4), 611–625.

McElroy, T.S., 2018. Recursive computation for block-nested covariance matrices. J. Time Ser. Anal. 39 (3), 299–312.

McElroy, T.S., 2021. A diagnostic for seasonality based upon polynomial roots of arma models. J. Off. Stat. 37 (2), 367–394.

McElroy, T.S., 2022. Casting vector time series: algorithms for forecasting, imputation, and signal extraction. Electron. J. Stat. 16 (2), 5534–5569.

McElroy, T.S., Monsell, B.C., 2015. Model estimation, prediction, and signal extraction for nonstationary stock and flow time series observed at mixed frequencies. J. Am. Stat. Assoc. 110, 1284–1303.

McElroy, T.S., Penny, R., 2019. Maximum entropy extreme-value seasonal adjustment. Aust. N. Z. J. Stat. 61, 152–174.

McElroy, T.S., Politis, D.N., 2020. Time Series: A First Course with Bootstrap Starter. CRC Press, New York.

McElroy, T.S., Roy, A., 2021. Testing for adequacy of seasonal adjustment in the frequency domain. J. Stat. Plan. Inference 221, 241–255.

McElroy, T.S., Roy, A., 2022a. Model identification via total frobenius norm of multivariate spectra. J. R. Stat. Soc. Series B: Stat. Methodol. 84 (2), 473–495.

McElroy, T.S., Roy, A., 2022b. A review of seasonal adjustment diagnostics. Int. Stat. Rev. 90 (2), 259–284.

McElroy, T.S., Trimbur, T., 2015. Signal extraction for non-stationary multivariate time series with illustrations for trend inflation. J. Time Ser. Anal. 36 (2), 209–227.

Pinheiro, J.C., Bates, D.M., 1996. Unconstrained parameterization for variance-covariance matrices. Stat. Comput. 6, 289–296.

Roy, A., McElroy, T.S., Linton, P., 2019. Estimation of causal invertible varma models. Stat. Sin. 29, 455–478.

Shapiro, S.S., Wilk, M.B., 1965. An analysis of variance test for normality (complete samples). Biometrika 591–611.

Tiao, G.C., Grupe, M.R., 1980. Hidden periodic autoregressive-moving average models in time series data. Biometrika 67, 365–373.

Tsay, R.S., 1986. Time series model specification in the presence of outliers. J. Am. Stat. Assoc. 81, 132–141.

Chapter 9

Combining information from multiple sources in official statistics

Changbao Wu*
University of Waterloo, Waterloo, Canada.
Corresponding author. e-mail address: cbwu@uwaterloo.ca

Chapter Outline

Abstract
We discuss issues with and techniques for combining information from multiple sources under settings involving probability samples, non-probability samples, multiple-frame survey samples, known population controls from a census or estimated population controls from other surveys. Two main principles, namely, *validity* and *efficiency*, for methodological developments are discussed under different scenarios. Several inferential techniques are presented under the general framework of calibration and pseudo empirical likelihood methods.

Handbook of Statistics, Vol. 53. https://doi.org/10.1016/bs.host.2025.04.002

1 Introduction

Official statistics conveys basic information about the state of society and the vibes of economy of particular region or jurisdictions. It plays a vital role in providing reliable information for decision making at various levels of government. It also supplies important and crucial data for scientific research and discovery. Over the past century and continuing to the present time, official statistics relies on conducting censuses for collecting basic population information. For more specialized data with measurements beyond census variables, sample surveys are the primary tool of data collection for both official statistics and scientific investigations. Design-based inference with probability samples has gained general acceptance by researchers and practitioners as well as government agencies due to its sound statistical foundations. Survey sampling theory and practice has become an integral part of statistical sciences.

The success of probability samples and design-based inference, however, has led to more frequent use of surveys and over-ambitious survey projects with measurements on hundreds of variables, and some requiring lab-based equipments for taking the measurements. Costs for conducting surveys have been skyrocketing while response rates have been declining in a dramatic fashion over the past few decades. Government statistical agencies have been making the shift in recent years from a "survey-first" mentality to using data from available sources. The trend was also emerging from the scientific world where data from multiple sources, especially those from non-probability samples, are combined and analyzed with suitable methods.

Statistical analyses using data from multiple sources face challenges in both theoretical developments and practical implementations. From a theoretical point of view, there are two fundamental principles that need to be followed for any methodological developments, namely, *validity* and *efficiency*. Validity refers to consistency of estimators under a suitable statistical framework, and efficiency is used to compare two or more competing valid estimators as measured by variances or mean squared errors. The two principles were discussed in detail by Wu and Thompson (2020) in the context of missing data analysis.

There are two broad scenarios in combining data from two or multiple sources. The first is that valid estimation and statistical inference is available by using only one of the sources. Combining data from multiple sources is for the purpose of increased efficiency. The second is that valid estimation and inference cannot be obtained from using any single source, and combining data from multiple sources makes it possible to achieve validity.

This chapter provides an overview of four specific scenarios in combining data from multiple sources (Sections 2–5). We first introduce the setting and data structure and discuss issues of *validity* and *efficiency* for each scenario. We then present some commonly used statistical methods for estimation and inference. In Section 6, we discuss the issue of under-coverage and show that some of the methods introduced in previous sections can be used to mitigate estimation biases due to under-coverage. Some additional remarks are given in Section 7.

2 Two probability survey samples

It often occurs in practice that two probability survey samples are selected from the same target population. Certain variables may be measured in both surveys in addition to survey-specific variables. Suppose that the finite population is denoted as $\mathcal{U} = \{1, 2, ..., N\}$, where N is the population size. Let $\mathcal{S}_1$ and $\mathcal{S}_2$ be two independent probability survey samples with respective sample sizes n_1 and n_2 selected from $\mathcal{U}$. Let the two survey datasets be represented by

$$\{(y_{1i}, y_i, \boldsymbol{x}_{1i}, z_i, d_{1i}), i \in \mathcal{S}_1\} \quad \text{and} \quad \{(y_{2i}, y_i, \boldsymbol{x}_{2i}, z_i, d_{2i}), i \in \mathcal{S}_2\},$$

where y_{1i}, y_{2i} and y_i are the so-called study variables with y_i being the common variable collected in both surveys, $\boldsymbol{x}_{1i}$, $\boldsymbol{x}_{2i}$ and z_i are the so-called auxiliary variables with z_i being the vector of common auxiliary variables available from both surveys. The d_{1i} and d_{2i} are respectively the survey weights for the two samples $\mathcal{S}_1$ and $\mathcal{S}_2$. For simplicity of presentation, we assume that d_{1i} and d_{2i} are the basic design weights obtained from the reciprocal of the first order inclusion probabilities.

Suppose that the goal is to estimate the population total or mean of a study variable (y_1, y_2 or y). Valid estimators are readily available using one of the survey samples for which the study variable is measured. The question then becomes "how to combine information from the two samples for more efficient estimation?". The answer depends on whether and how the auxiliary variables are used.

2.1 A combined estimator for the common study variable

Let $\mu_y = N^{-1}\sum_{i=1}^{N} y_i$ be the population mean of the common study variable y and assume that the population size N is known. A design unbiased estimator of μ_y can be computed from each of the two samples as

$$\hat{\mu}_y^{(1)} = \frac{1}{N}\sum_{i\in\mathcal{S}_1} d_{1i}y_i \quad \text{and} \quad \hat{\mu}_y^{(2)} = \frac{1}{N}\sum_{i\in\mathcal{S}_2} d_{2i}y_i.$$

A more efficient estimator of μ_y can be obtained by using a convex combination of the two design unbiased estimators in the form of

$$\hat{\mu}_y(\alpha) = \alpha\hat{\mu}_y^{(1)} + (1-\alpha)\hat{\mu}_y^{(2)}$$

for some $\alpha \in [0,1]$. The optimal choice of α which minimizes the design-based variance of $\hat{\mu}_y(\alpha)$ with respect to α is given by $\alpha_o = V_1^{-1}/(V_1^{-1} + V_2^{-1})$, where $V_1 = V(\hat{\mu}_y^{(1)})$ and $V_2 = V(\hat{\mu}_y^{(2)})$ which are the unknown theoretical variances of the two estimators. Let $\hat{V}_1$ and $\hat{V}_2$ be consistent variance estimators such that $\hat{V}_1 = V_1 + o_p(1)$ and $\hat{V}_2 = V_2 + o_p(1)$. The optimal estimator of μ_y used in practice is given by

$$\hat{\mu}_y^{(o)} = \hat{\mu}_y(\hat{\alpha}_o) = \hat{\alpha}_o\hat{\mu}_y^{(1)} + (1 - \hat{\alpha}_o)\hat{\mu}_y^{(2)},$$

where $\hat{\alpha}_o = \hat{V}_1^{-1}/(\hat{V}_1^{-1} + \hat{V}_2^{-1})$. Noting that $\alpha_o - \hat{\alpha}_o = o_p(1)$ and $\hat{\mu}_y^{(k)} - \mu_y = O_p(n_k^{-1/2})$ for $k = 1, 2$ under mild regularity conditions on the survey design and the finite population, we have $\hat{\mu}_y^{(1)} - \hat{\mu}_y^{(2)} = O_p(n^{-1/2})$ and

$$\hat{\mu}_y^{(o)} - \hat{\mu}_y(\alpha_o) = (\hat{\alpha}_o - \alpha_o)\{\hat{\mu}_y^{(1)} - \hat{\mu}_y^{(2)}\} = o_p(n^{-1/2}),$$

where $n = \min\{n_1, n_2\}$. The estimator $\hat{\mu}_y^{(o)} = \hat{\mu}_y(\hat{\alpha}_o)$ using the estimated α_o remains optimal asymptotically, i.e., the two versions $\hat{\mu}_y^{(o)}$ and $\hat{\mu}_y(\alpha_o)$ have the same asymptotic variance. Variance estimation for $\hat{\mu}_y^{(o)}$ can proceed by treating $\hat{\alpha}_o$ as a fixed quantity:

$$\hat{V}\{\hat{\mu}_y^{(o)}\} = \hat{\alpha}_o^2\hat{V}_1 + (1 - \hat{\alpha}_o)^2\hat{V}_2 = \frac{1}{\hat{V}_1^{-1} + \hat{V}_2^{-1}}.$$

Construction of the optimal estimator $\hat{\mu}_y^{(o)}$ involves the two variance estimators $\hat{V}_1$ and $\hat{V}_2$. For most survey designs and finite populations, we usually have $V_1 = O(n_1^{-1})$ and $V_2 = O(n_2^{-1})$, a popular "sub-optimal" estimator used in practice is to simply replace $\hat{\alpha}_o$ by $n_1/(n_1 + n_2)$. The resulting estimator is optimal if $V_1 = \sigma^2/n_1$ and $V_2 = \sigma^2/n_2$ for some common population quantity σ^2. The optimality result holds exactly under simple random sampling with replacement and approximately under simple random sampling without replacement with small sampling fractions.

2.2 Common auxiliary variables and internal consistency

The use of auxiliary information has been an active research topic in survey sampling over the past several decades. The term "auxiliary information" typically refers to certain population level information that is not part of the sample. For a single sample $\mathcal{S}$ with the sample dataset $\{(y_i, \boldsymbol{x}_i), i \in \mathcal{S}\}$, the observed auxiliary variables $\boldsymbol{x}_i$ for units in the sample do not provide additional information on the (marginal) distribution of the study variable y given the observed y_i, $i \in \mathcal{S}$. Popular statistical methods such as generalized regression and calibration to be discussed in Section 4 cannot be used in this scenario for more efficient estimation of μ_y.

When two samples $\mathcal{S}_1$ and $\mathcal{S}_2$ contain observations on a vector of common auxiliary variables $\boldsymbol{z}$, there is a research question on whether the information on $\boldsymbol{z}$ can be combined from the two samples for improved estimation of a study variable in one of the samples. A second research question with practical implications is the so-called "*internal consistency*". For the population mean $\mu_{\mathbf{z}} = N^{-1}\sum_{i=1}^{N} z_i$, we have two estimators available,

$$\hat{\mu}_{\mathbf{z}}^{(1)} = \frac{1}{N}\sum_{i\in\mathcal{S}_1} d_{1i}z_i \quad \text{and} \quad \hat{\mu}_{\mathbf{z}}^{(2)} = \frac{1}{N}\sum_{i\in\mathcal{S}_2} d_{2i}z_i.$$

These two estimators usually do not have the same value while estimating the same population quantity $\mu_{\mathbf{z}}$. It is desirable to create "calibrated weights" w_{1i} and w_{2i} for each of the two samples such that the weights provide valid estimators for any survey-specific variables while maintaining the "internal consistency" in the sense that two sets of weights lead to the same value of estimates for common variables in the two samples.

Zieschang (1990) and Renssen and Nieuwenbroek (1997) were among the first to address the problems. The proposed approach was to use the generalized regression techniques with an enlarged set of calibration equations. Unfortunately, the resulting regression-type weights can take negative values. Kim and Rao (2012) considered a model-assisted approach to combining two probability survey samples. In what follows, we describe the pseudo empirical likelihood methods of Wu (2004a), which guarantees positive weights with flexibilities of incorporating auxiliary information through a constrained maximization procedure.

Let $\boldsymbol{p} = (p_1, p_2, \ldots, p_{n_1})$ and $\boldsymbol{q} = (q_1, q_2, \ldots, q_{n_2})$ be two discrete probability measures over the sampled units in $\mathcal{S}_1$ and $\mathcal{S}_2$, respectively. The joint pseudo empirical likelihood (PEL) function in the spirit of Chen and Sitter (1999) is given by

$$\ell(\boldsymbol{p}, \boldsymbol{q}) = \sum_{i\in\mathcal{S}_1} d_{1i}\log(p_i) + \sum_{j\in\mathcal{S}_2} d_{2j}\log(q_j). \tag{1}$$

Maximizing $\ell(\boldsymbol{p}, \boldsymbol{q})$ given in (1) with respect to $\boldsymbol{p}$ and $\boldsymbol{q}$ under the two normalization constraints

$$\sum_{i\in\mathcal{S}_1} p_i = 1 \quad \text{and} \quad \sum_{j\in\mathcal{S}_2} q_j = 1 \tag{2}$$

leads to the maximizers

$$\hat{p}_i = d_{1i}/\sum_{k\in\mathcal{S}_1} d_{1k},\ i\in\mathcal{S}_1 \quad \text{and} \quad \hat{q}_j = d_{2j}/\sum_{k\in\mathcal{S}_2} d_{2k},\ j\in\mathcal{S}_2.$$

The maximum PEL estimator of a survey-specific variable is computed using $\hat{\boldsymbol{p}} = (\hat{p}_1, \hat{p}_2, \ldots, \hat{p}_{n_1})$ or $\hat{\boldsymbol{q}} = (\hat{q}_1, \hat{q}_2, \ldots, \hat{q}_{n_2})$. For instance, the maximum PEL estimator of $\mu_{y_1} = N^{-1}\sum_{i=1}^{N} y_{1i}$ is computed as

$$\hat{\mu}_{y_1} = \sum_{i\in\mathcal{S}_1} \hat{p}_i y_{1i} = \frac{\sum_{i\in\mathcal{S}_1} d_{1i}y_{1i}}{\sum_{i\in\mathcal{S}_1} d_{1i}},$$

where y_1 is a study variable measured specifically by the sample $\mathcal{S}_1$. Note that $\hat{\mu}_{y_1}$ is the Hájek estimator of μ_{y_1}. The maximum PEL estimator of a $\mathcal{S}_2$-specific variable y_2 is computed as $\hat{\mu}_{y_2} = \sum_{j\in\mathcal{S}_2} \hat{q}_j y_{2j}$.

For a common study variable y, the PEL approach allows seamless inclusion of the following constraint for the maximization procedure,

$$\sum_{i\in\mathcal{S}_1} p_i y_i = \sum_{j\in\mathcal{S}_2} q_j y_j. \tag{3}$$

The maximum PEL estimator of μ_y is computed as $\hat{\mu}_{yPEL} = \sum_{i\in\mathcal{S}_1} \hat{p}_i y_i = \sum_{j\in\mathcal{S}_2} \hat{q}_j y_j$, where $\hat{\boldsymbol{p}} = (\hat{p}_1, \hat{p}_2, ..., \hat{p}_{n_1})$ and $\hat{\boldsymbol{q}} = (\hat{q}_1, \hat{q}_2, ..., \hat{q}_{n_2})$ maximize $\ell(\boldsymbol{p}, \boldsymbol{q})$ given in (1) subject to constraints (2) and (3). The question of combining information from common auxiliary variables z along with the issue of "internal consistency" can be addressed simultaneously by including the set of constraints

$$\sum_{i\in\mathcal{S}_1} p_i z_i = \sum_{j\in\mathcal{S}_2} q_j z_j. \tag{4}$$

The PEL approach provides flexibilities for combining information from multiple probability survey samples. There have been established computational algorithms with R implementation for one-sample PEL problems available in the literature; see, for instance, the modified Newton Raphson procedure of Chen et al. (2002) and related R code in Wu (2005). PEL approaches for two or multiple samples require reformulations of the constrained maximization problem and use computational algorithms under stratified sampling as described in Wu (2004b). Technical details with an example involving three samples can be found in Wu (2004a).

3 Dual-frame and multiple-frame samples

In an ideal world for survey sampling, one has a single complete sampling frame covering the entire target population and a probability survey sample is drawn according to a sampling design. In a real world, unfortunately, such a complete sampling frame is often not available. Two or more separate incomplete frames might be available, and together they provide better coverage of the target population. For instance, one might have an area frame used in an existing survey conducted five years ago, but with rapid changes of the population the existing area frame becomes incomplete. A list frame of residential telephone numbers or a random digit dialing sampling frame can be used to increase coverage of the current population.

Dual-frame surveys are an early example of combining data from multiple sources for valid estimation and inference. The basic setting and estimation methods were first discussed by Hartley (1962). Following the notation of Rao and Wu (2010), let $\mathcal{A} = \{1, 2, ..., N_A\}$ and $\mathcal{B} = \{1, 2, ..., N_B\}$ be the two frame populations with sizes N_A and N_B, respectively. It is assumed that none of the frames is complete (i.e., $N_A < N$ and $N_B < N$), but together the two frames cover the entire target population. Let $\mathcal{S}_A$ and $\mathcal{S}_B$ be two probability survey samples with sample sizes n_A and n_B drawn respectively from frames $\mathcal{A}$ and $\mathcal{B}$, and the study variable y and auxiliary variables $\boldsymbol{x}$ are included in both samples. The goal is to estimate the mean $\mu_y = N^{-1}\sum_{i=1}^{N} y_i$ for the target population. It should be clear that a valid estimator of μ_y cannot be constructed using only one of $\mathcal{S}_A$ and $\mathcal{S}_B$ and one needs to combine information from the two dual-frame samples.

3.1 Dual-frame samples with post-stratification

The target population $\mathcal{U}$ may be divided into three mutually exclusive domains $a = \mathcal{A} \cap \mathcal{B}^c$, $b = \mathcal{A}^c \cap \mathcal{B}$, and $ab = \mathcal{A} \cap \mathcal{B} = \mathcal{B} \cap \mathcal{A} = ba$, where $\mathcal{A}^c$ and $\mathcal{B}^c$ denote the complement sets of $\mathcal{A}$ and $\mathcal{B}$. It follows that

$$\mathcal{A} = a \cup ab, \quad \mathcal{B} = b \cup ba, \quad \text{and} \quad \mathcal{U} = a \cup b \cup ab.$$

Let N_a, N_b and $N_{ab} = N_{ba}$ be the domain sizes of a, b and $ab = ba$. We have

$$N_A = N_a + N_{ab}, \quad N_B = N_b + N_{ba}, \quad \text{and} \quad N = N_a + N_b + N_{ab}.$$

The population mean μ_y can be written under a "stratified population" with three strata (i.e., domains), a, b and ab:

$$\mu_y = \frac{N_a}{N}\mu_{ya} + \frac{N_b}{N}\mu_{yb} + \frac{N_{ab}}{N}\mu_{yab}, \tag{5}$$

where μ_{ya}, μ_{yb} and μ_{yab} are the stratum population mean of y over a, b and ab, respectively.

Let $\{(y_i, \boldsymbol{x}_i, d_{Ai}), i \in \mathcal{S}_A\}$ and $\{(y_i, \boldsymbol{x}_i, d_{Bi}), i \in \mathcal{S}_B\}$ be the two datasets of the dual-frame samples, where d_{Ai} and d_{Bi} are the design weights. It is assumed that each unit in $\mathcal{S}_A$ can be identified as whether or not it also belongs to frame $\mathcal{B}$ and each unit in $\mathcal{S}_B$ can be identified as whether or not it also belongs to frame $\mathcal{A}$. This is a crucial part of conducting valid estimation and inference, and the information needs to be gathered at the data collection stage. The two samples $\mathcal{S}_A$ and $\mathcal{S}_B$ can be post-stratified as

$$\mathcal{S}_A = \mathcal{S}_a \cup \mathcal{S}_{ab} \quad \text{and} \quad \mathcal{S}_B = \mathcal{S}_b \cup \mathcal{S}_{ba},$$

where $\mathcal{S}_a = a \cap \mathcal{S}_A$, $\mathcal{S}_{ab} = \mathcal{S}_A \cap (ab)$, $\mathcal{S}_b = b \cap \mathcal{S}_B$, and $\mathcal{S}_{ba} = \mathcal{S}_B \cap (ab)$. The post-stratified sample $\mathcal{S}_a$ comes from stratum a and $\mathcal{S}_b$ from b. It is important to note that both $\mathcal{S}_{ab}$ and $\mathcal{S}_{ba}$ are samples from stratum ab, but $\mathcal{S}_{ab}$ is part of $\mathcal{S}_A$ from frame $\mathcal{A}$ and $\mathcal{S}_{ba}$ is part of $\mathcal{S}_B$ through frame $\mathcal{B}$. Let n_a, n_{ab}, n_b and n_{ba} be the sizes of the post-stratified samples $\mathcal{S}_a$, $\mathcal{S}_{ab}$, $\mathcal{S}_b$, and $\mathcal{S}_{ba}$, respectively.

Under the unrealistic scenario where the population stratum sizes N_a, N_b and N_{ab} are known, estimation of μ_y amounts to estimating μ_{ya}, μ_{yb} and μ_{yab} through Eq. (5). Valid estimators for μ_{ya} and μ_{yb} are readily available, namely,

$$\hat{\mu}_{ya} = \frac{1}{N_a}\sum_{i \in \mathcal{S}_a} d_{Ai} y_i \quad \text{and} \quad \hat{\mu}_{yb} = \frac{1}{N_b}\sum_{i \in \mathcal{S}_b} d_{Bi} y_i.$$

We have two independent samples $\mathcal{S}_{ab}$ and $\mathcal{S}_{ba}$ from the same population stratum ab. Estimation of the "population parameter" μ_{yab} can use the same techniques of combining information from two probability samples as described in Section 2.

A more realistic scenario is that the two frame population sizes N_A and N_B are known but the size N_{ab} of the population domain ab is unknown. We have two estimators of N_{ab} available from $\mathcal{S}_{ab}$ and $\mathcal{S}_{ba}$: $\hat{N}_{ab} = \sum_{i \in \mathcal{S}_{ab}} d_{Ai}$, and $\hat{N}_{ba} = \sum_{i \in \mathcal{S}_{ba}} d_{Bi}$. A pooled estimator of N_{ab}, denoted as $\hat{N}_{Pab}$, can be constructed once again using the methods of Section 2:

$$\hat{N}_{Pab} = \hat{\phi}\hat{N}_{ab} + (1 - \hat{\phi})\hat{N}_{ba},$$

with the optimal choice $\hat{\phi} = \hat{V}_B(\hat{N}_{ba}) / \{\hat{V}_A(\hat{N}_{ab}) + \hat{V}_B(\hat{N}_{ba})\}$, where $\hat{V}_A(\cdot)$ and $\hat{V}_B(\cdot)$ denote the estimated variance under the sampling designs for $\mathcal{S}_A$ and $\mathcal{S}_B$, respectively. A "sub-optimal" choice is to use $\hat{\phi} = n_{ab}/(n_{ab} + n_{ba})$. The other two population stratum sizes are estimated as $\hat{N}_a = N_A - \hat{N}_{Pab}$ and $\hat{N}_b = N_B - \hat{N}_{Pab}$. If the two frame population sizes N_A and N_B are also unknown, they can be estimated as $\hat{N}_A = \sum_{i \in \mathcal{S}_A} d_{Ai}$ and $\hat{N}_B = \sum_{i \in \mathcal{S}_B} d_{Bi}$.

3.2 Multiplicity-based approach for multiple-frame samples

Extensions of the post-stratification-based methods described in Section 3.1 for duel-frame samples become notationally awkward and practically difficult when there are three or more samples taken from different incomplete sampling frames. A more convenient approach to combining information from multiple-frame samples is to use the so-called "single-frame multiplicity-based" methods, which present a unified framework for dealing with duel-frame and multiple-frame samples.

Suppose that there are Q (≥ 2) frames, denoted by $\mathcal{A}_1, \ldots, \mathcal{A}_Q$. It is assumed that the Q frames together cover the target population $\mathcal{U} = \{1, 2, \ldots, N\}$. The term *multiplicity* is defined as *the number of frames* to which a particular unit belongs. It is a partial frame membership information since it does not require the identification of specific frames for which the unit belongs and can be obtained with some minor effort during the data collection process (Mecatti, 2007). Let $\mathcal{S}_1, \ldots, \mathcal{S}_Q$ be Q independent samples drawn from the Q frames. Let $\{(y_{qi}, \boldsymbol{x}_{qi}, d_{qi}), i \in \mathcal{S}_q\}, q = 1, 2, \ldots, Q$ be the Q-frame samples, where y_{qi} is the common response variable of interest attached to unit i in frame $\mathcal{A}_q$, $\boldsymbol{x}_{qi}$ is a vector of auxiliary variables that are not necessarily common across different frames, and d_{qi} is the design weight for unit i in the sample $\mathcal{S}_q$. Let m_{qi} be the *multiplicity* of unit i in frame q. We have $1 \leq m_{qi} \leq Q$ for all qi. For duel-frame samples, we have $m_{qi} = 1$ if $i \in a$ or $i \in b$ and $m_{qi} = 2$ if $i \in ab$.

The key to multiplicity-based methods is the observation that the population mean $\mu_y = N^{-1}\sum_{i=1}^{N} y_i$ can be rewritten as $\mu_y = N^{-1}\sum_{q=1}^{Q}\sum_{i\in\mathcal{A}_q} y_{qi}/m_{qi}$. A multiplicity-based design-unbiased estimator of μ_y is given by

$$\hat{\mu}_{yM} = \frac{1}{N}\sum_{q=1}^{Q}\sum_{i\in\mathcal{S}_q} d_{qi}\frac{y_{qi}}{m_{qi}}, \tag{6}$$

where the subscript M indicates "multiplicity". The population size N can be estimated as $\hat{N}_M = \sum_{q=1}^{Q}\sum_{i\in\mathcal{S}_q} d_{qi}/m_{qi}$. The multiplicity-based estimator $\hat{\mu}_y$ given in (6) not only has a simple form but also leads to a straightforward path for variance estimation:

$$Var(\hat{\mu}_y) = \frac{1}{N^2}\sum_{q=1}^{Q} Var\left(\sum_{i\in\mathcal{S}_q} d_{qi}\frac{y_{qi}}{m_{qi}}\right). \tag{7}$$

A variance estimator can be constructed using the survey design for each frame, treating $z_{qi} = y_{qi}/m_{qi}$ as a "shared variable" (i.e., the value of the response variable y_{qi} is shared by the m_{qi} frames for which unit qi belongs), and applying any available variance estimator for the Horvitz-Thompson estimator $\sum_{i\in\mathcal{S}_q} d_{qi} z_{qi}$.

3.3 Inference with pseudo empirical likelihood

The multiplicity-based approach to multiple-frame samples can be combined with pseudo empirical likelihood (PEL) for constructing confidence intervals or performing other analytic tasks. The (log) PEL function defined by Rao and Wu (2010) is given by

$$\ell_M(\boldsymbol{p}_1,\ldots,\boldsymbol{p}_Q) = \frac{n_M}{\hat{N}_M}\sum_{q=1}^{Q}\sum_{i\in\mathcal{S}_q}\frac{d_{qi}}{m_{qi}}\log(p_{qi}), \tag{8}$$

where $n_M = \sum_{q=1}^{Q} n_q$, and n_q is the size of sample $\mathcal{S}_q$, $\boldsymbol{p}_q = (p_{q1},\ldots,p_{qn_q})$ is a set of discrete probability measures assigned to the sample $\mathcal{S}_q$, $q = 1, \ldots, Q$. The factor $n_M/\hat{N}_M$ in the PEL function is for the purpose of scaling on the PEL ratio function as in Wu and Rao (2006) for single-frame surveys and has no impact on the constrained maximization process. Maximizing the PEL function given in (8) subject to the normalization constraint

$$\sum_{q=1}^{Q}\sum_{i\in\mathcal{S}_q} p_{qi} = 1 \tag{9}$$

leads to $\hat{p}_{qi} = (d_{qi}/m_{qi})/\hat{N}_M$. The maximum PEL estimator of μ_y is computed as $\hat{\mu}_{yPEL} = \sum_{q=1}^{Q}\sum_{i\in\mathcal{S}_q}\hat{p}_{qi}y_{qi}$, which is the Hájek estimator of μ_y obtained from (6) with N replaced by $\hat{N}_M$.

The PEL ratio confidence interval for the population mean μ_y is constructed through the PEL ratio function defined as

$$r_M(\theta) = -2\{\ell_M(\tilde{\boldsymbol{p}}_1(\theta),...,\tilde{\boldsymbol{p}}_Q(\theta)) - \ell_M(\hat{\boldsymbol{p}}_1,...,\hat{\boldsymbol{p}}_Q)\},$$

where $(\tilde{\boldsymbol{p}}_1(\theta),...,\tilde{\boldsymbol{p}}_Q(\theta))$ is the maximizer of $\ell_M(\boldsymbol{p}_1, \ldots, \boldsymbol{p}_Q)$ subject to the normalization constraint (9) and the parameter constraint given by

$$\sum_{q=1}^{Q} \sum_{i \in \mathcal{S}_q} p_{qi} y_{qi} = \theta \tag{10}$$

for a fixed θ. It can be shown that $r_M(\theta)$ follows asymptotically a scaled χ^2 distribution with one degree of freedom when $\theta = \mu_y$, with the scaling constant depending on the "design effect" of the multiple-frame samples (Rao and Wu, 2010). In practice, a bootstrap procedure may be used to approximate the asymptotic distribution of $r_M(\mu_y)$. Let b^*_α be the upper $\alpha \in (0, 1)$ quantile of the sampling distribution of $r_M(\mu_y)$ obtained through a bootstrap approximation. The $(1-\alpha)$-level confidence interval for μ_y is then constructed as $\{\theta | r_M(\theta) < b^*_\alpha\}$. Computational procedures with R implementation, which were developed for the PEL methods with single-frame surveys (Chen et al., 2002; Wu, 2004b, 2005), can be used with minor modifications for multiplicity-based PEL approach to multiple-frame survey samples.

4 Use of auxiliary population information

The use of auxiliary information on known population totals through calibration weighting is one of the earliest attempts in survey sampling and official statistics for combining information from multiple sources. The history goes back to Deming and Stephan (1940) on raking ratio estimation to incorporate known population counts for marginal totals. The general setting involves a probability survey sample from which valid estimation and inference are feasible. The known auxiliary population information is typically from census data or other external sources. The goal is to increase *efficiency* in estimation and maintain *internal consistency* for government and statistical agencies.

4.1 Traditional calibration techniques

The seminal paper by Deville and Särndal (1992) played a significant role in formalizing and popularizing the general concept and techniques of calibration weighting and estimation. Let $\{(y_i, \boldsymbol{x}_i, d_i), i \in \mathcal{S}\}$ be a survey dataset with $\mathbf{d} = (d_1, \ldots, d_n)$ being the initial set of survey weights. Let $T_{\mathbf{x}} = \sum_{i=1}^N \boldsymbol{x}_i$ be the known population totals of auxiliary variables.

The Horvitz-Thompson estimator for the population total $T_y = \sum_{i=1}^N y_i$ is given by $\hat{T}_{yHT} = \sum_{i \in \mathcal{S}} d_i y_i$. The traditional calibration estimator of T_y is computed as $\hat{T}_{yC} = \sum_{i \in \mathcal{S}} w_i y_i$, where the set of calibration weights $\mathbf{w} = (w_1, \ldots, w_n)$

minimizes a distance measure $D(\mathbf{w}, \mathbf{d})$ between the initial set of weights, $\mathbf{d}$, and the set of calibration weights, $\mathbf{w}$, subject to the set of benchmark constraints (also called calibration equations)

$$\sum_{i \in S} w_i \boldsymbol{x}_i = T_{\mathbf{x}}. \tag{11}$$

The left hand side of the calibration equations (11) is the calibration estimator for $T_{\mathbf{x}}$, which is forced to be the same as the known population totals to achieve "internal consistency" on auxiliary variables. However, the claim that the calibration estimator $\hat{T}_{yC}$ is more efficient than the Horvitz-Thompson estimator is not always true unless certain conditions are satisfied. The initial set of weights, the distance measure, and the form of population totals all play a role in the performance of calibration estimators. We provide some further discussions below.

(i) The initial set of weights. One crucial but often overlooked component in calibration estimation is that the initial set of weights provides valid weighted estimators for population quantities. This is not an issue for the setting used by Deville and Särndal (1992) where the design weights d_i are the initial weights. The calibration weights with constraints (11) can be expressed as $w_i = d_i g_i$ where the so-called g-weights satisfy $g_i = 1 + o_p(1)$. The result follows from the condition $\sum_{i \in S} d_i \boldsymbol{x}_i - T_{\mathbf{x}} = O_p(N/\sqrt{n})$, which holds for most commonly used survey designs and survey populations; see, for instance, Deville and Särndal (1992); Fuller (2009) and (Wu and Lu, 2016). The validity of calibration weights $\mathbf{w} = (w_1, \ldots, w_n)$ in terms of providing valid estimation depends on the validity of the initial weights $\mathbf{d} = (d_1, \ldots, d_n)$. Using a set of uniform (i.e., equal) weights as the initial weights for calibration weighting does not lead to valid inference unless the dataset is a simple random sample from the population.

(ii) The distance measure. Deville and Särndal (1992) considered a class of distance measures in the form of

$$D(\mathbf{w}, \mathbf{d}) = \sum_{i \in S} G_i(w_i, d_i),$$

where $G_i(w_i, d_i)$ depends on w_i and d_i and possibly other known quantities. A minimum requirement for the distance measure is that $G_i(w_i, d_i) \geq 0$ for all i and $G_i(w_i, d_i) = 0$ if $w_i = d_i$. While different distance measures all lead to first-order equivalent calibration estimators with large samples, the configurations of the calibration weights may differ considerably for finite samples. For instance, the chisquare distance may lead to negative weights, the generalized exponential titling distance may produce positive but very small weights, and the generalized pseudo empirical likelihood distance could have positive but very large weights; see (Wu and Lu, 2016) for a detailed discussion. The pseudo empirical likelihood (PEL) function (Chen and Sitter, 1999; Wu and Rao, 2006) is closely related to the generalized pseudo empirical likelihood distance as shown in (Wu and Lu, 2016). The maximum PEL estimators are suitable for population means while the traditional calibration estimators are designed for population totals. There is a gap between estimation of a population mean and estimation of a population total unless the population size N is known.

(iii) The known population totals. The validity of traditional calibration estimators relies on the result $\sum_{i\in S} d_i \boldsymbol{x}_i - T_{\mathbf{x}} = O_p(N/\sqrt{n})$. This latter result requires that the population totals $T_{\mathbf{x}}$ be accurate otherwise the calibration procedures would become a main potential source for biases. Population totals obtained from unreliable sources should not be used for calibration since the validity of estimators obtained from using the final set of calibration weights becomes questionable.

4.2 Model-calibration methods

The use of benchmark constraints (11) for traditional calibration weighting is inherent to a linear relationship between y and $\boldsymbol{x}$, and the gain of efficiency from using the resulting calibration estimator $\hat{\mu}_{yC}$ depends on the correlation between y and the linear predictor $\boldsymbol{x}'\boldsymbol{\beta}$ as demonstrated by Wu and Sitter (2001).

Suppose that the relation between y and $\boldsymbol{x}$ follows a general semiparametric model ξ through the first and second moment structures, and the finite population data $\{(y_i, \boldsymbol{x}_i), i = 1, 2, \ldots, N\}$ can be viewed as a random sample from the model ξ,

$$E_\xi(y_i|\boldsymbol{x}_i) = \mu(\boldsymbol{x}_i, \boldsymbol{\theta}), \quad V_\xi(y_i|\boldsymbol{x}_i) = v(\boldsymbol{x}_i)\sigma^2, \quad i = 1, 2, \ldots, N, \tag{12}$$

where $\mu(\cdot, \cdot)$ is the mean function and $v(\cdot)$ is the variance function, both with known forms and $v(\boldsymbol{x})$ depending on $\boldsymbol{x}$ sometimes through the mean function. The $\boldsymbol{\theta}$ and σ^2 are the so-called superpopulation parameters for the model ξ. The model-calibration estimator of μ_y proposed by Wu and Sitter (2001) replaces the conventional calibration equations (11) by the "size constraint" $\sum_{i\in S} w_i = N$ plus a single model-calibrated equation

$$\sum_{i\in S} w_i \mu(\boldsymbol{x}_i, \hat{\theta}) = \sum_{i=1}^{N} \mu(\boldsymbol{x}_i, \hat{\theta}), \tag{13}$$

where $\hat{\theta}$ is a survey-weighted estimator of the model parameters $\boldsymbol{\theta}$. While the one-dimensional constraint given in (13) reduces computational complexities of the constrained minimization problem for finding the final calibrated weights w_i, the right-hand side of equation (13) requires complete auxiliary information $\{\boldsymbol{x}_1, \boldsymbol{x}_2, \ldots, \boldsymbol{x}_N\}$ to be available if $\mu(\boldsymbol{x}_i, \boldsymbol{\theta})$ is nonlinear in $\boldsymbol{x}$.

The model-calibration approach has three theoretical implications. The first is on efficiency. The model-calibration estimator $\hat{\mu}_{yMC} = N^{-1}\sum_{i\in S} w_i y_i$ with w_i obtained using (13) is optimal among a class of calibration estimators in terms of the anticipated variance $E_\xi\{V_p(\hat{\mu}_{yMC})\}$, i.e., the model expectation $E_\xi(\cdot)$ of the design-based (asymptotic) variance $V_p(\cdot)$ of calibration estimators (Wu, 2003). The second is the confirmation of traditional calibration using (11) and its relation to a linear model, since the optimal model-calibration estimator reduces to the traditional calibration estimator under a linear model and the chisquared distance (Wu and Sitter, 2001). The third is the relationship between the types of population parameters and the types of calibration constraints for optimal estimation, as illustrated in Chen and Wu (2002) on the estimation of population distribution function and quantiles and in Sitter and Wu (2002) on the estimation of quadratic finite population functions such as population variance and covariance.

4.3 Calibration with estimated population controls

The model-calibration constraint (13) requires complete auxiliary information $\{\boldsymbol{x}_1, \boldsymbol{x}_2, \ldots, \boldsymbol{x}_N\}$ to be available when $\mu(\boldsymbol{x}_i, \boldsymbol{\theta})$ is nonlinear in $\boldsymbol{x}$. When such information is not available, the control total $\sum_{i=1}^{N} \mu(\boldsymbol{x}_i, \hat{\theta})$ may be replaced by a suitable estimate through a two-phase sampling strategy (Wu and Luan, 2003) or from an existing survey with information on auxiliary variables $\boldsymbol{x}$.

Suppose that $\mathcal{S}_0$ is a first-phase sample with size n_0 ($>n$) and information on $\boldsymbol{x}$. Let $d_i^{(0)}$ be the design weights for the first-phase sample. We can replace the model-calibration constraint (13) by

$$\sum_{i \in \mathcal{S}} w_i \mu(\boldsymbol{x}_i, \hat{\theta}) = \sum_{i \in \mathcal{S}_0} d_i^{(0)} \mu(\boldsymbol{x}_i, \hat{\theta}), \tag{14}$$

where $\hat{\theta}$ is obtained using the second-phase sample $\mathcal{S}$ with information on both y and $\boldsymbol{x}$. The model-calibration estimator remains valid under the constraint (14) as long as the control total on the right-hand size is estimated consistently from $\mathcal{S}_0$ or another source with information on $\boldsymbol{x}$. The efficiency of the resulting model-calibration estimator using w_i, however, depends on the accuracy of the estimated control total. In general, the sample size of the source used for the estimation needs to be much larger than the size of the sample $\mathcal{S}$.

Calibration techniques have been first developed in the context of survey sampling and official statistics for gaining efficiency in estimation and achieving internal consistency. The methodology has also found successful applications in other areas of statistics, including missing data analysis and causal inference. Zhang et al. (2023) provided a comprehensive review on the topic. Let $\mathcal{S}$ be a random sample of size n from an infinite population, with covariates $\boldsymbol{x}$ measured for all $i \in \mathcal{S}$. The study variable y, however, is subject to missingness. Let $R_i = 1$ if y_i is observed and $R_i = 0$ if y_i is missing. Let

$$\mathcal{S}_R = \{i | i \in \mathcal{S} \text{ and } R_i = 1\}, \quad \mathcal{S}_M = \{i | i \in \mathcal{S} \text{ and } R_i = 0\}$$

represent the set of respondents and the set of non-respondent (with missing y), respectively. It follows that $\mathcal{S} = \mathcal{S}_R \cup \mathcal{S}_M$. We have both y_i and $\boldsymbol{x}_i$ observed for $i \in \mathcal{S}_R$ but only $\boldsymbol{x}_i$ measured for $i \in \mathcal{S}_M$.

A powerful tool for dealing with missing data analysis is inverse probability weighting through the estimated propensity scores. It is one of the most successful applications of weighting through the Horvitz-Thompson estimator used in survey sampling to other areas of statistics. Unlike survey weights which are known from the sampling design, the propensity scores, defined as $\pi_i = P(R_i = 1|\boldsymbol{x}_i) = E(R_i|\boldsymbol{x}_i)$ under the so-called missing at random assumption, are unknown and need to be estimated under an assumed propensity score model. Suppose that the propensity score has a parametric form, $\pi_i = \pi(\boldsymbol{x}_i, \boldsymbol{\alpha})$, where the model parameters $\boldsymbol{\alpha}$ have the same dimensionality of the covariates $\boldsymbol{x}$ (including a constant 1 corresponding to an intercept term in the model). For instance, a logistic regression model is often used in practice where $\pi_i = 1 - \{1 + \exp(\boldsymbol{x}'\boldsymbol{\alpha})\}^{-1}$. The model parameters $\boldsymbol{\alpha}$ can be estimated through a set of calibration equations given by

$$\frac{1}{n}\sum_{i\in\mathcal{S}_R}\frac{\boldsymbol{x}_i}{\pi(\boldsymbol{x}_i,\boldsymbol{\alpha})} = \frac{1}{n}\sum_{i\in\mathcal{S}}\boldsymbol{x}_i. \tag{15}$$

The set of equations (15) has the same spirit of the set of equations (11). The factor $1/n$ on both side of (15) is redundant but the right-hand side of (15) can be viewed as an estimated mean for the covariates $\boldsymbol{x}$ as "population" controls. The simple sample means are a valid estimate of the population means due to the assumption that $\mathcal{S}$ is a random sample from the target population. The estimator $\hat{\boldsymbol{\alpha}}$ is the solution to (15). The inverse probability weighted (IPW) estimator of the population mean of y is given by $n^{-1}\sum_{i\in\mathcal{S}_R} y_i/\pi(\boldsymbol{x}_i, \hat{\boldsymbol{\alpha}})$, which has the exact same form of the calibration estimator $\hat{\mu}_{yC}$ with the inverse of the estimated propensity score as the calibration weights.

A parallel development can be applied to the estimation of the average treatment effect in causal inference with observational data. Let $\mathcal{S}$ be the initial random sample of n individuals with measurements on some baseline covariates $\boldsymbol{x}$ from the target population. Each individual in the sample receives either treatment or control but not both. Let $T_i = 1$ if individual i receives treatment and $T_i = 0$ otherwise. The initial sample can be divided into $\mathcal{S}_1 = \{i|i \in \mathcal{S} \text{ and } T_i = 1\}$ and $\mathcal{S}_0 = \{i|i \in \mathcal{S} \text{ and } T_i = 0\}$ with respective sizes n_1 and n_0. We have $\mathcal{S} = \mathcal{S}_1 \cup \mathcal{S}_0$ and $n = n_1 + n_0$. Let Y_{i1} and Y_{i0} be the values of the potential response variable Y under the treatment and the control, respectively, for individual i. The goal is to estimate the average treatment effect $\eta = \mu_1 - \mu_0$, where $\mu_1 = E(Y|T = 1)$ and $\mu_0 = E(Y|T = 0)$. Similar to missing data problems, the concept of propensity score $\pi_i = P(T_i = 1|\boldsymbol{x}_i)$ plays a central role in estimating η (Rosenbaum and Rubin, 1983). Under a parametric model $\pi_i = \pi(\boldsymbol{x}_i, \boldsymbol{\alpha})$, the model parameters $\boldsymbol{\alpha}$ can be estimated using the same set of calibration equations given in (15) where the set $\mathcal{S}_R$ is replaced by $\mathcal{S}_1$ for the treatment group. The inverse probability weighted estimator of η is given by $\hat{\eta} = n^{-1}\sum_{i\in\mathcal{S}_1} Y_{i1}/\hat{\pi}_i - n^{-1}\sum_{i\in\mathcal{S}_0} Y_{i0}/(1 - \hat{\pi}_i)$, where $\hat{\pi}_i = \pi(\boldsymbol{x}_i, \hat{\boldsymbol{\alpha}})$. A Hájek type estimator of η is often used in practice where the first and the second n in $\hat{\eta}$ are replaced respectively by $\hat{n} = \sum_{i\in\mathcal{S}_1}\hat{\pi}_i^{-1}$ and $\tilde{n} = \sum_{i\in\mathcal{S}_0}(1 - \hat{\pi}_i)^{-1}$.

5 Non-probability survey samples

Non-probability samples are not a new concept in survey sampling. However, the topic has been mostly ignored in research until recently partly because of the success of probability sampling and design-based inference widely adopted by survey organizations and official statistics. High costs and low response rates for surveying human populations and the emergence of new non-probability data sources have brought the topic back to the spotlight with increased research activities in recent years. Wu (2022) contains a comprehensive review of recent developments on analysis of non-probability survey samples.

Methodological developments for analyzing non-probability samples are built based on several basic assumptions. Let $\{(y_i, \boldsymbol{x}_i), i \in \mathcal{S}_A\}$ be a non-probability sample from the target population $\mathcal{U} = \{1, 2, ..., N\}$, where the sample $\mathcal{S}_A$ contains n_A units. At the heart of these assumptions is the issue on participation in the non-probability sample. While the exact reason(s) why a particular unit appears in the sample may never be known, it is common practice in statistics to use a model, denoted as q, to approximately describe the participation mechanism. Let $R_i = 1$ if $i \in \mathcal{S}_A$ and $R_i = 0$ otherwise, $i = 1, 2, \ldots, N$. Let $\pi_i^A = P(R_i = 1|\boldsymbol{x}_i, y_i) = \pi(\boldsymbol{x}_i, y_i)$ be the *participation probability*. The term *propensity score* was also used in this context by several authors; see, for instance, Chen et al. (2020). The estimated participation probabilities $\hat{\pi}_i^A$ play a crucial role for inference with non-probability survey samples. For instance, the inverse probability weighted (IPW) estimator of μ_y is computed as $\hat{\mu}_{yIPW} = (\hat{N}^A)^{-1} \sum_{i \in \mathcal{S}_A} y_i / \hat{\pi}_i^A$, where $\hat{N}^A = \sum_{i \in \mathcal{S}_A} (\hat{\pi}_i^A)^{-1}$.

5.1 The need for combining information from other sources

The fact that the unknown participation probabilities π_i are defined for $i = 1, 2, \ldots, N$ over the entire target population implies that they cannot be estimated without additional information on the population and assumptions on the participation mechanism. The scenario of the so-called ignorable participation serves as a starting point for methodological developments. More formally, the assumption is stated below.

1. (*Ignorability*) The participation indicator variable R_i and the study variable y_i are independent given the observed $\boldsymbol{x}_i$.

Under the ignorability assumption **A1**, we have $\pi_i^A = P(R_i = 1|\boldsymbol{x}_i, y_i) = P(R_i = 1|\boldsymbol{x}_i) = \pi(\boldsymbol{x}_i)$ which depends only on $\boldsymbol{x}_i$. The need for combining information from other sources becomes apparent once we look into the details of estimation for participation probabilities.

Consider a parametric form $\pi_i^A = \pi(\boldsymbol{x}_i, \boldsymbol{\alpha})$ with unknown model parameters $\boldsymbol{\alpha}$ for the participation probabilities. The full log-likelihood function for $\boldsymbol{\alpha}$ is given by (Chen et al., 2020)

$$\ell(\boldsymbol{\alpha}) = \sum_{i \in \mathcal{S}_A} \log\left(\frac{\pi_i^A}{1 - \pi_i^A}\right) + \sum_{i=1}^{N} \log(1 - \pi_i^A). \tag{16}$$

The likelihood function $\ell(\alpha)$ is not computable for a given α due to the second term in (16) which requires information at the population level.

The non-parametric Nadaraya-Watson kernel regression estimator (Nadaraya, 1964; Watson, 1964) of $\pi_i^A = P(R_i = 1|\boldsymbol{x}_i) = \pi(\boldsymbol{x}_i)$ is given by (Wu, 2022)

$$\tilde{\pi}(\boldsymbol{x}_i) = \frac{\sum_{j=1}^N K_h(\boldsymbol{x}_i - \boldsymbol{x}_j)R_j}{\sum_{j=1}^N K_h(\boldsymbol{x}_i - \boldsymbol{x}_j)} = \frac{\sum_{j\in\mathcal{S}_A} K_h(\boldsymbol{x}_i - \boldsymbol{x}_j)}{\sum_{j=1}^N K_h(\boldsymbol{x}_i - \boldsymbol{x}_j)}, \tag{17}$$

where $K_h(t) = K(t/h)$ is a chosen kernel function with bandwidth h. It is apparent that $\tilde{\pi}(\boldsymbol{x}_i)$ cannot be computed without additional information on $\boldsymbol{x}$ from the target population.

A popular setting for analysis of non-probability survey samples, first used by Rivers (2007) and later by many other authors in recent years, is to assume that there exists a probability survey sample containing the required auxiliary information. There are two datasets,

$$\{(y_i, \boldsymbol{x}_i), i \in \mathcal{S}_A\} \quad \text{and} \quad \{(\boldsymbol{x}_i, d_i^B), i \in \mathcal{S}_B\},$$

where $\mathcal{S}_A$ is the primary non-probability sample with measurements on both the study variable y and auxiliary variables $\boldsymbol{x}$ but with an unknown participation mechanism, and $\mathcal{S}_B$ is an existing reference probability sample with information on the survey weights d_i^B and auxiliary variables $\boldsymbol{x}$ but not on the study variable y. Valid estimation of the population mean μ_y cannot be achieved with any one of the datasets, and the goal of methodological developments is to combine the two data sources for valid and efficient estimation through inverse probability weighting using estimated participation probabilities or model-based prediction methods.

Under the two-sample setting and a parametric model $\pi_i^A = \pi(\boldsymbol{x}_i, \boldsymbol{\alpha})$ for the participation probabilities, Chen et al. (2020) proposed to estimate the model parameters $\boldsymbol{\alpha}$ by maximizing the pseudo log-likelihood function given by

$$\ell^*(\boldsymbol{\alpha}) = \sum_{i\in\mathcal{S}_A} \log\left(\frac{\pi_i^A}{1 - \pi_i^A}\right) + \sum_{i\in\mathcal{S}_B} d_i^B \log(1 - \pi_i^A). \tag{18}$$

The intuitive argument is that the computable pseudo log-likelihood function $\ell^*(\boldsymbol{\alpha})$ given in (18) is an unbiased estimator of the full log-likelihood function $\ell(\boldsymbol{\alpha})$ given in (16), i.e., $E_p\{\ell^*(\boldsymbol{\alpha})\} = \ell(\boldsymbol{\alpha})$, where $E_p(\cdot)$ refers to expectation under the probability sampling design for the reference sample $\mathcal{S}_B$. The validity of the pseudo maximum likelihood method of Chen et al. (2020) can be rigorously argued by noting that the corresponding pseudo score functions are unbiased under the joint randomization of the participation model q and the survey design p for $\mathcal{S}_B$, i.e., $E_{qp}\{\mathbf{U}(\boldsymbol{\alpha}_0)\} = \mathbf{0}$, where

$$\mathbf{U}(\boldsymbol{\alpha}) = \frac{\partial}{\partial\boldsymbol{\alpha}}\ell^*(\boldsymbol{\alpha}) = \sum_{i\in\mathcal{S}_A} \frac{\pi'(\boldsymbol{\alpha})}{\pi(\boldsymbol{x}_i, \boldsymbol{\alpha})\{1 - \pi(\boldsymbol{x}_i, \boldsymbol{\alpha})\}} - \sum_{i\in\mathcal{S}_B} d_i^B \frac{\pi'(\boldsymbol{\alpha})}{1 - \pi(\boldsymbol{x}_i, \boldsymbol{\alpha})},$$

$\pi'(\boldsymbol{\alpha}) = \partial\pi(\boldsymbol{x}_i, \boldsymbol{\alpha})/\partial\boldsymbol{\alpha}$, and $\boldsymbol{\alpha}_0$ is the true value of the model parameters $\boldsymbol{\alpha}$. The pseudo maximum likelihood estimator $\hat{\boldsymbol{\alpha}}$ is obtained by solving $\mathbf{U}(\boldsymbol{\alpha}) = \mathbf{0}$. If the nonparametric kernel regression estimator in the form of (17) is preferred, a computable version is given by (Wu, 2022)

$$\hat{\pi}(\boldsymbol{x}_i) = \frac{\sum_{j\in\mathcal{S}_A} K_h(\boldsymbol{x}_i - \boldsymbol{x}_j)}{\sum_{j\in\mathcal{S}_B} d_i^B K_h(\boldsymbol{x}_i - \boldsymbol{x}_j)}, \quad i \in \mathcal{S}_A. \tag{19}$$

The estimated participation probabilities in the form of $\hat{\pi}_i^A = \pi(\boldsymbol{x}_i, \hat{\boldsymbol{\alpha}})$ or $\hat{\pi}_i^A = \hat{\pi}(\boldsymbol{x}_i)$ for $i \in \mathcal{S}_A$ can then be used to construct valid inverse probability weighted estimators for parameters of the target population such as μ_y.

5.2 Calibration estimation of participation probabilities

The setting used in Section 5.1 for combining information from two sources, a non-probability sample and an existing reference probability sample, has a very restrictive requirement: The full vector of auxiliary variables $\boldsymbol{x}$ is observed in both samples. While the inclusion of all relevant auxiliary variables for modelling participation behaviour or for prediction purposes can be decided at the data collection stage for the non-probability sample, it is often impossible to find an existing probability survey with the same set of auxiliary variables. A calibration-based approach for estimating the participation probabilities offers a potential solution to combining information from multiple existing probability surveys.

Under a logistic regression model q for the participation probabilities with $\pi_i^A = \pi(\boldsymbol{x}_i, \boldsymbol{\alpha}) = 1 - \{1 + \exp(\boldsymbol{x}'\boldsymbol{\alpha})\}^{-1}$, the pseudo score functions become

$$\mathbf{U}(\boldsymbol{\alpha}) = \sum_{i\in\mathcal{S}_A} \boldsymbol{x}_i - \sum_{i\in\mathcal{S}_B} d_i^B \pi(\boldsymbol{x}_i, \boldsymbol{\alpha})\boldsymbol{x}_i.$$

It can be shown that a modified version of $\mathbf{U}(\boldsymbol{\alpha})$ with a user-specified function $\mathbf{h}(\boldsymbol{x}_i, \boldsymbol{\alpha})$ in the form of

$$\mathbf{G}(\boldsymbol{\alpha}) = \sum_{i\in\mathcal{S}_A} \mathbf{h}(\boldsymbol{x}_i, \boldsymbol{\alpha}) - \sum_{i\in\mathcal{S}_B} d_i^B \pi(\boldsymbol{x}_i, \boldsymbol{\alpha})\mathbf{h}(\boldsymbol{x}_i, \boldsymbol{\alpha})$$

remains as a set of unbiased estimating functions, i.e., $E_{qp}\{\mathbf{G}(\boldsymbol{\alpha}_0)\} = \mathbf{0}$. A valid estimator of $\boldsymbol{\alpha}$ can be obtained by solving $\mathbf{G}(\boldsymbol{\alpha}) = \mathbf{0}$. If we choose $\mathbf{h}(\boldsymbol{x}_i, \boldsymbol{\alpha}) = \boldsymbol{x}_i/\pi(\boldsymbol{x}_i, \boldsymbol{\alpha})$, then we have a special form of $\mathbf{G}(\boldsymbol{\alpha})$ denoted as $\mathbf{G}_C(\boldsymbol{\alpha}) = \sum_{i\in\mathcal{S}_A} \boldsymbol{x}_i/\pi(\boldsymbol{x}_i, \boldsymbol{\alpha}) - \sum_{i\in\mathcal{S}_B} d_i^B \boldsymbol{x}_i$. The corresponding estimator $\hat{\boldsymbol{\alpha}}$ which solves $\mathbf{G}_C(\boldsymbol{\alpha}) = \mathbf{0}$ becomes a calibration-based estimator since it is the solution to

$$\sum_{i\in\mathcal{S}_A} \frac{\boldsymbol{x}_i}{\pi(\boldsymbol{x}_i, \boldsymbol{\alpha})} = \sum_{i\in\mathcal{S}_B} d_i^B \boldsymbol{x}_i. \tag{20}$$

The right-hand side of (20) is the estimated population totals using the reference probability sample $\mathcal{S}_B$. The most important implication of the calibration-based approach to estimating $\boldsymbol{\alpha}$ is that the estimated population totals do not have to be from a single reference probability survey sample. It allows combining information from multiple existing probability samples for which certain auxiliary variables are included. The estimator $\hat{\boldsymbol{\alpha}}$ is obtained by solving a modified version of (20) where $\sum_{i\in\mathcal{S}_B} d_i^B \boldsymbol{x}_i$ is replaced by $\hat{T}_{\mathbf{x}}$ with components from any reliable sources.

In practice, the calibration-based approach requires an efficient computational procedure to solve $\mathbf{G}_C(\boldsymbol{\alpha}) = \sum_{i\in\mathcal{S}_A} \boldsymbol{x}_i/\pi(\boldsymbol{x}_i, \boldsymbol{\alpha}) - \hat{T}_{\mathbf{x}} = \mathbf{0}$. The Newton-Raphson iterative procedure can be used to find the solution $\hat{\boldsymbol{\alpha}}$:

$$\boldsymbol{\alpha}^{(m+1)} = \boldsymbol{\alpha}^{(m)} - \{\mathbf{H}(\boldsymbol{\alpha}^{(m)})\}^{-1}\mathbf{G}_C(\boldsymbol{\alpha}^{(m)}),$$

where $\mathbf{H}(\boldsymbol{\alpha}) = \partial \mathbf{G}_C(\boldsymbol{\alpha})/\partial \boldsymbol{\alpha}$. The parametric form $\pi(\boldsymbol{x}_i, \boldsymbol{\alpha})$ often falls within the framework of generalized linear models in the form of $\pi(\boldsymbol{x}_i, \boldsymbol{\alpha}) = g(\boldsymbol{x}'\boldsymbol{\alpha})$, where $g(\cdot)$ is a monotone increasing inverse link function. Let $k(t) = dg(t)/dt$. We have

$$\mathbf{H}(\boldsymbol{\alpha}) = -\sum_{i\in\mathcal{S}_A} \frac{k(\boldsymbol{x}'\boldsymbol{\alpha})}{[\pi(\boldsymbol{x}_i, \boldsymbol{\alpha})]^2}\boldsymbol{x}_i\boldsymbol{x}',$$

which is a negative definite matrix. The Newton-Raphson iterative procedure with the initial value $\boldsymbol{\alpha}^{(0)} = \mathbf{0}$ works very well for finding the solution $\hat{\boldsymbol{\alpha}}$ to $\mathbf{G}_C(\boldsymbol{\alpha}) = \mathbf{0}$.

5.3 Model-based prediction: Mass imputation

The two-sample setting described in Section 5.1 provides valid approaches to estimating the participation probabilities. It turns out that it also offers an opportunity of combining information from two sources for model-based prediction approach to parameters of the target population. Note that the reference probability sample dataset $\{(\boldsymbol{x}_i, d_i^B), i \in \mathcal{S}_B\}$ contains survey weights d_i^B. The dilemma is that the study variable y is not included, making it impossible to directly estimate population parameters related to y using the reference dataset.

Rivers (2007) was the first to explore the idea of treating y_i as missing for all i in the reference probability sample $\mathcal{S}_B$, and using $(y_i, \boldsymbol{x}_i)$ for i in the non-probability sample $\mathcal{S}_A$ to build an imputation model. He termed the approach as "sample matching", which has been developed into a general approach as "mass imputation" in recent years (Kim et al., 2021).

The mass imputation approach to combining information from the two samples provides valid estimation under the assumed imputation model and the so called transportability assumption. Recall that the ignorability assumption **A1** leads to $\pi_i^A = P(R_i = 1|\boldsymbol{x}_i, y_i) = P(R_i = 1|\boldsymbol{x}_i) = \pi(\boldsymbol{x}_i)$, i.e., the participation behaviour is characterized by the auxiliary variables included in the two sample datasets. Noting that imputation is prediction, and an imputation model ξ relates y to $\boldsymbol{x}_i$ through the first moment $E_\xi(y_i|\boldsymbol{x}_i)$. Under the assumption **A1**, we have

$$E_\xi(y_i|\boldsymbol{x}_i, R_i = 1) = E_\xi(y_i|\boldsymbol{x}_i), \tag{21}$$

which seems to indicate that the imputation model ξ can be built using the non-probability sample dataset $\{(y_i, \boldsymbol{x}_i), i \in \mathcal{S}_A\}$ (with $R_i = 1$), and the fitted model is then used (i.e., transported) to impute y_i for $i \in \mathcal{S}_B$. Unfortunately, the transportability from the fitted model using $\mathcal{S}_A$ to imputation using $\mathcal{S}_B$ requires another crucial assumption.

A2. (*Positivity*) The participation probabilities satisfy $\pi_i^A = P(R_i = 1|\boldsymbol{x}_i, y_i) > 0$ for $i = 1, 2, \ldots, N$.

The positivity assumption is a hidden part required for (21) since the relation does not extend to units with $P(R_i = 1) = 0$.

Rivers (2007) considered nearest neighbour imputation by matching the missing y_i for $i \in \mathcal{S}_B$ to an observed y_k with $k \in \mathcal{S}_A$. Kim et al. (2021) used a semiparametric model where $E_\xi(y_i|\boldsymbol{x}_i) = \mu(\boldsymbol{x}_i, \boldsymbol{\beta})$. The estimator $\hat{\boldsymbol{\beta}}$ for the model parameters is obtained by using the sample $\mathcal{S}_A$. The imputed values are simply computed as $y_i^* = \mu(\boldsymbol{x}_i, \hat{\boldsymbol{\beta}})$ for $i \in \mathcal{S}_B$. The mass imputation estimator of μ_y is then given by $\hat{\mu}_{yMI} = (\hat{N}^B)^{-1} \sum_{i \in \mathcal{S}_B} d_i^B y_i^*$, where $\hat{N}^B = \sum_{i \in \mathcal{S}_B} d_i^B$. The estimator $\hat{\mu}_{yMI}$ is consistent under the joint randomization of the imputation model and the sampling design for the reference probability sample. A valid variance estimator can also be constructed under the same framework Kim et al. (2021).

6 Mitigating under-coverage biases

Estimation biases due to under-coverage of the target population are a common but notoriously difficult problem to deal with in practice. In the context of non-probability survey samples, under-coverage issues are related to the violation of the positivity assumption **A2** stated in Section 5.3. Following the notation used in Section 5, let $\mathcal{U}_1 = \{i | i \in \mathcal{U} \text{ and } \pi_i^A > 0\}$ and $\mathcal{U}_0 = \{i | i \in \mathcal{U} \text{ and } \pi_i^A = 0\}$. Let the sizes of $\mathcal{U}_1$ and $\mathcal{U}_0$ be N_1 and N_0, respectively. We have $\mathcal{U} = \mathcal{U}_1 \cup \mathcal{U}_0$ and $N = N_1 + N_0$.

The $\mathcal{U}_1$ is the subpopulation represented by the non-probability sample $\mathcal{S}_A$, since every unit in $\mathcal{U}_1$ has a non-zero probability to participate in $\mathcal{S}_A$. The $\mathcal{U}_0$ is the subpopulation with no information being collected and hence is the uncovered part of the target population. The severity of under-coverage problems depends on the size N_0 and the difference between $\mathcal{U}_1$ and $\mathcal{U}_0$. This can be illustrated through the estimation of μ_y of the target population $\mathcal{U}$ using a simple random sample from $\mathcal{U}_1$. Let the means of the two subpopulations $\mathcal{U}_1$ and $\mathcal{U}_0$ be μ_{y1} and μ_{y0}, respectively. We have $\mu_y = W_1\mu_{y1} + W_0\mu_{y0}$, where

$W_1 = N_1/N$ and $W_0 = N_0/N$. Let $\bar{y}$ be the sample mean of a simple random sample from $\mathcal{U}_1$. It follows that $E(\bar{y}) = \mu_{y1}$. The bias due to under-coverage by using $\bar{y}$ as an estimator for the mean of the target population is given by

$$\text{Bias}(\bar{y}) = E(\bar{y}) - \mu_y = W_0(\mu_{y1} - \mu_{y0}).$$

The bias would be negligible if W_0 (or equivalently N_0) is small and/or the difference between μ_{y1} and μ_{y0} is small. Unfortunately, the under-covered part of the population often possesses very different characteristics among its units, leading to non-negligible under-coverage biases.

Under-coverage biases may be mitigated through combining information from multiple sources and suitable estimation strategies. We discuss two approaches as described in Chen et al. (2023) to dealing with under-coverage problems, both in the context of non-probability survey samples with the same notation used in Section 5.

6.1 The calibrated IPW approach

The calibration-based approach to estimating participation probabilities described in Section 5.2 is a useful tool for mitigating potential biases due to under-coverage. Let $\hat{\pi}_i^A = \pi(\boldsymbol{x}_i, \hat{\boldsymbol{\alpha}})$ be a parametric estimator of π_i^A with the model parameters $\boldsymbol{\alpha}$ estimated through the set of calibration equations given by (20). It is assumed that $\boldsymbol{x}_i$ contain the constant 1 as the first component and the first equation in (20) leads to

$$\hat{N}^A = \sum_{i \in \mathcal{S}_A} \{\pi(\boldsymbol{x}_i, \hat{\boldsymbol{\alpha}})\}^{-1} = \sum_{i \in \mathcal{S}_B} d_i^B = \hat{N}^B.$$

The resulting $\hat{\mu}_{yIPW} = (\hat{N}^A)^{-1} \sum_{i \in \mathcal{S}_A} y_i/\hat{\pi}_i^A$ is termed as the calibrated IPW estimator (Chen et al., 2020; Rao, 2021; Wu, 2022).

When the positivity assumption **A2** is violated, the IPW estimator is generally not a valid estimator for the mean of the target population due to under-coverage issues. However, the calibrated IPW estimator remains as a valid prediction-based estimator if $(y_i|\boldsymbol{x}_i)$ follows a linear regression model. More specifically, suppose that $E_\xi(y_i|\boldsymbol{x}_i) = \boldsymbol{x}'\boldsymbol{\beta}$, $i = 1, 2, \ldots, N$. It follows that $E_\xi(\mu_y) = N^{-1}T'\boldsymbol{\beta}$ conditional on the given $(\boldsymbol{x}_1, \boldsymbol{x}_2, \ldots, \boldsymbol{x}_N)$. We have

$$E_p E_\xi\left\{\frac{1}{\hat{N}^A}\sum_{i \in \mathcal{S}_A}\frac{y_i}{\hat{\pi}_i}\right\} = E_p\left\{\frac{1}{\hat{N}^A}\sum_{i \in \mathcal{S}_A}\frac{\boldsymbol{x}_i}{\hat{\pi}_i}\right\}'\boldsymbol{\beta} = E_p\left\{\frac{1}{\hat{N}^B}\sum_{i \in \mathcal{S}_B} d_i^B \boldsymbol{x}_i\right\}'\boldsymbol{\beta} \doteq \frac{1}{N}T_{\mathbf{x}}'\boldsymbol{\beta}.$$

In other words, we have $E_p E_\xi(\hat{\mu}_{yIPW} - \mu_y) \doteq 0$. Another important feature of the calibrated IPW estimator is that it is a valid model-based prediction estimator without involving the estimation of the parameters $\boldsymbol{\beta}$ in the outcome regression model $E_\xi(y_i|\boldsymbol{x}_i) = \boldsymbol{x}'\boldsymbol{\beta}$.

The right-hand side of (20) may be replaced by estimated population controls $\hat{T}_{\mathbf{x}}$ from multiple reliable sources such as existing large scale probability survey samples or census data. The calibrated IPW estimator $\hat{\mu}_{yIPW} = (\hat{N}^A)^{-1}\sum_{i\in\mathcal{S}_A} y_i/\pi(\boldsymbol{x}_i, \hat{\boldsymbol{\alpha}})$ with $\hat{\boldsymbol{\alpha}}$ solving $\sum_{i\in\mathcal{S}_A} \boldsymbol{x}_i/\pi(\boldsymbol{x}_i, \boldsymbol{\alpha}) = \hat{T}_{\mathbf{x}}$ remains a valid prediction estimator of μ_y if the target population follows a linear model, regardless of the positivity assumption.

A valid model-based prediction estimator of μ_y may also be constructed when the outcome regression model $E_\xi(y_i|\boldsymbol{x}_i) = \mu(\boldsymbol{x}_i, \boldsymbol{\beta})$ is nonlinear. It involves obtaining the initial weights $w_i^o = \{\pi(\boldsymbol{x}_i, \hat{\boldsymbol{\alpha}})\}^{-1}$ through an existing procedure in estimating $\boldsymbol{\alpha}$, and then computing the model-calibration weights w_i^C, $i \in \mathcal{S}_A$ by minimizing a chosen distance measure and using the model-calibration equations

$$\sum_{i\in\mathcal{S}_A} w_i^C = \hat{N}^B \quad \text{and} \quad \sum_{i\in\mathcal{S}_A} w_i^C \mu(\boldsymbol{x}_i, \hat{\boldsymbol{\beta}}) = \sum_{i\in\mathcal{S}_B} d_i^B \mu(\boldsymbol{x}_i, \hat{\boldsymbol{\beta}}),$$

where $\hat{\boldsymbol{\beta}}$ is obtained by fitting the outcome regression model to $\{(y_i, \boldsymbol{x}_i), i \in \mathcal{S}_A\}$. The model-calibrated weights w_i^C, $i \in \mathcal{S}_A$ are then used to construct estimators of parameters of the target population. The model-calibration estimators of μ_y in this scenario have a potential risk of extrapolation in computing $\mu(\boldsymbol{x}_i, \hat{\boldsymbol{\beta}})$ for $i \in \mathcal{S}_B$ using $\hat{\boldsymbol{\beta}}$ obtained from $\mathcal{S}_A$, which could result in an estimator with non-negligible bias. See a related discussion in the next section on the split population approach.

6.2 The split population approach

The root for under-coverage problems is the lack of information over certain part of the target population. A complete solution would require the identification of the subpopulation which is uncovered by the current sample data and additional information over the uncovered subpopulation. Under the two sample setting with the availability of a reference probability sample from the target population with information on auxiliary variables, the split population approach of Chen et al. (2023) offers a potential path to finding a solution.

Chen et al. (2023) introduced the concept of accessibility function for sample participation to define the two convex hulls $\mathcal{H}_1$ and $\mathcal{H}_0$, formed respectively by auxiliary variables in the two subpopulations, i.e., $\{\boldsymbol{x}_i, i \in \mathcal{U}_1\}$ and $\{\boldsymbol{x}_i, i \in \mathcal{U}_0\}$. The convex hull formulation may also be motivated by the concerns of extrapolation when using the fitted model from $\{(y_i, \boldsymbol{x}_i), i \in \mathcal{S}_A\}$ to do prediction for $\boldsymbol{x}_i$ with $i \in \mathcal{S}_B$. If $\mathcal{S}_B$ represents the entire target population, the convex hull $\mathcal{H}_B$ formed by $\{\boldsymbol{x}_i, i \in \mathcal{S}_B\}$ would approximate the convex hull $\mathcal{H}$ formed by $\{\boldsymbol{x}_i, i \in \mathcal{U}\}$, while the convex hull $\mathcal{H}_A$ formed by $\{\boldsymbol{x}_i, i \in \mathcal{S}_A\}$ would only approximate $\mathcal{H}_1$. It follows that we may break $\mathcal{S}_B$ into $\mathcal{S}_B = \mathcal{S}_{B,1} \cup \mathcal{S}_{B,0}$ through the convex hull formulation to identify the two subsamples $\mathcal{S}_{B,1}$ and $\mathcal{S}_{B,0}$ which represent $\mathcal{U}_1$ and $\mathcal{U}_0$, respectively.

The partition of $\mathcal{S}_B$ into $\mathcal{S}_B = \mathcal{S}_{B,1} \cup \mathcal{S}_{B,0}$ can be carried out through a simple procedure of convex hull membership identification. For each $j \in \mathcal{S}_B$, the verification of $\boldsymbol{x}_j \in \mathcal{H}_A$ (which is equivalent to verifying $j \in \mathcal{S}_{B,1}$) amounts to checking if there exists a sequence of constants $a_i \geq 0$ for $i \in \mathcal{S}_A$ such that

$$\sum_{i \in \mathcal{S}_A} a_i = 1 \quad \text{and} \quad \sum_{i \in \mathcal{S}_A} a_i \boldsymbol{x}_i = \boldsymbol{x}_j.$$

There are two possible scenarios after the convex hull partition of $\mathcal{S}_B$ into $\mathcal{S}_{B,1}$ and $\mathcal{S}_{B,0}$. The first is that the subsample $\mathcal{S}_{B,0}$ turns out to be empty. This is great news since the problem of extrapolation for model-based prediction is likely not an issue, and approaches such as the calibrated IPW estimators and mass imputation estimators can mitigate under-coverage biases. The second scenario is that the subsample $\mathcal{S}_{B,0}$ contains sizeable number of units. In principle, one would need to take a sub-subsample over $\mathcal{S}_{B,0}$ to obtain measurements on the study variable y in order to fully remove potential estimation biases due to under-coverage. What types of existing data sources could be used for mitigating under-coverage biases and under what conditions it can be achieved remain as a research topic to be pursued in the future.

7 Additional remarks

Data integration is the new terminology that has been frequently used in recent literature on combining information from multiple sources. Data from existing sources come with different quality at different time point for different purposes initially. Just because multiple data sources are available and can be combined does not necessarily mean that better estimation and inferential results are going to appear naturally. It is important to address data integration problems under a suitable statistical framework with guiding principles. Two of these principles are illustrated throughout this chapter under different practical scenarios on integrating data from multiple sources, namely, *validity* and *efficiency*. This is in line with the foundation of traditional statistics where theory is developed as precise as possible under assumed conditions while in applications methods are chosen to reflect the actual conditions as close as possible.

Modern statistical and machine learning tools may have an active role to play in data integration problems, especially when there are large and unstructured data sources available, such as streaming data and satellite image data. This is a wide open area that could see major developments in coming years.

Acknowledgements

This work was supported by a grant from the Natural Sciences and Engineering Research Council of Canada.

References

Chen, J., Sitter, R.R., 1999. A pseudo empirical likelihood approach to the effective use of auxiliary information in complex surveys. Stat. Sin. 9, 385–406.

Chen, J., Sitter, R.R., Wu, C., 2002. Using empirical likelihood methods to obtain range restricted weights in regression estimators for surveys. Biometrika 89, 230–237.

Chen, J., Wu, C., 2002. Estimation of distribution function and quantiles using the model-calibrated pseudo empirical likelihood method. Stat. Sin. 12, 1223–1239.

Chen, Y., Li, P., Wu, C., 2020. Doubly robust inference with non-probability survey samples. J. Am. Stat. Assoc. 115, 2011–2021.

Chen, Y., Li, P., Wu, C., 2023. Dealing with undercoverage for non-probability survey samples. Surv. Methodol. 49, 497–515.

Deming, W.E., Stephen, F.F., 1940. On a least squares adjustment of a sampled frequency table when the expected marginal totals are known. Ann. Math. Stat. 11, 427–444.

Deville, J.-C., Särndal, C.E., 1992. Calibration estimators in survey sampling. J. Am. Stat. Assoc. 87, 376–382.

Fuller, W.A., 2009. Sampling Statistics. John Wiley& Sons, Inc, Hoboken, New Jersey.

Hartley, H.O., 1962. Multiple frame surveys. In: Proceedings of the Social Statistics Section, American Statistical Association, pp. 203–206.

Kim, J.K., Park, S., Chen, Y., Wu, C., 2021. Combining non-probability and probability survey samples through mass imputation. J. R. Stat. Soc. Ser. A 184, 941–963.

Kim, J.K., Rao, J.N.K., 2012. Combining data from two independent surveys: a model-assisted approach. Biometrika 99, 85–100.

Mecatti, F., 2007. A single frame multiplicity estimator for multiple frame survey. Surv. Methodol. 33, 151–157.

Nadaraya, E.A., 1964. On estimating regression. Theory Proba. Its Appl. 9, 141–142.

Rao, J.N.K., 2021. On making valid inferences by integrating data from surveys and other sources. Sankhya B 83, 242–272.

Rao, J.N.K., Wu, C., 2010. Pseudo-empirical likelihood inference for multiple frame surveys. J. Am. Stat. Assoc. 105, 1494–1503.

Renssen, R.H., Nieuwenbroek, N.J., 1997. Aligning estimates for common variables in two or more sample surveys. J. Am. Stat. Assoc. 92, 368–374.

Rivers, D., 2007. Sampling for web surveys. In: Proceedings of the Survey Research Methods Section, Joint Statistical Meetings, American Statistical Association, Alexandria, VA, pp. 1–26.

Rosenbaum, P.R., Rubin, D.B., 1983. The central role of the propensity score in observational studies for causal effects. Biometrika 70, 41–55.

Sitter, R.R., Wu, C., 2002. Efficient estimation of quadratic finite population functions in the presence of auxiliary information. J. Am. Stat. Assoc. 97, 535–543.

Watson, G.S., 1964. Smooth regression analysis. Sankhya A 26, 359–372.

Wu, C., 2003. Optimal calibration estimators in survey sampling. Biometrika 90, 937–951.

Wu, C., 2004a. Combining information from multiple surveys through the empirical likelihood method. Can. J. Stat. 32, 15–26.

Wu, C., 2004b. Some algorithmic aspects of the empirical likelihood method in survey sampling. Stat. Sin. 14, 1057–1067.

Wu, C., 2005. Algorithms and R codes for the pseudo empirical likelihood method in survey sampling. Survey Methodology 31, 239–243.

Wu, C., 2022. Statistical inference with non-probability survey samples (with discussion). Surv. Methodol. 48, 283–311.

Wu, C., Lu, W.W., 2016. Calibration weighting methods for complex surveys. Int. Stat. Rev. 84, 79–98.

Wu, C., Luan, Y., 2003. Optimal calibration estimators under two-phase sampling. J. Off. Stat. 19, 119–131.

Wu, C., Rao, J.N.K., 2006. Pseudo empirical likelihood ration confidence intervals for complex surveys. Can. J. Stat. 34, 359–375.

Wu, C., Sitter, R.R., 2001. A model-calibration approach to using complete auxiliary information from survey data. J. Am. Stat. Assoc. 96, 185–193.

Wu, C., Thompson, M.E., 2020. Sampling Theory and Practice. Springer, Cham.

Zhang, S., Han, P., Wu, C., 2023. Calibration techniques encompassing survey sampling, missing data analysis and causal inference. Int. Stat. Rev. 91, 165–192.

Zieschang, K.D., 1990. Sample weighting methods and estimation of totals in the consumer expenditure survey. J. Am. Stat. Assoc. 85, 986–1001.

Chapter 10

The maturity structure of public debt: A granular approach using Indian data☆

Piyali Das[a,1], Chetan Ghate[b,c,*,2], and Subhadeep Halder[b,d,3]

[a]*National Institute of Public Finance and Policy, New Delhi, India,* [b]*Indian Statistical Institute, Delhi, India,* [c]*Centre for Applied Macroeconomic Analysis (CAMA), Australian National University (ANU), Canberra, Australia,* [d]*New York University, Abu Dhabi, United Arab Emirates*

**Corresponding author. e-mail address: cghate@isid.ac.in*

Chapter Outline

☆ We thank participants in the IEG Brown Bag Series, the CAFRAL Annual Conference, and the 62nd Meeting of the Philippine Economic Society (PES) for comments. We also thank the Editor, Arni Rao, and an anonymous referee for useful comments.

[1] Assistant Professor, National Institute of Public Finance and Policy (NIPFP), New Delhi, India. E-mail: piyali.das@nipfp.org.in

[2] Professor of Economics, Economics and Planning Unit, Indian Statistical Institute-Delhi Centre, India. Centre for Applied Macroeconomic Analysis (CAMA), Australia National University (ANU), Canberra, Australia. E-mail: cghate@isid.ac.in

[3] Ph.D. Student, Indian Statistical Institute-Delhi Centre, India and NYU-Abu Dhabi, UAE. E-mail: sh7991@nyu.edu

Handbook of Statistics, Vol. 53. https://doi.org/10.1016/bs.host.2025.04.008

Abstract

The traditional analysis of public debt sustainability uses the face value of securities and the number of remaining coupon payments to calculate aggregate debt. The aggregate approach ignores the granular security-wise breakup of public debt, the maturity structure of these securities, and the information about the path of future expected interest rates contained in bond prices. Hall and Sargent provide a framework to track the evolution of aggregate US public debt using the market value of debt. In this approach, the maturity structure of debt is explicitly tracked, and its role in the evolution of aggregate debt can be quantified. In this paper, we illustrate how the Hall-Sargent approach works through a series of examples. We then apply the Hall-Sargent approach using a novel dataset of outstanding Indian Centre government securities from 1999–2022. We show that the nominal return on marketable debt accounts for 55.8 % of the approximately 30 % of the change in aggregate debt over 1999–2022, a substantial amount. We show that the average maturity of debt using the Hall-Sargent approach is substantially lower than the average maturity calculated using the traditional approach. Our analysis builds on the application of the Hall-Sargent approach to Indian public debt analysis following earlier works of Das and Ghate. We discuss measures that can raise institutional support for fiscal capacity, and the data requirements needed to support sovereign debt analysis.

1 Introduction

After the COVID-19 pandemic (2020–2022), both advanced and emerging market economies have been grappling with high debt-to-GDP ratios. Global debt-to-GDP ratio reached around 125 % in 2021, a significant increase compared to pre-pandemic levels (IMF 2021). Emerging market economies like India also experienced a surge in debt primarily driven by increased borrowing to address the economic challenges posed by the pandemic. The strong macroeconomic response to COVID-19 involved substantial increases in the fiscal deficit for many economies as well as in the expansion of central bank's balance sheets.

It is well recognized that the lack of fiscal and monetary coordination leads to adverse macroeconomic outcomes.[1] The rise in sovereign debt however has led to policymakers and academics to reach very difficult conclusions about the role of fiscal policy (Blanchard, 2022). As noted by Blanchard, one aspect of the macro-stabilization debate has focused on the need for a steady decrease in debt through urgent fiscal consolidation. Another aspect of the debate has argued stabilizing debt at higher levels as an adequate policy goal. Even more radical solutions have been proposed such as the cancellation of debt held by central banks (Blanchard, 2022, page 1).

A missing aspect of this debate is *how* the sovereign debt burden should be measured in the first place. Traditional approaches report debt/GDP using the *face value* of debt, which uses information on the face value of securities and the number of remaining coupon payments. This approach ignores information

[1] See (Leeper, 1991) and (Bianchi et al., 2020).

about various economic conditions and the path of future expected interest rates contained in bond prices. Sovereign debt typically consists of debt of different maturities. There is short-term debt that is due within a year; debt that has a maturity between 2–10 years; and longer maturity debt that has a maturity greater than 10 years. Aggregate debt, therefore, can be "broken up" into debt with varying maturities. The value of a bond portfolio with different maturity debt changes due to various internal and external market forces, which tend to impact bond prices. Consequently, investors may experience capital gains or losses in each period as bond prices fluctuate. Thus, it is entirely possible for a bond with a longer tenure and higher face value to have a lower market value if it is not preferred by investors, potentially resulting in a capital loss. The standard calculation of average maturity does not capture bond price fluctuations in the secondary market, even for investor-held bonds. Additionally, a security-level analysis provides the flexibility of analyses at different maturity tranches.

Breaking away from the traditional approach, Hall and Sargent (see Hall and Sargent, 1997, Hall and Sargent, 2011) tracked the evolution of aggregate US public debt using the *market value* of debt. In this approach, the maturity structure of debt is explicitly tracked, and its role in the evolution of aggregate debt can be quantified. In Hall and Sargent(HS hereafter), each security is viewed as a zero-coupon bond, which is then priced using the yield-to-maturity (YTM) data to arrive at the market value of the security. Das and Ghate (2022) apply the HS approach to Indian public debt using detailed security-level data from 2000 to 2018. To the best of our knowledge, Das and Ghate (2022) is the first application of the HS approach to understand sovereign debt dynamics in a large emerging market economy. As shown by Das and Ghate (2022), unpacking aggregate debt into its constituent maturities leads to a more precise understanding of how the nominal interest rate, inflation, and growth rate affect aggregate debt dynamics, and the quantitative role different maturity buckets play in its evolution.[2]

The HS approach also changes how the average maturity of outstanding debt is calculated. This has important implications for roll-over risk. The traditional approach called *residual maturity*, which is reported in the Status Papers published by the Ministry of Finance, Government of India, measures average maturity using the information on the *face value* of securities and the number of remaining payments. The HS approach uses information on the market value of debt to create the weights in the calculation of the weighted average of maturities. Any change in the interest rate will be reflected in the market value of debt and **not** in the face value as the market value takes into account the price of the security which is a function of the interest rate. The

[2] Other studies that have used the (Hall and Sargent, 2011) accounting framework to look at public debt evolution include (Equiza-Goñi, 2016; Hilscher et al., 2022; Cochrane, 2022; De-Graeve and Mazzolini, 2023), and (Jiang et al., 2024).

price of security captures a host of factors that become important in analyzing the capital gains and losses that the government makes. Moreover, given that most of the securities around 95 % according to the Status paper.

Our paper makes several contributions.

First, we highlight that while data on aggregate debt is useful in providing a broad snapshot of the borrowing position and needs of the fiscal authority, it does not capture the granular security level break-up of sovereign debt. The underlying securities are important as they help track various aspects of different securities such as their risk and liquidity profiles, and the preferences of investors among a host of other factors that could drive the aggregate change in debt.

Second, as shown in Das and Ghate (2022), the HS approach leads to a *lower* average maturity of public debt, suggesting more roll-over risk compared to standard calculations of the average maturity using the face value of debt reported in the Status Papers. We propose that considering the market value of bonds is crucial when calculating a debt's average maturity. This underscores the importance of incorporating market values in average maturity calculations to provide a more accurate representation of the maturity profile of public debt and potential returns and prospects for roll-over risk in sovereign debt.

Third, our paper is related to a growing literature on state capacity and economic development (Besley and Persson, 2009; Johnson and Koyama, 2019). This literature argues that the role of the state is not just limited to security and stability, but also to the provision of market supporting institutions. State capacity is typically defined through three criterion: administrative capacity, legal capacity, and fiscal capacity. Since revenue generating capacity is usually accepted as the key to economic development, having several measures of fiscal capacity which includes public debt sustainability, the average maturity of debt, duration, and the drivers of public debt, helps with the understanding of state capacity.

Our paper outlines the data requirements for conducting an analysis of the drivers of public debt, and its stability. We suggest that the Treasury (Ministry of Finance in the Government of India), in consultation with the Central Bank (Reserve Bank of India), provide security level data at regular intervals as and when the securities are issued. There should be information about the holder of a particular bond, albeit at an aggregate sectoral level. Such information will reveal the preference for different securities across the maturity spectrum for different market participants. Information about the auction along with the identifiers of securities should also be provided. The government should make a dedicated portal to make such information available. This will make it convenient for research purposes, as well as to get a complete overview of the borrowing environment for the government. An independent Fiscal Council (if it is not present already) can be entrusted with such capacity-building measures. For an inflation targeting nation, this can simultaneously help with implementing flexible inflation targeting and sovereign debt management and thereby lead to better macroeconomic coordination and policy outcomes.

The appeal of the HS methodology lies in the fact that it deviates from the official (government's) calculations and this is true across economies. The official approach involves considering the nominal value required to roll-over/ retire the debt in upcoming period(s). By considering the *market value* of debt, the HS presents a viable alternative, one that is rooted in the public finance/ macroeconomics literature/discourse wherein debt is measured at its market value and interest payments are holding period returns.[3] Like any other methodology, however the HS approach has some limitations. While it focuses on the interest burden across different maturity tranches and helps understand how this affects the debt dynamics, it takes a careful researcher to see that there are other important aspects like primary surpluses (in other words, tax and spending policies), inflationary expectations, the maturity structure of debt, and monetary and fiscal policy interactions that affect debt dynamics. For emerging and less developed economies with less developed financial markets wherein securities of different maturities may not be issued (or liquid), getting complete yield data across the yield curve is a challenge. Since the HS methodology relies heavily on yield to maturity data, this can pose a problem in the analysis. Besides, the HS method is an accounting method. The constituent factors affecting the debt-GDP dynamics can therefore at best lend itself to a correlation analysis and make it difficult to extract any causal relationship.

2 An overview of Indian public debt and it's maturity profile

Fig. 1 plots the evolution of the debt-GDP ratio for India for the general government (centre and state; dotted line) and for the Centre government (solid blue line) from 1984 to 2024.[4] The trend increases broadly. The latest spike in sovereign debt after 2020 is due to the large borrowing undertaken by the government during the COVID-19 crisis.

The overall fiscal capacity of a government depends on the status of its borrowing needs and the cost of servicing the borrowings. Fig. 2 plots the evolution of two major deficit indicators of the Centre government: the fiscal deficit and the primary deficit, both as a percentage of GDP. The overall trend for both series are quite volatile with clear spikes after the global financial crisis of 2007–09 and the COVID-19 pandemic in 2020. Though there was a downward trend in both the series after the 1990s, which became even more pronounced after the passage of the Fiscal Responsibility and Budget Management (FRBM) Act in 2003, there has been an upward trend after the Global Financial Crisis (GFC) (2007–2009), highlighting the elevated fiscal

[3] The following podcast discusses the related issues in greater detail: https://www.mercatus.org/macro-musings/george-hall-history-us-national-debt-and-government-financing.

[4] General debt is the sum of Centre (Union) and States debt less loans and advances from the Union to the States and is not adjusted for 14-day treasury bills due to the lack of availability of granular data from source.

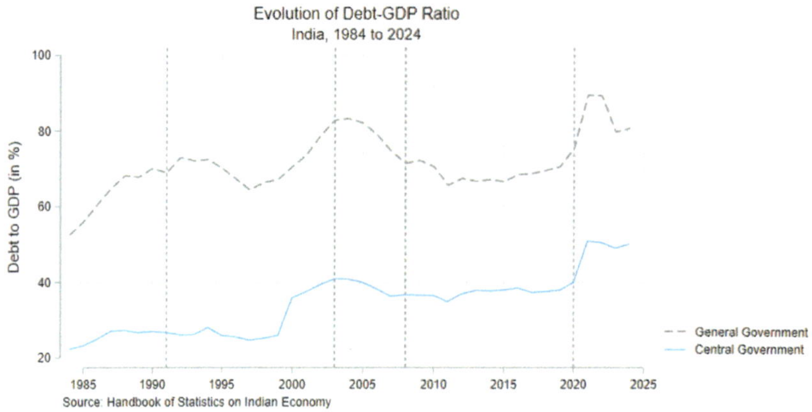

FIG. 1 Evolution of debt-GDP ratio, India. Note: General government consists of both the center (or union) as well as the state governments. The four vertical dotted lines denote the following episodes: the Liberalization Reforms (1991), the enactment of the FRBM Act (2003), the Global Financial Crisis (2008), and COVID-19 (2020). While the GFC is between 2007–2009, we take the mid-point, 2008. *The data is from the Handbook of Statistics on Indian Economy from 1984 to 2024.*

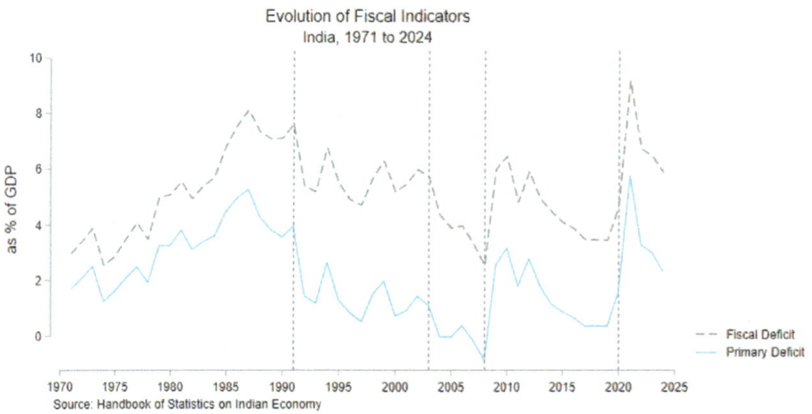

FIG. 2 Evolution of fiscal and primary deficit (as % of GDP) for centre government. Note: The four vertical dotted lines denote the following episodes: the Liberalization Reforms (1991), the enactment of the FRBM Act (2003), the Global Financial Crisis (2008), and COVID-19 (2020). For the Global Financial Crisis. While the GFC is between 2007–2009, we take the mid-point, 2008. *Data is for each financial year from 1971 to 2024 and is from the Handbook of Statistics on Indian Economy.*

requirements of the government and therefore the need for further borrowings in the last decade and a half. Ghate and Das (2024) show that the interest rate cost of servicing public debt (as a percentage of gross tax revenues) in India is about 25 % in 2024, a substantial amount. Fig. 3 plots the time path of interest payments to revenue receipts between 1983 and 2024 highlighting the elevated interest costs in the past two decades.

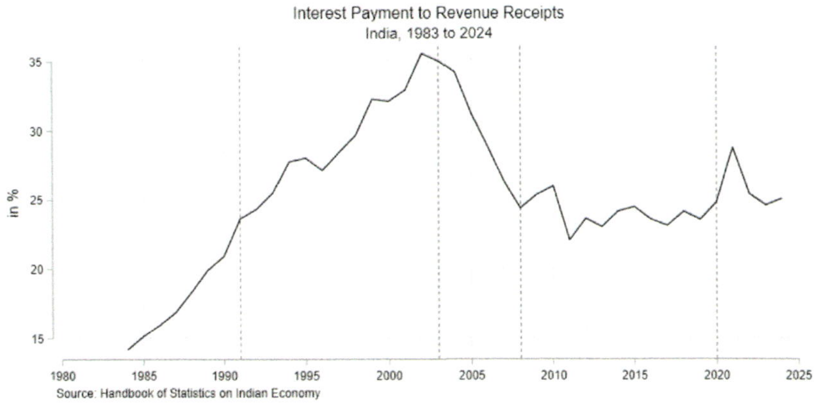

FIG. 3 Cost of servicing public debt. Note: The four vertical dotted lines denote the following episodes: the Liberalization Reforms (1991), the enactment of the FRBM Act (2003), the Global Financial Crisis (2008), and COVID-19 (2020). While the GFC is between 2007–2009, we take the mid-point, 2008. *The data is for each financial year from 1983 to 2023 and is taken from the Handbook of Statistics on Indian Economy.*

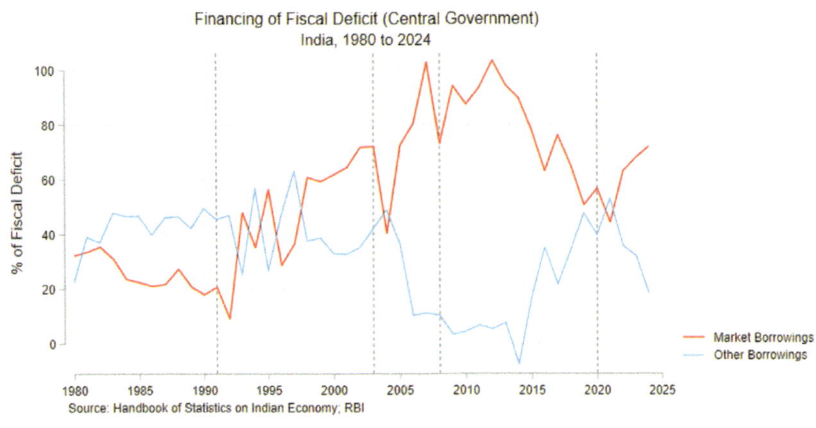

FIG. 4 Financing of gross fiscal deficit by Central Government. Note: The four vertical dotted lines denote the following episodes: the Liberalization Reforms (1991), the enactment of the FRBM Act (2003), the Global Financial Crisis (2008), and COVID-19 (2020).. While the GFC is between 2007–2009, we take the mid-point, 2008. *Data is for each financial year from 1980 to 2024 and is from the Handbook of Statistics on Indian Economy.*

How does the government of India finance its borrowing? Fig. 4 plots the time series evolution of the two major sources of borrowing: market borrowing (red solid line) and other forms of borrowings (blue solid line) which include small savings, the state provident fund, special deposits, reserve funds, 364-day treasury bills and the net impact of switching off of securities, from 1980 to 2022. The financing of fiscal deficit is also possible through external finance but its share is not more than 1 % of total financing (see RBI). As evident from the figure, the Indian government has financed more than 60 % of its total fiscal deficit from market borrowings, while the share of all other forms of borrowing has fallen below 20 % in the sample. This highlights the importance of market borrowing in meeting the borrowing needs of the government.

Fig. 5 plots the maturity profile of outstanding Centre government securities as a percentage of total outstanding securities up to 2022.[5] For each year we computed the share of securities within a particular maturity bracket as a proportion of the total outstanding securities for that particular year. We aggregated the maturity shares in five major bins: very short-term (within 1 year), short-term (between 2 to 5 years), medium-term (6 to 10 years), long-term (10 to 20 years), and very long-term (>20 years). A key takeaway from the figure is the steady increase in long-term securities (maturities exceeding 20 years) as a percentage of the total outstanding securities over the last 2 decades. Additionally, short-term debt (maturities less than 1 year) constitutes less than 10 % of the total outstanding debt. This figure underscores the government's commitment to extending the maturity structure of its debt. One noteworthy observation from this graph is the U-shaped curve for securities with maturities between 10 and 20 years, which experienced a sharp increase following the Global Financial Crisis (GFC). In India, the 10-year government security is the most actively traded, leading to a substantial liquidity premium, making it a cost-effective choice for the government to issue more to meet the demand. As a share of total borrowings, there has been an increase in the share of long-term securities (>10 years) in total borrowings in the past two decades, while relatively short-term securities have seen a decline in their shares, highlighting the increasing role of long-term securities in the total portfolio of borrowings. These trends are reported in Fig. 6 which plots the maturity pattern of borrowings for Centre government securities.

[5] We compile a comprehensive annual time series dataset on aggregate government debt from 1999 to 2022, including its constituent elements: nominal interest rate, inflation, real GDP growth rate, and primary deficit/surplus. We source security-level data from the Ministry of Finance, Government of India's Status Papers, which record the total outstanding value of Central government securities for each fiscal year and details of new issuances. We segment securities into zero-coupon bonds by separating coupons and creating distinct matrices for principal and coupon payments, focusing on Central Government securities from 1999 to 2022. We obtain yield information from the Reserve Bank of India's (RBI) Subsidiary General Ledger (SGL) transactions for government-dated securities across various maturities. Fig. 8, Section 5, plots the monthly yield to maturity (YTM) data for Indian Government securities across the term structure from 1996 till 2022. Using March-end yields for each year and maturity, we calculate corresponding bond prices. Fig. 8 shows month-end yields for government-dated securities. We use RBI yield curve data to calculate debt prices and market values. For more details refer to Section 5 and Appendix B.

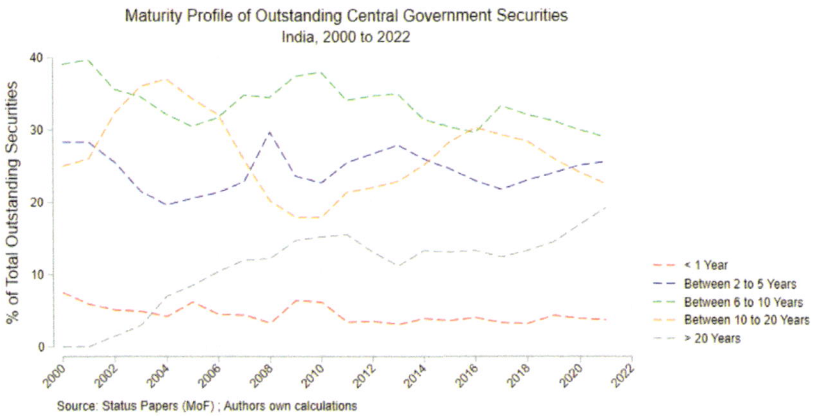

FIG. 5 Maturity profile of outstanding securities as % of total outstanding. *Note: Data is hand collected from the Ministry of Finance Status Papers for each financial year from 2000 to 2022.*

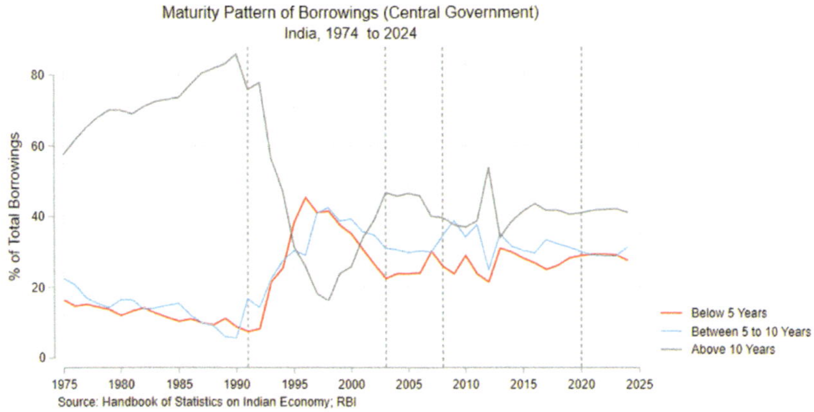

FIG. 6 Maturity pattern of rupee denominated borrowings by Central Government. Note: The four vertical dotted lines denote the following episodes: the Liberalization Reforms (1991), the enactment of the FRBM Act (2003), the Global Financial Crisis (2008), and COVID-19 (2020). While the GFC is between 2007–2009, we take the mid-point, 2008. *Data is from the Handbook of Statistics on Indian Economy for each financial year from 1974 to 2024.*

Fig. 7 plots the weighted average yields of Centre government securities for each year. The bars around each point shows the lower and upper bounds of yield for any particular year. Post 2000 there has been a significant drop in the yields for these securities. The length of each bar represents the range of yields provided.

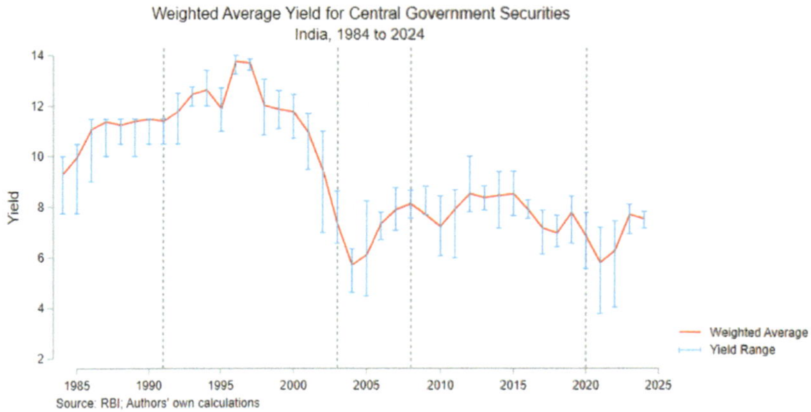

FIG. 7 Weighted average yields for Central Government securities. Note: The three vertical dotted lines denote the following episodes: the enactment of the FRBM Act (2003), the Global Financial Crisis (2008), and COVID-19 (2020). While the GFC is between 2007–2009, we take the mid-point, 2008. *Data is from the Handbook of Statistics on Indian Economy for each financial year from 1984 to 2024.*

3 The Hall-Sargent approach

To understand the factors that drive public debt using a granular approach, we start from a simple government budget constraint at time period t:

$$B_t + T_t = (1 + r_{t-1,t})B_{t-1} + G_t \tag{1}$$

where, B_t is nominal debt issued at time t, T_t is the tax revenue at time t, $r_{t,t-1}$ is the nominal interest rate on the past debt B_{t-1}, and G_t is government expenditure. We manipulate (1) to give:

$$\frac{B_t}{Y_t} - \frac{B_{t-1}}{Y_{t-1}} = (r_{t-1,t} - \pi_{t-1,t} - g_{t-1,t})\frac{B_{t-1}}{Y_{t-1}} + \frac{def_t}{Y_t} \tag{2}$$

where, Y_t is nominal GDP at time t, and $\pi_{t-1,t}$ and $g_{t-1,t}$ are the inflation rate, and growth rate of real GDP from period $t - 1$ to period t, respectively.

Eq. 2 shows that the change in debt-to-GDP between time period t and $t - 1$ is equal to the sum of the nominal interest rate payments net of growth and inflation (the first term on the RHS) and the primary deficit-to-GDP ratio of this time period. Whereas the nominal interest rate and primary deficit add to the change in debt, inflation and growth rates help reduce debt. A primary surplus also reduces debt.

Sovereign debt typically consists of debt of different maturities. Typically, there is debt that is due within a year, between 2–10 years, and those of more than 10 years maturity. Accordingly, we can manipulate Eq. (2) to represent aggregate debt into debt with varying maturities. As shown in Das and Ghate (2022), unpacking aggregate debt into its constituent maturities leads to a better understanding of how the nominal interest rate, inflation rate, and growth rate, affect debt dynamics via different maturity buckets. To account for the components, one needs to use security level debt data. Popularized by George Hall and Thomas Sargent (Hall and Sargent, 1997, 2011), this approach uses security-level data of the government to arrive at the aggregate market value of debt. Each security is viewed as a zero-coupon bond, which is then priced using the yield-to-maturity data to arrive at the market value of the security. Following the term structure of interest rates each security or coupon bond can be viewed as a bundle of zero coupon or discount bonds (Hall and Sargent, 1997). Each of these in turn can be priced by using the yield to maturity data and then aggregated to get the market value of debt. So that the bond is a weighted sum of pure discount bonds of maturities 1, 2, 3, ..., j.

We begin by first introducing the maturity structure in Eq. (2). Let $\tilde{B}_{t-1}^{j}$ be real values of nominal zero-coupon bonds of maturity j at $t-1$. $\tilde{r}_{t-1,t}^{j}$ is the net nominal holding period return between $t-1$ and t on nominal zero-coupon bonds of maturity j. Using these definitions, the re-defined budget constraint in Eq. (1) can be written as

$$\frac{\tilde{B}_t}{Y_t} = \sum_{j=1}^{n} \tilde{r}_{t-1,t}^{j} \frac{\tilde{B}_{t-1}^{j}}{Y_{t-1}} - (\pi_{t-1,t} + g_{t-1,t}) \frac{\tilde{B}_{t-1}}{Y_{t-1}} + \frac{def_t}{Y_t} + \frac{\tilde{B}_{t-1}}{Y_{t-1}} \tag{3}$$

Eq. 1 above differentiates between the contributions to the growth of the debt-GDP ratio that are contingent on the maturity of debt j, taking into account their nominal interest payments, and those that are not.

Next, at each date t we compute the number of rupees the government has promised to pay at each date $t+j$. The coupons are stripped from the coupon bonds and are valued as a weighted sum of zero coupon bonds. This is because any coupon bond can be decomposed into zero coupon bonds with varying maturity. Let s_{t+j}^{t} be the number of time $t+j$ rupees that the government has at time t promised to deliver while q_{t+j}^{t} is the number of time t rupees that it takes to buy a rupee at time $t+j$. q_{t+j}^{t}, the price of a bond, is given by

$$q_{t+j}^{t} = \frac{1}{(1+\rho_{jt})^{j}}$$

where ρ_{jt} is the time t yield to maturity (YTM) on bonds with j periods to maturity. We obtain the yield to maturity (YTM) data from the Reserve Bank of India (RBI) to compute prices. Since we are interested in the real market value, we define the deflator, $v_t = 1/p_t$, with p_t denotes the price level in the base year t.[6] These modifications give the total real value of government debt outstanding in period t as

$$v_t \sum_{j=1}^{n} q_{t+j}^{t} s_{t+j}^{t}$$

[6] This says that in order to convert 1 rupee to an equivalent amount of goods, it would be given by $v_t = 1/p_t$.

Hence, we can write the re-defined time t budget constraint in Eq. (4) as

$$v_t \sum_{j=1}^{n} q_{t+j}^{t} s_{t+j}^{t} = v_t \sum_{j=1}^{n} q_{t+j-1}^{t} s_{t+j-1}^{t-1} + def_t \tag{4}$$

Eq. (4) denotes the real counterpart of the government budget constraint taking into account the different maturities of the debt that the government has issued. It shows the real value of the interest-bearing debt at the end of period t, given by the LHS of Eq. (4), being equal to the real value of the government's outstanding debt at the beginning of period t and the primary deficit-GDP ratio in period t. Dividing both sides of Eq. (4) by Y_t, we can write

$$\left(\frac{v_t}{v_{t-1}} \frac{q_{t+j-1}^{t}}{q_{t+j-1}^{t-1}} \frac{Y_{t-1}}{Y_t} - 1 \right) \approx \tilde{r}_{t-1,t}^{j} - \pi_{t-1,t} - g_{t-1,t} \tag{5}$$

Defining terms in this way, we see a clear mapping between Eqs. (3) and (4).[7]

3.1 Introducing a term premium

We extend (Hall and Sargent, 2011) by adding a term premium, separating the short rate (or the safe real interest rate) and the term premium. Given that q_{t+j}^{t} is the number of time t rupees that it takes to buy a rupee at time $t+j$, this is like the price of a zero coupon bond with maturity j. Therefore, by definition, the net nominal holding period return between time t and $t-1$ on nominal zero-coupon bonds of maturity j is given as

$$\tilde{r}_{t-1,t}^{j} = \frac{q_{t+j-1}^{t}}{q_{t+j-1}^{t-1}} + 1$$

which can be interpreted as buying a j period bond at time $t-1$ and selling it as a $j-1$ year bond at time t. Following Cochrane and Piazzesi (2005) and Begenau et al. (2015) we can express this holding period return into two components: the short rate which is the 1-year YTM, and the excess return or term premium

$$\tilde{r}_{t-1,t}^{j} = \rho_{1t} + \widetilde{rx}_{t-1,t}^{j} \tag{6}$$

where ρ_{1t} is the short rate or the date t yield to maturity on a 1-period bond, and $\widetilde{rx}_{t-1,t}^{j}$ is the excess return holding for a j period bond from date $t-1$ to date t. Thus we can write Eq. (5) as

$$\left(\frac{v_t}{v_{t-1}} \frac{q_{t+j-1}^{t}}{q_{t+j-1}^{t-1}} \frac{Y_{t-1}}{Y_t} - 1 \right) \approx \left(\rho_{1t} + \widetilde{rx}_{t-1,t}^{j}\right) - \pi_{t-1,t} - g_{t-1,t} \tag{7}$$

where we have augmented the initial (Hall and Sargent, 2011) equation with an explicit term premium appearing in it.

7 Appendix (A) outlines in detail the derivation steps.

3.2 Contributions of inflation, growth and term premium for debt decomposition

To take into account the role played by inflation, growth, and nominal returns including the term premium component, we write Eq. (3) as

$$\frac{\widetilde{B}_t}{Y_t} = \sum_{j=1}^{n} \left(\rho_{1t} + \widetilde{rx}^{j}_{t-1,t}\right)\frac{\widetilde{B}^{j}_{t-1}}{Y_{t-1}} - \left(\pi_{t-1,t} + g_{t-1,t}\right)\frac{\widetilde{B}_{t-1}}{Y_{t-1}} + \frac{def_t}{Y_t} + \frac{\widetilde{B}_{t-1}}{Y_{t-1}} \tag{8}$$

We iterate the above equation backward to an initial time $t-\tau$ and re-write the above equation as,

$$\frac{\tilde{B}_t}{Y_t} - \frac{\tilde{B}_{t-\tau}}{Y_{t-\tau}} = \sum_{s=0}^{\tau-1}\left[\sum_{j=1}^{n}\left(\rho_{1t-s-1} + \widetilde{rx}^{j}_{t-s-1,t-s} - \pi_{t-s-1,t-s} - g_{t-s-1,t-s}\right)\frac{\widetilde{B}^{j}_{t-s-1}}{Y_{t-s-1}} + \frac{def_{t-s}}{Y_{t-s}}\right] \tag{9}$$

Eq. (9) illustrates that commencing from an initial time period τ, we can characterize the evolution of the debt-to-GDP ratio through the decomposition of the terms present on the RHS of Eq. (9). Note that the terms on the right side of the equation are actually weighted by the debt-to-GDP ratio of various maturity tranches. This suggests that the factors contributing to the debt-to-GDP evolution-specifically the nominal return (a combination of term premium and short rate) net of inflation and the growth rate-are influenced by the issuance of debt-to-GDP for each maturity within a given yearly tranche.[8]

In the next section, we use a newly assembled dataset, apply it to Eq. (9), and quantify the factors driving the change in the value of the debt-GDP ratio between any two arbitrary time periods.[9]

4 Maturity structure

The official statistics on government debt is provided by the Ministry of Finance in India through the Status Papers on Government Debt published annually since 2010. In the Status Papers the maturity structure of government bonds is analyzed using two key metrics: average maturity and duration. Average maturity refers to the average time it takes for bonds within a portfolio to reach maturity. This is calculated by considering individual bond maturity periods and deriving a weighted average based on their respective values. This straightforward measure, which takes into account the face value of bonds, provides a broad overview of a portfolio's maturity profile. A more nuanced measure is duration, also called Macaulay Duration, which represents the time taken to recover the initial investment in present value terms. In its simplest

[8] To understand how the decomposition works in practice, in Appendix A we provide notation with respect to two times to clarify how the algebra works.

[9] In Appendix A we show, using a simple example with two time periods, how Eq. (9) can be written and is decomposed into different components.

form, duration can be thought of as the payback period for a bond to break even, or the time it takes for a bond to repay its purchase price.

Both measures play an important role in shaping market dynamics and investment strategies. They help investors and policymakers assess the overall risk and liquidity profile of the government debt market and are instrumental in determining the interest rate sensitivity of bond portfolios. The Central Bank (Reserve Bank of India or RBI in short) utilizes these metrics to manage the maturity profile of government debt and influence market liquidity. The metrics also provide valuable insights into the structure and behavior of the bond market, allowing for better risk management and strategic planning.

4.1 A simple example

As mentioned in the Introduction, the Status Papers calculate the average maturity using the residual maturity approach, where the average maturity is simply the weighted average of all outstanding securities over a period of time t with the weights being their relative share in the total face value.

As an example, following Hall and Sargent (1997), we calculate the average maturity of a security based on the market value of that security in the total portfolio as opposed to the face value of that security. For the purpose of calculating the average maturity, we follow the initial steps of the decomposition equation mentioned in the previous section in converting the bonds into a series of zero coupon bonds by stripping and allocating its coupon payments and the principal amount into a zero coupon bond based on the year in which the coupons are due.

We illustrate this by using a simple example. Suppose at time period t there are two bonds, namely bond A and bond B with maturities in $t+1$ and $t+2$ respectively. The coupon rate and face value are given below:

	Bond A	**Bond B**
Face Value	100	300
Coupon Rate	10%	10%
Interest rate (ρ_{jt})	5%	10%
Time to Maturity (years)	1	2

The calculation of the average maturity is given by:

$$\text{Average Maturity} = w_A * Mat_A + w_B * Mat_B \tag{10}$$

where w_A and w_B are the respective weights of the two bonds and Mat_A and Mat_B are the maturities of the bonds.

$$w_A = \frac{100}{100+300}$$

$$w_B = \frac{300}{100+300}$$

Therefore,

$$\text{Average Maturity} = w_A * Mat_A + w_B * Mat_B = 0.25 * (1) + 0.75 * (2) = 1.75 \text{ years} \tag{11}$$

We call $AvgMat_{fisc} = 1.75$ years as the average maturity calculated by the fiscal authority.

We now calculate the average maturity by the Hall and Sargent (2011) methodology using the numbers above. We first have to determine the prices of the two bonds as well as the promised repayments. The HS approach does so by breaking the two bonds into zero coupon bonds and pricing them separately.

Let q^t_{t+j} be the price of a zero coupon bond at time t with years to maturity $t + j$, then we have

$$q^t_{t+j} = \frac{1}{(1 + \rho_{jt})^j}$$

In this example

$$q^t_{t+1} = \frac{1}{(1 + 0.05)} = 0.952$$

$$q^t_{t+2} = \frac{1}{(1 + 0.1)^2} = 0.826$$

we also define s^t_{t+j} as the sum of the coupon, c^t_{t+j} and the principal or face value, b^t_{t+j}.

The market value is given by

$$\sum_{j=1}^{2} q^t_{t+j} s^t_{t+j}$$

$$\Rightarrow q^t_{t+1}(c^t_{t+1} + b^t_{t+1}) + q^t_{t+2}(c^t_{t+2} + b^t_{t+2})$$
$$= 0.952(10 + 30 + 100) + 0.826(30 + 300) = 405.86$$

Now to calculate average maturity, we have

$$\text{Average Maturity} = m_A * Mat_A + m_B * Mat_B \tag{12}$$

where m_A and m_B are the market shares of the two bonds. In order to calculate them, we first need to calculate the market value of the bonds at time t.

$$m_A = \frac{s^t_{t+1} q^t_{t+1}}{\sum_{j=1}^{2} q^t_{t+j} s^t_{t+j}} = \frac{140 * 0.952}{405.86} = 0.33$$

$$m_B = \frac{s^t_{t+2} q^t_{t+2}}{\sum_{j=1}^{2} q^t_{t+j} s^t_{t+j}} = \frac{330 * 0.826}{405.86} = 0.67$$

$$AvgMat_{hs} = 0.33 * (1) + 0.67 * (2) = 1.67 \tag{13}$$

The market value calculation takes into account the government,™s outstanding promises to pay coupons that are economically indistinguishable from promises to pay principal, as well as any change in the interest rate (ρ_{jt}) will also have a bearing on the average maturity. As can be seen the average maturity under the residual approach is higher (1.75 years) versus the HS approach (1.67 years).

As shown in the example, there could be differences between the calculation of average maturity based on the way the weights are used in the calculation. The Status Papers uses the face value of securities while the (Hall and Sargent, 2011) approach uses the market value for the calculation of weights. Since the market value of securities incorporates the movement of the price of securities, it gives a better reflection of the "true" average maturity as perceived by the market. In the above example, the market values the short-term security (long-term security) with a higher (lower) weight compared to the calculation done using the face value approach. Consequently, the overall average maturity is lower when calculated using the market value approach as opposed to the face value approach.

5 Applying the Hall-Sargent approach to Indian security level data

Building on the dataset used in Das and Ghate (2022), we compile a comprehensive annual time series dataset on aggregate government debt from 1999 to 2022, including its constituent elements: nominal interest rate, inflation, real GDP growth rate, and primary deficit/surplus. We source security-level data from the Ministry of Finance, Government of India's Status Papers on Government Debt, which record the total outstanding value of Central government securities for each fiscal year and details of new issuances. We segment securities into zero-coupon bonds by separating coupons and creating distinct matrices for principal and coupon payments, focusing on Central Government securities from 1999 to 2022.

We obtain yield information from the Reserve Bank of India's (RBI) Subsidiary General Ledger (SGL) transactions for government-dated securities across various maturities. Fig. 8 plots the monthly yield to maturity (YTM) data for Indian Government securities across the term structure from 1996 till 2022. The policy rate set by the Reserve Bank of India is shown by the solid black line. As can be seen, the yields decreased after the global financial crisis in 2008 and post-COVID-19 when monetary policy loosened. Using March-end yields for each year and maturity, we calculate corresponding bond prices.

We undertake the debt decomposition following Hall and Sargent (2011). Other essential variables for decomposition, such as GDP data, come from various Economic Survey issues. CPI inflation data is sourced from the Organization for Economic Cooperation and Development (OECD) CPI and WPI

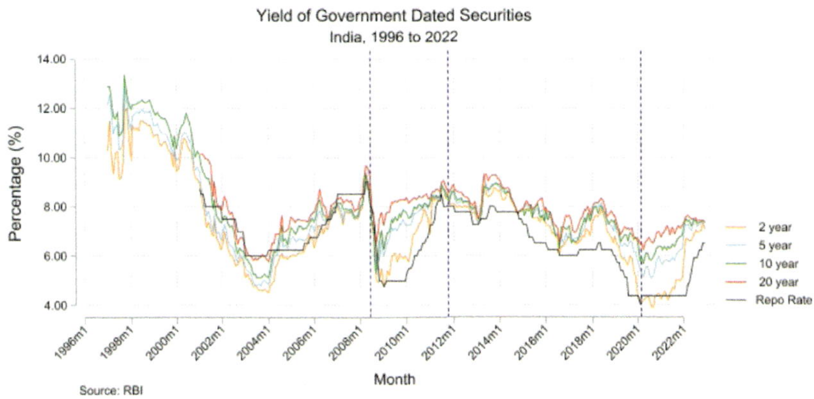

FIG. 8 Month end yields of Government dated securities. Note: The three vertical dotted lines denote the following episodes: the enactment of the FRBM Act (2003), the Global Financial Crisis (2008), and COVID-19 (2020). While the GFC is between 2007–2009, we take the mid-point, April 2008. *Data is from the Reserve Bank of India, monthly from 1996 to 2022.*

series, while primary deficit/surplus data is from the RBI. Next, we outline the steps for constructing the dataset in order to perform the decomposition exercise.

We begin by collecting data on all Central government securities issued from 1999 onwards using the Status Papers of government debt. The RBI, acting as the debt manager for the government, issues these securities on behalf of the Ministry of Finance. We provide additional details on the process of security issuance by the RBI in Appendix B. Our approach treats coupon bonds as bundles of zero-coupon bonds with different maturities. We strip each bond into its component zero-coupon bonds, corresponding to each date when a coupon or principal payment is due, and value each promised payment separately. This method allows us to construct two key matrices: the coupon matrix (**C**) and the principal matrix (**P**). The **C** matrix captures coupon payments for all securities across all years and maturities, while the **P** matrix records principal payments over the entire maturity horizon.

By stripping coupons from each bond, we effectively price bonds as weighted sums of zero-coupon bonds of various maturities. Combining the **P** and **C** matrices yields the total payments outstanding for the Central government, weighted by various maturity tranches. Using yield-to-maturity (YTM) data from SGL transactions in government-dated securities for various maturities, we derive the "price" of each security. In our context, this price represents the number of time t rupees required to purchase one rupee at time $t + j$. It's worth noting that all securities in our analysis are rupee-denominated.

We calculate the value of currency measured in goods per rupee as the inverse of the price in the base year, denoted as v_t. By multiplying q^t_{t+j} (the price of the security) with v_t and s^t_{t+j} (the payment amount), we obtain the real value of the marketable bond in year t. Summing these values across all maturities provides the total real value of government debt outstanding in period t.

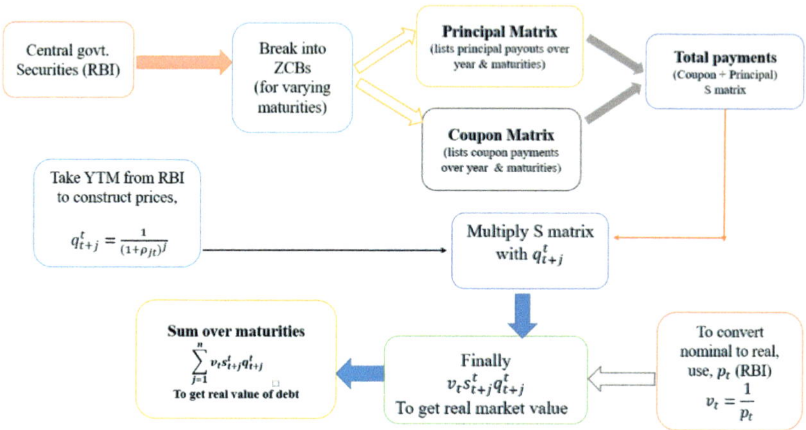

FIG. 9 Flow chart showing calculation of the real value of government debt.

A schematic chart illustrating the calculation process for the real value of marketable debt is provided in Fig. 9. This comprehensive approach to data collection and processing enables us to conduct a thorough accounting decomposition of India's government debt, providing insights into its composition and dynamics over the studied period.

5.1 Decomposition results

Table 1 presents our decomposition findings based on the compiled data based on Eq. (9). It outlines the evolution of the debt-GDP ratio for each specified time period, accompanied by an analysis of the contributing factors that have influenced the changes observed in the debt-GDP ratio. Across the entire sample (1999–2022), the debt-GDP ratio grew by 30.8 % points. This increase was driven predominantly by the nominal return component, particularly at the short rate segment, which exerted a significant upward influence on debt. Conversely, inflation and economic growth contributed to mitigating the rise, thereby reducing the debt-GDP ratio. Notably, bondholders realized a net positive real return, as is evident from the positive real return component in the first column.

Upon dissecting the decomposition for various sub-periods, a recurring pattern emerges. Across most sub-periods, the dominant factors influencing the change in the debt-GDP ratio remain the nominal return and the inflation components. With the exception of the 2004–2014 period, the debt-GDP ratio increases in all other sub-periods. Notably, in the years following the COVID pandemic, the debt-GDP increased by approximately 13 % points in two years (2020–2022). Once again nominal return and inflation are the major components, the negative growth rate of GDP leads to an *increase* in the real value of debt-GDP.[10] The key takeaway from our decomposition result is the important role played by the nominal interest rates and inflation in the evolution of debt-GDP ratio in India.

Given the important role played by the maturity structure of debt in our study, we extend our analysis across the same time periods as indicated in Table 1, while also explicitly incorporating the maturity structure. This approach enables us to discern the contributions of different components within various maturity tranches, thereby influencing the overall debt-GDP ratio. We categorize the analysis into three maturity ranges: 1–2 years, 2–10 years, and 10 + years. A salient observation from Table 2 is that inflation and the short rate component collectively contribute to more than half of the changes in the debt-GDP ratio across the entire sample period and within the delineated sub-periods. This shows that short rate plays an important role in impacting the maturity structure of debt. Notably, in the aftermath of the global financial crisis (2009–2014), holders of long-term bonds encountered negative realreturns-a trend that also emerged in the post-COVID era (2020–2022). Following the de-facto adoption of inflation-targeting in 2014, between 2014–2020 the contribution of the inflation component in reducing debt-GDP reduces, but it regains prominence in the immediate aftermath of the COVID pandemic (2020–2022).

The results highlight the substantial role played by the short rate component in the above mentioned maturity tranches, contributing approximately 12.2 % of the total shift in the debt-GDP ratio. The term premium component, accounting for about 12 % of the total change in debt-GDP during the entire sample period, underscores the influence of market expectations, risk considerations, and liquidity factors that impact the valuation of long term bonds. It is pertinent to note that due to data limitations, the explicit separation of these contributing factors isn't feasible; thus, these results show the collective contribution of all of these components.

The key takeaway from these decomposition results underscores the substantial influence exerted by inflation and nominal interest rates. While inflation significantly contributes to reducing the debt-GDP ratio, its impact becomes more pronounced under a longer maturity structure of debt. Nominal interest rates, represented by the short rate component, which we use as a proxy for the central bank,s policy rate, also assume a pivotal role in driving up the debt-GDP ratio. Consequently, the role of inflation becomes vital in maintaining the real value of debt at a sustainable level.

However, when the central bank is committed to containing inflation within a specified range, a mere escalation of price levels may not suffice to ensure debt sustainability. Instead, the Central Bank can use the maturity structure of debt to defer the surge in prices to the future (leading to future unexpected inflation) and use the market value of debt to keep debt at a sustainable level in the current time period.

[10] The component of growth rate is positive which denotes that negative growth during these years actually led to an increase in the debt-GDP rather than reducing it.

TABLE 1 Debt decomposition for Central Government securities.

		Period					
Start End		**2000–2022**	**2000–2004**	**2004–2009**	**2009–2014**	**2014–2020**	**2020–2022**
Debt-GDP							
	Start	19.70	19.70	35.60	34.90	34.70	37.30
	End	50.50	35.60	34.90	34.70	37.30	50.50
	Change	30.80	15.90	−0.70	−0.20	2.60	13.20
Marketable debt							
	Nominal return	55.80	15.10	8.40	9.40	13.60	9.40
	Short rate	45.60	9.10	8.50	12.10	9.90	5.90
	Ex_Return(TP)	14.10	6.90	−0.10	−2.70	4.20	3.50
	Inflation	−41.30	−5.60	−7.70	−15.70	−6.40	−6.00
	Real return	14.50	9.10	0.60	−6.30	6.80	3.40
	Growth rate	−32.00	−8.90	−9.00	−7.00	−8.10	1.00
Non-marketable debt							
	Nominal return	29.70	13.30	7.40	4.10	3.10	1.70
	Inflation	−5.40	−0.40	−1.10	−2.10	−0.70	−1.10
	Growth rate	−3.60	−0.80	−1.30	−0.90	−0.90	0.30
Primary Deficit/ GDP		26.00	3.40	1.90	10.80	2.30	7.70

Note: Data for Centre government securities used in this calculation are from the Ministry of Finance Status Papers on Government Debt, ranging from 2000 to 2022. Data for all other variables are from the Reserve Bank of India.

TABLE 2 Debt decomposition for Central Government securities by maturity structure.

			Period					
Start End			2000–2022	2000–2004	2004–2009	2009–2014	2014–2020	2020–2022
Debt-GDP								
		Start	19.70	19.70	35.60	34.90	34.70	37.30
		End	50.50	35.60	34.90	34.70	37.30	50.50
		Change	30.80	15.90	−0.70	−0.20	2.60	13.20
Marketable debt								
	Nominal return		55.80	15.10	8.40	9.40	13.60	9.40
		1–2 years	13.80	3.00	2.60	3.20	3.10	1.90
		TP(1–2 years)	1.60	0.60	0.30	−0.2	0.40	0.50
		2–10 years	30.01	8.80	4.40	5.00	6.90	4.90
		TP(2–10 years)	8.20	4.40	0.40	−1.1	2.00	2.50
		10 + years	12.20	3.30	1.30	1.20	3.60	2.60
		TP(10 + years)	4.30	1.90	−0.20	−0.7	1.70	1.60
	Inflation		−41.3	−5.60	−7.70	−15.70	−6.40	−6.00

Continued

Table 2 Debt decomposition for Central Government securities by maturity structure. —Cont'd

			Period					
Start **End**			**2000–2022**	**2000–2004**	**2004–2009**	**2009–2014**	**2014–2020**	**2020–2022**
		1–2 years	−12.30	−1.60	−2.40	−4.70	−1.90	−1.70
		2–10 years	−21.60	−3.10	−3.90	−8.30	−3.30	−3.00
		10 + years	−7.50	−1.00	−1.40	−2.60	−1.20	−1.20
	Growth rate		−32.00	−8.90	−9.00	−7.00	−8.10	1.00
		1–2 years	−9.00	−2.40	−2.60	−2.00	−2.3 0	0.30
		2–10 years	−16.80	−4.80	−4.60	−3.80	−4.20	0.5
		10 + years	−6.20	−1.80	−1.80	−1.20	−1.60	0.2
Non-marketable debt								
	Nominal return		29.70	13.30	7.40	4.10	3.10	1.70
	Inflation		−5.40	−0.40	−1.10	−2.10	−0.70	−1.10
	Growth rate		−3.60	−0.80	−1.30	−0.90	−0.90	0.30
	Primary Deficit to GDP		26.00	3.40	1.90	10.80	2.30	7.70

Note: Data for Centre government securities used in this calculation is from the Ministry of Finance Status Papers on Government Debt, ranging from 2000 to 2022. Data for all other variables are from the Reserve Bank of India.

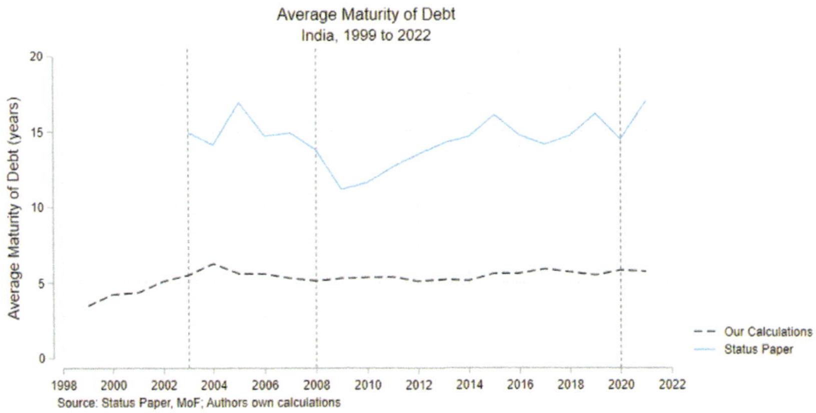

FIG. 10 Average maturity. Note: The three vertical dotted lines denote the following episodes: the enactment of the FRBM Act (2003), the Global Financial Crisis (2008), and COVID-19 (2020). While the GFC is between 2007–2009, we take the mid-point, 2008. *Data is from the Status Papers for each financial year from 1999 to 2022. For our calculations, we used the (Hall and Sargent, 2011) approach.*

5.2 Average maturity

Fig. 10 illustrates the evolution of average maturity based on how the Status Papers calculate it, and how average maturity is calculated in Hall and Sargent (2011). The solid blue line represents the average maturity as calculated in the Status Papers, while the dotted orange line depicts the average maturity based on the (Hall and Sargent, 2011) approach, which uses market values as weights. The figure clearly shows a substantial gap between the average maturity measures derived from the two approaches. The Status Papers indicate an upward trend in average maturity after the global financial crisis, reflecting the fiscal authority's efforts to extend the debt maturity profile. However, while the fiscal authority is elongating the maturity profile, the market continues to place greater value on shorter-term securities, as evidenced by the lower average maturity calculated using the (Hall and Sargent, 2011) approach. Furthermore, the average maturity based on market values (dotted orange line) remains nearly flat, indicating that market participants have preferred securities with maturities that are less than eight years. This suggests that the risks associated with a relatively shorter average maturity of debt, particularly rollover risk, remain elevated, even as the fiscal authority issues long-term securities to extend the maturity profile.

6 Capacity building and policy implications

While data on aggregate debt is useful in providing a broad snapshot of the borrowing position and needs of the fiscal authority (e.g., Treasury/Ministry of Finance), it does not capture the heterogeneous partitions that aggregate is made up of. Our paper argues that using the underlying securities constituting overall debt is important as it helps to keep track of various aspects of different securities such as risk, liquidity, and preferences of investors amongst a host of other factors that could drive the aggregate change in debt. A granular approach can also help in tracking the capital gains or losses from any security that accrues to the government and therefore could guide the government on current sentiments and future expectations in the bond market.

Importantly, the maturity structure of debt could be utilized when we work with security-level data, which is important in order to "discover" the yield curve for the economy as well as to get a sense of the term premium that investors expect. The maturity structure also guides policymakers about the features of future securities to be issued in the market. By treating debt not as an aggregate, but rather as an aggregate entity made up of a combination of different securities with different maturities, the government gets an accurate sense of the interest cost of debt in terms of coupon payments every period. According to the Hall-Sargent approach, these are not simply coupon payments but the summation of coupon and principal payments (Hall and Sargent, 2011).

Given the advantages of using security-level data as compared to aggregate data, the government should provide such data at regular intervals in any calendar year as and when the securities are issued. This would involve making information available on other statistics on all the major indicators like the issue/re-issue/buyback of securities; the rate of inflation; the borrowing requirement; information about the auction along the identifiers of securities. There should be information about the holder of any particular bond, albeit at an aggregate sectoral level. Such information reveals the preference of different securities across the maturity spectrum for different market participants.

Providing such a dataset and using the (Hall and Sargent, 2011) accounting framework to understand the evolution of debt is important, and adds to state and fiscal capacity. The government should have a dedicated portal to make the granular security-wise decomposition of debt publicly available. This will facilitate complete transparency of the borrowing environment for the government. For easy tracking of securities, the unique ISINs of each security should be revealed so as to map any existing security with its past issuance/buyback. This enables tracking the government borrowings accurately and helps assess the sustainability of debt including calculations of average maturity.

The monitoring of these requirements can be entrusted to an independent institution, such as a Fiscal Council. While there has been a lot of discussion for a Fiscal Council in India, its setting up has been delayed. The urgency of setting up an independent Fiscal Council cannot be emphasized more. Such an institution will assist the Central Bank (Reserve Bank of India) and the Fiscal Authority (Ministry of Finance) in strengthening rule-based monetary and fiscal policy. This will help with implementing flexible inflation targeting and undertaking sovereign debt management more effectively which will lead to better macroeconomic coordination and policy outcomes.

7 Conclusion

In recent years, the global economy has witnessed a substantial increase in sovereign debt, a trend exacerbated by the global financial crisis and the COVID-19 pandemic. High levels of public debt worry policy makers because high debt levels are typically associated with high interest costs of servicing the debt and unstable debt-dynamics. Policy makers would also like to assess what the average maturity of debt is to gauge roll-over risk, and quantify the drivers of the increase in public debt.

To provide answers to the above questions, the traditional analysis of public debt sustainability uses the face value of securities and the number of remaining coupon payments to calculate aggregate debt. Such an an aggregate approach, while widely employed, ignores the granular security-wise breakup of public debt, and the maturity structure of these securities. Following two path-breaking papers, Hall and Sargent (1997), and Hall and Sargent, (2011), we illustrate how the approach in these papers can help with a more accurate measure of a country's sovereign debt burden. We first discuss the Hall-Sargent approach through a series of simple examples. We then apply it to a novel dataset of outstanding Indian Centre government securities from 1999–2022. Our paper builds on earlier published work by Das and Ghate (2022) and Ghate and Das (2024) by extending the dataset used in these papers. We also extend the Hall-Sargent framework by breaking up the interest component on public debt to separate the short rate component from the term premium. This extension allows us to quantify the short rate and term premium component contributions to debt changes.

We show that the average maturity of public debt in India using the Hall-Sargent approach is substantially lower than the average maturity calculated using the traditional approach (or residual maturity). Since the amount of roll-over risk in public debt is an important parameter from a macro-stabilization perspective, our results suggest that there is substantially more roll-over risk in Indian public debt than calculated in official statistics publications, such as the Status Papers. Using a debt-decomposition exercise, we show that the nominal return on marketable debt accounts for 55.8 % of the approximately 30 % of the change in aggregate debt over 1999–2022, a substantial amount. While the debt-GDP ratio rose by 13.2 % in COVID (2020–2022), approximately 71 % of this increase $\left(\frac{9.4}{13.20}\right)$ was driven by the interest rate component from the nominal return on marketable debt.

From a policy prescription, Central Banks and Treasuries/Ministries of Finance in EMDEs can support state and fiscal capacity by using alternative approaches for public debt analyses. In this paper, we outline one such approach, as in Hall-Sargent. The requirements to conduct an analysis such as Hall-Sargent requires detailed granular security level data. The security-wise breakup of public debt should be made available in official publications like the Status Papers. The undertaking of analyses using the data can be entrusted to an independent institution such as a Fiscal Council. Such a Council would ensure that a broader and more transparent analysis on macro-fiscal issues is conducted. Incorporating these changes would allow for better implementation of rule-based monetary and fiscal policy in EMDEs like India which is a key criterion for macroeconomic stability.

Appendix A. Derivation of the decomposition equation

In this appendix, we derive Eq. (5) of the paper detailing all the steps in the process. We start by writing the government budget constraint, which is given as

$$\frac{B_t}{Y_t} = (r_{t-1,t} - \pi_{t-1,t} - g_{t-1,t})\frac{B_{t-1}}{Y_{t-1}} + \frac{def_t}{Y_t} + \frac{B_{t-1}}{Y_{t-1}} \tag{14}$$

To bring some of the consequences of interest rate risk and the maturity structure of the debt for the evolution of the Debt-GDP ratio, the budget constraint is redefined.

Let $\tilde{B}^j_{t-1}$ be real values of nominal zero-coupon bonds of maturity at $t-1$.

$\tilde{r}^j_{t-1,t}$ is the net nominal holding period return between t and $t-1$ on nominal zero-coupon bonds of maturity j.

Then using these definitions the new redefined budget constraint is given as

$$\frac{\tilde{B}_t}{Y_t} = \sum_{j=1}^{n} \tilde{r}^j_{t-1,t}\frac{\tilde{B}^j_{t-1}}{Y_{t-1}} - (\pi_{t-1,t} + g_{t-1,t})\frac{\tilde{B}_{t-1}}{Y_{t-1}} + \frac{def_t}{Y_t} + \frac{\tilde{B}_{t-1}}{Y_{t-1}} \tag{15}$$

The above equation distinguishes contributions to the growth of the debt-GDP ratio that depend on debt maturity j from those that don't.

Accounting details.

To carry out the accounting, at each date, t, compute the number of rupees the government has promised to pay at each date $t+j$. The coupons are stripped from the coupon bonds and are valued as a weighted sum of zeroes as any coupon bond can be decomposed into zero coupon bonds with varying maturity. Let.

s^t_{t+j} is the number of time $t+j$ rupees that the government has at time t promised to deliver.

q^t_{t+j} be the number of time t rupees that it takes to buy a rupee at time $t+j$, so this is like the price of a bond.

$$q^t_{t+j} = \frac{1}{(1+\rho_{jt})^j}$$

where ρ_{jt} the time t yield to maturity on bonds with j periods to maturity.

To convert t rupees to goods, use $v_t = 1/p_t$, with p_t being the price level.

The total real value of government debt outstanding in period t equals

$$v_t \sum_{j=1}^{n} q^t_{t+j} s^t_{t+j}$$

Thus now we can define the time t budget constraint as

$$v_t \sum_{j=1}^{n} q_{t+j}^{t} s_{t+j}^{t} = v_t \sum_{j=1}^{n} q_{t+j-1}^{t} s_{t+j-1}^{t-1} + def_t \tag{16}$$

The left-hand side of Eq. (16) is the real value of the interest bearing debt at the end of period t. The right-hand side of Eq. (16) is the sum of the real value of the primary deficit and the real value of the outstanding debt that the government owes at the beginning of the period.

Now to attain the government budget constraint that we have in (3), we will proceed as follows. Using Eq. (16),

$$v_t \sum_{j=1}^{n} q_{t+j}^{t} s_{t+j}^{t} = v_t \sum_{j=1}^{n} q_{t+j-1}^{t} s_{t+j-1}^{t-1} + def_t$$

Divide by Y_t on both sides to get

$$\frac{v_t \sum_{j=1}^{n} q_{t+j}^{t} s_{t+j}^{t}}{Y_t} = \frac{v_t \sum_{j=1}^{n} q_{t+j-1}^{t} s_{t+j-1}^{t-1}}{Y_t} + \frac{def_t}{Y_t}$$

Now add and subtract $\frac{\sum_{j=1}^{n} v_{t-1} q_{t+j-1}^{t-1} s_{t+j-1}^{t-1}}{Y_{t-1}}$ on the RHS of the above equation, we get

$$\frac{v_t \sum_{j=1}^{n} q_{t+j}^{t} s_{t+j}^{t}}{Y_t} = \frac{\sum_{j=1}^{n} v_t q_{t+j-1}^{t} s_{t+j-1}^{t-1}}{Y_t} - \frac{\sum_{j=1}^{n} v_{t-1} q_{t+j-1}^{t-1} s_{t+j-1}^{t-1}}{Y_{t-1}} + \frac{def_t}{Y_t} + \frac{\sum_{j=1}^{n} v_{t-1} q_{t+j-1}^{t-1} s_{t+j-1}^{t-1}}{Y_{t-1}}$$

$$\frac{v_t \sum_{j=1}^{n} q_{t+j}^{t} s_{t+j}^{t}}{Y_t} + = \left(\frac{\frac{\sum_{j=1}^{n} v_t q_{t+j-1}^{t} s_{t+j-1}^{t-1}}{Y_t}}{\frac{\sum_{j=1}^{n} v_{t-1} q_{t+j-1}^{t-1} s_{t+j-1}^{t-1}}{Y_{t-1}}} - 1 \right) \frac{\sum_{j=1}^{n} v_{t-1} q_{t+j-1}^{t-1} s_{t+j-1}^{t-1}}{Y_{t-1}} + \frac{\sum_{j=1}^{n} v_{t-1} q_{t+j-1}^{t-1} s_{t+j-1}^{t-1}}{Y_{t-1}} + \frac{def_t}{Y_t}$$

Canceling the terms and rearranging, we get

$$\frac{v_t \sum_{j=1}^{n} q_{t+j}^{t} s_{t+j}^{t}}{Y_t} = \sum_{j=1}^{n} \left(\frac{v_t}{v_{t-1}} \frac{q_{t+j-1}^{t}}{q_{t+j-1}^{t-1}} \frac{Y_{t-1}}{Y_t} - 1 \right) \frac{v_{t-1} q_{t+j-1}^{t-1} s_{t+j-1}^{t-1}}{Y_{t-1}} + \frac{def_t}{Y_t} + \frac{v_{t-1} \sum_{j=1}^{n} q_{t+j-1}^{t} s_{t+j-1}^{t-1}}{Y_{t-1}} \tag{17}$$

To recognize the terms in the above equation and see the analogy with Eq. (3), we see that

$$v_{t-1} q_{t+j-1}^{t-1} s_{t+j-1}^{t-1} = \tilde{B}_{t-1}^{j}$$

$$\tilde{B}_{t-1} = \sum_{j=1}^{n} \tilde{B}_{t-1}^{j}$$

Now since $v_t = 1/p_t$, thus we will have

$$\left(\frac{v_t}{v_{t-1}} \frac{q_{t+j-1}^{t}}{q_{t+j-1}^{t-1}} \frac{Y_{t-1}}{Y_t} - 1 \right) = \tilde{r}_{t-1,t}^{j} - \pi_{t-1,t} - g_{t-1,t} \tag{18}$$

A.1 Example: Two time periods.

To see things in a much more simpler way and to see an example of how Eq. (9) will be implemented, let's take a concrete example.

Suppose we want to see how the debt to GDP ratio evolved during the time period 2005–09. For this purpose the appropriate values of t and τ will be 2009 and 04 respectively as we are measuring the change in the Debt-GDP for 4 years. We will have $n = 30$ as the maximum maturity period bond available to us was for 30 years.

Thus with this formulation, Eq. (9) will become

$$\begin{aligned} \frac{\tilde{B}_{2009}}{Y_{2009}} - \frac{\tilde{B}_{2005}}{Y_{2005}} = \sum_{s=0}^{3} \Bigg[\sum_{j=1}^{30} \big(\rho_{1,2009-s} + \widetilde{rx}_{2009-s-1,t-s}^{j} - \pi_{2009-s-1,2009-s} \\ - g_{2009-s-1,2009-s}\big) \frac{\tilde{B}_{2009-s-1}^{j}}{Y_{2009-s-1}} + \frac{def_{2009-s}}{Y_{2009-s}} \Bigg] \end{aligned} \tag{19}$$

Now opening the outer summation of the RHS of the above equation to see the impact of each year, we will have

$$\begin{aligned} \frac{\tilde{B}_{2009}}{Y_{2009}} - \frac{\tilde{B}_{2005}}{Y_{2005}} = & \left[\sum_{j=1}^{30} (\rho_{1,2008} + \widetilde{rx}_{2008,2009}^{j} - \pi_{2008,2009} - g_{2008,2009}) \frac{\tilde{B}_{2008}^{j}}{Y_{2008}} + \frac{def_{2009}}{Y_{2009}} \right] \\ & + \left[\sum_{j=1}^{30} (\rho_{1,2007} + \widetilde{rx}_{2007,2008}^{j} - \pi_{2007,2008} - g_{2007,2008}) \frac{\tilde{B}_{2007}^{j}}{Y_{2007}} + \frac{def_{2008}}{Y_{2008}} \right] \\ & + \left[\sum_{j=1}^{30} (\rho_{1,2006} + \widetilde{rx}_{2006,2007}^{j} - \pi_{2006,2007} - g_{2006,2007}) \frac{\tilde{B}_{2006}^{j}}{Y_{2006}} + \frac{def_{2007}}{Y_{2007}} \right] \\ & + \left[\sum_{j=1}^{30} (\rho_{1,2005} + \widetilde{rx}_{2005,2006}^{j} - \pi_{2005,2006} - g_{2005,2006}) \frac{\tilde{B}_{2005}^{j}}{Y_{2005}} + \frac{def_{2006}}{Y_{2006}} \right] \end{aligned} \tag{20}$$

So Eq. (20) decomposes the total change in the debt to GDP between the period 2005 to 2009 into the individual changes pertaining to each year.

A.2 A simple example with two types of bonds with different maturities. In order to get a concise picture of how terms in the decomposition interact, we provide a simple illustration consisting of just two types of bonds with maturities in 1 and 2 time periods. Suppose we are in time period $t = 1$ and there are bonds of two types of maturities (one period and two period). The maturity profile of the bond is therefore $j = 1,2$, where we have a short-duration bond that matures in one time period forward and a long duration bond that matures in two time periods. Thus we have $t = 0,1,2,3$ and $j = 1,2$.

So the consolidated budget constraint of the government in period 1 is given by:

$$\frac{B_1}{Y_1} = (r_{0,1} - \pi_{0,1} - g_{0,1})\frac{B_0}{Y_0} + \frac{def_1}{Y_1} + \frac{B_0}{Y_0} \tag{21}$$

For sake of simplicity we assume that there are only nominal zero coupon bonds and not inflation-indexed. So let $\tilde{B}_0^j$ be real values of nominal zero-coupon bonds of maturity j at time *0*.

If we denote the holding period return (HPR) between time *0 and 1* by $\tilde{r}_{0,1}^j$, then we can write the new budget constraint at time $t = 1$ as

$$\frac{\tilde{B}_1}{Y_t} = \left(\tilde{r}_{0,1}^1 \frac{\tilde{B}_0^1}{Y_0} + \tilde{r}_{0,1}^2 \frac{\tilde{B}_0^2}{Y_0}\right) - (\pi_{0,1} - g_{0,1})\frac{\tilde{B}_0}{Y_0} + \frac{def_1}{Y_1} + \frac{\tilde{B}_0}{Y_0} \tag{22}$$

Again to carry out the decomposition, we have taken the same route as before and thus we can write the total real value of outstanding debt in period 1 as

$$v_1\left(q_2^1 s_2^1 + q_3^1 s_3^1\right)$$

where s_2^1 and s_3^1 are the number of time 2 and time 3 rupees that the government has at *time 1* promised to deliver.

q_2^1 and q_3^1 are the number of time 1 rupee that it takes to buy a rupee at time 2 and time 3 respectively, akin to price of a bond and thus

$$q_2^1 = \frac{1}{1+\rho_{11}}$$

$$q_3^1 = \frac{1}{(1+\rho_{21})^2}$$

where ρ_{11} and ρ_{21} are the YTM at time 1 for bonds maturing with 1 and 2 periods to maturity.

v_1 is simply to convert time 1 dollars into goods, so $v_1 = 1/p_1$.

Thus we can define the time 1 budget constraint as

$$v_1\left(q_2^1 s_2^1 + q_3^1 s_3^1\right) = v_1\left(q_1^1 s_1^0 + q_2^1 s_2^0\right) + def_1 \tag{23}$$

Note that the price of bond that will mature in the same period, q_1^1, is simply 1, as the number of time 1 rupee required to buy a rupee in time 1 is simply 1 itself.

To arrive at the equation, that we have in the paper, divide by Y_1 on both sides of Eq. (23) to get

$$\frac{v_1\left(q_2^1 s_2^1 + q_3^1 s_3^1\right)}{Y_1} = \frac{v_1\left(q_1^1 s_1^0 + q_2^1 s_2^0\right)}{Y_1} + \frac{def_1}{Y_1}$$

After doing the similar manipulation like we have done in the paper, we will arrive at the following equation which is given as

$$\begin{aligned}\frac{v_1\left(q_2^1 s_2^1 + q_3^1 s_3^1\right)}{Y_1} &= \left(\frac{v_1}{v_0}\frac{q_1^1}{q_1^0}\frac{Y_0}{Y_1} - 1\right)\frac{v_0 q_1^0 s_1^0}{Y_0} + \left(\frac{v_1}{v_0}\frac{q_2^1}{q_2^0}\frac{Y_0}{Y_1} - 1\right) \\ &\quad \frac{v_0 q_2^0 s_2^0}{Y_0} + \frac{def_1}{Y_1} + \frac{v_0\left(q_1^0 s_1^0 + q_2^0 s_2^0\right)}{Y_0}\end{aligned} \tag{24}$$

Now to recognize the terms and see the analogy, we have

$$\begin{aligned}v_0 q_1^0 s_1^0 &= \tilde{B}_0^1 \\ v_0 q_2^0 s_2^0 &= \tilde{B}_0^2 \\ \tilde{B}_0 &= \tilde{B}_0^1 + \tilde{B}_0^2\end{aligned}$$

Also, by taking the relevant approximations,

$$\begin{aligned}\left(\frac{v_1}{v_0}\frac{q_1^1}{q_1^0}\frac{Y_0}{Y_1} - 1\right) &\approx \tilde{r}_{0,1}^1 - \pi_{0,1} - g_{0,1} \\ \left(\frac{v_1}{v_0}\frac{q_2^1}{q_2^0}\frac{Y_0}{Y_1} - 1\right) &\approx \tilde{r}_{0,1}^2 - \pi_{0,1} - g_{0,1}\end{aligned}$$

Since by definition, $\tilde{r}_{0,1}^1$ is the net nominal holding period return of a one period to maturity bond between periods 0 and 1, hence it is actually nothing but $\tilde{r}_{0,1}^1 = \rho_{10}$, where ρ_{10} is the YTM of a one period to maturity bond at time 0.

In order to decompose the term premia (excess return) component from the nominal returns, we need to keep in mind the fact that by definition, term premium or excess return is the excess holding period return on holding a zero coupon bond for different maturities. To be more precise, it is the excess return of a long bond over a short bond. In our case, we will have:

$$\begin{aligned}\tilde{r}_{0,1}^1 &= \rho_{10} \\ \tilde{r}_{0,1}^2 &= \rho_{10} + \widetilde{rx}_{0,1}^2\end{aligned} \tag{25}$$

where $\widetilde{rx}_{0,1}^2$ is the excess return of holding a 2 period to maturity bond from periods 0 to 1.

Excess returns or Term Premium is in general the compensation demanded by investors for holding duration risk. So it's the compensation demanded for holding the longer duration asset as against rolling the investment with short-term instruments.

Note that $\tilde{r}^1_{0,1}$ will have no excess return as its the HPR of a one period to maturity bond between time 0 and 1.

Thus, from Eq. (25) we have

$$\widetilde{rx}^2_{0,1} = \tilde{r}^2_{0,1} - \rho_{10}$$

As can be seen from this equation that the excess returns or term premium can be negative or positive depending upon the preference of the investors and other market information like monetary policy, inflation expectations, macro scenario-like any shock or the supply side dynamics of the bond market to name some.

In principle the excess returns or term premium can be negative as well if suppose the investor prefer to secure a fixed return for long period (say for institutional pension funds), then the investor may be willing to accept a negative term premium than taking over the rolling over risk.

Now plugging all these into Eq. 24, we will have

$$\frac{v_1\left(q_2^1 s_2^1 + q_3^1 s_3^1\right)}{Y_1} = \left(\frac{v_1}{v_0}\frac{q_1^1}{q_1^0}\frac{Y_0}{Y_1} - 1\right)\frac{v_0 q_1^0 s_1^0}{Y_0} + \left(\frac{v_1}{v_0}\frac{q_2^1}{q_2^0}\frac{Y_0}{Y_1} - 1\right)\frac{v_0 q_2^0 s_2^0}{Y_0}$$
$$+ \frac{def_1}{Y_1} + \frac{v_0\left(q_1^0 s_1^0 + q_2^0 s_2^0\right)}{Y_0}$$

$$\Rightarrow \frac{\widetilde{B}_1}{Y_1} = (\widetilde{r}^1_{0,1} - \pi_{0,1} - g_{0,1})\frac{\widetilde{B}^1_0}{Y_0} + (\widetilde{r}^2_{0,1} - \pi_{0,1} - g_{0,1})\frac{\widetilde{B}^2_0}{Y_0} + \frac{def_1}{Y_1} + \frac{\widetilde{B}_0}{Y_0}$$

$$\Rightarrow \frac{\widetilde{B}_1}{Y_1} - \frac{\widetilde{B}_0}{Y_0} = (\widetilde{r}^1_{0,1})\frac{\widetilde{B}^1_0}{Y_0} + (\widetilde{r}^2_{0,1})\frac{\widetilde{B}^2_0}{Y_0} - (\pi_{0,1} + g_{0,1})\left(\frac{\widetilde{B}^1_0}{Y_0} + \frac{\widetilde{B}^2_0}{Y_0}\right) + \frac{def_1}{Y_1} \quad (26)$$

$$\Rightarrow \frac{\widetilde{B}_1}{Y_1} - \frac{\widetilde{B}_0}{Y_0} = (\rho_{10})\frac{\widetilde{B}^1_0}{Y_0} + (\rho_{10} + \widetilde{rx}^2_{0,1})\frac{\widetilde{B}^2_0}{Y_0} - (\pi_{0,1} + g_{0,1})\left(\frac{\widetilde{B}_0}{Y_0}\right) + \frac{def_1}{Y_1}$$

Eq. (26) is the decomposition equation, which is actually taking into account the fact that we have a term premium term explicitly floating in it.

Appendix B. Background information on Indian securities

How does the RBI issue securities?

How does the Reserve Bank of India (RBI) issue Centre government securities?[11] The RBI acts as the banker and the debt-manager to the government. A Government Security (G-Sec) is a tradable instrument issued by the Central Government or the State Governments acknowledging the obligation of the

[11] https://rbidocs.rbi.org.in/rdocs/Publications/PDFs/PRI85CEC73A987F41AC89068EE7607A8BEC.PDFDetails can be found here.

government's debt. In India, the Central Government issues both treasury bills and bonds or dated securities while the State Governments issue only bonds or dated securities, which are called State Development Loans (SDLs). The *Public Debt Office (PDO)* of the Reserve Bank of India acts as the registry / depository of G-Secs and deals with the issue, interest payment and repayment of principal at maturity. Most of the *dated securities* are **fixed coupon** securities.

There are various types of bonds issued. Most government bonds in India are issued as fixed rate bonds. Floating-rate bonds (FRBs) were first issued in September 1995 in India and have a variable coupon and can carry the coupon, which will have a base rate plus a fixed spread, to be decided by way of an auction mechanism. Government had last issued a zero coupon bond in 1996. Inflation Indexed Bonds (IIBs) are bonds wherein both coupon flows and principal amounts are protected against inflation. STRIPS are essentially zero coupon bonds and they are created out of existing securities only and unlike other securities, are not issued through auctions.

Besides banks, insurance companies and other large investors, smaller investors like Co-operative banks, Regional Rural Banks, Provident Funds are also required to statutorily hold G-Secs.

How are G-secs issued?

G-Secs are issued through auctions conducted by RBI on its electronic platform. Participants include Commercial Banks (CBs), Primary Dealers (PD), Insurance companies and Provident funds, who maintain funds accounts (current account) and securities accounts (Subsidiary General Ledger (SGL) accounts) with RBI. All non-members including non-scheduled CBs can participate in the primary auction through scheduled commercial banks or PDs. The RBI, in consultation with the Government of India, issues an indicative half-yearly auction calendar. Auction for *dated securities* is conducted on Friday for settlement on a T + 1 basis (i.e. securities are issued on next working day i.e., Monday). The Reserve Bank of India conducts auctions usually every Wednesday to issue T-bills of 91 day, 182 day and 364 day tenors. Settlement for the *T-bills* auctioned is made on T + 1 day.

An auction may either be yield-based or price-based. A **yield-based auction** is generally conducted when a **new** G-Sec is issued. Investors bid in yield terms up to two decimal places. Bids are arranged in ascending order and the cut-off yield is arrived at the yield corresponding to the notified amount of the auction and the cut-off yield is then fixed as the coupon rate for the security. Bids that are higher than the cutoff yield are *rejected.*

A **price-based auction** is conducted when the Government of India re-issues **securities** which have already been issued earlier. Bidders quote in terms of price per `100 of face value of the security. Bids are organized in *descending order* of the price offered, and successful bidders are those who have bid at or above the cut-off price. Depending upon the method of allocation to successful bidders, the auction may be conducted on a uniform price basis or multiple price basis.

In a **competitive bidding**, an investor bids at a specific price / yield and is allotted securities if the price / yield quoted is within the cut-off price / yield. These are carried out by well-informed institutional investors such as banks, financial institutions, PDs, mutual funds, and insurance companies.

How does one get information about the price of a G-Sec?

The return on a security is a combination of two elements (i) coupon income and (ii) the gain / loss on the security due to price changes. Information on the traded prices of securities is available on the FBIL website. The Clearing Corporation of India Limited (CCIL) is the clearing agency for G-Secs. In effect, during settlement, the CCP becomes the seller to the buyer and the buyer to the seller of the actual transaction. CCIL also guarantees settlement of all trades in G-Secs.

References

Begenau, J., Monika P., Martin S., 2015. Banks' Risk Exposures, Technical Report, National Bureau of Economic Research.

Besley, T., Persson, T., 2009. The origins of state capacity: property rights, taxation, and politics. Am. Economic Rev. 99 (4), 1218–1244.

Bianchi, F., Renato F., Melosi, L., 2020. Monetary and Fiscal Policies in Times of Large Debt: Unity is Strength, Technical Report, National Bureau of Economic Research.

Blanchard, O., 2022. Fisc. Policy Low. Interest. Rates. MIT Press.

Cochrane, Piazzesi, M., 2005. Bond risk premia. Am. Economic Rev. 95 (1), 138–160.

Cochrane, J.H., 2022. A fiscal theory of monetary policy with partially-repaid long-term debt. Rev. Economic Dyn. 45, 1–21.

Das, P., Ghate, C., 2022. Debt decomposition and the role of inflation: a security level analysis For India. Econ. Model 113, 105855.

De-Graeve, F.D., Mazzolini, G., 2023. The maturity composition of government debt: a comprehensive database. Eur. Economic Rev. 154, 104438.

Equiza-Goñi, J., 2016. Government debt maturity and debt dynamics in Euro area countries. J. Macroecon 49, 292–311.

Ghate, C., Das, P., 2024. Fiscal requirements for debt reduction targets in India. Economic Political Wkly 59 (39).

Hall, Sargent, 2011. Interest rate risk and other determinants of post-WWII US Government Debt/ GDP dynamics. Am. Economic Journal: Macroecon. 3 (3), 192–214.

Hall, G.J., Sargent, T.J., 1997. Accounting for the Federal Government's cost of funds. Economic Perspect. - Fed. Reserve Bank. Chic. 21, 18–28.

Hilscher, J., Alon, R., Reis, R., 2022. Inflating away the public debt? An empirical assessment. Rev. Financial Stud. 35 (3), 1553–1595.

Jiang, Z., Hanno, L., Stijn, V.N., Mindy, Z.X., 2024. The US public debt valuation puzzle. Econometrica 92 (4), 1309–1347.

Johnson, N.D., Koyama, M., 2019. Persecution & Toleration: The Long Road to Religious Freedom. Cambridge University Press.

Leeper, E.M., 1991. Equilibria under 'Active' and 'Passive' monetary and fiscal policies. J. Monetary Econ. 27 (1), 129–147.

Chapter 11

The importance of accurate, timely, credible crime data to inform crime and justice policy

John K. Roman, Stephanie Kennedy, John Buntin, and Alex R. Piquero*
Sociology & Criminology, University of Miami, Coral Gables, FL, United States
Corresponding author. e-mail address: apiquero@miami.edu

Chapter Outline

Abstract

Patterns of crime began changing dramatically in the summer of 2020 as the nation responded to the COVID-19 pandemic and the police killing of George Floyd. The Council on Criminal Justice (CCJ) began tracking the trends, first in support of our National Commission on COVID-19 and Criminal Justice and then for our Violent Crime Working Group. By providing snapshots of trends in large cities that make data readily available, the reports have filled a yawning gap in the nation's knowledge about what was actually happening – and not happening – with crime in real time. In April 2023, CCJ launched the Crime Trends Working Group to continue the important work of grounding debates about crime and criminal justice reform in facts. Its mission is to explore and explain current crime trends, while building consensus for significant improvements in the nation's capacity to produce timely, accurate, and complete crime data. With a diverse membership of expert producers and consumers of criminal justice statistics, the Working Group collects real-time crime data, identifies gaps in data collection and availability, and develops strategies to improve crime reporting nationwide as law enforcement agencies transition to a new federal reporting system.

Handbook of Statistics, Vol. 53. https://doi.org/10.1016/bs.host.2025.05.001

1 Introduction

Patterns of crime began changing dramatically in the summer of 2020 as the nation responded to the COVID-19 pandemic and the police killing of George Floyd. The Council on Criminal Justice (CCJ) began tracking the trends, first in support of our National Commission on COVID-19 and Criminal Justice and then for our Violent Crime Working Group. By providing snapshots of trends in large cities that make data readily available, the reports have filled a yawning gap in the nation's knowledge about what was actually happening – and not happening – with crime in real time.

In April 2023, CCJ launched the Crime Trends Working Group to continue the important work of grounding debates about crime and criminal justice reform in facts. Its mission is to explore and explain current crime trends, while building consensus for significant improvements in the nation's capacity to produce timely, accurate, and complete crime data. With a diverse membership of expert producers and consumers of criminal justice statistics, the Working Group collects real-time crime data, identifies gaps in data collection and availability, and develops strategies to improve crime reporting nationwide as law enforcement agencies transition to a new federal reporting system.

2 Background

On the first Friday of each month, the U.S. Bureau of Labor Statistics releases estimates of employment and unemployment for the nation, states, and selected local areas (U.S. Department of Labor Statistics, n.d.). Comprehensive, official national statistics on crime, on the other hand, are released annually and lag by nearly a year. Data collected by the federal government on homicide and other serious violent and property crimes for 2023, for example, will not be published until the fall of 2024.

The FBI's Uniform Crime Reporting (UCR) Program collects data on crimes known to law enforcement from over 19,000 law enforcement agencies. Agencies participate voluntarily and submit their crime data either directly or through a state UCR program. The FBI's program provides crime data going back to 1960 to researchers, the media, and the public. In the 1980s, the FBI announced plans to transition from the UCR's Summary Reporting System, which reports total crimes by month, to the National Incident-Based Reporting System (NIBRS), which reports data on each crime individually. In January 2021, the FBI sunsetted the Summary Reporting System and announced that it would only accept data through NIBRS. Given that a significant number of agencies had not made the transition (discussed below) by the sunset date, the FBI decided to accept data reported through both the Summary Reporting System and NIBRS in 2022 and 2023, but will not do so thereafter.

The National Crime Victimization Survey (NCVS), developed by the Bureau of Justice Statistics and fielded annually by the Census Bureau, is the nation's primary source of information on criminal victimization. Each year, data are obtained from a nationally representative sample of about 240,000 people in about 150,000 U.S. households. People aged 12 or older (who are not unhoused or in a facility) are interviewed on the frequency, characteristics, and consequences of criminal victimization they have experienced. The NCVS collects information on non-fatal personal crimes (e.g., assault, robbery, rape and sexual assault, and personal larceny) and household property crimes (e.g., burglary, trespassing, motor vehicle theft, and other types of theft), including whether these crimes were reported to law enforcement. Respondents provide demographic details about themselves and the offender (when known), characteristics of the crime, whether the crime was reported to law enforcement (and reasons for not reporting), the consequences of the victimization, and victim experiences with the criminal justice system and victim services.

The National Violent Death Reporting System (NVDRS), was established in 2002 by the Centers for Disease Control and Prevention to provide detailed information on the circumstances surrounding violent deaths, including information about weapon(s) used, victim and offender characteristics, and other contextual information. NVDRS compiles information from frontline investigators, including law enforcement, coroners, and medical examiners, on all violent deaths, including firearms and homicide deaths (Centers for Disease Control, n.d.b). In 2008, NVDRS became freely accessible online. As of April 2024, 48 states, the District of Columbia, and Puerto Rico were participating in NVDRS.

Timely, accurate, complete, and usable crime trends data are critical to guiding the nation's efforts to control and prevent crime. Real-time national data on crime trends–which crimes are rising, which are falling, and by how much–are needed to help policymakers spot emerging issues that require urgent attention and to guide the development of appropriate interventions. Timely national data can also ground politically charged policy conversations and public perceptions about crime in facts rather than anecdotes, and, in so doing, can help shape data-driven policymaking. This chapter thus focuses on gaps in crime data infrastructure that limit the nation's ability to track trends in real time and respond quickly with informed policy solutions.

3 Timeliness

Timely data are essential to meet the demands of policymakers and members of the media who want real-time information, particularly during periods when crime rates are changing rapidly.

Data from the Federal Bureau of Investigation's (FBI) Uniform Crime Reporting (UCR) Program and the Bureau of Justice Statistics' (BJS) National Criminal Victimization Survey (NCVS) have been published annually for decades. The UCR and NCVS reports are typically published in September or October and present information on crime during the previous year, although the two systems do not have identical periods of coverage.[1] The FBI recently began to release limited quarterly crime statistics from NIBRS through the agency's Crime Data Explorer; however, media outlets and policymakers remain cued to the more comprehensive annual reports.

The 9-to-10-month lag in the release of comprehensive UCR data on crimes reported to the police limits the value of the data to law enforcement agencies, policymakers, members of the media, and the public. This has become more problematic in recent years as the internet, social media, and other technologies permit the rapid spread of new types of crime. In 2021, for example, a viral video showed viewers how to steal Kia and Hyundai model cars quickly and easily, resulting in an immediate and dramatic rise in thefts of these vehicles in Wisconsin and Colorado. Data indicate that claims and losses in Wisconsin were more than 30 times higher in 2021 than in 2019, and more than 10 times higher in Colorado over the same time period (Highway Loss Data Institute, 2021). The rapid rise of Kia and Hyundai thefts in the wake of the video has been most evident in Milwaukee, WI, often cited as the epicenter of the trend. In August 2021, for example, Kia and Hyundai thefts comprised nearly three-quarters (73 %) of all reported vehicle thefts in Milwaukee (Fig. 1).

At the time Kia and Hyundai thefts were surging in the summer of 2021, national data on the crime were only available through 2019. Given the age of the data, the national numbers were of little help as decision-makers sought to respond to the surge in motor vehicle thefts and develop effective changes to policy and practice in a rapidly changing environment. In the absence of a near real-time crime surveillance system, similar to those used in the public health sector, it is difficult, if not impossible, to distinguish a crisis (like the homicide spike in 2020, or the Kia and Hyundai vehicle theft spike in 2021) from what may not be a crisis (including conflicting narratives on shoplifting), or to determine whether any such crisis is local or national in scope (Council on Criminal Justice, 2023c; Lopez, Boxerman and Cundiff, 2023). The intervening information vacuum during periods of extreme flux provides an opportunity for anecdotal evidence and political rhetoric, rather than data, to steer the national narrative on crime, and the resulting policies and practices are likely to be piecemeal at best.

Nongovernmental organizations have tried to fill the void by collecting and releasing timely data from relatively small samples of law enforcement agencies from which data are available. The Major Cities Chiefs Association

[1] UCR data reflect crime incidents known to police that occurred in a calendar year (January through December). NCVS data reflect victimizations experienced during the six months prior to the interview, excluding the month of the interview; annual reports cover victimizations experienced from July 1 of one year to November 30 of the next. Victimizations in annual NCVS reports are classified by the year of the survey and not by the year of the victimization. See, page 18: (Thompson and Tapp, 2023).

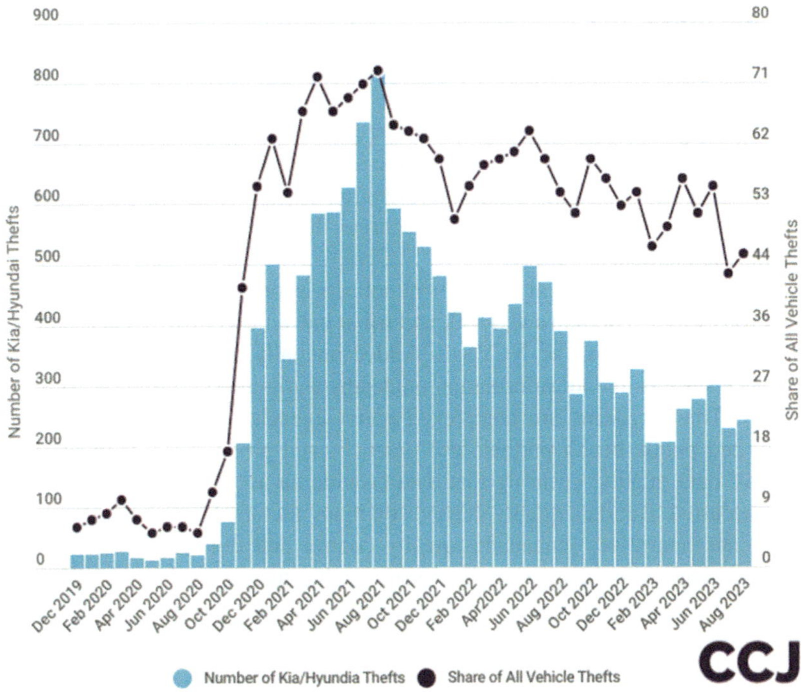

FIG. 1 Number and share of Kia/Hyundai vehicle thefts in Milwaukee, 2019–2023. *Source*: Motherboard (n.d.).

publishes a survey of violent crime reports in about 70 American and nine Canadian cities; reflecting the desire for more timely data, the survey recently became quarterly (Major Cities Chiefs Assocation, n.d.). A private firm, AH Datalytics, scrapes homicide data from more than 200 cities to generate a "murder dashboard" that updates the most recent numbers available, often within days or a few weeks of new incidents (Datalytics, n.d.). Since crime patterns began shifting early in the COVID-19 pandemic, the Council on Criminal Justice has been producing periodic reports on crime trends in about three dozen cities that provide monthly incident-level data on 10 to 12 types of violent and property crimes (Council on Criminal Justice, n.d.). In May 2024, NORC at the University of Chicago (2024) launched the Live Crime Tracker, which scrapes and reports law enforcement agency website data from 56 cities on all serious crimes (e.g., homicide, rape, robbery, burglary, simple and aggravated assault, larceny, and motor vehicle theft) in real time.

While these nongovernmental efforts help paint a timelier picture of national crime trends, they are no substitute for official government figures that are national in scope.

4 Accuracy

Quality data are dependent on a number of factors, some of which are common to large data enterprises (enhancing quality control and reducing errors), while others emerge when comparing data sources that measure crimes in different ways as well as reports that measure crime across different time periods. These factors can pose challenges to the interpretation and use of the data. In addition, data involving firearms are often inaccurate due to the miscoding of some assaults as incidents.

As indicated, the federal government collects information on crime trends in numerous ways: federal, state, local, and tribal law enforcement agencies report to the UCR Program, the NCVS is fielded annually, and public health surveillance systems record information on deaths and injuries caused by criminal actions. These systems are complex. Federal agencies and their state, local, tribal, and public health partners use practices and processes to maintain and improve data quality, ensuring that data provide an accurate representation of what occurred. But data quality depends on many factors. Some are common to large data enterprises (enhancing quality control and reducing errors), while others emerge when comparing data sources that measure crimes in different ways or reports that measure crime across different time periods. These factors can pose challenges to the interpretation and use of the data.

Working Group interviews and focus groups with local law enforcement officials and state UCR Program managers conducted for this chapter revealed concerns about the quality of crime trends information reported to NIBRS (Council on Criminal Justice, 2023b). The transition from the Summary Reporting System to the more complex incident-based reporting system increased the burden for under-resourced law enforcement agencies, including the front-line police officers charged with recording the data that are ultimately reported. As part of its transition to NIBRS, for example, the New York Police Department increased the length of its incident report form by 50 %–from four to six pages.

Other law enforcement officials shared concerns about the quality of data submitted by patrol officers and the difficulty of securing staff and resources for quality assurance and data infrastructure at a time when department leaders and elected officials were eager to put more officers on the street. Data gleaned from focus groups and a survey conducted with state UCR program managers indicated that the overwhelming majority of programs lacked the staff or resources to provide much quality control assistance to local law enforcement agencies (Council on Criminal Justice, 2023b).

While official crimes reported to law enforcement and published through the UCR Program have typically been the leading barometer of crime trends, most crimes are not reported to the police. For example, 2022 NCVS data indicate that about 42 % of violent crimes were reported to law enforcement; rape and sexual assault were the least likely to be reported, at only about 1 in 5 (Thompson and Tapp, 2023). About 32 % of property crimes were reported.

The UCR and NCVS measures indicate very different total volumes of crime. But over the past three decades, they have generally shown fairly consistent year-to-year rates of change for violent and property crime. When they don't agree, however, there can be significant confusion about the most fundamental question: Did crime go up or down?

This was especially the case in 2022, as noted by the founding chair of this Working Group, Richard Rosenfeld, and Group member Janet Lauritsen. In an October 2023 brief, the pair pointed out that law enforcement data reported through the UCR Program indicated that all types of violent crime fell 2 % from 2021 to 2022, while NCVS victimization data showed an increase of 75 %, a discrepancy primarily due to a much higher rate of aggravated assaults in the victimization survey (Fig. 2; Rosenfeld and Lauritsen, 2023). Potential explanations for the divergence include increased police response time due to reduced staffing levels; declining public trust in law enforcement, leading to fewer public calls for police service; data reporting periods that differ by a few months in an environment of changing trends; or potential issues encountered with data collected during the COVID-19 pandemic period.

Beyond the big-picture methodological questions are the more mundane but critical issues of accuracy in the recording and reporting of law enforcement data. Many law enforcement agencies present their crime trends data in multiple formats, including reporting to the UCR Program, posting on agency websites and data portals, and publishing data and analysis in official reports.

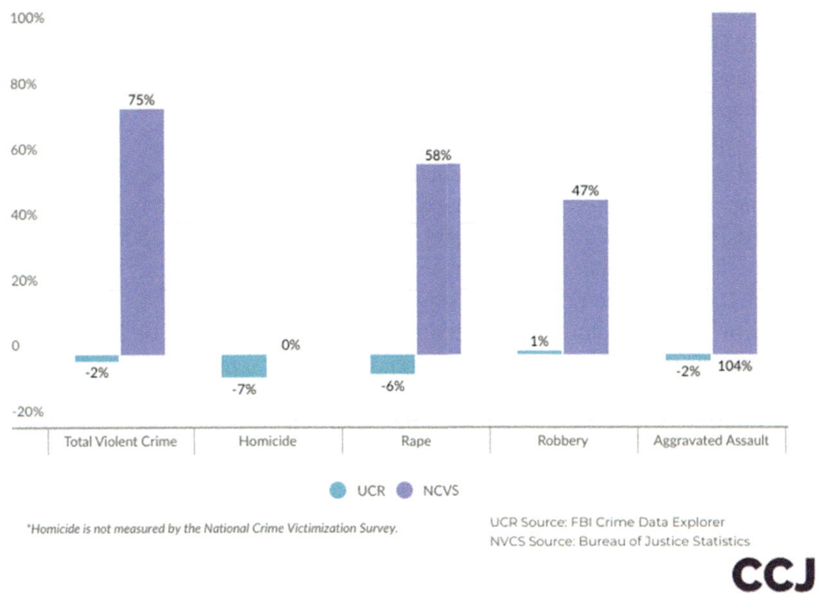

FIG. 2 Percentage change in UCR program and NCVS violent crime rates, 2021–2022.

Reporting to NIBRS is more complicated than reporting to the Summary Reporting System, and officers don't always record incident data correctly or completely. As noted above, law enforcement agencies and state UCR programs often lack the staff to do much quality control on the backend. The FBI does provide state UCR programs and law enforcement agencies with a variety of quality assurance tools, but they generally focus on ensuring that data are complete and properly formatted; they cannot ensure that data were entered correctly.

As with any data enterprise, inaccuracies and inconsistencies within and between data sources are inevitable due to various types of errors, including simple data entry mistakes. Additionally, data published on agency websites may be preliminary and updated over time. Discrepancies between values published on websites and those shared by the FBI, however, may be substantial and lead to confusion and concern about what the actual volume of crime might be in a city or jurisdiction (See, for example: Nguyen and Diaz, 2024).

Surveys such as the NCVS are also subject to errors because they rely on respondents to report on all crimes experienced by those in their household aged 12 and older (National Archive of Criminal Justice Data, n.d.). Errors occur when survey respondents may be unaware of or may not accurately recall the nature or timing of all victimization events—or may be unwilling to tell a researcher about them.

Even small discrepancies can have big consequences in the policy and political environment. This is especially true when an inaccuracy makes the difference between a particular crime trend going up or down. In a jurisdiction that had 50 murders in a prior year, for example, a count that's off by five could mean homicide is either up 10 % or down 10 %.

Perhaps such differences shouldn't have an outsized impact on public debate, especially when base rates are low compared to historical levels or are recorded in sparsely populated jurisdictions. But they do. When a crime type is declining, citizens, the media, and political opponents tend toward contentment. Rising crime, on the other hand, brings political attacks and clamor for changes. The potential fallout can be huge—in terms of crime tactics and strategies, whether the police chief stays on the job or gets fired, even whether the mayor or prosecutor is re-elected or driven from office. The high-stakes consequences of getting data on crime trends right put a premium on accuracy.

There are limitations of the current federal data collection systems that leave significant gaps in our understanding of levels of gun crime, particularly non-fatal firearms assaults and injuries (Roman, 2020). There are currently no national crime statistics counting the number or rate of non-fatal shooting victimizations; data are sometimes published by law enforcement agencies but are, in general, sparsely available (Piquero and Wheeler, 2024). NIBRS does have a data element for "assault with a firearm," which records whether a firearm was used in an incident, but this element does not allow specification of whether the gun was fired, a person was shot, a gun was used as a blunt instrument (e.g., pistol-whipping), a gun was brandished (i.e., unlawfully displayed), or a gun was otherwise used in the assault.

Existing data on non-fatal firearms assaults and injuries come from public health surveillance systems: the Firearm Injury Surveillance Through Emergency Rooms (FASTER) program, which runs the Advancing Violence Epidemiology in Real-Time (FASTER: AVERT) initiative (Centers for Disease Control and Prevention, 2021a); the Firearm Injury Surveillance System of the National Electronic Injury Surveillance System (NEISS-FISS); and the Healthcare Cost and Utilization Project (HCUP; Agency for Healthcare Research and Quality, 2022).

FASTER: AVERT is funded by the CDC and administered in partnership with state health departments and participating hospitals, collects near real-time information on infectious diseases, firearms and other assaults and injuries, and mental health conditions from a broad swath of hospitals in 12 states. The NEISS-FISS is funded by the CDC and administered by the U.S. Consumer Product Safety Commission, collects data from what was intended to be a nationally representative sample of 100 hospital emergency departments across the nation. Currently, roughly 85 hospitals participate. NEISS-FISS pays hospitals to review medical records to determine whether injuries are intentional (assault), unintentional (accident), or intentionally self-inflicted, which reduces the miscoding of firearms assaults as accidental injuries. Estimates of non-fatal firearms assaults and injuries from NEISS-FISS, however, have been highly variable due to sample size and composition issues.

While these systems provide valuable information, they also suffer from a significant drawback: Data involving firearms are often inaccurate due to the miscoding of some assaults as accidents (Campbell et al., 2018).

Hospitals record the diagnoses assigned and treatment provided to patients using codes established in the International Classification of Diseases, Ninth and Tenth Revisions. Public health surveillance systems (along with health services researchers) use these codes to record the nature of firearms injuries, determining how many non-fatal firearms injuries occur and whether those injuries stemmed from an accident or an assault. Evidence suggests, however, that coding practices often result in firearms assaults being incorrectly coded as accidental injuries, which obscures understanding about how often firearms are used in assaults (Barber, Cook, and Parker, 2022; Miller et al., 2022). The miscoding of data may be complicated by overlapping but not identical categorization of these types of incidents between data sources; in the public health data, a firearms assault is called an intentional injury, and in the crime data, they are categorized as an aggravated assault. One result of this miscoding is that many people incorrectly believe that the accidental discharge of a firearm in their home is a larger risk than a firearm-related homicide or suicide (Barber et al., 2022; Miller, Azrael, and Hemenway, 2002; Miller et al., 2007). For example, of the roughly 49,000 people in the U.S. who died by firearms in 2021, only 1 % died by unintentional gun injury (Davis, Kim and Crifasi, 2023).

All of these databases provide policymakers and the public with valuable insights but fall short of what's needed: a full, accurate national portrait of the number, circumstances, and the health outcomes of non-fatal firearm assaults.[2] Without accurate information on these and other aspects of non-fatal firearms assaults and injuries, policymakers and the public miss opportunities to shape effective strategies. In many shootings, chance dictates whether a gunshot is fatal (Beard et al., 2017), so the lack of clear understanding about how firearms are used in crimes that do not result in death—and whether these crimes were intended to result in death—also significantly hampers homicide prevention and intervention strategies.

5 Completeness of data

There has been significant progress in transitioning law enforcement agencies to NIBRS, but substantial obstacles to generating a more comprehensive view of American crime remain. These include achieving full NIBRS coverage, effectively tracking non-fatal firearms assaults and injuries data, and expanding coverage of other important types of crime, such as white-collar, environmental, and cybercrimes, that are not captured well by existing systems.

Under the UCR Summary Reporting System, national reporting of crime data by law enforcement agencies typically covered 98 % of the total population (Federal Bureau of Investigation Uniform Crime Reporting Program, n.d.). While the percentage of Americans living in jurisdictions that report to NIBRS has risen significantly in recent years, from 57 % in 2020 to 83 % at the end of 2024 (Barnett-Ryan and Berzofsky, 2022), NIBRS coverage is still below levels reached by the Summary Reporting System. Substantial obstacles to generating a more comprehensive view of American crime remain, including the challenges related to the NIBRS transition, effectively tracking non-fatal firearms assault and injury data, and expanding data collection to cover other important types of crime, such as white collar, environmental, and cybercrimes, that are not captured well by existing systems.

In 2012, BJS funded the creation of the National Crime Statistics Exchange (NCS-X) to explore whether it was possible to use NIBRS data to produce accurate, national estimates of reported crime that included incident-level details and characteristics (Bureau of Justice Statistics, 2021). Working closely with the FBI, BJS determined that it was possible to produce accurate results on crime trends from a sample of 400 carefully selected jurisdictions (Bureau of Justice Statistics, 2021). A joint effort was launched to help these 400 agencies transition to NIBRS with the ultimate goal of constructing a weighted

[2] For a discussion of how changes in coding practices and changes in hospitals participating in NEISS-FISS can distort trends in nonfatal firearm injuries, see: Cook, Rivera-Aguirre, Cerdá, and Wintemute (2017) and Roman and Cook (2021).

sample that would allow for the generation of nationally representative estimates of crime trends.[3] A few years later, however, the FBI appeared to have changed its emphasis.

In December 2015, then-FBI Director James Comey announced that the FBI would transition all reporting from the Summary Reporting System to NIBRS by January 1, 2021 (Federal Bureau of Investigation, 2015). The switch to NIBRS aimed to enhance the completeness, as well as the timeliness and accuracy, of available crime data nationwide and provide the FBI and other users with a greater ability to identify, understand, and address evolving crime issues (Bureau of Justice Statistics, n.d.b). But the announcement also created new challenges. Instead of focusing on helping 400 agencies transition to NIBRS as part of the NCX-S effort, the task became to transition all U.S. law enforcement agencies to NIBRS.

Unlike the aggregate-level data available through the Summary Reporting System, NIBRS provides more detailed information, including expanded crime types and offense categories, characteristics of each crime and whether multiple crimes occurred during the same incident, profiles of both victims and offenders, the type and value of any property lost, and whether the case was cleared by law enforcement. Like earlier data from the Summary Reporting System, NIBRS data are standardized to facilitate comparisons across jurisdictions, enhancing the capacity for users to conduct strategic analyses to amplify prevention and intervention strategies (Bureau of Justice Statistics, n.d.b).

When the transition was announced in 2015, NIBRS covered approximately 36 % of the population. Early in the transition, large cities experienced the lowest compliance rates; by 2022, however, smaller cities were still struggling with the transition (Fig. 3; Asher, 2024).

While there has been significant progress in onboarding agencies to the new system, particularly in the past year, the transition to NIBRS remains incomplete. The New York Police Department and the Los Angeles Police Department recently began reporting to NIBRS; high-population states like California, Florida, and Pennsylvania, however, continue to lag. In April 2024, NIBRS reporting covered 83 % of the population.

To better understand the obstacles to sustaining the transition to NIBRS, the Working Group held a series of interviews with the executive committee of the Association of State UCR Programs, state UCR Program managers, and local law enforcement representatives from seven states that varied in geography, population size, and the types of active law enforcement agencies (Council on Criminal Justice, 2023a, 2023b). States that successfully transitioned to NIBRS, in general, provided adequate funding to their state UCR programs and used federal grants to provide local agencies with technical assistance and

[3] To learn about the methodology, see the "NCS-X Sample: 400 State and Local Law Enforcement Agencies" tab: Bureau of Justice Statistics (2021).

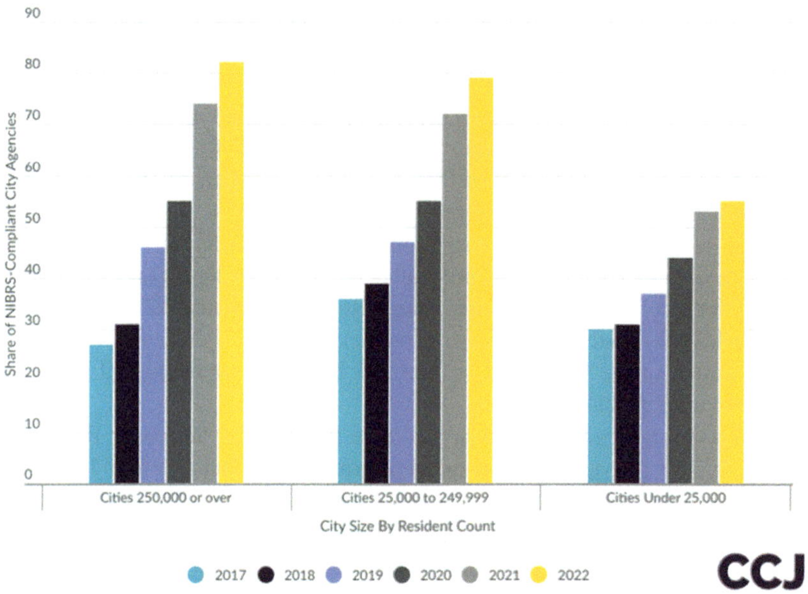

FIG. 3 Share of city agencies that were NIBRS-compliant by Population, 2017–2022.

innovative tools such as a low- or no-cost records management system. Many of the states that transitioned to NIBRS also enacted statutes that required local law enforcement agencies to submit incident-level data on a timely basis. States that had been less successful in the transition to NIBRS typically underfunded state UCR programs, failed to help local law enforcement agencies navigate a sometimes confusing software vendor landscape, and lacked statutes mandating NIBRS reporting.

As was the case with the Summary Reporting System before it, most of the offenses in NIBRS are physical crimes against people, property, and society.[4] Other types of offenses that have a significant impact on American families, society, and communities receive less attention. These include intellectual property theft, crimes committed online, crimes against animals, environmental crimes, campaign finance and elections offenses, securities fraud, embezzlement, and other forms of white-collar crime.[5]

[4] NIBRS organizes crimes into three categories: crimes against people (e.g., murder, rape, and assault), crimes against property (e.g., robbery, bribery, and burglary, where the goal of the crime is to obtain money), and crimes against society (e.g., prostitution and gambling). See: Federal Bureau of Investigation (2018), and National Academies of Sciences, Engineering, and Medicine (2016), and National Academies of Sciences, Engineering, and Medicine (2018).

[5] The United Nations identified 11 Level 1 crimes that encompass a range of offenses, from those causing or intending death to acts against public safety and the environment. (See: United Nations Office on Drugs and Crime, 2015).

Varying amounts of information about these categories of crime are available from other sources, but their exclusion from the two major Justice Department reports, *Crime in the Nation* (Federal Bureau of Investigation, 2023) and *Criminal Victimization* (Bureau of Justice Statistics, 2023a), limits understanding of the full nature and extent of crime in the country. That, in turn, can distort enforcement and prevention priorities across state, local, tribal, and federal agencies.

6 Usability

Law enforcement agencies tend to see national crime data reporting as an administrative task; few make full and effective use of the insights that the rich NIBRS data can provide. The primary federal online data tool is cumbersome and does not always allow even expert users to easily extract the statistics they seek.

Many local and state law enforcement agencies skillfully use the data they collect to make real-time decisions about where to deploy officers and other tactics. But few law enforcement agencies make full and effective use of the insights that the rich data reported to NIBRS can provide, such as gleaning relevant information about the relationships between victims and offenders or exploring the circumstances surrounding incidents where multiple crimes occur. Instead of seeing NIBRS as a tool that can help them be more effective, many officers instead see reporting to NIBRS as a burdensome administrative task. Working Group members expressed the need to identify ways to enhance the usability of NIBRS for law enforcement agencies, public officials and policymakers, researchers, and members of the media through the creation of "playbooks," or user guides, to unlock the value of NIBRS as a tool that can add narratives to criminal incidents and lead to more effective responses.

Statistical Analysis Centers (SACs) exist in 49 states (every state but Texas). They were developed to use data from all facets of the justice system to conduct research and analysis and support the administration of criminal and juvenile justice (Brick, 2023). The majority of SACs (specifically, 32 of them) are administered within a State Administering Agency, although 10 states house their SACs in other state agencies, and seven are housed in universities (Fig. 4) (Justice Information Resource Network, n.d.).

SACs are authorized to collect, analyze, and report crime and criminal justice statistics to serve all levels of government and the public and support BJS and national priorities. Analyses and dissemination focus on core areas, including incident-based crime data and criminal justice system performance, as well as data accessibility and special projects, such as using criminal history records for research and conducting targeted analyses with administrative data (Bureau of Justice Statistics, 2021).

However, a 2022 survey of SAC directors and designees representing 44 states suggests the centers are an underutilized resource. Nearly half (45 %) of the respondents reported being unable to conduct research and analysis at

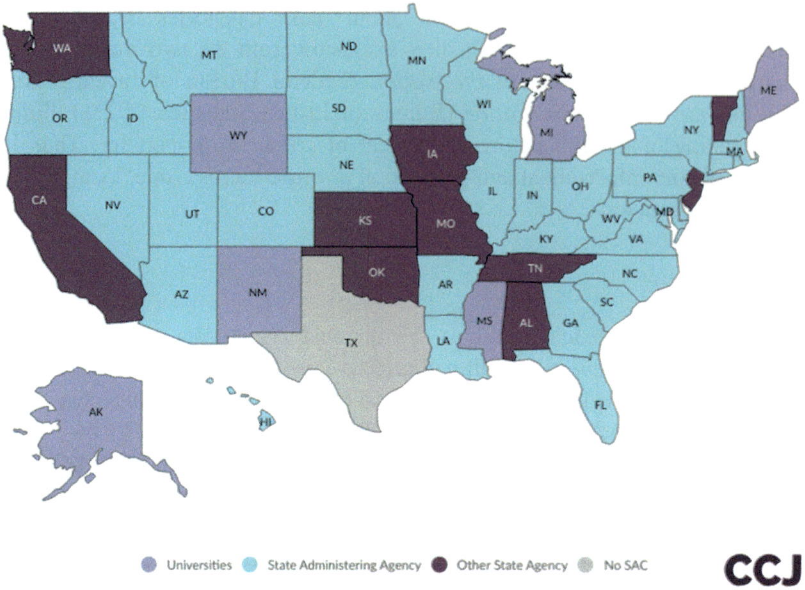

FIG. 4 Organizational structure of state statistical analysis centers. Map lines delineate study areas and do not necessarily depict accepted national boundaries.

the expected levels due to a lack of research staff, with comments indicating that salary constraints were a barrier to hiring and retention (Brick, 2023). More than a third of respondents (38 %) noted challenges due to poor data quality, while 33 % indicated problems accessing relevant data. As a result, SACs have played a limited role in unlocking the value of NIBRS data.

In addition to the need to enhance the utility of national crime data to criminal justice agencies and leaders, such data should be more easily accessible to the public and researchers. National reports made available to the public should offer clear, simple, and complete time series data that can be used to generate crime trends according to user specifications. Currently, the online tool maintained by the FBI is cumbersome, and even expert users sometimes face difficulty in easily extracting the statistics they seek.

The FBI has made improvements to its online tool, the Crime Data Explorer (CDE; Federal Bureau of Investigation, n.d.a), but several shortcomings remain and limit the usefulness of crime trends data to policy and practice (Council on Criminal Justice, 2023a). While historical crime data extend back through 1960, CDE provides access only as far back as 1985. Data on the CDE are not available according to FAIR principles (Findable, Accessible, Interoperable, and Re-Usable) (FORCE11, n.d.), and restrictions are placed on how users can access and query the data. Comparing data between states, cities, and cities of different sizes is difficult to execute, an issue amplified by the elimination of these comparisons from the Crime in the U.S. reports published by the FBI beginning in 2021. Note: Other online tools – notably, the Federal Reserve Bank of St. Louis' Federal Reserve Economic Data (FRED) platform – allow users to explore trends and make comparisons across data sources and time periods (Federal Reserve Bank of St. Louis, n.d.). Further, in the CDE, data sources are not always properly documented, nor are there sufficient resources available to support user navigation of the platform.

In addition to the FBI's CDE, BJS created the Law Enforcement Agency Reported Crime Analysis Tool (LEARCAT), a dashboard that enables users to examine crime information reported to NIBRS by participating law enforcement agencies (Bureau of Justice Statistics, n.d.a). This tool provides interactive data visualizations and filtering tools to explore crime and victimization metrics based on user-defined parameters. Users can generate incident and victimization counts, rates, and percentages by incident and victim characteristics. It is important to note that while LEARCAT and the CDE both present NIBRS data, the tools use different definitions for some offense categories, such that counts between the two sources may differ (Bureau of Justice Statistics, 2023b).

BJS also created the National Crime Victimization Survey Data Dashboard (N-DASH) tool (Bureau of Justice Statistics, 2024). It presents estimates from the NCVS as interactive data visualizations, providing users with a range of views and analysis types to explore questions of interest. Users can explore victimization by crime type, reporting to police, victim-offender relationship, weapon use, and victim services use, and can assess trends in victimization over time across various demographic factors through drop-down menus.

7 Governance and funding

A variety of federal agencies collect crime-related data across multiple departments, but the Executive Branch lacks a structure for coordinating these efforts. Years of underfunding have made it difficult for the Bureau of Justice Statistics (BJS) to perform its mandated functions.

The FBI currently collects information on crimes reported to law enforcement from state UCR program offices and, in some cases, directly from federal, state, local, and tribal law enforcement agencies. It then shares this information through the CDE, press releases, and an annual report. The FBI has deep relationships with state, local, and tribal law enforcement agencies and with state UCR programs. As a result, it is well-positioned to support the collection of NIBRS data. However, as the FBI is primarily an investigative and intelligence agency, and not a recognized federal statistical agency, it is less suited to data analysis and reporting.

As the federal statistical agency for criminal justice, BJS is statutorily empowered to "compile, collate, analyze, publish, and disseminate uniform national statistics concerning all aspects of criminal justice and … crime…."(34 U.S. Code § 10132). Its staff have the expertise to gather and conduct sophisticated statistical analyses of data from NIBRS, NCVS, and other sources. In addition, unifying the public presentation of crime trends data at BJS would minimize the chances of the federal government publishing conflicting information while creating more of a one-stop shop for understanding the full picture of crime in America.

The lack of a federal governance structure for justice data collection and reporting has hampered the nation's access to timely, accurate, complete, and usable crime trends data and analysis. The vast majority of national crime trends statistics are collected and distributed by two federal agencies: the FBI and BJS. The FBI collects and reports through the UCR Program/NIBRS and BJS reports on NIBRS and collects and reports on NCVS. The CDC shares information on intentional injuries and firearm-related homicide from NVDRS and from other health surveillance tools like FASTER: AVERT and NEISS-FISS. The result is that information is siloed: The media and policymakers focus on UCR Program data, criminological and other social science researchers use NIBRS and NCVS, and public health researchers focus on violent injury and death data from CDC sources.

Other federal agencies also collect data and produce statistics about crime trends:

- The Environmental Protection Agency reports on violations of regulations and environmental crimes.
- The Securities and Exchange Commission reports on financial crimes and violations of securities law.
- The U.S. Secret Service reports on financial crimes including counterfeiting and identity theft.
- The Financial Crimes Enforcement Network, a branch of the U.S. Treasury Department, reports on money laundering tied to criminal and terrorist activity.
- The Federal Trade Commission reports on consumer fraud and identity theft.
- The U.S. Postal Inspection Service reports on mail fraud and related crimes.

No federal agency currently compiles time series data from these various sources.

In their 2018 report, the National Academies of Sciences, Engineering, and Medicine's National Research Council[6] concluded that there is "neither a strong coordinating body nor a defined governance structure for the whole U.S. crime statistics enterprise." The National Research Council recommended that the Office of Management and Budget "explore the range of coordination and governance processes for the complete U.S. crime statistics enterprise—including the 'new' crime categories—and then establish such a structure."

[6] In 2013, the National Academies of Sciences, Engineering, and Medicine's National Research Council convened a panel to "assess and make recommendations for the development of a modern set of crime measures in the United States and the best means for obtaining them." For details, see: National Academies of Sciences, Engineering, and Medicine (2016).

To date, no such structure has been put in place. A variety of federal agencies across multiple departments currently collect crime-related data, but the Executive Branch still lacks a structure for coordinating these efforts. This creates significant hurdles for those seeking to link crime, health, and economic data to inform policy and budget decisions.

Although BJS is the principal federal agency responsible for collecting, analyzing, publishing, and disseminating information on crime, criminal offenders, victims of crime, and the operations of the criminal justice system at all levels of government, it receives one of the smallest base appropriations from Congress of any principal statistical agency. In FY 2024, BJS received a $35- million base appropriation from Congress, significantly less than the President's $78 million request and a decrease from FY 2023′s $42 million. For FY 2025, the President's budget proposed $42 million in base funding for BJS (COSSA, 2024). In contrast, the Bureau of Labor Statistics received an appropriation of $698 million in FY 2024. And that same year, Congress appropriated $307 million to the National Center for Education Statistics, $187 million to the National Center for Health Statistics, and $188 million to the National Agricultural Statistics Service (American Statistical Association, 2024).

Years of underfunding—punctuated by occasional budget cuts—have made it difficult for BJS to perform its mandated functions.[7] BJS has not conducted the Survey of Local Jail Inmates since 2002 or the Survey of Prison Inmates since 2016. These surveys provide valuable insights into crime trends and how people acquire firearms used in crime.

The Office of Justice Programs, administered under the U.S. Department of Justice, has attempted to compensate for these cuts by transferring funds to BJS to support their work.[8] Yet rising costs, particularly for the implementation of the labor-intensive NCVS, which now exceed $60 million a year (Federal Bureau of Investigation, n.d.b; Office of Management and Budget, 2023), have limited the ability of BJS to support state and local criminal justice agencies and forced it to pull back from fielding surveys of people in prisons, which provide valuable insights into emerging crime trends and the operations of the

[7] The BJS budget has restricted its ability to expand crime data offerings. In 2009, the National Research Council of the National Academies of Science, Engineering, and Medicine conducted a sweeping review of the bureau's responsibilities and the nation's crime data needs. The National Research Council identified major gaps in crime data collection, including for white-collar crimes such as corporate fraud and money laundering; internet crimes; juvenile victimization, offending, law enforcement experiences; and the interactions of drugs and crime. Fifteen years later, gaps such as these remain. For more details, see: National Research Council, 2009.

[8] Every year, BJS and the National Institutes of Justice receive a transfer from the Office of Justice Programs. The amount ranges from 1 % to 3 % of the Office of Justice Program's direct programming budget. For FY 2025, the Office of Justice Programs plans to transfer 2.5 % of its direct programming funds to the National Institutes of Justice and BJS, which will result in $30 million in additional funds for BJS. In addition, BJS is expected to receive $15 million from the Office for Victims of Crime.

justice system. Budget limitations also have impeded BJS's efforts to gather information on the operations of the courts and other important parts of the criminal justice system. Funding levels must increase for the nation to have access to high-quality, rigorous analyses of timely, accurate, and complete crime trends data.

NIBRS also faces governance challenges, especially the update of NIBRS. Proposals to edit or expand NIBRS must be considered and reviewed by the FBI's Criminal Justice Information Services (CJIS) Division's Advisory Policy Board (APB). The APB is responsible for reviewing several programs, including the UCR Program, for policy, technical, and operational issues, and making recommendations to the FBI director (Federal Bureau of Investigation, n.d.b). The overwhelming majority of APB recommendations, including recommendations to update NIBRS, are approved by the FBI director. The FBI then develops specifications that it shares with state UCR programs, which, in turn, provide guidance to records management system software vendors. All of this requires coordination across multiple federal and state entities.[9] Based on the complexity of the process, the movement from proposal to implementation is slow, often spanning years.

Since 2020, the APB has made more than 30 NIBRS-related recommendations that have been approved by the FBI director but have not yet resulted in technical specifications (Council on Criminal Justice, 2023a). How the FBI is prioritizing these recommendations is unclear. Some state UCR program managers told the Working Group that they were unsure which APB recommendations were moving forward and when new data elements would be included in future iterations of the technical specifications (Council on Criminal Justice, 2023a). For example, the APB recently approved a measure to add additional data elements to distinguish between shootings and other firearms assaults (e.g., pistol-whipping), identify whether the victim(s) sustained an injury, and determine whether a firearm was brandished. But even if the FBI prioritized producing technical specifications to implement this recommendation, the change would not be reflected in records management systems for years.

[9] With 35 representatives from criminal justice and national security organizations throughout the country, the APB provides critical stakeholder input on FBI CJIS programs, which go far beyond simply collecting crime data. Proposals to modify NIBRS start with a "topic paper." These can be submitted to the APB by anyone. Topic papers are then considered by the APB's UCR subcommittee, which meets every six months. Approved measures then go to the full APB, which sends them to the FBI Director for approval. Once approved, the FBI must develop technical specifications for the new data elements to share with states. Most states use NIBRS as the basis for a customized, state-specific incident-reporting system. As a result, state UCR programs must also develop technical specifications for records management system software vendors so that vendors can update the NIBRS-specific portions of their platforms. Records management system vendors typically have two years to update their systems. Many states have dozens of vendors, some of whom struggle to update NIBRS. For details, see: Federal Bureau of Investigation (n.d.b).

While some long-term participants in the APB see the process for updating NIBRS as one that is necessary to thoroughly vet proposals, the current process for updating NIBRS has limited policymakers' ability to develop and implement responsive, data-driven violence prevention strategies in a timely fashion. Between 2019 and 2023, for example, the rate of firearm assaults rose 32 % in 11 American cities, yet national or nationally representative data on the nature of these assaults–including whether anyone was shot during them–were not available to local law enforcement, policymakers, the media, or the public.

Adding data elements to the NIBRS data collection can pose significant challenges to law enforcement agencies. While many local law enforcement agencies are now including provisions for NIBRS upgrades in their contracts with records management system vendors, the costs to large vendors for customizing upgrades for multiple states can be significant. Upgrading records management systems is just one of the costs associated with NIBRS reporting. Unlike some other complex national data collection series, NIBRS reporting is voluntary (for those not in federal agencies) and uncompensated (Centers for Disease Control and Prevention, 2021b).

8 Recommendations

The following are recommendations and implementation strategies to improve the collection and use of crime-related statistics.

8.1 Improve the nation's capacity to collect and analyze crime data

1. The Attorney General should request a plan from the FBI to increase the number of law enforcement agencies reporting to the NIBRS to cover 90 % of the U.S. population by the end of 2025, 95 % by the end of 2026, and 98 % by the end of 2027 to match the level of Summary Reporting System reporting in 2020. The Attorney General should also request that the plan include action steps to bring the 58 noncompliant federal entities (the only law enforcement agencies subject to a federal mandate to transition to NIBRS) into compliance with statutory requirements (see page 7: Department of Justice, n.d.)
2. The U.S. Department of Justice should create a direct working relationship between the FBI and BJS to produce timely, accurate, complete, and usable crime trends reports. The FBI should continue to collect NIBRS data; beginning in 2025, BJS should be the lead agency in releasing the official U.S. crime trends statistics.
3. The Office of Justice Programs should issue a request for proposals to review funding and staffing levels for state UCR Programs and, in coordination with state and local justice stakeholders, identify standards for funding and staffing. The request for proposals should require the estimation of the ongoing costs to states and state, local, and tribal agencies for maintaining and expanding NIBRS.

4. Congress should provide BJS with the resources to support state efforts to develop low- or no-cost records management systems for smaller law enforcement agencies.
5. Congress should provide permanent funding to the Office of Justice Programs' Crime and Corrections Analyst in Residence Program, which helps small and mid-sized law enforcement agencies with training and resources to analyze crime data in a way that informs operations and provides greater transparency to residents (Bureau of Justice Assistance, 2021).
6. Congress should provide the BJS with funds to establish and support a National Justice Data Analysis Center. The Center should work with key stakeholders to establish national best practices for law enforcement agencies to visualize and share data that adhere to FAIR principles (ensuring data are Findable, Accessible, Interoperable, and Re-Usable), provide training and technical assistance to law enforcement agencies to help them deploy local and national data to enhance crime analysis and prevention, and enhance crime trends data science by identifying avenues to link data across the criminal justice and public health systems (FORCE11, n.d.).
7. The Bureau of Justice Assistance should incentivize law enforcement agencies to become NIBRS-compliant by changing Byrne-JAG funding rules to require that state, local, and tribal law enforcement agencies applying directly for grants, but not currently reporting to NIBRS, commit to reporting to NIBRS within a year of receiving the grant. Agencies that receive grants and fail to meet the deadline should be required to describe their obstacles to reporting and how they intend to overcome them. Failure to comply should result in a 5 % funding reduction until the non-reporting agency begins the NIBRS transition process.
8. States not already reporting monthly should enact or strengthen statutes to require local law enforcement agencies to report NIBRS-compliant crime data to state UCR programs on a monthly basis.
9. The FBI should provide its Criminal Justice Information Services (CJIS) division with resources to hire experienced data analysts and scientists to perform quality control and provide additional technical assistance on quality control to federal, state, local, and tribal law enforcement agencies and state UCR programs.
10. The CJIS division should clearly communicate anticipated updates and additions to NIBRS, as well as the timing of those updates and additions, to UCR programs and other key stakeholders.
11. The White House Office of Science and Technology should request that the Office of Management and Budget convene the U.S. Chief Information Officers Council, n.d. to create an Executive Branch governance structure to coordinate the collection and distribution of crime and justice statistics across the federal government. During this process, the Chief Information Officers Council should work with the Advisory

Committee on Data and Evidence Building (U.S. Bureau of Economic Assistance, 2022), which was established under HR 4174—the Foundations for Evidence-Based Policymaking Act of 2018 (Congress.gov, n.d.).

8.2 Publish monthly crime trends reports

1. BJS should create a monthly crime trends dashboard using data from NIBRS. Jurisdictions with available data should be chosen and weighted to create a representative national sample. BJS should publish crime statistics for this sample of jurisdictions on a regular monthly schedule, similar to how the Bureau of Labor Statistics publishes employment statistics on the first Friday of every month.
2. BJS should overhaul its website to focus on providing clear, simple, and complete time series data from NIBRS and the NCVS that can be used to generate crime trends according to user specifications. BJS should also seek to aggregate data from a wide variety of sources like the Federal Reserve Bank of St. Louis' Federal Reserve Economic Data (FRED) platform (Federal Reserve Bank of St. Louis, n.d). Where BJS is unable to aggregate data, links to other federal agencies that present criminal justice-related data, statistics, and tools should be provided on its website so that users can more easily access public information about the operation of the criminal justice system.
3. The CDC should work with BJS and law enforcement partners to explore ways for the CDC to access partial police investigative notes before the police investigation is completed. Earlier access to police information would result in the release of National Violent Death Reporting System (NVDRS) information in a timelier fashion (Centers for Disease Control and Prevention, 2021b).

8.3 Create a comprehensive portrait of crime in America

1. In partnership with the FBI, the CDC, and other federal agencies, BJS should publish a comprehensive annual compendium of crime in the U.S. The first such report should be published in 2025. Law enforcement data collected through NIBRS and victim survey data collected through the NCVS should be published simultaneously, and the combined publication should seek to explain any significant difference in the reported trends. This report should include information on major violent and property offenses (Group A and Group B offenses in NIBRS), as well as on other important types of crime, including intellectual property theft, cybercrime, hate crime, crimes against animals, environmental crimes, campaign finance and elections offenses, securities fraud, embezzlement, and other forms of white-collar crime (Bureau of Justice Statistics, 2024; United Nations Office on Drugs and Crime, 2015). In subsequent years, the compendium should include regular reports on firearms crime trends and seek to integrate additional sources of data from public health surveillance systems and other stakeholders.

2. To enhance access to information in between the publication of annual reports, BJS should create an updated website that provides access to downloadable crime trends data and visualizations dating back to 1960. The site should comply with FAIR principles (data should be Findable, Accessible, Interoperable, and Re-Usable; FORCE11, n.d.). Data should be described and footnoted in ways that make it clear where information was incomplete or where estimate techniques were used, important tables should have unique URLs, and data should be downloadable in ASCII and.csv format to maximize accessibility.
3. The FBI's Criminal Justice Information Services (CJIS) division should prioritize recommendations to improve how NIBRS captures information on crimes involving firearms. Specifically, CJIS should prioritize producing technical specifications for recently approved recommendations related to firearm assaults.[10]
4. The CDC should improve the accuracy and completeness of the health surveillance systems that collect information on violent deaths and firearms assaults and injuries by prioritizing the following actions:
   - The CDC should direct the National Center for Health Statistics to develop a plan to address the systematic miscoding of firearms assaults and injuries in hospital discharge coding.
   - The CDC should double the number of state health departments participating in the Firearm Injury Surveillance Through Emergency Rooms (FASTER) program's Advancing Violence Epidemiology in Real-Time (FASTER: AVERT) initiative from 12 to 24 as a first step to making FASTER: AVERT available in all 50 states. The CDC should also expand opportunities for states and emergency departments to provide timely access to select FASTER: AVERT data to researchers and work with state health departments to increase access to non-fatal firearm injury data.
   - The CDC should continue to support the U.S. Consumer Product Safety Commission in its efforts to increase the number of hospitals participating in the National Electronic Injury Surveillance System-Firearm Injury Surveillance System (NEISS-FISS), including by providing support to vendors assisting with hospital recruitment and selection.
5. The Office of Justice Programs and the CDC should fund pilot projects to explore ways to integrate crime trends and public health data.[11] As part of

[10] These include adding "Firearm Stolen" data value to Type of Criminal Activity (Data Element 12); "Gunshot Wound" data value to Type of Injury (Data Element 33); "Firearm Discharge" data value to Type of Weapon/Force Used (Data Element 13); and "Type of Injury" to a victim of a Homicide Offense (Data Element 33).

[11] Camden, NJ's Camden Arise project provides a good example of the benefits of such linkages: https://camdenhealth.org/work/camden-arise/.

this effort, the CDC and BJS should "crosswalk" criminal justice and public health definitions where competing descriptions could lead to confusion, link data sources, and engage in cross-agency analysis of violent crimes and victimizations.

6. BJS, in collaboration with the CDC and other federal agencies, should produce regular reports on firearm-related crime and injury trends. The reports should be published at least every other year (currently, these data are published only in special reports), and also should provide trends across demographic groups (Kena and Truman, 2022; Piquero and Roman, 2024). As part of this work, BJS should develop standards and processes for capturing consistent estimates of firearm fatalities; non-fatal firearm assaults, including whether a gun was used as a blunt instrument (e.g., pistol-whipping) or was brandished (i.e., unlawfully displayed) during an assault; the use of firearms in intimate partner violence and domestic abuse incidents; and the use of stolen firearms in criminal incidents. Disaggregating information by sex, race, and other demographic characteristics should be prioritized to examine how gun crime affects different population subgroups.

8.4 Funding considerations

BJS is the principal federal agency responsible for collecting, analyzing, publishing, and disseminating information on crime, criminal offenders, victims of crime, and the operations of the justice system. But years of insufficient funding have hampered the bureau's ability to provide timely, accurate, complete, and usable information on crime trends and the justice system, impeding the ability of policymakers to adjust public safety strategies rapidly and effectively.

Based on the findings and recommendations presented above, the Crime Trends Working Group recommends that Congress increase appropriations for statistical collections to $75 million for FY 2025, scaling up to $93 million in FY 2026. BJS should be funded at a level that allows it to perform its mandated functions and assume a new role in analyzing and publishing national crime trends data, without having to rely on transfers from the Office of Justice Programs and the Office for Victims of Crime. While far greater increases are warranted in the short and longer term, the Working Group believes these amounts can be administered responsibly and will tangibly improve the nation's crime data infrastructure.

Increasing BJS funding to $93 million would allow the agency to conduct the following new initiatives:

1. Enhance its website, adding a dashboard that shows monthly crime trends for a sample of cities and that aggregates historical time series data for multiple data sets, at a cost of roughly $4 million annually

2. Support the publication of a comprehensive annual compendium of crime in the U.S. and regular reports on firearms crime trends with an additional $3 million for staff and resources
3. Support states with grants for low- or no-cost records management software for use among state, local, and tribal law enforcement agencies, with $4 million in one-time additional funding
4. Resume the Survey of Local Jail Inmates and the Survey of Prison Inmates every five years, at a cost of $4 million per administrative year
5. Support the development of a National Justice Data Analysis Center to enhance data analysis and data science across the criminal justice and public health systems, at a cost of $3 million annually

In addition to its recommendation that Congress increase funding for BJS, the Working Group makes several funding recommendations regarding other federal agencies:

- Hold the FBI harmless in any shifting of functions to BJS, allowing the FBI to reinvest savings to bolster staffing at its Criminal Justice Information Services (CJIS) division, with a focus on data collection and quality control
- Provide permanent operational funding for the Bureau of Justice Assistance's Crime and Corrections Analyst in Residence program, at a cost of $3 million annually
- Double the number of state health departments participating in the Firearm Injury Surveillance Through Emergency Rooms (FASTER) program's Advancing Violence Epidemiology in Real-Time (FASTER: AVERT) initiative, administered by the CDC, from 12 to 24, at a cost of $1.8 million annually, as the next step in moving toward implementing FASTER: AVERT in all 50 states.
- Expand participation in the U.S. Consumer Product Safety Commission's National Electronic Injury Surveillance System-Firearm Injury Surveillance System (NEISS-FISS) from roughly 85 hospitals to the full sample of 100 hospitals, incurring a one-time cost of approximately $300,000
- Provide funding to support the integration and linking of crime trends and public health data, at a cost of $500,000 annually for five years.

Acknowledgements

The work of this group would not have been possible without the contributions of the Crime Trends Working Group's Founding Chair, Richard Rosenfeld, who passed away on January 8, 2024. An eminent criminologist, Rick spearheaded the formation of the working group, was the original lead author of CCJ's crime trends report series, and helped shape these recommendations to maximize their relevance, impact, and utility. His contributions to this group and to the criminal justice field are inestimable, and he is deeply missed.

This chapter was adapted from a report produced by the Council on Criminal Justice's Crime Trends Working Group. The report and its recommendations reflect the contributions and insights of working group members, who are among the nation's leading experts on crime data. We are also grateful to the Council on Criminal Justice staff for their support in developing and refining the analysis presented here.

Conflict of interest and neutrality statement

The Council on Criminal Justice is a nonpartisan criminal justice think tank and national invitational membership organization. Its mission is to advance understanding of the criminal justice policy choices facing the nation and build consensus for solutions based on facts, evidence and fundamental principles of justice.

The Council does not take policy positions. As part of its array of activities, the Council conducts research and convenes independent task forces composed of Council members who produce reports with findings and policy recommendations on matters of concern. The findings and conclusions in this research report are those of the authors alone. They were not subject to the approval of the Council's Board of Directors or its Board of Trustees.

For more information about the Council, visit counciloncj.org.

References

34 U.S. Code § 10132. https://www.law.cornell.edu/uscode/text/34/10132.

Agency for Healthcare Research and Quality, 2022. *Healthcare costs and utilization project.* https://hcup-us.ahrq.gov/overview.jsp.

Datalytics, A.H., n.d. *YTD murder comparison.* https://www.ahdatalytics.com/dashboard/ytd-murder-comparison/.

American Statistical Association, 2024. Unpublished analysis provided to the Working Group.

Asher, J., 2024. *NIBRS noncompliance has become more of a small agency problem.* AH Analytics. https://jasher.substack.com/p/nibrs-noncompliance-has-become-more.

Barber, C., Cook, P.J., Parker, S.T., 2022. The emerging infrastructure of US firearms injury data. Preventive Med. 165, 107129. https://doi.org/10.1016/j.ypmed.2022.107129

Barnett-Ryan, C., Berzofsky, M., 2022. *Establishing new methods for estimating crime in the U.S: the transition to incident-based crime reporting through NIBRS.* Presentation delivered by the Bureau of Justice Statistics. https://bjs.ojp.gov/sites/g/files/xyckuh236/files/media/document/Establishing%20New%20Methods%20for%20Estiimating%20Crime%20in%20the%20U.S.–The%20Transition%20to%20Incident-Based%20Crime%20Reporting%20through%20NIBRS.pdf.

Beard, J.H., Morrison, C.N., Jacoby, S.F., Dong, B., Smith, R., Sims, C.A., et al., 2017. Quantifying disparities in urban firearm violence by race and place in Philadelphia, Pennsylvania: a cartographic study. Am. J. Public. Health 107 (3), 371–373. https://www.ncbi.nlm.nih.gov/pmc/articles/PMC5296702/.

Brick, B.T., 2023. *Annual statistical analysis center survey: 2022 results.* Justice Research and Statistics Association. https://justiceresearch.dspacedirect.org/collections/a618488f-41a9–468b-afa0–782ccd7345?cp.page=1.

Bureau of Justice Assistance, 2021. *Crime and corrections analyst in residence program.* https://bja.ojp.gov/funding/awards/15pbja-21-gk-04010-jagp.

Bureau of Justice Statistics, 2021. *National crime statistics exchange (NCS-X).* https://bjs.ojp.gov/programs/national-crime-statistics-exchange.

Bureau of Justice Statistics, 2023a. *Criminal victimization*, 2022. https://bjs.ojp.gov/press-release/criminal-victimization-2022.

Bureau of Justice Statistics, 2023b. *Law enforcement agency reported crime analysis tool: Methodology*. https://bjs.ojp.gov/learcat-methodology.

Bureau of Justice Statistics, 2024. *Group "A" offenses, group "B" offenses*. https://bjs.ojp.gov/sites/g/files/xyckuh236/files/media/document/offensea_offenseb.pdf.

Bureau of Justice Statistics, 2024. *National crime victimization survey data dashboard (N-DASH)*. https://ncvs.bjs.ojp.gov/Home.

Bureau of Justice Statistics, n.d.-a. *LEARCAT: Law enforcement agency reported crime analysis tool*. https://learcat.bjs.ojp.gov/IncidentsCrime?Dat%20Year=2022&Unit%20of%20Analysis=Count.

Bureau of Justice Statistics, n.d.-b. *National incident-based reporting system (NIBRS)*. https://bjs.ojp.gov/national-incident-based-reporting-system-nibrs.

Campbell, S., Nass, D., Nguyen, M., 2018. The CDC says gun injuries are on the rise. But there are big problems with this data. *The Trace*. https://www.thetrace.org/2018/10/cdc-nonfatal-gun-injury-data-estimate-problems/.

Centers for Disease Control and Prevention, 2021a. *Funded surveillance: advancing violence epidemiology in real-time (AVERT) CDC-RFA-CE-23-007*. https://www.cdc.gov/violenceprevention/firearms/funded-surveillance.html.

Centers for Disease Control and Prevention, 2021b. *WISQARS national violent death reporting system (NVDRS) help menu*. https://www.cdc.gov/injury/wisqars/nvdrs/index.html.

Congress.gov., n.d. *HR 4174: Foundations for evidence-based policymaking act of 2018*. https://www.congress.gov/bill/115th-congress/house-bill/4174.

Cook, P.J., Rivera-Aguirre, A.E., Cerdá, M., Wintemute, G.C., et al., 2017. Respond. Am. J. Public Health. 107 (12), e23. https://doi.org/10.2105/AJPH.2017.304101

COSSA, 2024 April 2. COSSA's analysis of the President's FY 2025 budget request for social and behavioral science. https://cossa.org/cossas-analysis-of-the-presidents-fy-2025-budget-request-for-social-and-behavioral-science/.

Council on Criminal Justice, n.d. *Crime trends reports*. https://counciloncj.org/crime-trends-reports/.

Council on Criminal Justice, 2023a. *Meeting #5: crime trends data tool usability and obstacles to NIBRS adoption*. https://counciloncj.org/meeting-5-crime-trends-data-tool-usability-and-obstacles-to-nibrs-adoption/.

Council on Criminal Justice, 2023b. *National incident-based reporting system focus groups and interviews: Key themes*. https://counciloncj.org/wp-content/uploads/2024/05/NIBRS-Interviews.pdf.

Council on Criminal Justice, 2023c. *Trends in homicide: What we need to know*. https://counciloncj.org/homicide-trends-report/.

Davis, A., Kim, R., Crifasi, C.K., 2023. *A year in review: 2021 gun deaths in the U.S.* John Hopkins Center for Gun Violence Solutions. https://publichealth.jhu.edu/sites/default/files/2024–01/2023-june-cgvs-u-s-gun-violence-in-2021-v3.pdf.

Department of Justice, n.d. *The report of the attorney general pursuant to section 18(a) of executive order 14074: department of Justice review of the transition of law enforcement agencies to the national incident-based reporting system (NIBRS)*. https://www.justice.gov/opa/speech/file/1563061/dl.

Federal Bureau of Investigation, 2015. The transition to NIBRS. *FBI This Week*. https://www.fbi.gov/audio-repository/news-podcasts-thisweek-the-transition-to-nibrs.mp3/view.

Federal Bureau of Investigation, 2018. *Crimes against persons, property, and society*. https://ucr.fbi.gov/nibrs/2018/resource-pages/crimes_against_persons_property_and_society-2018.pdf.

Federal Bureau of Investigation, 2023. *FBI releases 2022 crime in the nation statistics*. https://fbi.gov/news/press-releases/fbi-releases-2022-crime-in-the-nation-statistics.

Federal Bureau of Investigation. n.d.a. *Crime data explorer*. https://cde.ucr.cjis.gov/LATEST/webapp/#/pages/home.

Federal Bureau of Investigation, n.d.b. *The CJIS advisory process*. https://le.fbi.gov/cjis-division/the-cjis-advisory-process.

Federal Bureau of Investigation Uniform Crime Reporting Program, n.d. *Summary of the uniform crime reporting program*. https://ucr.fbi.gov/leoka/2013/standard-ucr-info/about_ucr_-2013-_final.

Federal Reserve Bank of St. Louis. n.d. *FRED economic data: St. Louis FRED*. https://fred.stlouisfed.org.

FORCE11, n.d. *The FAIR data principles*. https://force11.org/info/the-fair-data-principles/.

Highway Loss Data Institute, 2021. Hyundai and Kia theft losses. Bulletin 38 (28).

Kena, G., Truman, J.L., 2022. *Trends and patterns in firearm violence, 1999–2018* (NCJ 251663). Bureau of Justice Statistics. https://bjs.ojp.gov/content/pub/pdf/tpfv9318.pdf.

Lopez, E., Boxerman, R., Cundiff, K., 2023. *Shoplifting trends: What you need to know*. Council on Criminal Justice.

Major Cities Chiefs Assocation, n.d. l *Resources: Violent crime reports*. https://majorcitieschiefs.com/resources/.

Miller, M., Azreal, D., Yenduri, R., Barber, C., Bowen, A., MacPhaul, E., et al., 2022. Assessment of the accuracy of firearm injury intent coding at 3 US hospital. JAMA Netw. Open 5 (12), e2246429. https://pubmed.ncbi.nih.gov/36512356/.

Miller, M., Azrael, D., Hemenway, D., 2002. Firearm availability and unintentional firearm deaths, suicide, and homicide among 5-14 year old. J. Trauma. & Acute Care Surg 52 (2), 267–275. https://journals.lww.com/jtrauma/abstract/2002/02000/firearm_availability_and_unintentional_firearm.11.aspx.

Miller, M., Lippmann, S.J., Azreal, D., Hemenway, D., 2007. Household firearm ownership and rates of suicide across the 50 United States. J. Trauma. & Acute Care, Surgery 6 (24), 1029–1035. https://journals.lww.com/jtrauma/abstract/2007/04000/household_firearm_ownership_and_rates_of_suicide.31aspx.

Motherboard, n.d. *Motherboard VICE news Kia Hyundai theft data*. https://docs.google.com/spreadsheets/d/1–6614t_Ata5k7oESrm-HAdF1uOe8aPGRaQEC144GYeE/edit#gid=0.

National Academies of Sciences, Engineering, and Medicine, 2016. *Modernizing crime statistics: Report 1: Defining and classifying crime*. https://doi.org/10.17226/23492.

National Academies of Sciences, Engineering, and Medicine, 2018. *Modernizing crime statistics: Report 2: New systems for measuring crime*. https://doi.org/10.17226/25035.

National Archive of Criminal Justice Data, n.d. *National crime victimization survey: Accuracy of NCVS estimates*. https://www.icpsr.umich.edu/web/pages/NACJD/NCVS/accuracy.html.

National Research Council, 2009. Ensuring the quality, credibility, and relevance of U.S. justice statistics. The National Academies Press, Washington, DC. https://doi.org/10.17226/12671

Nguyen, J., Diaz, M., 2024 March 6. Houston police department's crime stats don't match federal numbers. *KPRC 2*. https://www.click2houston.com/news/local/2024/03/06/houston-police-departments-crime-stats-dont-match-federal-numbers/.

NORC at the University of Chicago, 2024. *Live crime tracker*. https://livecrimetracker.norc.org.

Office of Management and Budget, 2023. *ICR documents: NCVS redesign split sample supporting statement A 09.19.23*. https://www.reginfo.gov/public/do/PRAViewDocument?ref_nbr=202309–1121-001.

Piquero, A.R., Roman, J.K., 2024. Firearm homicide demographics before and after the COVID-19 pandemic. JAMA Netw. Open 7 (5), e2412946. https://doi.org/10.1001/jamanetworkopen.2024.12946

Piquero, A.R., Wheeler, A., 2024. Disaggregating fatal and non-fatal shootings in Boston, 2015-2023. Int. J. Police Sci. 3 (1). https://doi.org/10.56331/20240122/IJPS3

Roman, J.K., 2020. A blueprint for a U.S. firearms data infrastructure. NORC at the University of Chicago. https://www.norc.org/content/dam/norc-org/pdfs/A%20Blueprint%20for%20a%20U.S.%20Firearms%20Dat%20Infrastructure_NORC%20Expert%20Panel%20Final%20Report_October%202020.pdf.

Roman, J.K, and Cook, P., 2021. Improving data infrastructure to reduce firearms violence. Final Report. NORC: Chicago.

Rosenfeld, R., Lauritsen, J., 2023. *Did violent crime go up or down? Yes, it did.* Council on Criminal Justice. https://counciloncj.org/did-violent-crime-go-up-or-down-last-year-yes-it-did/.

Thompson, A., Tapp, S.N., 2023. *Criminal victimization, 2022* (NCJ 307089). Bureau of Justice Statistics. https://bjs.ojp.gov/document/cv22.pdf.

United Nations Office on Drugs and Crime, 2015. *International classification of crime statistical purposes (ICCS).* https://unstats.un.org/unsd/statcom/doc15/BG-ICCS-UNODC.pdf.

U.S. Bureau of Economic Assistance, 2022. *Advisory committee on data for evidence building.* https://www.bea.gov/evidence.

U.S. Chief Information Officers Council, n.d. *Purpose and vision: what we do.* https://www.cio.gov/.

U.S. Department of Labor Statistics, n.d. *U.S. Bureau of Labor Statistics.* https://www.bls.gov/.

Index

Note: Page numbers followed by "*f*" indicate figures and "*t*" indicate tables.

D

N

T

U

V

W

X

Y

Z

9780443314223